Edited by
Richard A. Layfield and Muralee Murugesu

Lanthanides and Actinides in Molecular Magnetism

Related Titles

de Bettencourt-Dias, A. (ed.)

Luminescence of Lanthanide Ions in Coordination Compounds and Nanomaterials

2014

Print ISBN: 978-1-119-95083-7

(Also available in a variety of electronic formats)

Gatteschi, D., Benelli, C.

Introduction to Molecular Magnetism

From Transition Metals to Lanthanides

2015

Print ISBN: 978-3-527-33540-4

(Also available in a variety of electronic formats)

Hilzinger, R., Rodewald, W.

Magnetic Materials

Fundamentals, Products, Properties, Applications

2013

Print ISBN: 978-3-895-78352-4

Edited by
Richard A. Layfield and Muralee Murugesu

Lanthanides and Actinides in Molecular Magnetism

WILEY-VCH
Verlag GmbH & Co. KGaA

The Editors

Dr. Richard A. Layfield
The University of Manchester
School of Chemistry
Oxford Road
M13 9PL Manchester
United Kingdom

Prof. Dr. Muralee Murugesu
University of Ottawa
Department of Chemistry/D'Iorio Hall
Marie Curie 10
ON K1N 6N NK
Canada

Cover
Periodic table: iStockphoto © Tomacco
Magnet: iStockphoto © jgroup

Library of Congress Card No.: applied for

British Library Cataloguing-in-Publication Data
A catalogue record for this book is available from the British Library.

Bibliographic information published by the Deutsche Nationalbibliothek
The Deutsche Nationalbibliothek lists this publication in the Deutsche Nationalbibliografie; detailed bibliographic data are available on the Internet at <http://dnb.d-nb.de>.

Print ISBN: 978-3-527-33526-8
ePDF ISBN: 978-3-527-67350-6
ePub ISBN: 978-3-527-67349-0
Mobi ISBN: 978-3-527-67348-3
oBook ISBN: 978-3-527-67347-6

Typesetting Laserwords Private Limited, Chennai, India
Printing and Binding Markono Print Media Pte Ltd, Singapore

Printed on acid-free paper

Contents

Preface

A huge surge in the number of publications dealing with the magnetic properties of coordination compounds based on highly anisotropic lanthanide ions occurred in the past decade. This body of work provided the inspiration for the current book. In light of the recent trends, one could be forgiven for assuming that f-block magnetism is a new field of research: it is not, but there is no doubt that certain types of lanthanide and actinide compounds have breathed new life into an established field of molecular magnetism. In compiling *Lanthanide and Actinides in Molecular Magnetism*, our aim has been to set the scene by covering the important fundamental aspects of f-element electronic structure and magnetism and then to use this as a basis for understanding the most eye-catching recent developments and emerging cutting-edge topics. This aim has been achieved by the contributing authors, who address their chosen topics using a range of experimental and theoretical considerations, much of which is based on their own pioneering contributions to the field.

In Chapter 1, Sorace and Gatteschi deal with the fundamental aspects of lanthanide electronic structure and magnetism, and in Chapter 2, Clemente-Juan, Coronado and Gaita-Ariño develop a picture of how the crystal field theory can be used to design single-ion magnets and spin qubits. Tang and Zhang condense a huge volume of literature in Chapter 3 in order to cover the bewildering array of polynuclear lanthanide single-molecule magnets (SMMs), and extended molecular networks of lanthanide complexes are dealt with by Sessoli and Bernot in Chapter 4. The experimental aspects of SMM physics are covered by Pedersen, Woodruff, Bendix and Clérac in Chapter 5, and a detailed description of theoretical models of lanthanide magnetism is provided by Chibotaru and Ungur in Chapter 6. The promising role of lanthanide complexes in quantum computing is presented by Aromí, Luis and Roubeau in Chapter 7. Chapter 8 is a tour-de-force of lanthanide-phthalocyanine chemistry and physics authored by Lan, Klyatskaya and Ruben, from synthetic supramolecular chemistry to applications in molecular spintronic devices. The role of lanthanide complexes in the magnetocaloric effect is covered by Sharples and Collison in Chapter 9, and in

Chapter 10, Liddle and van Slageren cover yet another emerging topic – namely, the applications of actinide elements in single-molecule magnetism.

One of the most exciting aspects of the material covered in this book is its multi-disciplinary nature: chemistry, physics, materials science and nanoscience all play their part. The potential applications in advanced technologies such as quantum computing add to the excitement. The beauty of this science is that it does not respect 'traditional' subject boundaries, which, the reader will find, is clearly demonstrated throughout. The target audience should span the full range of career stages. Established experts in the field will find the book to be an invaluable summary, while new research students will find it to be an excellent entry point to the wider literature. We hope that the book will become an indispensable guide for all scientists with interests in magnetism and in the magnetic properties of f-element compounds.

The editors owe a great debt of gratitude to the contributing authors, all of whom have invested significant time and effort in order to share their wisdom and insight.

The University of Manchester, UK — *Richard Layfield*
University of Ottawa, Canada — *Muralee Murugesu*

List of Contributors

Guillem Aromí
Universitat de Barcelona
Departament de Química
Inorgànica
Grup de Magnetisme i Molècules
Funcionals (GMMF)
Diagonal 645
08028 Barcelona
Spain

Jesper Bendix
University of Copenhagen
Department of Chemistry
Universitetsparken 5
2100 Copenhagen
Denmark

Kevin Bernot
Université Européenne de
Bretagne
INSA, SCR, UMR 6226
20 Avenue des buttes de
Coësmes CS70839
35708 Rennes Cedex
France

Liviu F. Chibotaru
Katholieke Universiteit Leuven
Theory of Nanomaterials Group
Celestijnenlaan 200F
3001 Heverlee
Belgium

Juan M. Clemente-Juan
Universidad de Valencia
Instituto de Ciencia Molecular
c/Catedrático José Beltrán, 2
46980 Paterna
Spain

Rodolphe Clérac
CNRS, CRPP
UPR 8641
33600 Pessac
France

and

University of Bordeaux
CRPP, UPR 8641
33600 Pessac
France

David Collison
The University of Manchester
School of Chemistry
Oxford Road
Manchester M13 9PL
UK

Eugenio Coronado
Universidad de Valencia
Instituto de Ciencia Molecular
c/Catedrático José Beltrán, 2
46980 Paterna
Spain

Alejandro Gaita-Ariño
Universidad de Valencia
Instituto de Ciencia Molecular
c/Catedrático José Beltrán, 2
46980 Paterna
Spain

Dante Gatteschi
Università degli studi di Firneze
Laboratory for Molecular
Magnetism
Dipartimento di Chimica
"Ugo Schiff" and UdR INSTM
Via della Lastruccia 3
50019 Sesto Fiorentino
Italy

Svetlana Klyatskaya
Karlsruher Institut für
Technologie (KIT)
Institut für Nanotechnologie
76344 Eggenstein-Leopoldshafen
Germany

Yanhua Lan
Karlsruher Institut für
Technologie (KIT)
Institut für Nanotechnologie
Hermann-von-Helmholtz-Platz 1
76344 Eggenstein-Leopoldshafen
Germany

Stephen T. Liddle
University of Nottingham
School of Chemistry
University Park
Nottingham NG7 2RD
UK

Fernando Luis
CSIC-Universidad de Zaragoza
Instituto de Ciencia de
Materiales de Aragón
Pedro Cerbuna 12
50009 Zaragoza
Spain

Kasper S. Pedersen
CNRS, CRPP
UPR 8641
33600 Pessac
France

and

Univ. Bordeaux
CRPP, UPR 8641
33600 Pessac
France

and

University of Copenhagen
Department of Chemistry
Universitetsparken 5
2100 Copenhagen
Denmark

Olivier Roubeau
CSIC-Universidad de Zaragoza
Instituto de Ciencia de
Materiales de Aragón
Pedro Cerbuna 12
50009 Zaragoza
Spain

Mario Ruben
Karlsruher Institut für
Technologie (KIT)
Institut für Nanotechnologie
Hermann-von-Helmholtz-Platz 1
76344 Eggenstein-Leopoldshafen
Germany

and

Université de Strasbourg
IPCMS-CNRS
23 Rue du Loess
67034 Strasbourg
France

Roberta Sessoli
Università degli Studi di Firenze
Laboratory of Molecular
Magnetism
Department of Chemistry and
INSTM
Via della Lastruccia 3
50019 Sesto Fiorentino
Italy

Joseph W. Sharples
School of Chemistry
The University of Manchester
Oxford Road
Manchester M13 9PL
UK

Joris van Slageren
Universität Stuttgart
Institut für Physikalische Chemie
Pfaffenwaldring 55
70569 Stuttgart
Germany

Lorenzo Sorace
Università degli studi di Firneze
Laboratory for Molecular
Magnetism
Dipartimento di Chimica
"Ugo Schiff" and UdR INSTM
Via della Lastruccia 3
50019 Sesto Fiorentino
Italy

Jinkui Tang
Chinese Academy of Sciences
State Key Laboratory of Rare
Earth Resource Utilization
Changchun Institute of Applied
Chemistry
Changchun, 130022
P. R. China

Liviu Ungur
Katholieke Universiteit Leuven
Theory of Nanomaterials Group
Celestijnenlaan 200F
3001 Heverlee
Belgium

Daniel N. Woodruff
CNRS, CRPP
UPR 8641
33600 Pessac
France

and

University of Bordeaux
CRPP, UPR 8641
33600 Pessac
France

Peng Zhang
Chinese Academy of Sciences
State Key Laboratory of Rare
Earth Resource Utilization
Changchun Institute of Applied
Chemistry
Changchun, 130022
P. R. China

1 Electronic Structure and Magnetic Properties of Lanthanide Molecular Complexes

Lorenzo Sorace and Dante Gatteschi

1.1 Introduction

The first studies on the magnetic and electronic properties of compounds containing lanthanide ions date back to the beginning of the twentieth century [1]. However, detailed investigation on these systems only began in the 50s[2] and helped set up an appropriate theoretical framework for the analysis of their properties [2–5]. Most of the studies reported in the early literature, which involved optical spectroscopy, magnetism, or electron paramagnetic resonance (EPR), were however concerned with inorganic systems in which the lanthanide occupied high symmetry sites, and paramagnetic ions were often doped in diamagnetic host lattices [6, 7].

On the other hand, the number of molecular complexes (which usually show a low point symmetry at the lanthanide site) whose magnetic properties had been well characterized remained quite small even in 1993, when Kahn [8] wrote his landmark book entitled *Molecular Magnetism*. The field of lanthanide molecular magnetism has indeed really boomed only in the last 15 years, when the availability of powerful theoretical and experimental techniques allowed deep insight into these systems. As a result, some more specific applications of the theory that was developed for inorganic systems to the molecular magnets case were needed. The purpose of this chapter is to describe the fundamental factors affecting the electronic structure of lanthanide complexes, with some specific focus on the symmetry, and the way this is related to their static magnetic properties (dynamic magnetic properties being the focus of a subsequent chapter).

Lanthanide atoms in the electronic ground state are characterized by the progressive filling of 4f shells, with the general configuration $[Xe]4f^n6s^2$ (with the exception of La, Ce, Gd, Lu, for which the ground configuration is $[Xe]4f^n5d^16s^2$). For this reason, the most stable lanthanide ions are the tripositive ones, obtained by loss of the 5d and 6s electrons (notable exceptions are Eu^{2+}, Ce^{4+} and Tb^{4+},

Lanthanides and Actinides in Molecular Magnetism, First Edition.
Edited by Richard A. Layfield and Muralee Murugesu.

which have stable electronic configurations). In the following, we discuss the paramagnetic properties of rare earth compounds arising from the unpaired 4f electrons: since these are effectively shielded by the completely filled 5s and 5p orbitals, their behaviour is much less affected by the coordination environment of the ion compared to the 3d transition metal series. Consequently, optical spectra consist of very sharp, weak lines due to formally forbidden 4f–4f transitions, while the magnetic properties can, to a first approximation, be expressed as those of a free Ln^{3+} ion. This means that rare earth ions present an essentially unquenched orbital momentum, since the core-like character of 4f orbitals (compared, e.g. to 3d ones) prevents the crystal field (CF) from quenching the orbital momentum.[1] For this reason, in the early days of magnetochemistry, lanthanides were studied as a model of free ions, much more accessible than the paramagnetic gases [9].

The first attempts to rationalize the magnetic properties of rare earth compounds date back to Hund [10], who analysed the magnetic moment observed at room temperature in the framework of the 'old' quantum theory, finding a remarkable agreement with predictions, except for Eu^{3+} and Sm^{3+} compounds. The inclusion by Laporte [11] of the contribution of excited multiplets for these ions did not provide the correct estimate of the magnetic properties at room temperature, and it was not until Van Vleck [12] introduced second-order effects that agreement could be obtained also for these two ions.

The effect of the coordinating ligands over the magnetic properties of lanthanides becomes important in lowering the temperature, as the ground multiplets are split by an amount comparable to thermal energy: as a consequence, depopulation of the sublevels occurs, and deviation from the Curie law is observed. This, in turn, complicates the interpretation of magnetic properties of systems in which the lanthanide(III) ion interacts with another paramagnetic species. Indeed, effects due to the magnetic exchange are very small – because the unpaired electrons are in the well-shielded f orbitals – and may be hidden by ligand field effects at low temperature. It is, then, of paramount importance to appropriately determine the split components of the lowest lying multiplet and to understand the factors on which this depends.

In the following sections, we start by discussing in some detail the electronic structure of the free ion, following the classic treatment of Wybourne [3], and we successively analyse the effect of the ligand field. The relation between the Stevens' formalism [2], to which the molecular magnetism community is more used, and Wybourne's notation is presented. Indeed, the latter takes more easily into account the effect of the excited multiplets, and its use might facilitate interchange and data comparison with results from luminescence and absorption spectroscopy. The resulting magnetic properties and EPR spectra are discussed,

1) Exceptions to this behaviour are Eu^{2+} and Gd^{3+}, which – as a consequence of their $4f^7$ electronic configuration – present an orbitally non-degenerate ground state.

with some examples from more recent literature. Finally, we briefly discuss the way exchange coupling effects are treated in molecular systems containing anisotropic lanthanides.

1.2
Free Ion Electronic Structure

We start our description of the electronic structure of complexes of lanthanides by the analysis of the free ion energy structure. The relevant Hamiltonian is written as

$$\hat{H} = \sum_{k=1}^{n} \left\{ \frac{1}{2m}\boldsymbol{p}_k^2 - \frac{Ze^2}{r_k} \right\} + \sum_{k=1}^{n} \zeta(r_k)\boldsymbol{l}_k \cdot \boldsymbol{s}_k + \sum_{k<\lambda}^{n} \frac{e^2}{r_{k\lambda}} \tag{1.1}$$

In Equation 1.1, the first term is the sum of hydrogen-like terms for single electrons and the second one is the sum of the single-electron spin–orbit interaction, while the third term contains the interelectronic repulsion. By applying the central field approximation [13], each single electron can be considered as moving in an average, spherically symmetric field due to the nucleus and to the remaining electrons:

$$\hat{H}_0 = \sum_{k=1}^{n} \left\{ \frac{1}{2m}\boldsymbol{p}_k^2 + U\left(r_k\right) \right\} \tag{1.2}$$

This allows separation of the corresponding Schrödinger equation in n independent equations, one for each electron, so that the solution of Equation 1.2 will be a product of functions of the type:

$$\Psi(k) = r^{-1}R_{nl}\,(r)Y_l^m(\theta, \varphi) \tag{1.3}$$

In Equation 1.3, the radial function $R_{nl}\,(r)$ is defined by the quantum numbers n and l and the spherical harmonics Y_l^m depend on the quantum numbers l and m_l. When the spin of the electron is taken into account, the normalized antisymmetric function is written as a Slater determinant. The corresponding eigenvalues depend only on n and l of each single electron, which determine the electronic configuration of the system.

The difference $\hat{H}_1$ between the Hamiltonians (1.1) and (1.2):

$$\hat{H}_1 = \sum_{k=1}^{n} \left\{ -\frac{Ze^2}{r_k} - U\left(r_k\right) \right\} + \sum_{k=1}^{n} \zeta(r_k)\boldsymbol{l}_k \cdot \boldsymbol{s}_k + \sum_{k<\lambda}^{n} \frac{e^2}{r_{k\lambda}} \tag{1.4}$$

can now be treated as a perturbation, the first term of which only causes a global shift of the energies without affecting the relative differences. Both the second and the third terms split each configuration into separate multiplets, since their effect is different for different states of the same configuration.

Within this framework it is assumed that the energy differences between each configuration are much higher than the splitting induced through Equation 1.4, so that each configuration is treated separately. The calculation of the matrix element of $\widehat{\boldsymbol{H}}_1$ is obtained after defining a set of basis states in a specified angular momenta coupling scheme: this corresponds to assigning a different relative importance to the second and third terms of Equation 1.4, that is, respectively the spin–orbit and the interelectronic repulsion interactions. If the former dominates over the latter, the j–j coupling scheme is applied, in which the spin $\boldsymbol{s}_i$ and the angular momenta $\boldsymbol{l}_i$ of each electron are first coupled to provide a global momentum $\boldsymbol{j}_i$. This is done following the rules of angular momentum addition, so that $|\boldsymbol{l}_i - \boldsymbol{s}_i| < \boldsymbol{j}_i < \boldsymbol{l}_i + \boldsymbol{s}_i$. After this, the $\boldsymbol{j}_i$ values of each electron are coupled to obtain a global $\boldsymbol{J}$.

The former scheme is usually applied to heavy atoms, while for rare-earth ions the LS coupling scheme (also known as *Russell–Saunders coupling*) is normally used. In this approach, interelectronic repulsion is considered to be dominant over spin–orbit coupling. As a consequence, the spins of all the electrons are first coupled together to obtain a global spin $\boldsymbol{S} = \sum_i \boldsymbol{s}_i$, and the same is done with angular momenta $\boldsymbol{L} = \sum_i \boldsymbol{l}_i$. The corresponding matrix elements of the interelectronic repulsion Hamiltonian are then diagonal with respect to S, L, M_S, M_L and are independent of the latter, providing a $[(2S+1)(2L+1)]$-fold degenerate set of states. This set of states is called a *term*, and is characterized by n, l, S and L quantum numbers (where the first two values are fixed to 4 and 3 for 4f electrons), and the term is indicated by a ^{2S+1}L symbol, with $S, P, D, F, \ldots$ etc. corresponding to $L = 0, 1, 2, 3, \ldots$ etc. However, terms characterized by the same $|S, L, M_S, M_L\rangle$ may appear more than once in a given configuration, so that this set of quantum numbers is not always sufficient to characterize the term unambiguously. For this purpose, Racah introduced the irreducible representations of the groups R_7 and G_2, indicated by W and U, respectively [14].

The interelectronic repulsion Hamiltonian also commutes with $\boldsymbol{J} = \boldsymbol{S} + \boldsymbol{L}$, so that its matrices are diagonal also with respect to J and M_J, the corresponding eigenvalues being independent of M_J. Since states of $|J, M_J\rangle$ are linear combinations of $|S, L, M_S, M_L\rangle$ states with the same S, L in the scheme $|S, L, J\rangle$ the energy is also independent of J. There then exist two sets of functions depending on the quantum numbers ($|S, L, M_S, M_L\rangle$ and $|S, L, J, M_J\rangle$) in which the interelectronic repulsion is diagonal, whose energies depend only on S and L.

The labelling of terms as $|S, L, J, M_J\rangle$ is preferable when one takes into account the effect of spin–orbit coupling, since J and M_J remain good quantum numbers even after this perturbation is accounted for. In detail, the effect of spin–orbit coupling over a many-electron atomic term is evaluated by writing the spin–orbit operator in terms of the total angular and spin momentum, $\boldsymbol{L}$ and $\boldsymbol{S}$:

$$\widehat{H}_{s.-o.} = \sum_{k=1}^{n} \zeta(r_k)\boldsymbol{l}_k \cdot \boldsymbol{s}_k = \lambda \boldsymbol{L} \cdot \boldsymbol{S} \tag{1.5}$$

In Equation 1.5, λ is the spin–orbit coupling within a given Russell–Saunders multiplet, which is related to the spin–orbit coupling constant of the ion, ζ, by the relation $\lambda = \pm\zeta/2S$, the + sign applying for $n < 7$ and the – sign for $n > 7$ [15]. The effect of spin–orbit coupling is to split the terms in multiplets with same L, S and different values of J, with energies $E(^{2S+1}L_J) = (\lambda/2)[J(J+1) - L(L+1) - S(S+1)]$, so that the ground state is $J = L - S$ for $n < 7$ and $J = L + S$ for $n > 7$. The important point is that the ground state corresponds to different orientation of the spins with respect to the angular orbital momentum: antiparallel for $n < 7$ and parallel for $n > 7$. As we will see, this has paramount importance for the magnetic properties throughout the lanthanide series. The remaining $2J + 1$ (Figure 1.1) degeneracy can only be removed by a further perturbation: either by a magnetic field, or by the ligand field, the effect of which is considered in the following sections.

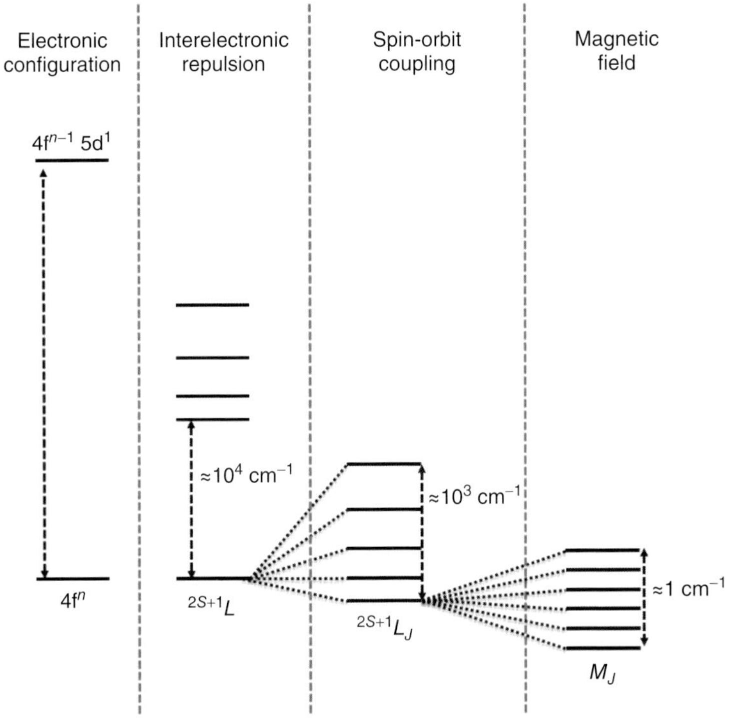

Figure 1.1 Effect of interelectronic repulsion, spin–orbit coupling and magnetic field on the energy levels arising from a given $4f^n$ configuration for a free-ion Ln^{3+}. The magnetic field effect is estimated assuming a 1 T field.

1.2.1 Free Ion Magnetism

By applying a magnetic field, the degeneracy of the $(2J + 1)$ levels in each $^{2S+1}L_J$ multiplet is removed, providing a series of levels identified by their M_J value, with $-J \leq M_J \leq +J$. The corresponding field-dependent energy can be written in terms of a Landé splitting factor, g_J, depending on J, L and S. The value of this factor as a function of J, L and S can be obtained by considering that the corresponding magnetic moment will be

$$\mu_J = \mu_B g_J \boldsymbol{J} = \mu_B(g_L \boldsymbol{L} + g_S \boldsymbol{S}) \tag{1.6}$$

One can thus write the following equality for the diagonal matrix elements of $\mu_J \cdot \boldsymbol{J}$:

$$\left\langle J, M_J \right| \mu_J \cdot \boldsymbol{J} | J, M_J \rangle = \mu_B g_J \langle J, M_J | \boldsymbol{J} \cdot \boldsymbol{J} | J, M_J \rangle \tag{1.7}$$

The second term of Equation 1.7 may be rewritten, making use of Equation 1.6, as

$$\mu_B \left(g_L \boldsymbol{L} \cdot \boldsymbol{J} + g_S \boldsymbol{S} \cdot \boldsymbol{J} \right) = \mu_B \left[g_L \left(\boldsymbol{L}^2 + \frac{1}{2} \left(\boldsymbol{J}^2 - \boldsymbol{L}^2 - \boldsymbol{S}^2 \right) \right) + g_S \left(\boldsymbol{S}^2 + \frac{1}{2} \left(\boldsymbol{J}^2 - \boldsymbol{L}^2 - \boldsymbol{S}^2 \right) \right) \right] \tag{1.8}$$

which finally yields

$$g_J = \frac{g_L(L(L+1) + J(J+1) - S(S+1)) + g_S(S(S+1) + J(J+1) - L(L+1))}{2J(J+1)} \tag{1.9}$$

Since $g_L = 1$ and assuming for the sake of simplicity $g_S = 2$, one obtains

$$g_J = \frac{3}{2} + \frac{S(S+1) - L(L+1)}{2J(J+1)} \tag{1.10}$$

For $J = 0$ there is obviously no first-order Zeeman splitting: however, application of a magnetic field can result in second-order splitting. As such, it is necessary to evaluate the corresponding g_J factor, which is $g_0 = 2 + L(2 + S)$.

The magnetic susceptibility of the free ion will follow the Curie law:

$$\chi^M = \frac{N_A g_J^2 \mu_B^2}{3kT} J(J+1) \tag{1.11}$$

where χ^M is the molar magnetic susceptibility and N_A is the Avogadro number. This situation is indeed experimentally observed for Ln^{3+} complexes at room temperature, where the measured magnetic moment coincides with what is expected for the free-ion Curie behaviour, provided that the ground J multiplet is well isolated from the excited states (see Table 1.1). This is not the case both for Eu^{3+} – whose first excited state $^7F_1$ is lying only at ca 350 cm^{-1} above the non-magnetic $^7F_0$ ground state – and Sm^{3+}, for which the first excited state $^6H_{7/2}$ is lying about 700 cm^{-1} above the ground $^6H_{5/2}$ state [10].

Table 1.1 Relevant magnetic information for free Ln^{3+} ions.

Ion	Ground multiplet	*S*	*L*	*J*	*g*	χT value calculated	χT value experimental
Ce^{3+}	$^2F_{5/2}$	1/2	3	5/2	6/7	0.80	0.82
Pr^{3+}	$^3H_4$	1	5	4	4/5	1.60	1.54
Nd^{3+}	$^4I_{9/2}$	3/2	6	9/2	8/11	1.64	1.57
Pm^{3+}	$^5I_4$	2	6	4	3/5	0.90	Not measured
Sm^{3+}	$^6H_{5/2}$	5/2	5	5/2	2/7	0.09 (0.31)	0.27
Eu^{3+}	$^7F_0$	3	3	0	0	0 (1.5)	1.40
Gd^{3+}	$^8S_{7/2}$	7/2	0	7/2	2	7.87	8.10
Tb^{3+}	$^7F_6$	3	3	6	3/2	11.82	11.33
Dy^{3+}	$^6H_{15/2}$	5/2	5	15/2	4/3	14.17	13.91
Ho^{3+}	$^5I_8$	2	6	8	5/4	14.07	13.52
Er^{3+}	$^4I_{15/2}$	3/2	6	15/2	6/5	11.48	11.28
Tm^{3+}	$^3H_6$	1	5	6	7/6	7.15	6.51
Yb^{3+}	$^2F_{7/2}$	1/2	3	7/2	8/7	2.57	2.49

For Sm^{3+} and Eu^{3+} the χT values (given in emu kelvin per mole) obtained by including the Van Vleck contribution are reported in brackets. The experimental values refer to the $Ln_2(SO_4)_3 \cdot 8H_2O$ series, obtained as an average of the different measurements reported by Van Vleck [9].

For the latter two ions, the presence of low-lying excited states makes the inclusion of both the first-order contribution of the excited states and the second-order effects due to coupling of the ground J state with the excited states crucial in a correct estimation of the room-temperature values of χT. Indeed, second-order contribution in Van Vleck [9] expansion of the susceptibility is inversely proportional to the energy difference between the ground and the excited states:

$$\chi_{VV} = -2N \sum_{M_J=-J}^{J} \sum_{J'=-M_J'}^{-M_J'} \frac{\langle J, M_J | \mu_B (\boldsymbol{L} + g_S \boldsymbol{S}) | J', M_J' \rangle}{E_{J,M_J} - E_{J',M_J'}} = \frac{2N\mu_B^2 (g_J - 1)(g_J - 2)}{3\lambda} \tag{1.12}$$

1.3 Electronic Structure of Lanthanide Ions in a Ligand Field

When a lanthanide ion is placed in a ligand environment with symmetry lower than spherical, the energies of its partly filled 4f orbitals are split by the electrostatic field of the ligand. The result is a splitting of the $2J + 1$ degeneracy of the free ion states (see Figure 1.2).

This is an immediate consequence of the lowering of the symmetry as, even in the regular octahedral geometry, group theory tells us that the highest dimension of the irreducible representation is three. This is the basis of Crystal Field Theory, whose deeply symmetry-based formalism was developed by Bethe in 1929 [16].

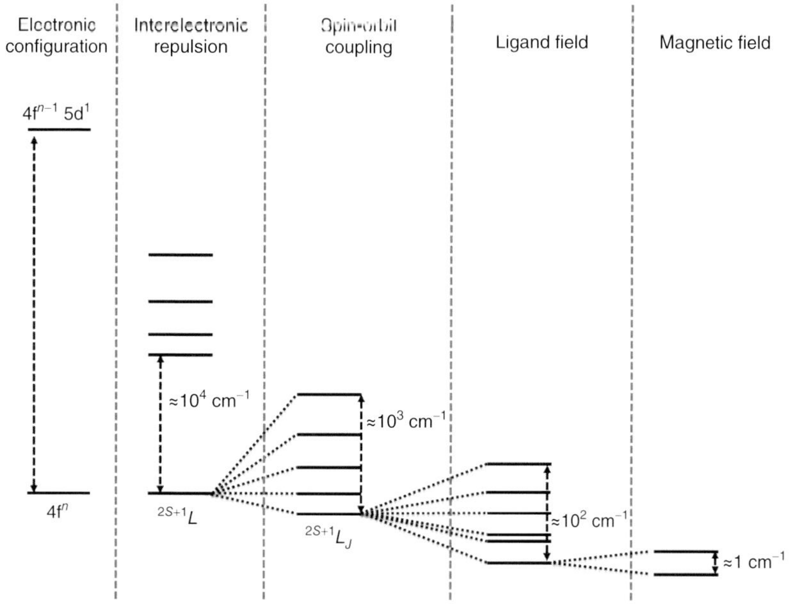

Figure 1.2 Energetic structure of a Kramers lanthanide ion in a ligand field evidencing the effect of progressively weaker perturbation. The magnetic field effect is estimated assuming a 1 T field.

Within this framework, the effect of the ligand can be described by an operator U_{CF}, which is the sum of one-electron operators for all the $4f^n$ electrons of the lanthanide, which accounts for the potential created by a charge distribution $\rho(\boldsymbol{R})$ at a distance r_i:

$$\boldsymbol{U}_{\mathrm{CF}} = -e\sum_{j=1}^{nel.} \boldsymbol{U}_j = -\,e\sum_{j=1}^{nel.} \int \frac{\rho(\boldsymbol{R})}{|\boldsymbol{R} - r_i|} dv \tag{1.13}$$

The problem of evaluating the effect of the perturbation created by the ligands thus reduces to the solution of the secular determinant with matrix elements of the type $\langle \varphi_l | \boldsymbol{U}_{\mathrm{CF}} | \varphi_k \rangle$, where $|\varphi_l\rangle$ and $|\varphi_k\rangle$ identify the eigenfunctions of the free ion. Since $|\varphi_l\rangle$ and $|\varphi_k\rangle$ are spherically symmetric, and can be expressed in terms of spherical harmonics, the potential is expanded in terms of spherical harmonics to fully exploit the symmetry of the system in evaluating these matrix elements. In detail, two different formalisms have been developed in the past to deal with the calculation of matrix elements of Equation 1.13 [2, 3]. Since U_{CF} is the sum of one-electron operators, while $|\varphi_l\rangle$ and $|\varphi_k\rangle$ are many-electron functions, both the formalisms require decomposition of free ion terms in linear combinations of monoelectronic functions.

1.3.1
Stevens' Formalism

Stevens' [2] first noted that the electrostatic potential fulfilling the symmetry requirements of the lanthanide site can be conveniently expressed as the product of a radial function and of Legendre polynomials, and thus transforms according to an irreducible representation of the rotation group. It is then possible to find a suitable constant to express such matrix elements by using an operator equivalent approach, in which the potential is expanded in a sum of equivalent angular momentum operators. In this approach, the matrix elements can be easily computed if mixing between different J multiplets is neglected. The resulting expression for Equation 1.13 using the Stevens' formalism is

$$\boldsymbol{U}_{\mathrm{CF}} = \widehat{\boldsymbol{H}}_{\mathrm{CF}}^{\mathrm{Stev}} = \sum_{k=2,4,6} \rho^k \sum_{q=-k}^{k} A_k^q \langle r^k \rangle \widehat{\boldsymbol{O}}_k^q \tag{1.14}$$

In Equation 1.14, $A_k^q\langle r^k\rangle$ is a parameter, $\widehat{\boldsymbol{O}}_k^q$ is the operator equivalent of the crystal field potential and ρ^k *is* a number which is different for the different f^n configurations and for the different k values. This accounts for the proportionality between the electrostatic potential, expressed as a spherical harmonic of order k, and the corresponding operator equivalent for that configuration [2, 6]. We note here that the number of q terms with $q \neq 0$ to be included is limited by the point group of the rare-earth site, since the CF Hamiltonian has to be invariant under all symmetry operations of the point group. The non-vanishing CF parameters in sites with commonly occurring point symmetries are reported in Table 1.2 [17, 18].

It is worth noting here that negative q values correspond to complex operators, while Stevens' parameters are always real [19]. The forms of the operator equivalents are reported in Table 1.3 [20], and the corresponding matrix elements are found tabulated in books by Abragam and Bleaney and by Altshuler [6, 21].

1.3.2
Wybourne's Formalism

Stevens' formalism turned out to be very powerful, and works easily as long as only the ground ${}^{2S+1}L_J$ multiplet of the lanthanide ion is considered. As such, it has been widely used in studies on EPR properties of lanthanide-based inorganic systems [6, 22], while it is not well suited for optical spectroscopy. Indeed, when starting to include excited multiplets the Stevens' formalism becomes much too involved. This is the reason why a more general formalism, developed by Wybourne [3], is of widespread use in optical studies – naturally dealing with excited multiplets – and

Table 1.2 Non-vanishing crystal field terms (Stevens' formalism) for common lanthanide point symmetries.

k	$\|q\|$	D_{3h} D_{6h} C_{6v}	C_6 C_{3h} C_{6h}	D_{2d} D_{4h} C_{4v}	S_4 C_{4h}	D_{3d} C_{3v}	C_3 S_6	D_{2h} C_{2v}	C_{2h} C_s
2	2							+	±
4	2							+	±
4	3					+	±		
4	4			+	±			+	±
6	2							+	±
6	3					+	±		
6	4			+	±			+	±
6	6	+	±			+	±	+	±

When only $q > 0$ are non-vanishing, the entry is indicated by '+', while ± indicates that both $q > 0$ and $q < 0$ terms are non-zero.

Table 1.3 Stevens' operators expressed in terms of J_+, J_-, J_z polynomials.

k	q	$\hat{O}_k^q$
2	0	$3\hat{J}_z^2 - J(J+1)$
2	±1	$\frac{\{\hat{J}_z,(\hat{J}_+ \pm \hat{J}_-)\}_+}{2c_\pm}$
2	±2	$\frac{\hat{J}_+^2 \pm \hat{J}_-^2}{c_\pm}$
4	0	$35\hat{J}_z^4 - [30J(J+1) - 25]\hat{J}_z^2 + 3J^2(J+1)^2 - 6J(J+1)$
4	±1	$\frac{\{[7\hat{J}_z^3 - 3J(J+1)\hat{J}_z - \hat{J}_z],(\hat{J}_+ \pm \hat{J}_-)\}_+}{2c_\pm}$
4	±2	$\frac{\{[7\hat{J}_z^2 - J(J+1) - 5)],(\hat{J}_+ \pm \hat{J}_-)\}_+}{2c_\pm}$
4	±3	$\frac{\{\hat{J}_z,(\hat{J}_+ \pm \hat{J}_-)^3\}_+}{2c_\pm}$
4	±4	$\frac{\hat{J}_+^4 \pm \hat{J}_-^4}{c_\pm}$
6	0	$231\hat{J}_z^6 - [315J(J+1) - 735]\hat{J}_z^4 + [105J^2(J+1)^2 - 525J(J+1) + 294]\hat{J}_z^2 - 5J^3(J+1)^3 + 40J^2(J+1)^2 - 60J(J+1)$
6	±1	$\frac{\{33\hat{J}_z^5 - [30J(J+1) - 15]\hat{J}_z^3 + [5J^2(J+1)^2 - 10J(J+1) + 12]\hat{J}_z,(\hat{J}_+ \pm \hat{J}_-)\}_+}{2c_\pm}$
6	±2	$\frac{\{33\hat{J}_z^4 - [18J(J+1) + 123]\hat{J}_z^2 + J^2(J+1)^2 + 10J(J+1) + 102,(\hat{J}_+^2 \pm \hat{J}_-^2)\}_+}{2c_\pm}$
6	±3	$\frac{\{11\hat{J}_z^2 - J(J+1) + 38,(\hat{J}_+^4 \pm \hat{J}_-^4)\}_+}{2c_\pm}$
6	±4	$\frac{\{11\hat{J}_z^2 - J(J+1) - 38,(\hat{J}_+^4 \pm \hat{J}_-^4)\}_+}{2c_\pm}$
6	±5	$\frac{\{\hat{J}_z,(\hat{J}_+ \pm \hat{J}_-)^5\}_+}{2c_\pm}$
6	±6	$\frac{\hat{J}_+^6 \pm \hat{J}_-^6}{c_\pm}$

$\{A, B\}_+$ is a shorthand notation for the product $(AB + BA)$, $c_+ = 2$, $c_- = 2i$.

is becoming increasingly applied also to molecular magnetism studies [23–25]. We thus find it appropriate to discuss it in this section.

In this approach, the ligand field potential is expressed as

$$\begin{aligned} \boldsymbol{U}_{\mathrm{CF}} &= \widehat{\boldsymbol{H}}_{\mathrm{CF}}^{\mathrm{Wyb}} \\ &= \sum_{k=0}^{\infty}\left[B_0^0 C_0^0(i) + \sum_{q=1}^{k} B_q^k (C_{-q}^k(i) + (-1)^q C_q^k(i)) + iB'^k_q (C_{-q}^k(i) - (-1)^q C_q^k(i))\right] \end{aligned} \tag{1.15}$$

In Equation 1.15, B_q^k and B'^k_q are the crystal field coefficients, which are all real, and $C_q^k(i)$ are tensor operators, related to the spherical harmonics $Y_q^k(i)$ by

$$C_q^k(i) = \sqrt{\frac{4\pi}{2k+1}}\, Y_q^k(i) \tag{1.16}$$

In much the same way as Stevens' operators, the summation in Equation 1.15 is limited to well-defined values: for f-electrons, the restriction $k \leq 7$ holds, while q is limited to those values consistent with the point symmetry of the site. Finally, the even part ($k = 0, 2, 4, 6$) is responsible for the CF splitting, while the odd part ($k = 1, 3, 5, 7$) is responsible for the intensity of induced electric dipole transitions in optical spectroscopy [5b, 26].

Crystal field energy levels can be found by diagonalizing the corresponding matrix, which is made up by elements of the type:

$$\langle l^n \tau SLJM_J | \widehat{\boldsymbol{H}}_{\mathrm{CF}}^{\mathrm{Wyb}} | l^{n'} \tau' S' L' J' M_J' \rangle \tag{1.17}$$

For low enough symmetries, both B_q^k and B'^k_q coefficients will be present in Equation 1.15, so that Equation 1.17 will be, in those cases, complex quantities. We finally note that the coefficients are transformed into CF parameters by multiplying them by the radial parts of the wave functions, represented by $R_{nl}(r)$, on which the tensor operators do not act.

Calculation of the angular part of the matrix elements thus remains, which can be performed exactly using tensor algebra techniques based on group theory. Since the calculation of the matrix elements is not straightforward, we provide here some details on it for the interested reader. The treatment follows the procedure described in Ref. [17].

The matrix element

$$\langle l^n \tau SLJM_J | \sum_i C_q^k(i) | l^{n'} \tau' S' L' J' M_J' \rangle \tag{1.18}$$

can be rewritten in terms of a unit tensor U_q^k as

$$\langle l^n \tau SLJM_J | U_q^k | l^{n'} \tau' S' L' J' M_J' \rangle \langle l \| C^k \| l' \rangle \tag{1.19}$$

The second term of the product is a reduced matrix element which contains the l-state dependence of the f electrons: since only $4f^n$ configuration, for which $l = 3$, is considered, it can be rewritten making use of Wigner–Eckart theorem and 3-j symbols for the coupling of angular momenta as

$$\langle l \| C^k \| l \rangle = (-1)^l [(2l+1)(2l'+1)]^{\frac{1}{2}} \begin{pmatrix} l & k & l' \\ 0 & 0 & 0 \end{pmatrix} = -7 \begin{pmatrix} 3 & k & 0 \\ 0 & 0 & 0 \end{pmatrix} \quad (1.20)$$

Using a similar approach, the first member of the product (Equation 1.19) may be rewritten as

$$\langle l^n \tau SLJM_J | U_k^q | l^{n'} \tau' SL'J'M_J' \rangle = (-1)^{J-M_J} \begin{pmatrix} J & k & J' \\ -M_J & q & M_J' \end{pmatrix} \langle l^n \tau SLJ \| U^k \| l^{n'} \tau' SL'J' \rangle \quad (1.21)$$

where the last term is a reduced matrix element, and is independent of M_J, q and M_J. Equation 1.21 can be further simplified by making use of Wigner 6-j symbols and a doubly reduced matrix element, obtaining:

$$\langle l^n \tau SLJM_J | U_k^q | l^{n'} \tau' SL'J'M_J' \rangle = (-1)^{S+L'+J+k} [(2J+1)(2J'+1)]^{\frac{1}{2}} \begin{Bmatrix} J & J' & k \\ L & L & S \end{Bmatrix} \langle l^n \tau SL \| U^k \| l^{n'} \tau' SL' \rangle \quad (1.22)$$

The last term on the right in Equation 1.22 represents a doubly reduced matrix element, which can be calculated by recursive formula in terms of the *coefficients of fractional parentage* [4, 14], tabulated in the work of Nielson and Koster [27]. Finally, Equation 1.18 is rewritten as

$$\begin{aligned} &\langle l^n \tau SLJM_J | \sum_i C_q^k(i) | l^{n'} \tau' S'L'J'M_J' \rangle \\ &= (-1)^{S+L'+2J-M_J+k+1} 7 [(2J+1)(2J'+1)]^{\frac{1}{2}} \begin{pmatrix} 3 & k & 3 \\ 0 & 0 & 0 \end{pmatrix} \\ &\times \begin{pmatrix} J & k & J' \\ -M_J & q & M_J' \end{pmatrix} \begin{Bmatrix} J & J' & k \\ L & L & S \end{Bmatrix} \langle l^n \tau SL \| U^k \| l^{n'} \tau' SL' \rangle \end{aligned} \quad (1.23)$$

Inspection of Equation 1.23 and consideration of the properties of 3-j and 6-j symbols confirm that only even k-values contribute to crystal field splitting. Further, it indicates that mixing between levels belonging to different J multiplets can only occur if terms with $k \leq J + J'$ and $-M_J + q + M_J' = 0$ are allowed by the site symmetry of the lanthanide, in much the same way as discussed above for the Stevens' formalism.

The advantage in using the Wybourne's approach in molecular magnetism is that it may provide a direct comparison with data obtained by optical spectroscopy, which is the most accurate technique for determination of the electronic structure of lanthanide ions. The application of this technique to molecular systems is just beginning, and while usually only the ground J multiplet is considered in the interpretation of the magnetic properties, the coupling of excited states can provide important effects on the wavefunction composition, thus affecting both static and dynamic magnetic properties of lanthanide-containing systems. In this respect, complementing the results of magnetic and EPR spectroscopy with those obtained by luminescence spectroscopy can provide detailed information on the energy pattern of 4f-containing systems. This approach is still seldom applied to the study of molecular magnets, but it has been shown in past few years to provide a much improved understanding of the relation between the electronic structure and the peculiar magnetic behaviour of these systems [28–30].

1.3.3 Standardization

When treating CF parameters in any of the two formalisms, non-specialists often overlook that the coefficients of the expansion of the CF potential (i.e. the values of CF parameters) depend on the choice of the coordinate system, so that conventions for assigning the correct reference framework are required. The conventional choice in which parameters are expressed requires the z-direction to be the principal symmetry axis, while the y-axis is chosen to coincide with a twofold symmetry axis (if present). Finally, the x-axis is perpendicular to both y- and z-axes, in such a way that the three axes form a right-handed coordinate system [31]. For symmetry in which no binary axis perpendicular to principal symmetry axis exists (e.g. C_{3h}, C_{4h}), y is usually chosen so as to set one of the B'^k_q (in Wybourne's approach) or A^q_k with $q < 0$ (in Stevens' approach) to zero, thereby reducing the number of terms providing a non-zero imaginary contribution to the matrix elements of the ligand field Hamiltonian. Finally, for even lower symmetry (orthorhombic or monoclinic), the correct choice is such that the ratio of the Stevens' parameter is restrained to $\lambda' = A^0_2/A^2_2 \in (0, \pm 1)$ and equivalently $\kappa = B^0_2/B^2_2 \in (0, \pm 1/\sqrt{6})$ in the Wybourne's notation [19, 32]. If literature-reported or experimentally determined parameters do not conform to this convention, rotation of the reference system should be applied, resulting in a standardized form of CF parameters [33]. This is of fundamental importance if different sets of parameters are to be compared to derive magnetostructural correlations and the direction of the quantization axis, and thus of the principal anisotropy axis, appropriately defined.

1.3.4 Calculation of Crystal Field Parameters

It is evident that the approach described so far to derive the electronic structure of lanthanide ions, based on perturbation theory, requires a large number of parameters to be determined. While state-of-the-art *ab initio* calculation procedures, based on complete active space self consistent field (CASSCF) approach, are reaching an extremely high degree of accuracy [34–37], the CF approach remains widely used, especially in spectroscopic studies. However, for low point symmetry, such as those commonly observed in molecular complexes, the number of CF parameters to be determined is as high as 27 (for C_i symmetry), which is clearly too large to be meaningful. To reduce the number of free parameters in the fit of either spectroscopic or magnetic data, some kind of *ab initio* calculation on the CF parameters is clearly necessary and we present here the basic concepts underlying some of them. Such calculations usually require the use of structural parameters (e.g. length and orientation of lanthanide–ligand atom distance with respect to a defined reference frame), and the determination of parameters describing the charge distribution on the ligand (or the ligand–lanthanide interaction in a molecular orbital oriented description), which should be consistent with chemical intuition.

The simplest approximation is that of considering a point-charge electrostatic model (PCEM), which parameterizes the crystal field effect generated by the n atoms coordinated to the lanthanide by using n point charges placed at the corresponding atomic positions. In other words, one may express Equation 1.13 as

$$\mathcal{U}_{\mathrm{CF}} = \sum_{j=1}^{nel.} \sum_{i=1}^{nlig} \frac{Z_i e^2}{r_{ij}} \tag{1.24}$$

The corresponding parameters obtained by this assumption do not provide, with some exceptions [38], acceptable fits to the experimentally available data for lanthanide-based molecular magnets. Indeed, such a model neglects the degree of covalent interaction of the metal–ligand configuration, and is thus only justified for ionic compounds. The situation for lanthanide(III) ions in a molecular compound with organic ligands can be rather different, and it has been already evidenced that for such compounds the CF can be strongly affected by covalency. Among the applications of models which take into account covalence to molecular magnets, it is worth mentioning the determination of CF parameters by using the simple overlap model. In this model, the ligand field is calculated by considering effective charges, located in the middle of the Ln–ligand bond, which are proportional to the total overlap between lanthanide and ligand wave functions and to charge factors [39]. Among other systems, this model has been successfully applied to rationalize the magnetic properties of holmium–nitrone complexes, which were then used as a starting point for the analysis of the exchange coupling in Ho-nitronyl nitroxide systems [40].

More recently, Coronado *et al.* [41] suggested a radial effective charge (REC) model, in which the effect of the ligating atom is modelled through an effective point charge situated along the lanthanide–ligand axis at a distance R, which is a parametric distance smaller than the real metal–ligand distance. At the same time, the charge value (q) is scanned in order to achieve the minimum deviation between calculated and experimental data. This was shown to work quite well for halide ligands, due to the spherical character of the electron density in the coordinating atom, while a lone pair covalent effective charge (LPEC) model was needed in the case of nitrogen ligands. In the latter case, a further displacement parameter was needed in order to account for the mismatch between the orientation of the nitrogen lone pair and the Ln–N direction. This model was used to explain the results obtained in one of the most common classes of single-ion magnet (SIM), namely lanthanide phthalocyanine sandwich complexes [41b]. Promising results on these systems have been obtained also by Klokishner *et al.* [42] by using the exchange-charge model [43], which considers both the interaction of the 4f electrons with the point charges of the surrounding ligands, and the overlap of the 4f orbitals with the ligand orbitals to contribute to the ligand field.

Among the most relevant approaches in the description of electronic properties of lanthanides is surely the angular overlap model (AOM) [44, 45]. The key idea underlying the model is best explained by looking at the main assumptions characterizing the AOM: (i) The energy of any f orbital E_f is obtained as a perturbation, which is proportional to the squares of metal–ligand overlap integrals. (ii) If the basis of the f-orbitals is defined relative to a coordinate system *xyz*, then the perturbation matrix due to a ligand placed on *z* is diagonal. (iii) Contributions arising from different ligands are additive.

The parameters describing the interaction between the ligand and the lanthanide are directly related to the σ-, π-, δ- and φ-bonding ability of the ligands, which may be influenced by the synthetic chemist. Moreover it is, in principle, possible to calculate the effect of the ligand by using the real coordination geometry around the metal ion site, thus including effects due to the low symmetry of the ligand field. The relations between angular overlap parameters and CF parameters was first reported by Urland for the case of isotropic π-ligands [44b]. This was successfully improved by including anisotropic π-ligands and applied to the interpretation of susceptibility, electronic and EPR spectra of $[Ph_4As]_2[Yb(NO_3)_5]$ [46] and of the extremely well-characterized series of $[Ln(H_2O)_9][EtOSO_3]$, which showed that electrostatic effects due to neighbouring complexes become relevant in the determination of axial second-order ligand field parameters [47]. More recently, Flanagan *et al.* [45] provided relations which also include anisotropic π-ligands, and applied the AOM to a global fit of CF parameters derived by polarized luminescence spectroscopy. This approach was used for the series of homologous compounds Ln(trensal) (where H_3trensal = $2, 2', 2''$-tris(salicylideneimino)triethylamine), where the variations of the AOM parameters were related to the small structural changes observed along the series. In particular, it was concluded that the metal–ligand overlap decreases on moving across the series as the result of the contraction of the f orbitals, which overcomes the effect of the decreasing bond lengths.

Finally, we note that even if in these studies only e_σ and e_π parameters were used for the description of the ligand field, we have recently shown, by applying AOM to the interpretation of the magnetic properties of Dy^{3+} derivatives of both polyoxometallate and phthalocyaninate, that e_δ cannot always be considered as negligible [48].

1.4 Magnetic Properties of Isolated Lanthanide Ions

1.4.1 Effect of a Magnetic Field

The application of a magnetic field to the wavefunctions obtained by the procedure described in the previous sections results in the complete removal of the degeneracy of the J multiplet, either pertaining to Kramers or non-Kramers ions, and yields a temperature-dependent population of the different $2J + 1$ components (Figure 1.2) Thus, at low temperatures, large deviations from the Curie law are observed. The effect of the magnetic field is described by the Zeeman Hamiltonian:

$$\hat{H}_{\text{Zeem}} = -\hat{\boldsymbol{\mu}} \cdot \boldsymbol{B} \tag{1.25}$$

where

$$\hat{\boldsymbol{\mu}} = -\mu_{\text{B}}[(\hat{L}_x + 2\hat{S}_x)\boldsymbol{i} + (\hat{L}_y + 2\hat{S}_y)\boldsymbol{j} + (\hat{L}_z + 2\hat{S}_z)\boldsymbol{k}] \tag{1.26}$$

For a general orientation α of the magnetic field B, with intensity $|B|$ and components (B_x, B_y, B_z), Equation 1.25 can be rewritten as

$$\hat{H}_{\text{Zeem},\alpha} = \mu_{\text{B}}[(\hat{L}_x + 2\hat{S}_x)B_x + (\hat{L}_y + 2\hat{S}_y)B_y + (\hat{L}_z + 2\hat{S}_z)B_z] \tag{1.27}$$

The effect of $\hat{H}_{\text{Zeem},\alpha}$ is evaluated by applying it to the eigenfunctions resulting from the crystal field analysis, which provides the eigenvectors $|n_\alpha\rangle$ and the energies $E_{n,\alpha}$. The resulting magnetization in a magnetic field applied along the direction α is written as

$$M_\alpha = N_{\text{A}} \frac{\sum_{n,\alpha} -\frac{\partial E_{n,\alpha}}{\partial B} \exp\left(-\frac{E_{n,\alpha}}{k_{\text{B}}T}\right)}{\sum_{n,\alpha} \exp\left(-\frac{E_{n,\alpha}}{k_{\text{B}}T}\right)} \tag{1.28}$$

where the values of $\partial E_{n,\alpha}/\partial B$ can be calculated by applying the operator $\hat{H}_{\text{Zeem},\alpha}/B$ on the eigenvectors $|n_\alpha\rangle$. As a consequence of the crystal field effects, $\partial E_{n,\alpha}/\partial B$ can be strongly dependent on α, thus providing a magnetization which can be largely anisotropic even at room temperature [49]. Single crystal measurements, of magnetization, susceptibility or torque, are then very powerful to obtain an accurate determination of CF parameters. On the other hand, magnetic properties of powder samples are much less sensitive to CF parameters, especially the transverse ones (i.e. $q \neq 0$), so that reliable estimates can only be obtained in these cases when additional information is available (i.e. homologous series of lanthanides, further spectroscopic information). Quite often, susceptibility measurements are indeed interpreted by assuming an idealized site symmetry neglecting most of the supposedly small transverse term of the crystal field [24, 50]. This has to be accurately considered when trying to correlate the low temperature dynamic behaviour of the magnetization with the eigenvalues and eigenvectors pattern arising by static magnetic measurements.

1.4.2
EPR Spectroscopy of Lanthanide Complexes

EPR spectroscopy has been heavily applied in the past to the analysis of the magnetic properties of compounds containing ions of the rare-earth group [6]. Results obtained by this technique were indeed one of the driving forces in the derivation of a consistent theory of the crystal field. In particular, some degree of admixture of low-lying excited states in the ground state had often to be taken into account to fit the experimental data, still within a model of ionic character. Due to the unquenched orbital contribution, EPR spectra of lanthanide ions are usually observed only at low temperatures, since the resulting fast spin–lattice relaxation times hamper signal detection at higher temperatures. As a consequence, they are normally interpreted by considering that only the lowest levels arising from the ligand field splitting of the ground J multiplet are EPR active.[2)]

If the Ln^{3+} centre is a Kramers ion, the spectra can be interpreted in terms of a doublet with largely anisotropic effective g-values. If one neglects the admixture of higher lying J multiplets and considers an axial symmetry, the effective g values will be

$$\begin{aligned} g_{//} &= 2g_J\langle\Psi_+|\hat{J}_z|\Psi_+\rangle \\ g_{\perp} &= g_J\langle\Psi_+|\hat{J}_+|\Psi_-\rangle \end{aligned} \tag{1.29}$$

The wavefunctions $\Psi_\pm$ can be written as

$$\Psi_\pm = \sum c_{M_J}|M_J, J\rangle \tag{1.30}$$

The number of terms retained in Equation 1.30 obviously depends on the symmetry of the ligand field: if an axis of q-fold symmetry is present, only M_J values differing by $\pm q$ will contribute to Equation 1.30. Note that for rhombic and lower symmetries, two different effective values of g are expected for the x and y directions. In this respect, EPR is extremely sensitive to transverse terms of the CF,

2) To further reduce the linewidth of the spectra, doping in isostructural Y^{3+}, La^{3+} or Lu^{3+} host is usually performed to minimize dipolar broadening and spin–spin relaxation.

since they not only alter c_{M_J} coefficients, and thus the effective g values, but may provide the only mechanism to obtain a non-zero transition probability within the doublet.

In more general terms, considering also the mixing of higher lying J states, the ground doublet wavefunction will be of the type

$$\Psi_{\pm} = \sum c_{M_J} |M_J, J\rangle + \sum c'_{M_J} |M_J, J'\rangle \tag{1.31}$$

While this generally has only small effects on the static magnetic properties, it may play a relevant role in providing a quantitative agreement with experimental resonance data, and in explaining quantum tunnelling relaxation of the magnetization at low temperature [51].

For non-Kramers lanthanide ions, the ligand field splitting may provide, depending on the symmetry, either singlets, doublets or pseudo-doublets. In this situation, spectra are usually analysed, depending on the symmetry, either as pseudo-doublets or pseudo-triplets (if the energy difference between singlet and doublet or pseudo-doublet groups of levels is relatively small) with very anisotropic g-value and a sizeable zero field splitting (ZFS) term of the effective spin Hamiltonian:

$$\hat{H}_{\text{nK}} = \mu_{\text{B}} \boldsymbol{B} \cdot \boldsymbol{g} \cdot \boldsymbol{S}_{\text{eff}} + \boldsymbol{S}_{\text{eff}} \cdot \boldsymbol{D} \cdot \boldsymbol{S}_{\text{eff}} \tag{1.32}$$

Despite the historical relevance of EPR spectroscopy in the elucidation of the electronic structure of lanthanide compounds, relatively few results have been reported in the domain of molecular magnetism. The vast majority of these studies, as will be seen hereafter, were focused on complexes of the second half of the lanthanide series, to gain more information about the slow relaxation of the magnetization in these systems. Among the few exceptions, our group reported a few powder EPR spectra of the family $Ln(trp)(HBPz_3)_2$ (where trp, tropolonate, $HBPz_3$, hydrotrispyrazolylborate), of which only $Yb(trp)(HBPz_3)_2$ could be fully interpreted and simulated. Despite the rhombic structural symmetry, this system provided a near-axial spectrum, indicating that the best description of the ligand field geometry is square antiprismatic with fourfold symmetry. The observed effective g-values were rationalized on the basis of this idealized geometry, and the ground doublet of $Yb(trp)(HBPz_3)_2$ turned out to be composed by $|J, M_J\rangle = |7/2, \pm 5/2\rangle$ for 98% [52].

A more complex situation was encountered in the analysis of single-crystal EPR spectra of the member of $[Ce(dmf)_4(H_2O)_3(\mu\text{-}CN)Co(CN)_5]$ (dmf, N,N'-dimethylformamide, Co = Co^{3+}) [53]. In this case, the investigation provided the orientation of the principal g-values of Ce^{3+} ground doublet, which, quite interestingly, did not show any peculiar relation to the axis of the bicapped trigonal prism formed by the ligands (Figure 1.3). Furthermore, on an analysis based on orthorhombic symmetry and the lowest $J = 5/2$ value only, Equation 1.30 yielded calculated g-values which agree with the observed values only within 12%. The observed discrepancy was attributed both to a small amount of mixing

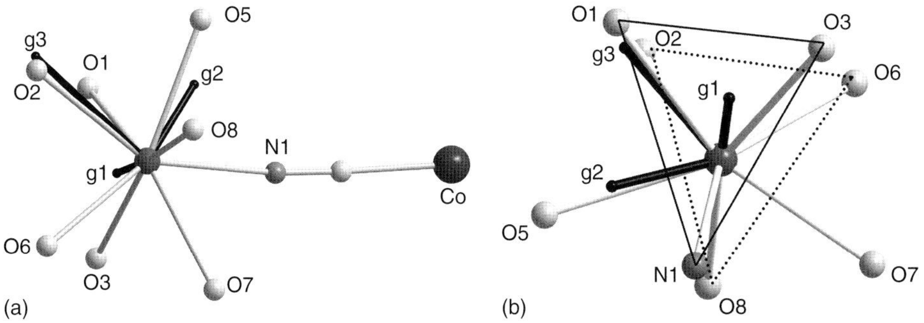

Figure 1.3 Orientation of the g tensor of Ce^{3+} in $[Ce(dmf)_4(H_2O)_3(\mu\text{-}CN)Co(CN)_5]$ with respect to the bicapped trigonal prism of the coordination sphere of Ce^{3+} centre (a), and with respect to the Co–C–N–Ce plane (b). For both figures, each component of the tensor is proportional to its respective modulus. (After Ref. [53], WIley-VCH.)

of the $^2F_{7/2}$ state in the ground doublet and to the simplifying assumption of orthorhombic symmetry.

The importance of EPR spectroscopy in evidencing low-symmetry terms of the CF was reported by one of us, in collaboration with the group of Boskovic and Kogerler [24], in the investigation of a Dy^{3+} polyoxometallate, based on the $[As_2W_{19}O_{67}(H_2O)]^{14-}$ anion. X-band EPR spectra were strongly temperature dependent, with a signal reaching a maximum intensity around 40 K: on the basis of the fit of magnetic susceptibility, which neglected the transverse CF terms, this was assigned to a doublet with prevailing $M_J = \pm 9/2$ component, which, according to ligand field calculations, lies 34 K above the ground state. The observation of an EPR spectrum with $g_\perp \neq 0$ demonstrated unequivocally that the transverse fourth-order ligand-field components are non-zero, since these terms provide the only means to observe a non-zero perpendicular component of the spectrum. Calculation based on Equations 1.29–1.30 suggested that a rather large mixing is induced by these terms, indicating the importance of spectroscopic information complementing the thermodynamic ones in providing an accurate picture of the electronic structure of these systems. A similar phenomenon was more recently observed by Schelter and coworkers [54], who performed X-band EPR spectroscopy on two of the Dy^{3+} complexes, showing that, while the ground state was essentially $M_J = \pm 13/2$ in nature, non-negligible mixing due to low symmetry components of the ligand field occurred.

Finally, we note that, to the best of our knowledge, only one report exists about EPR spectra of non-Kramers lanthanide ions in molecular magnets. In 2012, Hill and coworkers [51] performed a multifrequency study on powder and single crystal samples of $Na_9Ho(W_5O_{18})_2 \cdot nH_2O$, in both the pure form and when doped into the isostructural Y^{3+} derivative. While crystallizing in a triclinic unit cell, the symmetry of the lanthanide ion in this family is very close to D_{4d}. For this reason, susceptibility data had been previously fitted by a purely axial Hamiltonian [50].

However, consistent explanation of the EPR properties of the Ho^{3+} derivative required the inclusion of at least transverse fourth-order terms. The parameters obtained by accurate single-crystal EPR simulations confirmed the dominant $M_J = \pm 4$ character of the ground pseudo-doublet, which is separated by about $16\,cm^{-1}$ from the first excited state, of essentially $M_J = \pm 5$ character. More important, however, is that the two components of the ground state are partially split both by a fourth-order transverse term and by hyperfine interaction with the $I = 7/2$ Ho nucleus, resulting in the pattern outlined in Figure 1.4. In particular, the former terms are the only ones capable of explaining the actual observation of non-zero intensity in the EPR spectrum. Furthermore, they are very efficient (to just the second-order of perturbation) in mixing the two components of the ground state, thus explaining the field-dependent spin dynamics of this system. Such studies clearly highlight the relevance of EPR techniques in unveiling subtle details of the electronic structure of lanthanide-based molecular magnets, which are otherwise inaccessible by magnetic characterization. This is of paramount importance in determining the key factors affecting their low temperature spin dynamics. In particular, the possibility of having access to details of hyperfine interaction may be of much relevance to the explanation of hyperfine mediated tunnelling relaxation [55, 56].

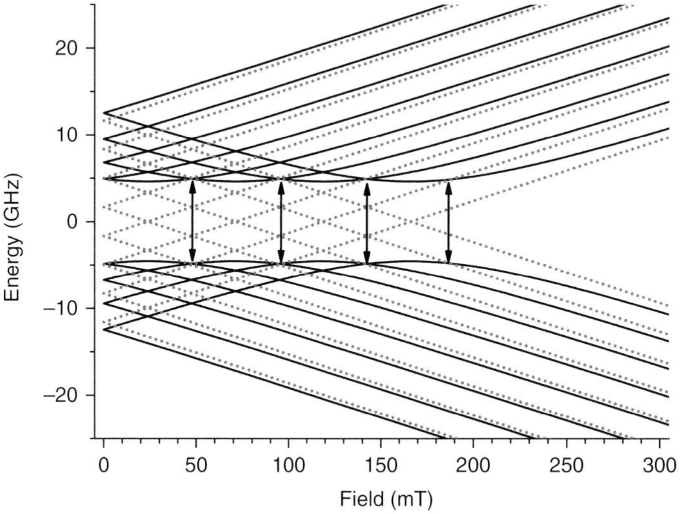

Figure 1.4 Plot of the lowest hyperfine split pseudo-doublet for field applied along the easy magnetization axis of $Na_9Ho(W_5O_{18})_2 \cdot nH_2O$. Grey dotted lines are obtained in the assumption of purely axial symmetry while black continuous lines result after inclusion of symmetry allowed fourth-order transverse anisotropy. Black arrows represent the calculated resonance position in the latter assumption for X-band frequency. (Redrawn using original data reported in Ref. [51], Royal Society of Chemistry.)

1.5
Exchange Coupling in Systems Containing Orbitally Degenerate Lanthanides

Up to now, we have only treated systems containing a single lanthanide ion. We wish to introduce here some of the problems which are encountered when dealing with exchange coupling in systems containing orbitally degenerate ions. The main problem in evaluating the exchange coupling in such compounds lies in the fact that one cannot use the isotropic spin Hamiltonian approach which is normally adopted to parameterize the magnetic interactions in compounds containing orbitally non-degenerate centres [57]. Indeed, when systems with unquenched angular orbital momentum are considered, S is no longer a good quantum number, and the energies of the levels depend also on the value of M_S, with $-S \leq M_S \leq S$, that is, there will be some preferential orientation of the magnetic moment even in zero field. This is the main reason for the peculiar difficulties arising in the analysis of the magnetic behaviour of the lanthanides. The first detailed and quantitative treatment of the interactions involving one orbitally non-degenerate ion, such as Fe^{3+}, and an orbitally degenerate lanthanide ion was performed by Levy [58] in an attempt to rationalize the exchange interactions in Yb^{3+} garnets, and was later employed by Yamaguchi and Kamimura [59] to analyse the behaviour of Ho^{3+} garnets. The assumptions on which this approach relies are (i) that the spin dependence of the superexchange interaction between a 4f electron and the orbitally non-degenerate state ions is well described by the Dirac–Van Vleck Hamiltonian and (ii) that the dependence of this interaction on the orbital state of the lanthanide is accounted for by an exchange potential. The exchange interaction is then described by an anisotropic exchange Hamiltonian exploiting the formalism of irreducible tensor operators $T_q^{[k]}$:

$$\widehat{\boldsymbol{H}}_{\text{exc}} = \sum_{k=0}^{2l} \sum_{q=-k}^{k} \alpha_{kq} \boldsymbol{T}_q^{[k]}(i)\boldsymbol{s}(i) \cdot \boldsymbol{S}(TM) \tag{1.33}$$

where $l = 3$ for lanthanides, i indicates the i-th electron of the lanthanide ion, $S(TM)$ is the spin operator of the transition metal ion and α_{kq} are the exchange coupling parameters, α_{00} being the isotropic one. Furthermore, it is necessary to know the wavefunctions of the lanthanide (and thus the relative coefficient of S values entering Equation 1.33 in the absence of exchange, which can be obtained by the CF parameters as discussed in Section 1.3). It is evident that the number of adjustable parameters is now huge, especially in the low site symmetry usually characteristic of molecular complexes of Ln^{3+}, and a meaningful fit of the experimental parameters becomes extremely difficult. Furthermore, it is in principle absolutely necessary to perform orientation-dependent measurements because the global Hamiltonian, including both exchange and single ion contributions, is intrinsically anisotropic. For this reason, there have been no reports of the application of this approach to molecular complexes, after the series of papers by the Florence University group [60], which synthesized and magnetically characterized different copper–lanthanide molecular complexes. CF parameters were obtained by an independent fit of the powder magnetic susceptibility of $[Ln(hfac)_3(H_2O)_2]$ (hfac, hexafluoroacetylacetonate) and fixed in order to reduce the parameterization of the system. It was then possible to find out that the isotropic term of the coupling Hamiltonian was antiferromagnetic while the anisotropic ones were ferromagnetic: to reduce the number of parameters, however, only second-order anisotropic exchange was considered.

More often, a simple qualitative approach has been used to get information on the nature of the interaction (ferro- or antiferromagnetic) in lanthanides coupled to organic radicals or to copper(II). Within this approach, the exchange interaction in coupled systems is made apparent by subtracting from χT of the complex the contribution arising from the thermal depopulation of the M_J sublevels of Ln^{3+}, $\chi_{Ln} T$, which is obtained by measuring an isostructural Ln^{3+} complex with a diamagnetic surrounding. This approach has been successfully applied both to the investigation of 4f–3d couples, for example, by replacing the paramagnetic copper(II) by a square-planar nickel(II) or zinc(II), and to 4f–2p complexes, where the organic radical was substituted by a diamagnetic ligand analogue carrying the same charge [52, 61–63].

Extension of this approach to quantitative estimation of the exchange coupling has also been reported by different groups [40, 42] by fitting the data of the diamagnetic substitute analogue with one of the CF calculation models described in Section 1.3.4. This allows derivation of the eigenfunctions and eigenvalues of the ground multiplet of the isolated lanthanide. The exchange interaction is subsequently evaluated in the assumption of isotropic coupling, which is often sufficient to yield a reasonable fit of the powder susceptibility.

As powerful as it can be, the diamagnetic substitution approach does not provide reliable information about the anisotropic features of the exchange between the species involved if only powder measurements are considered. We have recently demonstrated this by an integrated single crystal EPR and magnetic study on the $[Ln(dmf)_4(H_2O)_3(\mu\text{-}CN)M(CN)_5]$ family discussed above, with M = Co^{3+}, Fe^{3+} and Ln = La, Ce. Investigation of the two members of the family containing one paramagnetic centre resulted in a detailed picture of the low-lying levels of Ce^{3+} and Fe^{3+} ions. Analysis of the coupled species performed using this piece of information clearly indicated that a sound explanation of the data required the inclusion of isotropic, anisotropic and antisymmetric terms to describe the exchange interaction among the doublets [53].

Acknowledgements

We are grateful to Prof. R. Sessoli, Prof. C. Benelli, Mrs. E. Lucaccini and Mr. M. Perfetti for stimulating discussions.

References

1. Becquerel, J. (1908) *Le Radium*, **5**, 5–17.
2. Stevens, K.W.H. (1952) *Proc. Phys. Soc. London, Sect. A*, **65**, 209–215.
3. Wybourne, B.G. (1965) *Spectroscopic Properties of Rare Earths*, John Wiley & Sons, Inc., New York.
4. Judd, B.R. (1963) *Operator Techniques in Atomic Spectroscopy*, McGraw-Hill, New York.
5. (a) Ofelt, G.S. (1963) *J. Chem. Phys.*, **38**, 2171–2180; (b) Ofelt, G.S. (1962) *J. Chem. Phys.*, **37**, 511–520.
6. Abragam, A. and Bleaney, B. (1970) *Electron Paramagnetic Resonance of Transition Ions*, Clarendon Press, Oxford.
7. Dieke, G.H. and Crosswhite, H. (1968) *Crosswhite Spectra and Energy Levels of Rare Earth Ions in Crystals*, Interscience Publishers, New York.
8. Kahn, O. (1993) *Molecular Magnetism*, Wiley-VCH Verlag GmbH, New York.
9. Van Vleck, J.H. (1932) *The Theory of Electric and Paramagnetic Susceptibilities*, Oxford University Press, Oxford.
10. Hund, F. (1925) *Z. Phys.*, **33**, 855–859.
11. Laporte, O. (1928) *Z. Phys.*, **47**, 761–769.
12. (a) Van Vleck, J.H. and Frank, A. (1929) *Phys. Rev.*, **34**, 1494–1496; (b) Van Vleck, J.H. and Frank, A. (1929) *Phys. Rev.*, **34**, 1625–1625.
13. (a) Slater, J.C. (1929) *Phys. Rev.*, **34**, 1293–1322; (b) Condon, E.U. and Shortley, G.H. (1951) *The Theory of Atomic Spectra*, Cambridge University Press, Cambridge.
14. Racah, G. (1949) *Phys. Rev.*, **76**, 1352–1365.
15. Casey, A.T. and Mitra, S. (1976) in *Theory and Applications of Molecular Paramagnetism* (eds E.A. Boudreaux and L.N. Mulay), Wiley-Interscience, New York, 271–316.
16. Bethe, H. (1929) *Ann. Phys.*, **395**, 133–208.
17. Gorller-Walrand, C. and Binnemans, K. (1996) Rationalization of crystal-field parametrization, in *Handbook On the Physics and Chemistry of Rare Earths*, vol. 23 (eds K.A. Gschneidner and L. Eyring), Elsevier, Amsterdam.
18. Newman, D.J. and Ng, B. (2000) Empirical crystal fields, in *Crystal Field Handbook* (eds D.J. Newman and B. Ng), Cambridge University Press, Cambridge.
19. Rudowicz, C. (1986) *J. Chem. Phys.*, **84**, 5045–5058.
20. Rudowicz, C. and Chung, C.Y. (2004) *J. Phys. Condens. Matter*, **16**, 1–23.
21. Altshuler, S.A. and Kozyrev, B.M. (1974) *Electron Paramagnetic Resonance in Compounds of Transition Elements*, 2nd edn, John Wiley & Sons, Inc., New York.
22. (a) Elliott, R.J. and Stevens, K.W.H. (1952) *Proc. R. Soc. London, Ser. A*, **215**, 437–453; (b) Elliott, R.J. and Stevens, K.W.H. (1953) *Proc. R. Soc. London, Ser. A*, **218**, 553–566; (c) Elliott, R.J. and Stevens, K.W.H. (1953) *Proc. R. Soc. London, Ser. A*, **219**, 387–404.
23. Schilder, H. and Lueken, H. (2004) *J. Magn. Magn. Mater.*, **281**, 17–26.
24. Ritchie, C., Speldrich, M., Gable, R.W., Sorace, L., Kogerler, P. and Boskovic, C. (2011) *Inorg. Chem.*, **50**, 7004–7014.
25. Reu, O., Palii, A., Ostrovsky, S., Wallace, W., Zaharko, O., Chandrasekhar, V., Clerac, R. and Klokishner, S. (2013) *J. Phys. Chem. C*, **117**, 6880–6888.
26. Judd, B.R. (1962) *Phys. Rev.*, **127**, 750–761.
27. Nielson, C.W. and Koster, G.F. (1963) *Spectroscopic Coefficients for the* p^n,

d^n and f^n Configurations, MIT press, Boston, MA.
28. Cucinotta, G., Perfetti, M., Luzon, J., Car, P.-E., Mael, E., Caneschi, A., Calvez, G., Bernot, K. and Sessoli, R. (2012) *Angew. Chem. Int. Ed.*, **51**, 1606–1610.
29. Rinehart, J.D. and Long, J.R. (2012) *Dalton Trans.*, **41**, 13572–13574.
30. (a) Lucaccini, E., Sorace, L., Perfetti, M., Costes, J.-P. and Sessoli, R. (2014) *Chem. Commun.*, **50**, 1648–1651; (b) Pedersen, K.S., Ungur, L., Sigrist, M., Sundt, A., Schau-Magnussen, M., Vieru, V., Mutka, H., Rols, S., Weihe, H., Waldmann, O., Chibotaru, L.F., Bendix, J. and Dreiser, J. (2014) *Chem. Sci.*, **5**, 1650.
31. Prather, J.L. (1961) *Atomic Energy Levels in Crystals*, National Bureau of Standards Monograph, U.S. National Bureau of Standards, Washington, DC.
32. Rudowicz, C., Chua, M. and Reid, M.F. (2000) *Physica B*, **291**, 327–338.
33. Rudowicz, C. (1985) *J. Phys. C: Solid State Phys.*, **18**, 1415–1430.
34. Chibotaru, L.F., Ungur, L. and Soncini, A. (2008) *Angew. Chem. Int. Ed.*, **47**, 4126–4129.
35. Luzon, J., Bernot, K., Hewitt, I., Anson, C., Sessoli, R. and Powell, A.K. (2008) *Phys. Rev. Lett.*, **100**, e247205.
36. Ungur, L. and Chibotaru, L.F. (2011) *Phys. Chem. Chem. Phys.*, **13**, 20086–20090.
37. Boulon, M.E., Cucinotta, G., Luzon, J., Degl'Innocenti, C., Perfetti, M., Bernot, K., Calvez, G., Caneschi, A. and Sessoli, R. (2013) *Angew. Chem. Int. Ed.*, **52**, 350.
38. Baldoví, J.J., Cardona-Serra, S., Clemente-Juan, J.M., Coronado, E. and Gaita-Ariño, A. (2012) *Inorg. Chem.*, **51**, 12565–12574.
39. Porcher, P., Dos Santos, M.C. and Malta, O. (1999) *Phys. Chem. Chem. Phys.*, **1**, 397.
40. Kahn, M.L., Ballou, R., Porcher, P., Kahn, O. and Sutter, J.P. (2002) *Chem. Eur. J.*, **8**, 525.
41. (a) Baldoví, J.J., Cardona-Serra, S., Clemente-Juan, J.M., Coronado, E., Gaita-Ariño, A. and Palii, A. (2013) *J. Comput. Chem.*, **34**, 1961–1967; (b) Baldovi, J.J., Clemente-Juan, J.M., Coronado, E. and Gaita-Ariño, A. (2012) *Dalton Trans.*, **41**, 13705–13713.
42. Reu, O.S., Palii, A.V., Ostrovsky, S.M., Tregenna-Piggott, P.L.W. and Klokishner, S.I. (2012) *Inorg. Chem.*, **51**, 10955–10965.
43. Popova, M.N., Chukalina, E.P., Malkin, B.Z. and Saikin, S.K. (2000) *Phys. Rev. B*, **61**, 7421.
44. (a) Jørgensen, C.K., Pappalardo, R. and Schmidtke, H.-H. (1963) *J. Chem. Phys.*, **39**, 1422; (b) Urland, W. (1976) *Chem. Phys.*, **14**, 393–401.
45. Flanagan, B.M., Bernhardt, P.V., Luthi, S.R., Riley, M.J. and Krausz, E.R. (2001) *Inorg. Chem.*, **40**, 5401.
46. Urland, W. and Kremer, R. (1984) *Inorg. Chem.*, **23**, 1550–1553.
47. Urland, W. (1979) *Chem. Phys. Lett.*, **62**, 525–528.
48. Sorace, L., Benelli, C. and Gatteschi, D. (2011) *Chem. Soc. Rev.*, **40**, 3092–3104.
49. Mironov, V.S., Galyametdinov, Y.G., Ceulemans, A., Gorller-Walrand, C. and Binnemans, K. (2002) *J. Chem. Phys.*, **116**, 4673–4685.
50. AlDamen, M.A., Clemente Juan, J.M., Coronado, E., Martì-Gastaldo, C. and Gaita-Ariño, A. (2008) *J. Am. Chem. Soc.*, **130**, 8874–8875.
51. Ghosh, S., Datta, S., Friend, L., Cardona-Serra, S., Gaita-Ariño, A., Coronado, E. and Hill, S. (2012) *Dalton Trans.*, **41**, 13697–13704.
52. Caneschi, A., Dei, A., Gatteschi, D., Poussereau, S. and Sorace, L. (2004) *Dalton Trans.*, 1048–1055.
53. Sorace, L., Sangregorio, C., Figuerola, A., Benelli, C. and Gatteschi, D. (2009) *Chem. Eur. J.*, **15**, 1377–1388.
54. Williams, U.J., Mahoney, B.D., DeGregorio, P.T., Carroll, P.J., Nakamaru-Ogiso, E., Kikkawa, J.M. and Schelter, E.J. (2012) *Chem. Commun.*, **48**, 5593–5595.
55. Giraud, R., Wernsdorfer, W., Tkachuk, A.M., Mailly, D. and Barbara, B. (2001) *Phys. Rev. Lett.*, **87**, 057203.
56. Vincent, R., Klyatskaya, S., Ruben, M., Wernsdorfer, W. and Balestro, F. (2012) *Nature*, **488**, 357–360.

57. Borras-Almenar, J.J., Clemente-Juan, J.M., Coronado, E., Palii, A.V. and Tsukerblat, B.S. (1998) *J. Phys. Chem. A*, **102**, 200–213.
58. (a) Levy, P.M. (1964) *Phys. Rev.*, **135**, A155–A165; (b) Levy, P.M. (1966) *Phys. Rev.*, **147**, 311–319.
59. Kamimura, H. and Yamaguchi, T. (1970) *Phys. Rev. B*, **1**, 2902–2911.
60. (a) Benelli, C., Caneschi, A., Gatteschi, D., Guillou, O. and Pardi, L. (1990) *J. Magn. Magn. Mater.*, **83**, 522–524; (b) Benelli, C., Caneschi, A., Gatteschi, D., Guillou, O. and Pardi, L. (1990) *Inorg. Chem.*, **29**, 1750–1755.
61. Costes, J.P., Dahan, F., Dupuis, A. and Laurent, J.P. (1998) *Chem. Eur. J.*, **4**, 161–1620.
62. Kahn, M.L., Sutter, J.P., Golhen, S., Guionneau, P., Ouahab, L., Kahn, O. and Chasseau, D. (2000) *J. Am. Chem. Soc.*, **122**, 3413–3421.
63. Figuerola, A., Diaz, C., Ribas, J., Tangoulis, V., Lloret, F., Mahia, J. and Maestro, M. (2003) *Inorg. Chem.*, **42**, 641–649.

2
Mononuclear Lanthanide Complexes: Use of the Crystal Field Theory to Design Single-Ion Magnets and Spin Qubits

Juan M. Clemente-Juan, Eugenio Coronado, and Alejandro Gaita-Ariño

2.1
Introduction

The magnetic behaviour of single-molecule magnets (SMMs) is characterized by the presence of slow relaxation of the magnetization at low temperatures (superparamagnetic blocking), which gives rise to a magnetic memory effect similar to that found in hard magnets, together with quantum tunnelling effects. Contrary to conventional bulk magnets and molecule-based magnets, collective long-range magnetic ordering of magnetic moments is not necessary and the molecule exhibits its characteristic magnetic behaviour even at high dilutions in a diamagnetic matrix. This property of some coordination complexes makes them promising candidates for high-density memory storage. In quantum computing terms one considers these two-state molecules as the fundamental unit of quantum information (qubits) [1]. There is a large family of SMMs based on transition metal clusters. The most quoted example is $[Mn_{12}O_{12}(O_2CMe)_{16}(H_2O)_4]$ (Mn_{12}ac) [2] a molecule that combines a high axial anisotropy and a high spin ground state [3].

In the last few years a new approximation has been proposed to design SMMs that use lanthanide (and even actinide) coordination complexes as the source of anisotropic magnetic moments. Even mononuclear metal complexes, instead of a magnetic cluster, have been shown to provide the basic ingredients needed for detecting SMM behaviour, namely, axial anisotropy and large magnetic moment. This class of nanomagnets is known as single-ion magnets (SIMs). The first SIMs were the double-decker phthalocyaninato lanthanide complexes, in short $LnPc_2$, first reported by Ishikawa *et al.* [4] in 2003 (Figure 2.1). In these $LnPc_2$ complexes, the lanthanide ion is 8-coordinate and submitted to a pseudo-axial crystal field (CF) of D_{4d} symmetry. This CF splits the ground state, described by a total magnetic moment J, into $\pm M_J$ doublets. When this splitting is such that the lowest lying M_J doublet is such that the ground doublet has a high $|M_J|$ value, a thermal energy barrier, which often exceeds several hundreds of wave numbers [5], may appear for the reversal of the magnetization, thus leading to SMM behaviour.

Lanthanides and Actinides in Molecular Magnetism, First Edition.
Edited by Richard A. Layfield and Muralee Murugesu.

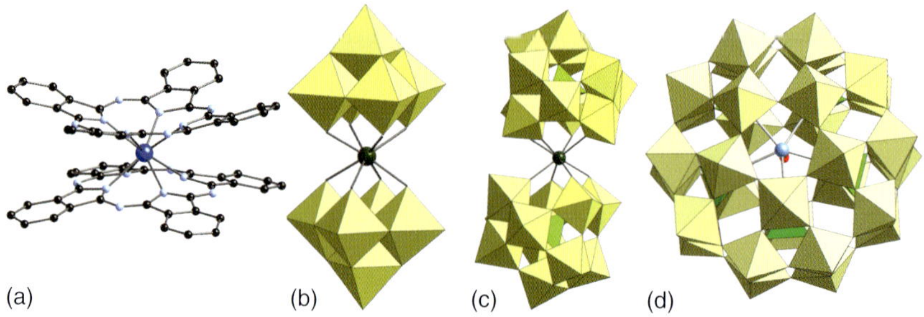

Figure 2.1 Structure of (a) terbium phthalocyaninato derivative, (b) $[Ln(W_5O_{18})_2]^{9-}$ ion, (c) $[Ln(\beta_2\text{-}SiW_{11}O_{39})_2]^{13-}$ ion and (d) Preyssler anion $[LnP_5W_{30}O_{110}]^{12-}$.

Recognizing that the key factor to obtain SMM behaviour is the CF symmetry around the Ln(III) ion, we explored the possibility of getting such behaviour in the mononuclear lanthanide complexes based on polyoxometalates (POMs), and in 2008 we reported the first example of this kind [6–8]. Thus, the erbium derivative of the series $[Ln(W_5O_{18})_2]^{9-}$ (in short LnW_{10}) was found to behave as an SMM. As observed in the $LnPc_2$ complexes, this polyanion presents a 'sandwich' type structure where the lanthanide ion also has D_{4d} coordination symmetry (Figure 2.1). Here, the lanthanide has a compressed environment so that, as we will discuss later, the sublevel scheme caused by the CF splitting is very different from that of $LnPc_2$ complexes for the same metals. Still, symmetries other than D_{4d} have also shown SMM behaviour. This is the case of the series $[LnP_5W_{30}O_{110}]^{12-}$ (in short LnW_{30}) (Figure 2.1), which have fivefold symmetry. This unusual symmetry gives rise to remarkably large off-diagonal anisotropy parameters [9]. The spin dynamics, especially at low temperatures, are then dominated by fast tunnelling processes [10].

After our report, many other examples of SIMs based on mononuclear lanthanide complexes have appeared. As relevant examples, we should mention the erbium-organometallic double-decker complexes [11] and the dysprosium–acetylacetonate complexes [12], and the dysprosium-DOTA (H4DOTA, 1,4,7,10-tetraazacyclododecane-1,4,7,10-tetraacetic acid) complexes reported by Sessoli *et al.* [13].

The CF theory deals with the effect of coordination on the ion spin states. In the case of lanthanide complexes, this model provides a useful theoretical framework to rationalize their magnetic properties [14]. In this chapter, we will describe a general theoretical approach that considers a point charge approximation to calculating the energy spectrum and the wave function of the low-lying magnetic levels of a lanthanide ion submitted to the CF created by the surrounding ligands. With this rough geometrical information, the non-vanishing CF parameters can be estimated and general trends about the most suitable coordination geometries needed to design SIMs can be obtained. Finally, more quantitative information about parameters, levels and magnetic properties can be extracted using an effective description of each donor atom, which takes into account effective charges and metal–ligand distances. With the help of this electrostatic model we will draw the key structural and electronic features that are needed to be taken into account in the design of SMMs and spin qubits based on these mononuclear lanthanide-based complexes.

2.2 Modelling the Magnetic Properties of Lanthanide Single-Ion Magnets: The Use of the Crystal Field Model

2.2.1 Theoretical Background

In a lanthanide ion spin–orbit coupling dominates the CF. Hence, even at room temperature it is commonly assumed that only the ground multiplet J, which accounts for the total angular momentum of the lanthanide, is populated. This multiplet is equal to $J = |L - S|$ for early lanthanides and $J = L + S$ for late lanthanides. Thus, the magnetic properties of a mononuclear lanthanide-based complex depend on the splitting of the ground state of the lanthanide, J, caused by the CF created by the ligands. The traditional CF Hamiltonian, $\widehat{H}_{\mathrm{cf}}$, parameterizes the electric field effect caused by the surrounding ligands acting over the central ion.

$\widehat{H}_{\mathrm{cf}}$ can be expressed in terms of extended Stevens operators, O_k^q [15, 16], and for a given J-multiplet it takes the general form:

$$\widehat{H}_{\mathrm{cf}}(J) = \sum_{k=2,4,6} \sum_{q=-k}^{k} B_k^q O_k^q = \sum_{k=2,4,6} \sum_{q=-k}^{k} a_k (1 - \sigma_k) A_k^q \langle r^k \rangle O_k^q \tag{2.1}$$

where k is the operator order, $\langle r^k \rangle$ is the expectation value of r^k, a_k are the α, β, γ Stevens equivalent coefficients for $k = 2, 4, 6$, respectively and σ_k are the shielding parameters of the 4f electronic shell [17]. α, β, γ are tabulated for the ground state of each lanthanide ion. Hence, the CF parameters, A_k^q and B_k^q, refer to the ground state as well.

Taking into account that B_k^q parameters represent the coefficient of an operator related to the spherical harmonic Y_{kq} then the ranges of k and q are limited to a maximum of 27 parameters (26 independent): B_k^q with $k = 2, 4, 6$ and $q = 0, 1, \ldots, k$. The B_k^0 values are real and the rest are complex. Due to the invariance of the CF Hamiltonian under the operations of the symmetry groups, the number of parameters is also limited by the point symmetry of the lanthanide site. Notice that for some groups, the number of parameters will depend on the choice of axes. In Table 2.1, the effect of site symmetry is illustrated for some common ion site symmetries.

Table 2.1 Non-forbidden crystal field parameters that a group of point charges may contribute collectively, as a function of their point-group symmetry.

Symmetry	CF parameters (real)	CF parameters (complex)
C_1, C_i	$B_2^0, B_2^1, B_4^0, B_6^0$	$B_2^2, B_4^1, B_4^2, B_4^3, B_4^4,$ $B_6^1, B_6^2, B_6^3, B_6^4, B_6^5, B_6^6$
C_2, C_{2h}, C_s	$B_2^0, B_2^2, B_4^0, B_6^0$	$B_4^2,\ B_4^4,\ B_6^2,\ B_6^4,\ B_6^6$
D_2, D_{2h}, C_{2v}	$B_2^0, B_2^2, B_4^0, B_4^2, B_4^4, B_6^0, B_6^2, B_6^4, B_6^6$	
C_3, S_6	$B_2^0, B_4^0, B_4^3, B_6^0$	$B_6^3,\ B_6^6$
C_4, C_{4h}, S_4	$B_2^0, B_4^0, B_4^4, B_6^0$	B_6^4
C_5	$B_2^0, B_4^0, B_6^0, B_6^5$	
D_3, D_{3d}, C_{3v}	$B_2^0, B_4^0, B_4^3, B_6^0, B_6^3, B_6^6$	
$D_{4h}, D_{2d}, C_{4v}, D_4$	$B_2^0, B_4^0, B_4^4, B_6^0, B_6^4$	
$D_{6h}, D_{3h}, C_{3h}, D_6, C_{6v}, C_{6h}, C_6$	$B_2^0, B_4^0, B_6^0, B_6^6$	
O_h, T_d, O, T_h, T a)	$B_4^0, B_4^4, B_6^0, B_6^4$	
I, I_h a)	B_6^0, B_6^5	
$D_{4d}, C_{5h}, D_{5h}, D_{6d}, C_n, C_{nv}, C_{nh},$ $D_{nh}, D_{nd} \forall n > 6$	B_2^0, B_4^0, B_6^0	

a) For cubic and icosahedral groups not all parameters are independent.

2.2.2
How to Determine the Crystal-Field Parameters: 1. The Ishikawa Approach

Ishikawa *et al.* proposed an approach for the determination of the ligand-field (LF) parameters of a set of isostructural lanthanide complexes. This method consists of a simultaneous fit of the temperature dependence of magnetic susceptibilities and NMR spectra for the whole isostructural series [18]. In order to avoid over-parametrization a key restriction is imposed: each parameter is expressed as a linear function of the number of f electrons, n:

$$A_k^q \langle r^k \rangle (n) = a_k^q + b_k^q (n - n_0)$$

where the value of n_0 is set to 10.5, the mean value of n for the series of complexes to fit. Under this condition, a unique set of parameters can simultaneously reproduce experimental data of a whole series of complexes.

This method was proposed to study LF parameters of anionic bis(phthalocyaninato)lanthanides, [TBA][LnPc_2] (TBA, tetrabutylammonium). They present an idealized molecular symmetry D_{4d}, where the two equivalent Pc rings are placed parallel to each other with a skew angle of 45° (Figure 2.2). For this ideal symmetry, only LF terms with $q = 0$ do not vanish, and the LF Hamiltonian is

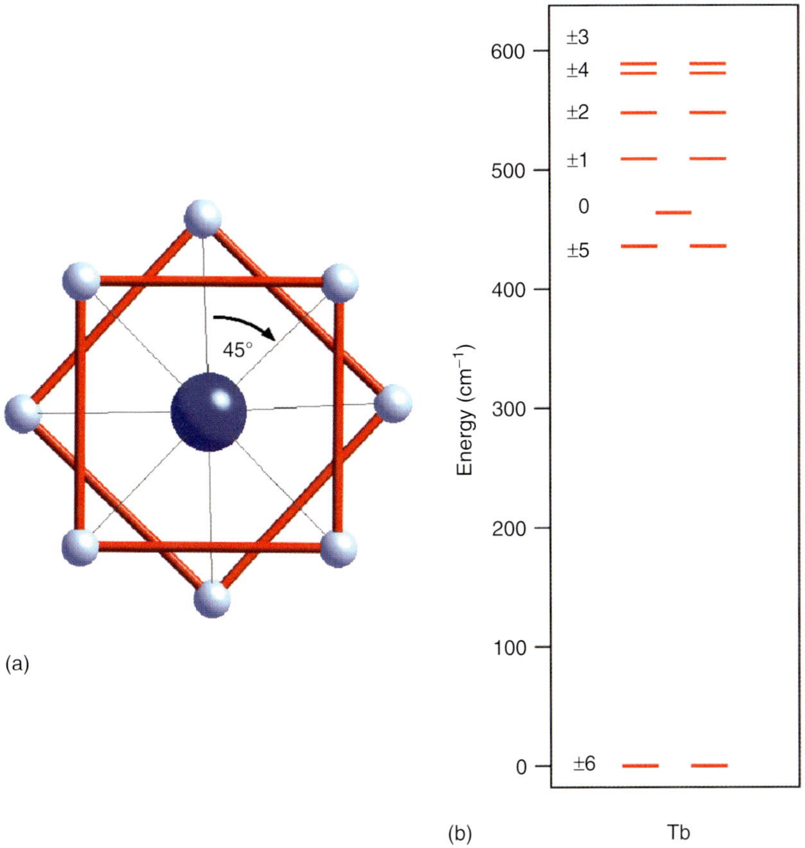

Figure 2.2 (a,b) Projection of terbium phthalocyaninato derivative showing the twist angle in the Tb(III) ion square-antiprismatic coordination site and energy level scheme for the ground $J = 6$ multiplet.

$$\hat{H} = A_2^0\langle r^2\rangle\alpha\hat{O}_2^0 + A_4^0\langle r^4\rangle\beta\hat{O}_4^0 + A_6^0\langle r^6\rangle\gamma\hat{O}_6^0$$

The best fit obtained is dominated by a large and positive $A_2^0\langle r^2\rangle$ parameter that varies linearly from 413 cm^{-1} for Tb(III) to 319 cm^{-1} to Yb(III) [14]. This fit leads to a ground magnetic doublet for terbium that corresponds to that with the maximum M_J value (= ±6), being separated from the first excited level ($M_J = \pm5$) by more than 400 cm^{-1} (Figure 2.2). This feature allows one to explain the SMM behaviour observed in [TBA][$LnPc_2$].

The same procedure was performed for two families of POMs, $[Ln(W_5O_{18})_2]^{9-}$ (in short LnW_{10}) and $[Ln(\beta_2\text{-}SiW_{11}O_{39})_2]^{13-}$ (in short LnW_{22}) [7] (Figure 2.1b,c).

Table 2.2 Determined LF parameters for the LnW_{10} and LnW_{22} families (in cm^{-1}).

Ln	$[LnW_{10}]^{9-}$ series			$[LnW_{22}]^{13-}$ series		
	$A_2^0r^2$	$A_4^0r^4$	$A_6^0r^6$	$A_2^0r^2$	$A_4^0r^4$	$A_6^0r^6$
Tb	−36.8	−89.0	−5.2	−11.3	−137.4	−7.1
Dy	−153.6	−145.5	16.3	−88.3	−134.4	7.0
Ho	−270.4	−202.0	37.7	−165.2	−131.3	21.1
Er	−387.2	−258.5	59.2	−242.2	−128.2	35.2
Tm	—	—	—	−319.2	−125.2	49.2
Yb	—	—	—	−396.1	−122.1	63.3

In these two families the coordination site around the Ln(III) also exhibits an approximate D_{4d} symmetry. In both families, an excellent agreement between the experimental values and the theoretical expectations was found using the CF parameters reported in Table 2.2. If we compare these two sets of LF parameters with those derived from the $[LnPc_2]^-$ complexes, the most important difference is the sign of the axial zero field splitting (ZFS) parameter of order 2, $A_2^0r^2$, which is negative in the two rare-earth POM families but positive in the phthalocyaninato-lanthanide family. We can associate this change in the sign with differences in the axial distortion of the coordination site. In fact, while the POM derivatives exhibit an axial compression, axial elongation of the square antiprism is observed in the phthalocyaninato-lanthanide complexes. This different sign has important consequences in the structure of the low-lying energy levels (Figure 2.3) and in the dynamic magnetic properties of these complexes. Thus, while in POM complexes, the erbium derivatives behave as SMMs, in the phthalocyaninato complexes, they do not show any sign of SMM behaviour. In turn, the terbium derivative behaves as an SMM in the phthalocyaninato complexes but not in the POM complexes. As a consequence of the different structural distortions, the energy levels for erbium and terbium are very different, and they are almost reversed in the two classes of complexes. In fact, while in the case of Er the POM ligands stabilize larger $|M_J|$ values, the phthalocyaninato ligand stabilizes the lowest value, $M_J = \pm 1/2$. In a similar way, in the terbium case, the POM derivatives stabilize the singlet $M_J = 0$, while in the phthalocyaninato derivative the doublet with the largest J value, $M_J = \pm 6$, becomes the ground state.

This procedure has been applied to other symmetries such as the well-known Preyssler anion $[LnP_5W_{30}O_{110}]^{12-}$, LnW_{30} (Figure 2.1d) [9]. This POM has a central coordination site with an unusual, fivefold axial symmetry. In spite of the low symmetry provided by this anion, we have observed that the dysprosium and holmium derivatives behave as SMMs, exhibiting slow relaxation of the magnetization. The remarkable difference between this system and the previous

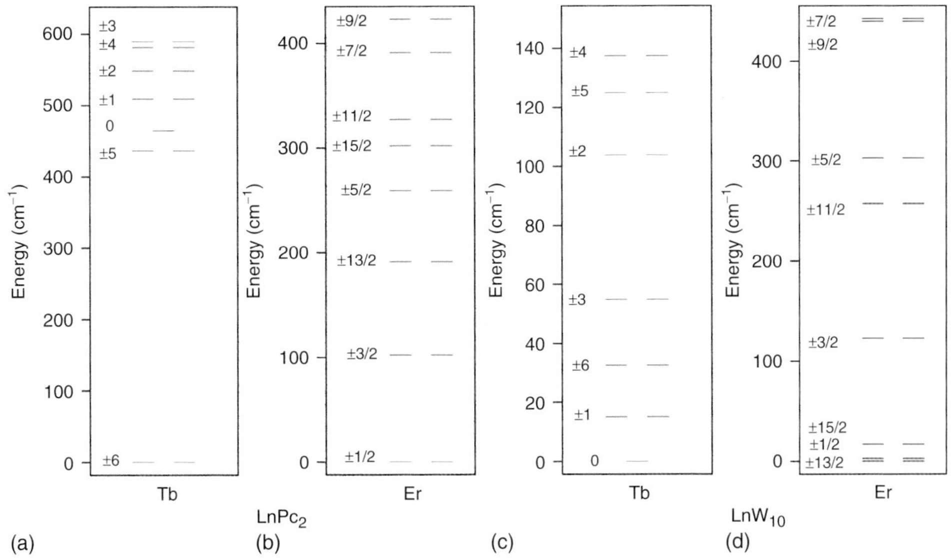

Figure 2.3 (a–d) Energy level scheme derived from Ishikawa's model for the ground $J = 6$ multiplet for terbium and $J = 15/2$ for erbium in $LnPc_2$ and LnW_{10} complexes.

Table 2.3 CF parameters determined for the LnW_{30} series (in cm^{-1}).

Ln^{3+}	f^n	$A_2^0r^2$	$A_4^0r^4$	$A_6^0r^6$	$A_6^5r^6$
Tb^{3+}	8	69.3	34.3	6.7	-1.08×10^3
Dy^{3+}	9	49.7	43.6	11.3	-1.16×10^3
Ho^{3+}	10	30.1	52.9	15.9	-1.24×10^3
Er^{3+}	11	10.5	62.3	20.6	-1.32×10^3
Tm^{3+}	12	−9.1	71.6	25.3	-1.40×10^3
Yb^{3+}	13	−28.68	80.9	29.9	-1.47×10^3

one is the presence of a large off-diagonal anisotropy parameter A_6^5 due to the fivefold symmetry, which mixes magnetic states with different M_J values. Although a simultaneous fit for the whole series is possible, the presence of an extra parameter makes it difficult to obtain a unique set (Table 2.3 and Figure 2.4). This is the reason why the Ishikawa procedure would be usable only in systems of high symmetry and with a significantly reduced number of CF parameters.

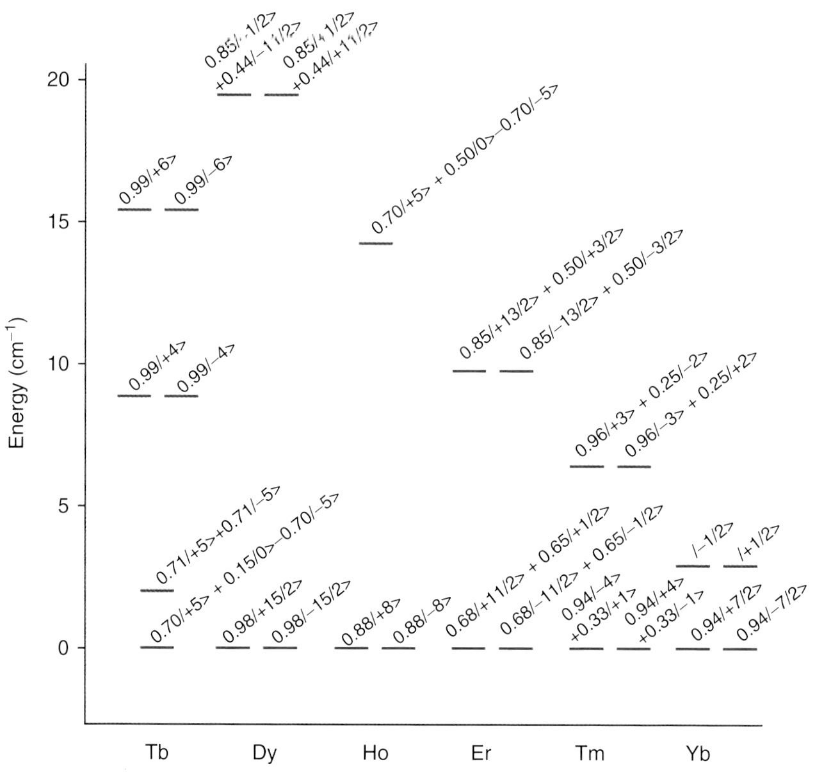

Figure 2.4 Low lying levels of the different complexes in the LnW_{30} series (Ln = Tb, Dy, Ho, Er, Tm and Yb).

2.2.3
How to Determine the Crystal-Field Parameters: 2. The Point Charge Electrostatic Model

In order to avoid overparametrization, we can use the point charge electrostatic model (PCE model) [19, 20], in which N ligands are represented by their point charges (Z_i). The $\hat{H}_{cf}$ generated by a charge distribution can be written in its most primitive form as a sum of Coulomb fields created by the charges. The A_k^q CF parameters can then be calculated by the following expression:

$$A_k^q = \frac{4\pi}{2k+1} c_{kq} (-1)^q \sum_{i=1}^{N} \frac{Z_i e^2 Y_{k-q}(\theta_i, \varphi_i)}{R_i^{k+1}} \qquad (2.2)$$

where θ_i, φ_i, R_i are the spherical coordinates of the point charge Z_i which represent the ith ligand with the lanthanide at the origin, e is the electron charge and

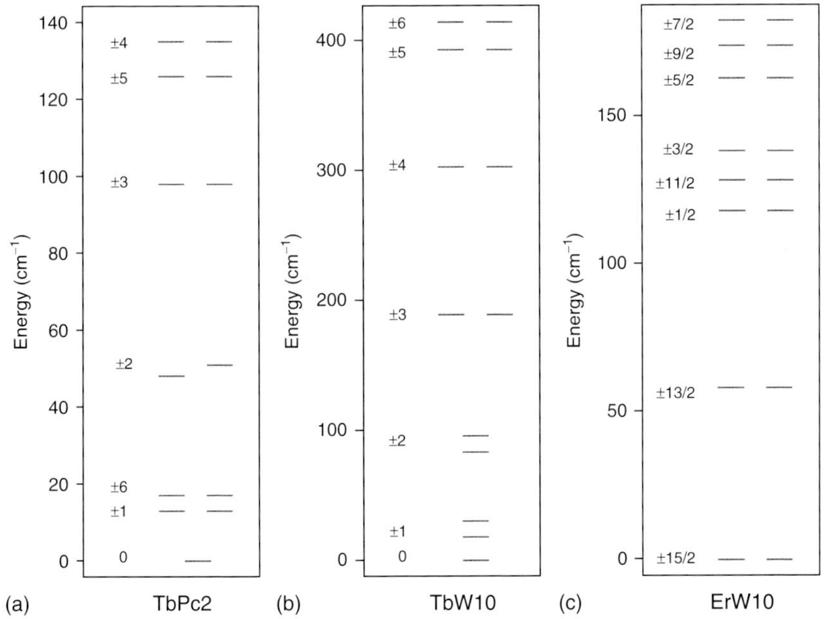

Figure 2.5 (a–c) Energy level scheme derived from PCE model for the ground $J = 6$ multiplet for Tb and $J = 15/2$ for Er in $Lnpc_2$ and LnW_{10} complexes.

c_{kq} is a tabulated numerical factor that relates spherical harmonics and Stevens equivalent operators. Note that, unless otherwise specified, Y_{kq}, A_k^q, B_k^q are complex numbers. As they have the same angular nodes, they can sometimes be used interchangeably.

The strengths and weaknesses of a purely PCE approach can be analysed with help of Figure 2.5. The magnetic behaviour of the POM lanthanide complexes can be at least qualitatively predicted from a purely geometrical basis. For example, TbW_{10} has a diamagnetic ground state and ErW_{10} has a high-spin Kramers doublet as the ground state. This geometrical prediction coincides with the energy level scheme obtained a posteriori by means of a fit of the magnetic properties. It is easy to see that such a tool is simple and fast as the magnetic properties can be predicted by a calculation that takes a few seconds on a desktop computer. Moreover, the M_J labels are simplified in Figure 2.5, but the PCE model produces – with no free parameters – a full CF Hamiltonian with all diagonal and extradiagonal parameters, so that the full wavefunction information is available for each level. The resulting mixing of functions produces a tunnelling splitting, which is clearly visible in the case of TbW_{10}. Still, this simple approach has an important limitation since it is valid only when dealing with ionic complexes. In fact, when the covalent effects are important, this model may lead to a wrong CF splitting. This happens, for example, in the case of phthalocyaninato double decker complexes for which the PCE model predicts a diamagnetic ground state $M_J = 0$, while the experimental results support a magnetic ground state with $M_J = \pm 6$.

2.2.4
How to Determine the Crystal-Field Parameters: 3. The Effective Point Charge Model

When the nature of the ligand and the orientation of its electronic lone-pairs are taken into account, the simple picture provided by a pure PCE model is not realistic and the model needs to be improved to provide quantitative predictions. An effective point-charge model should be introduced to overcome this limitation. From direct fitting of energy levels coming from spectroscopic measurements or from a fitting of the magnetic properties, it is possible to extract these effective point charges. In order to do that, two models have been distinguished depending on the ligand character: a radial effective charge (REC) model for those ligands having a spherical character (such as halides, hydroxide or alkoxide ligands) and a lone pair covalent effective charge (LPEC) model for those ligands interacting with the metal through lone pairs (such as nitrogen from amino groups, or oxygen from a water molecule) [21].

In the REC model, the ligand is modelled through an effective point charge situated in the axis described by the lanthanide-coordinated atom axis, at a distance R, which is smaller than the real metal–ligand distance (Figure 2.6). To account for the effect of covalent electron sharing, a radial displacement vector (D_r) is defined, in which the polar coordinate R is varied. At the same time, the charge value (q) is scanned in order to achieve the minimum deviation between calculated and experimental data, whereas θ and φ remain constant.

In the LPEC model the electron lone pair of the coordinating atom (nitrogen or oxygen) is not located along the radial Ln–ligand direction, making it necessary to define a new displacement vector (D_h) referred to as the *horizontal displacement* (Figure 2.6). Both vectors, D_h and D_r, are applied to the original position of each nucleus, determining the position of the effective centre of charge. Vector D_r is a segment of the distance between the position of the lone pair and the lanthanide

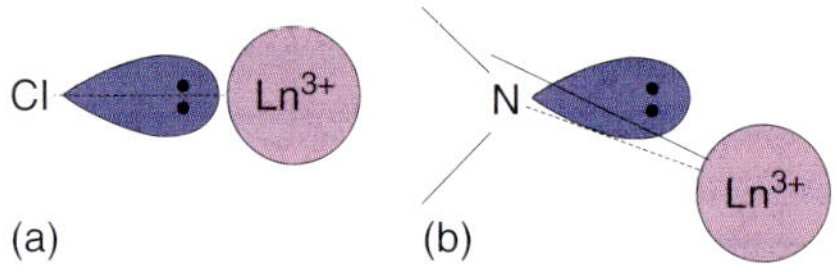

Figure 2.6 Types of orientations between the electronic pair of a ligand and a lanthanide cation. (a) The lone pair is directly oriented towards the lanthanide cation. (b) The lone pair is not directly oriented towards the lanthanide cation. The node in B_2^0 is shown as a solid line.

nucleus, and it is defined by the polar coordinate R. Note that this latter displacement reflects, as in the REC model, the effective charge resulting from the sharing of the ligand electron density by the lanthanide ion. This correction does possess physical meaning due to the fact that the nearest part of the electron cloud pointing towards the lanthanide induces a more marked effect than the areas placed further away.

As commented above, this effect is especially easy to see in the double-decker bis-phthalocyaninato complexes, by the position of the nuclei of nitrogen. PCE or REC models could never explain the sign of B_2^0 to justify that the $[TbPc_2]^-$ complex has SIM behaviour. However, when considering the direction in which the nitrogen lone pairs point, the B_2^0 term changes in sign naturally. An LPEC fit situates the effective charge on the other side of the B_2^0 node (Figure 2.6). This means a change in the sign of the second order uniaxial anisotropy, with dramatic consequences for the magnetic energy level scheme (Figure 2.7). Now one can see that the levels again correspond to the experimentally observed behaviour, that is, $[TbPc_2]^-$ has a high-spin ground state. Again, for simplicity only the main M_J component is displayed, but with only three LPEC parameters the full CF Hamiltonian is produced, with the whole set of diagonal and extradiagonal parameters. Thus, tunnelling splittings and minor M_J contributions are directly obtained.

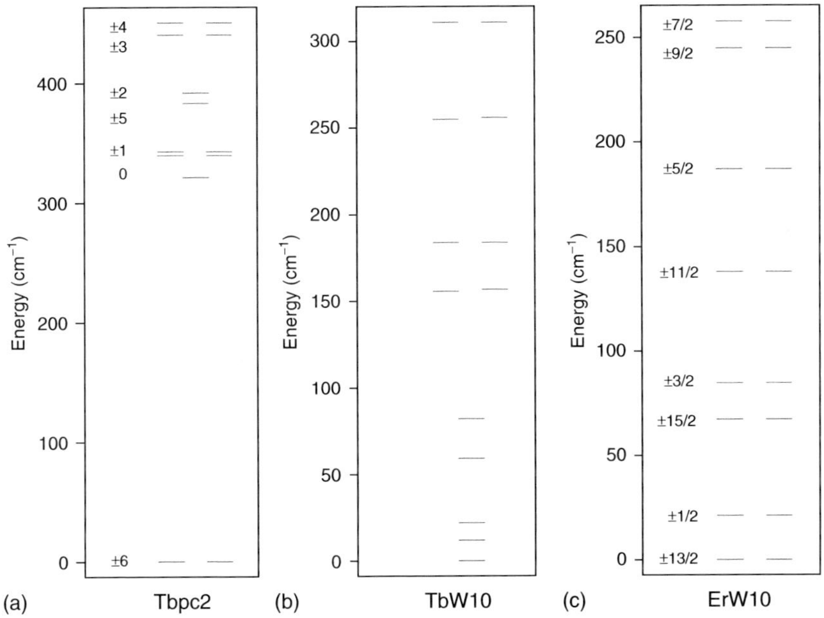

Figure 2.7 (a–c) Energy level scheme derived from REC and LPEC models for the ground $J = 6$ multiplet for terbium and $J = 15/2$ for erbium in $LnPc_2$ and LnW_{10} complexes.

This effect of the position of the lone pair is relevant not only in the case of phthalocyaninato ligands, but also can be a clue to the intriguing behaviour of the [Dy(DOTA)] complex, where the rotation of a water molecule changes the magnetic properties [13]. A PCE or an REC model cannot account for the effect of such a rotation, but an LPEC model would predict a dramatic effect, since the change in the position of the lone pair effectively means a completely different geometry.

The predictive power of REC and LPEC models is due to the possibility of associating each type of ligand with a reduced set of effective parameters (two or three), which are specific to the ligands and therefore can be used for many lanthanide complexes. Series of analogous compounds, such as those provided by POM chemistry, are of special interest for the goal of building a general reusable library of ligands for the rational design of SIMs based on lanthanide complexes. In order to obtain a trustable library upon which one can firmly found further work, the desiderata are as follows:

1) High-quality experimental data to be fitted, for example, magnetic data (χT vs T curves) and spectroscopic data, in order to obtain an accurate picture of the energy level scheme.
2) Simultaneous fits of a series of related complexes, as mentioned above. In this case, a handful of parameters can be used to reproduce the experimental behaviour of a large quantity of compounds. This reduces the possibilities of numerical artefacts, and also the propagation of experimental errors into the final parameters.
3) The use of homoleptic compounds, where all coordinating atoms are as chemically indistinguishable as possible. Indeed, with heteroleptic complexes it can easily be the case that the error in the parameterization of the effect of a type of ligand can be compensated and thus shadowed by another error in the parameterization of a different ligand, resulting in flawed parameters for both of them.

Finally, the REC model will be in general more trustable than the LPEC model, as it is simpler. Of course, in complexes where the effect of the lone pair is crucial, for example, in the case of the phthalocyaninate sandwiches, REC has to be abandoned in favour of LPEC.

Figure 2.8 is a partial snapshot of the recent progress in the construction of this general library, taken from Ref. [23]. Only the homoleptic complexes which can be fitted by the REC model are included. In the case of the halides F^-, Cl^-, Br^-, energy levels were used [21]. For oxygen, two series of a total of 10 POM complexes were used, but in this case only χT data were available. For nitrogen, a combination of simultaneous fits, χT data and spectroscopic data, was used [22, 24]. In this comparison we chose to depict the position

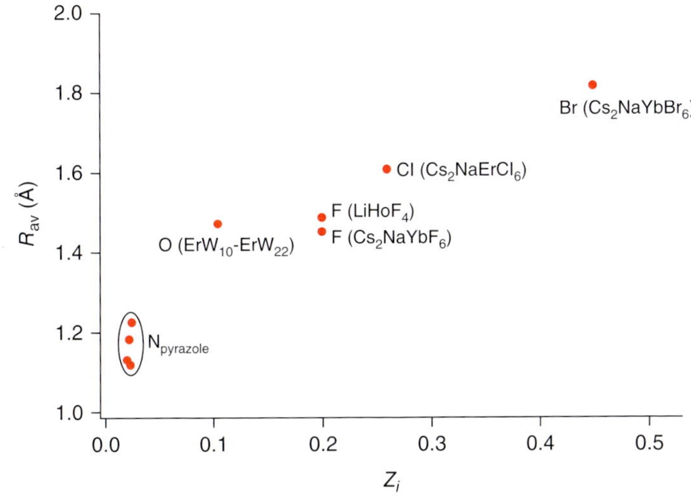

Figure 2.8 Average radius R_{av} and effective charge Z_i resulting from the REC correction of different homoleptic compounds: nitrogenated ligands of the type bis-pyrazole or trispyrazole [22], oxygen atoms of the POM series LnW_{10} and LnW_{22}, [23] and halogenides F^-, Cl^-, Br^- anions [21].

of each ligand in the plane defined by the effective charge and the average distance of the said effective charge from the nucleus of the lanthanide. This allows us to quantify, at any given molecular geometry, the different relative tendency of each kind of ligand to produce CF splittings that are dominated by a simple double well (B_2^0, qualitatively equivalent to the ZFS D), or the more complex energy level structures typical of higher level CF operators.

Notice that, with the current data, one cannot draw immediate parallels with the spectrochemical series (nor with the nephelauxetic series). At any rate, these comparisons are not trivial: depending on the symmetry, CF parameters of orders 4 and 6 are more or less important compared with those of order 2. Thus, the ratio charge/distance cannot generally quantify the strength of the LF exerted by a kind of ligand, or at least not in a way that is totally independent of the geometry of the complex.

It is expected that progress along this path will be successful for the rational design strategy for obtaining lanthanide complexes with the desired magnetic properties. Indeed, to be useful, geometrical comparisons such as those presented in this section and in Ref. [25] need to take into account the properties of the ligands to be used. This is only possible with the systematic study of a large number of complexes.

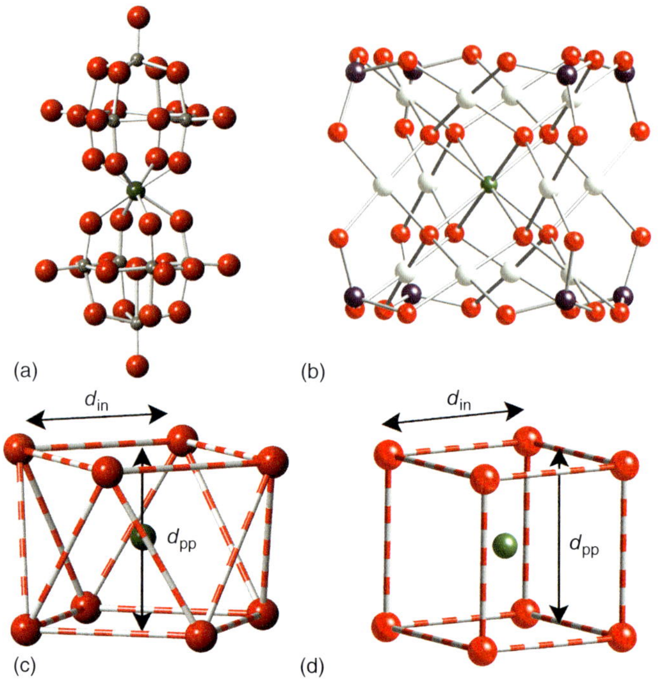

Figure 2.9 Schematic structure of two different polyoxometalate complexes with interesting coordination symmetry (a) LnW_{10} and (b) $[LnPd^{II}_{12}(As^{V}Ph)_8O_{32}]^{5-}$, and their coordination polyhedra (c) square antiprism and (d) cube. The d_{pp} parameter defines the average distance between the two oxygen-based square planes. The d_{in} is the average O–O distance within the oxygen-based square planes.

2.3
Magneto-Structural Correlations for Some Typical Symmetries

Herein, we compare the well-known square antiprism geometry (D_{4d} ideal symmetry) and the cubic geometry (O_h ideal symmetry). POM chemistry provides nice examples of these two 8-coordinate environments (Figure 2.2). In fact, these molecular metal oxides have rigid and stable structures that can impose highly symmetric CFs on the lanthanide. In Figure 2.9, the structure of the square antiprism $[Ln(W_5O_{18})_2]^{9-}$ (LnW_{10}) (**1**) is compared to the cubic structure of $[LnPd^{II}_{12}(As^{V}Ph)_8O_{32}]^{5-}$ (**2**) [26]. Experimentally, the magnetic behaviour of these two series shows that while for **1** some derivatives exhibit SMM behaviour, for **2** all the derivatives exhibit fast spin relaxation at low temperatures (U. Kortz and E. Coronado, *et al.*, unpublished results). It seems therefore that the D_{4d} symmetry is more suitable to design SIMs than the O_h symmetry.

The square antiprism is a common lanthanide coordination environment. Using this geometry, the main CF parameters to be considered are B_2^0, B_4^0, B_6^0 (those allowed by an ideal D_{4d} system), but due to the φ distortion of the site non-negligible values for B_4^4 and B_6^4 appear (Figure 2.2). In the case of a cubic environment, due to the well-known relations for this O_h symmetry, $B_2^0 = 0, B_4^4/B_4^0 = 5$ and $B_6^4/B_6^0 = -21$, this system may be defined only by two CF parameters, B_4^0 and B_6^0. This is better understood by performing a magneto-structural study in which the D_{4d} symmetry is taken as a distorted case of the O_h symmetry when the two squares are staggered rather than eclipsed. This situation is defined by the torsion angle between upper and lower squares, φ, which goes from 0° in the O_h symmetry to 45° in the D_{4d} symmetry. The calculations indicate that the CF parameters B_4^0 and B_6^0 are invariant with respect to φ. In addition, B_4^4 and B_6^4 are non-zero, except for $\varphi = 45°$, evolving as a function of φ (Figure 2.2). Thus, exact cancellation of these two parameters occurs for D_{4d} symmetry, while the maximum values are reached for the O_h one. As a conclusion, one can anticipate that for an ideal D_{4d} symmetry the wavefunctions will be described by pure $\pm M_J$ values, while for a cubic symmetry the presence of the B_4^4 and B_6^4 CF parameters will result in an extensive mixing of functions with different M_J values (in particular those differing by $\Delta M_J = 4$). On the other hand, as we have seen before, the B_2^0 parameter will be zero either for a cubic symmetry or for a non-axially distorted antiprism symmetry. This is the key parameter for creating a gap between the M_J ground doublet and the excited states (and therefore a barrier), which will only appear by axial distortion of the antiprism (elongation or compression). Using such a procedure, one can analyse real structures that are intermediate between these two ideal geometries, as determined by shape analysis (Figure 2.10) [27].

Two other common lanthanide coordination environments are the triangular dodecahedron and the trigonal prism. For the first case, a calculated extensive mixture of M_J is a consequence of the presence of non-negligible extradiagonal parameters B_4^4 and B_6^4. In particular, B_4^4 and B_6^4 parameters are one and two orders of magnitude larger than the corresponding diagonal terms (B_4^0 and B_6^0), respectively [28]. In general, this means that the triangular dodecahedron, similarly to cubic environments, is not ideal for generating SIM behaviour. For the second case, an ideal trigonal prism can only present $B_2^0, B_4^0, B_6^0, B_6^6$ terms. The presence of the extradiagonal CF parameter B_6^6 means mixing between M_J values differing by ± 6. It is crucial to note that this will produce direct tunnelling only for doublets involving states with $M_J = \pm 3$ or ± 6 and, even in these cases, the compound may show SMM behaviour if B_2^0 is large enough.

Energy splitting will be dependent on the lanthanide and on the symmetry and distortion of the coordination environment. As far as the lanthanide is concerned, we notice that J as well as the Stevens coefficients (α, β and γ) [29] are fixed and are specific for each lanthanide. As discussed above, symmetry and distortion decide which CF parameters are not cancelled and thus the mixing conditions between

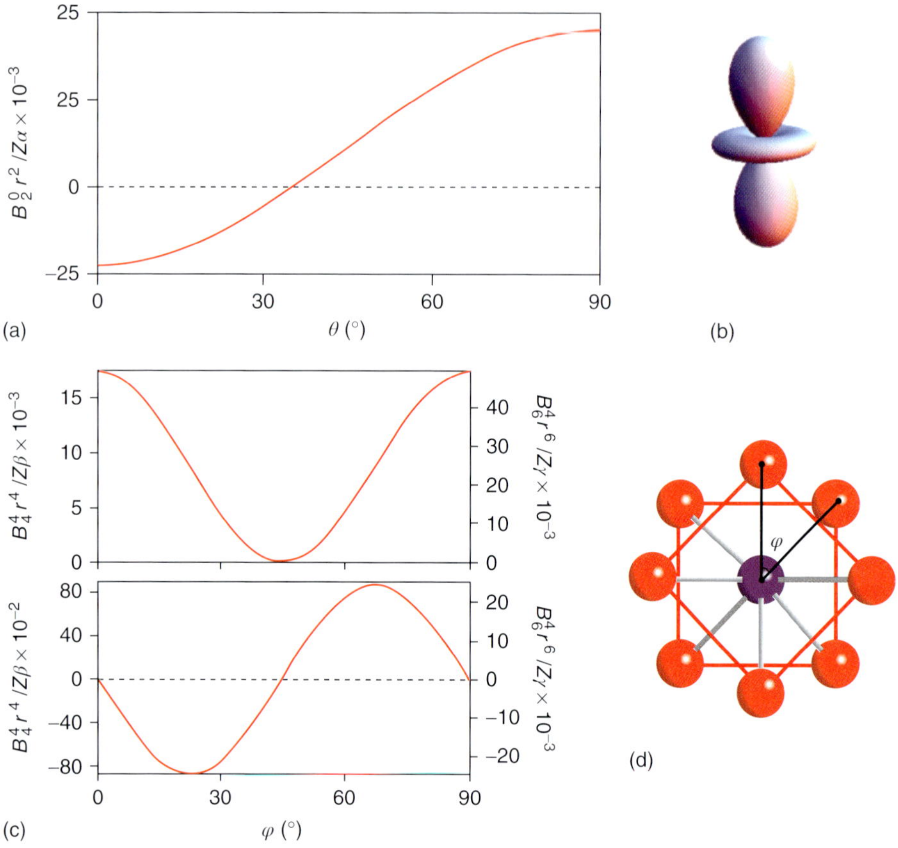

Figure 2.10 (a) Magnetostructural correlation of B_2^0/α with the θ polar angle in the case of a D_{4d} and (b) shape of Y_{20}; (c) Magnetostructural correlation of the real and imaginary parts of B_4^4/β and B_6^4/γ with φ and (d) scheme depicting the torsion angle that describes the rotation of a cube from O_h to D_{4d} symmetries.

M_J-sublevels. Taking into account these two factors some conditions can favour mononuclear SIMs behaviour [25]:

1) As a general rule, SIMs require a high M_J ground state, to create an energy barrier leading to slow spin relaxation, and low mixing in the wave functions, to minimize the fast spin relaxation through quantum tunnelling processes. In the simplest case, this can be achieved with an ideal pseudoaxial symmetry such as D_{4d}, C_{5h}, D_{6d}, or any symmetry of order 7 or higher. In all these symmetries, the most suitable case is achieved when the second-order uniaxial anisotropy, accounted by the B_2^0 parameter, is maximized. Depending on the

metal, this favourable situation can be reached either by increasing the electron density near the uniaxial axis (in the case of Tb, Dy and Ho), or near the basal plane (in the particular case of erbium, but also for thulium and ytterbium). For an axially distorted D_{4d} symmetry, axially elongated sites are favourable for Tb^{3+}, Dy^{3+} and Ho^{3+}, as exemplified by the double-decker bis-phthalocyaninato complexes, while axially compressed sites are favourable for Er^{3+}, as exemplified by POMs.

2) Not all the molecular symmetries are equally suited for favouring a SIM behaviour. Cubic O_h symmetry, for example, in the case of 8-coordinate complexes, does not have second-order uniaxial anisotropy, B_2^0, and this geometry does not favour a large energy barrier for magnetization reversal.
3) The presence of an extensive mixture of M_J values in the ground state doublet does not always lead to quantum tunnelling. In fact, if the two wave functions do not present any overlap, the relaxation through tunnelling is forbidden. For example, lanthanide complexes exhibiting a D_{2d} triangular dodecahedron geometry typically have extradiagonal CF terms of the type B_4^4 and B_6^4, whose values are higher than those of the diagonal terms. In some cases the mixing generated by these terms, which involve functions with M_J-values differing by four units, does not prevent the presence of an energy barrier.

Many SIMs will also be useful as spin qubits as discussed in the final section of this chapter. For the rational design of either SIMs or molecular spin qubits we have to exert control on two levels: the CF geometric effect and the interaction with further electron spins, nuclear spins and phonons. From the effect of the CF around the lanthanide, these two levels require (i) controlled mixing of the wave functions in a well-defined level subset and (ii) sufficient isolation of this subset from the rest of the spectrum. However, lanthanide complexes can present a large tunnelling gap in the ground state, Δ, which can minimize decoherence through nuclear spins [30]. This intense tunnelling is useful for qubits but usually kills the SMM magnetic behaviour. In this sense, the tunnelling gap in lanthanide SIMs can be much higher than that exhibited by cluster-type SMMs (by a factor larger than 10^6). Hence, this favourable effect for minimizing decoherence is expected to be much more pronounced in the case of mononuclear SIMs. Taking into account this last effect, the three conditions that favour the design of a qubit will be (i) to use a non-Kramers ion (for example, terbium or holmium); (ii) to choose a geometry for this lanthanide allowing for a large extradiagonal CF term and (iii) to have a ground state doublet $\pm M_J$ such that $2M_J$ is a multiple of the order of this extradiagonal operator, q. As the q-value coincides, in general, with the order of the main symmetry axis of the molecule, an easy way to satisfy the third requirement is that of finding a molecular geometry having a q-order axis equal to $2M_J$. Two examples that illustrate this point are provided by the POMs $[Ho(W_5O_{18})_2]^{9-}(C_{4v}, M_J = \pm 4, \Delta \approx 0.3\,cm^{-1})$

and $[TbW_{30}O_{110}]^{12-}$(C_{5v}, $M_J = \pm 5$, $\Delta \approx 2\,cm^{-1}$), both of which have been experimentally found to show unusually long decoherence times.

On the other hand, lanthanides with 100% isotopical purity such as terbium or holmium are preferred to simplify the operation and minimize decoherence in spin qubits. In this respect, the existence, for some lanthanides, of a manifold of electronuclear states can provide additional resources for the implementation of multiple qubit states within the same molecule [31]. All atoms in the first coordination sphere should be oxygen, and the sample should be deuterated if the compound contains hydrogen, to avoid interaction with other nuclei spins. Again, POM chemistry has been shown to provide ideal examples of this kind.

2.4 Impact of Lanthanide Complexes in Quantum Computing

A qubit, in its most ample definition, is a quantum two-level system. In a molecular spin qubit, the two levels are embodied by the 'up' and 'down' states of a spin in a magnetic molecule. As mentioned above, lanthanide complexes are good candidates for spin qubits, as in their ground state they often bear an M_J doublet which is well separated from the next energy level. A crucial, but often overlooked, issue is that any given magnetic system will be more or less apt as a molecular spin qubit depending on how its properties fit in the chosen quantum computing approach. Thus, the next section are devoted to presenting a few of these computational approaches: the quantum circuit model, the global control paradigm, measurement-based quantum computing and adiabatic control. In each case, comments and suggestions are given for their implementation with lanthanide complexes.

The main hardware types offered by physics are mentioned, namely: trapped ions (or trapped atoms), quantum dots, quantum optical cavities, rf superconducting quantum interference devices (SQUIDs) and nitrogen-vacancy (NV) defects on diamond. Some are important simply as a benchmark to evaluate the quality of the implementations offered by chemistry, whereas others might be combined with lanthanide complexes to produce heterogeneous quantum information processors which combine the advantages of different hardware types.

The final section deals with known examples of molecular spin qubits based on lanthanide SIMs. Distinction is made between single-qubit molecules and molecules which embody more than one qubit. This section includes some comments about decoherence in these molecular systems and strategies to control it.

2.4.1 Quantum Computing Paradigms and Design Criteria

Quantum computing consists in the explicit use of quantum mechanical phenomena – such as quantum superposition and entanglement – for the purposes of information technology [32]. As mentioned above, lanthanide complexes present quantum effects more readily than their transition-metal analogues, thus their interest as model systems in this field. It is well known that the coherent manipulation of quantum systems on a large scale would enable quantum information processing (QIP) to be qualitatively more efficient than classical information processing [33]. On the other hand, even in the absence of QIP systems with a computational power exceeding that of classical computers, the coherent manipulation of quantum systems is a formidable tool for the study of decoherence. Decoherence – the loss of quantum coherence of a system because of an uncontrolled entanglement with its environment – is a process of fundamental importance in physics, with deep philosophical consequences, notably a possible role in the arrow of time, the measurement problem in quantum mechanics and the transition from quantum to classical mechanics [34].

Quantum systems of any kind can in principle be candidates for quantum hardware, including different kinds of spin qubits; we briefly review some of these in the next section. Much effort has been expended on the question of which physical systems are best suited for use in QIP, but no ultimate answer has been found so far. A much quoted list of conditions to build computers was established by DiVincenzo [35], but one has to note that some of these restrictions are specific to the 'quantum circuit' paradigm.

In the *quantum circuit model* any quantum algorithm is expressed as a series of quantum gates (i.e. unitary quantum operations), each acting on one or several qubits. In the case of lanthanide complexes, this can be visualized as complex electron paramagnetic resonance (EPR) pulse sequences on systems with several coupled magnetic ions. The quantum circuit starts with all qubits in a known state, for example, $|000\ldots00\rangle$ and ends with a measurement of the state of one or several of the qubits. In a pulsed EPR setup, this would mean extreme cooling before the start of the computation to ensure full population of the ground state, and a careful analysis of the free induction decay signal after the experiment. During the computation, the spins are manipulated as needed, including the possibility of conditioning the quantum gate on a 'target' spin on the state of a 'control' spin. A typical example is the Controlled-NOT gate, where the target qubit is flipped from $|0\rangle$ to $|1\rangle$ and vice versa if and only if the value of the control qubit is $|1\rangle$ (Figure 2.11). This operation on dinuclear lanthanide complexes is described in the chapter by Aromí *et al.* in the present book.

There are several other ways of approaching QIP that do not use the quantum circuit model. The *global control paradigm*, embodied in the seminal proposal of

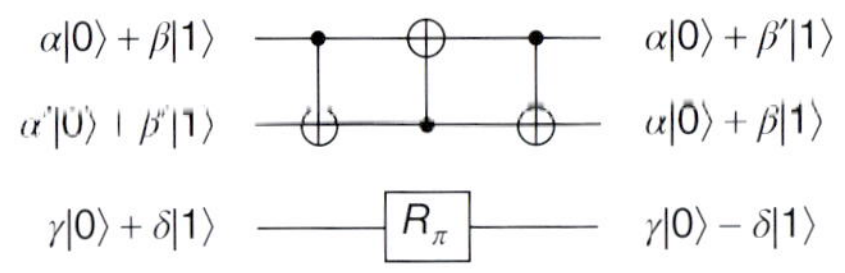

Figure 2.11 Sample fragment of a quantum circuit illustrating the effect of some typical quantum gates: a 'SWAP' operation between the upper and centre qubits is effected through three consecutive 'controlled-NOT' operations, while the lower qubit suffers a simple phase rotation. Dark dots mark control qubits and crossed circles mark target qubits.

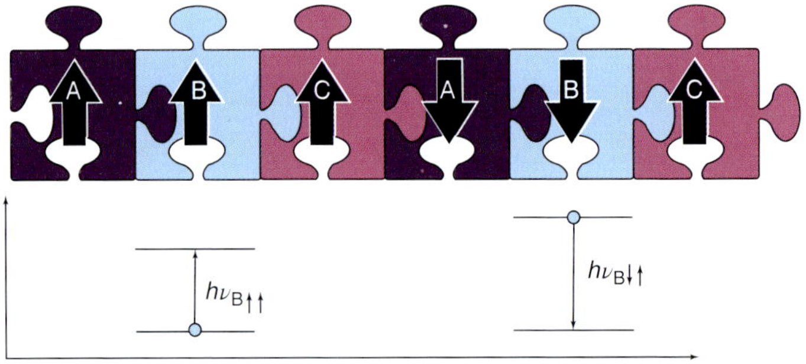

Figure 2.12 Section of an $(-ABC-)_n$ magnetic heteropolymer, or periodic array of spin qubits of three different types. Note than while both B sites are chemically equivalent, the effect of the magnetic exchange with their nearest neighbours modulates the energy of the up-down transition.

'a potentially realizable quantum computer' by Lloyd [36], assumes a body of interacting qubits, such as a loosely coupled spin chain, or possibly a 2D or 3D array of SMMs. The major conditions are that the spins are distributed periodically and that they are of more than one kind, meaning they can be independently manipulated by spectroscopy. Lloyd's proposal was based on three kinds of qubits A, B, C, but more recent proposals proved that the principle is sustainable with just two [37]. Requiring neither local one- nor two-qubit gates, all the qubits of the same kind are then addressed simultaneously, for example, by switching the state of all qubits of type B from $|0\rangle$ to $|1\rangle$ and vice versa, with the following critical exceptions:

- At least one qubit is different from all others and can be addressed individually; this can be simply the final spin of a spin chain.
- The interaction with the (nearest) neighbours is such that it allows differential addressability depending on the surroundings.

As we can see in Figure 2.12, the fact that chemically equivalent centres can have a different magnetic environment enables a selective manipulation, for example, 'flip all spins of kind A, which have as neighbours a spin C pointing down and a qubit B pointing up'. The procedure is reminiscent of cellular automata, and indeed some extensions of this proposal are periodic not only in space but also in time and are called *Quantum Cellular Automata* [38, 39].

Chemically, global control would in principle proceed using magnetic coordination polymers. These do not need to be single-chain magnets of transition metal complexes, and in fact weakly coupled chains of lanthanides or actinides might be excellent candidates. Promising systems include lanthanide-substituted metal-organic frameworks (MOFs) [40]. The accuracy required for full QIP is outside the technical capabilities of EPR equipment, and yet it is likely that small-scale but valuable quantum simulations can be achieved with global control on lanthanide-based systems.

The other two major paradigms for quantum computing are probably farther from an implementation using molecular systems. *Measurement-based quantum computing* requires the ability to prepare a starting so-called 'resource state' which is massively entangled. By this we mean that the state of each qubit depends non-trivially on the state of many other qubits. In a context of molecular magnetism, the simplest case would be the antiferromagnetic coupling of two effective spins $S = 1/2$: the up/down orientation of spin A is not determined, but we know that it will be opposite to the orientation of spin B. This entanglement may be achieved, for example, using a self-assembled monolayer of lanthanide Kramers' doublets on a conductive surface, which results in non-local coupling between the spins. Indeed, it has been shown that even a thermal state with always-on interaction can be used as a 'resource state' [41]. While the computation requires the ability to perform single-qubit measurements in a variety of bases, the preparation of a versatile resource state is a more realistic project and would be by itself an interesting result.

The paradigm of *adiabatic control* appears to be even more exotic. And yet, one has to note that it is through adiabatic quantum computing that the record-smashing factorization of the number 143 into (11×13) was achieved [42]. Adiabatic computation does not proceed in discrete steps, but through a slow and continuous manipulation of the Hamiltonian, so that the system is kept in a highly nontrivial ground state. Note that a limited adiabatic control of the Hamiltonian describing a multinuclear lanthanide–lanthanide complex is conceivable not only by using magnetic fields, but also electrically, that is, by the means of molecular spintronics, as the CF can in principle be manipulated with additional electric fields. In this scenario the position of the ligands can vary continuously, and thus the CF felt by each centre, as well as the magnetic exchange between them, could also be manipulated adiabatically. It is possible to simulate this kind of procedure with the help of Single Ion Magnet magnetic PREdiction (SIMPRE), a software package that calculates the electronic level structure of lanthanide complexes based on an effective point-charge approach [21, 25, 43].

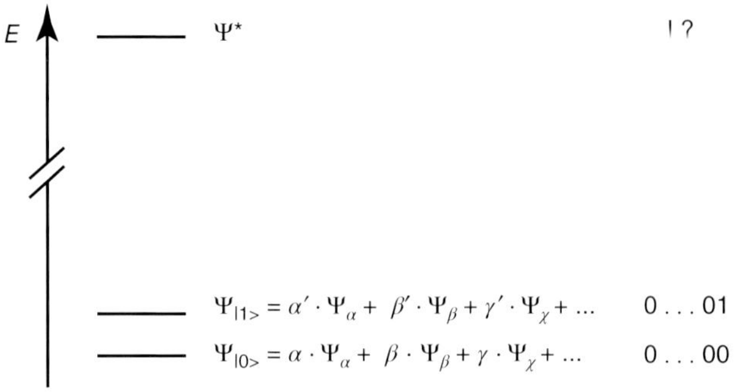

Figure 2.13 The first 2^n magnetic states of the system are assigned to combinations of n qubits. The rest of the spectrum is outside of the computational basis.

Considering any of these paradigms, a minimal goal for toy models would be 'to manipulate the quantum dynamics of a small number of spin levels', and that requires a known and controlled composition of the wavefunction, sufficient isolation and a method for coherent manipulation. As illustrated in Figure 2.13, the first few magnetic states of the system are labelled and thus assigned qubit values. The rest of the spectrum is outside of the computational basis, so one needs to ensure that these levels are not populated during the coherent manipulation.

2.4.2
Combining Physical Qubit Implementations with Lanthanide Complexes

It is important to remember that quantum paradigms are not necessarily limited to a particular qubit implementation: qubit hardware can often be used for different quantum computing paradigms, and vice versa. Over the past few decades, physics has offered a large range of quantum bits such as trapped ions or trapped atoms, quantum dots, quantum optical cavities, SQUIDs and NV defects on diamond (Table 2.4). Some are relevant as benchmarks against which it is possible to evaluate the quality of the implementations offered by chemistry. Others, more interestingly, may actually be combined with lanthanide complexes functioning as molecular spin qubits. This would produce heterogeneous quantum information processors which combine the advantages of different systems. It is now commonly accepted that the eventually successful quantum hardware will be some kind of combination of hardwares, so this is naturally an attractive path.

The use of trapped ions, or trapped atoms, as qubits [46, 47] is one of the most mature techniques. They have been used to achieve remarkable feats, mainly in the field of quantum simulation [48]. However, there is no clear link between this technology and molecular systems.

Table 2.4 Some examples of physical qubits, with an estimation of their transversal relaxation times and error rate in one- or two-qubit gates (for details see [33] and references therein).

Qubit candidate	T_2 (s)	Error % one (two)-qubit gate
Trapped ion	15	0.48 (0.7)
GaAs/Si quantum dot [44]	2	5 (n.d.)
Infrared photon	10^{-4}	0.016 (1)
SQUID	4×10^{-6}	0.7 (10)
NV centre in diamond [45]	1	2 (5)

Quantum dots – the manipulation of electrons on a semiconducting surface [49] – constitute another promising qubit candidate that has advanced enormously in recent years [50]. From a theoretical point of view, quantum dots, either as spin or as charge qubits, are arguably the most interesting scheme for chemists, as there are several analogous phenomena between them and magnetic or mixed-valence molecules. Indeed, in quantum dot qubits there are g-tensors and spin–orbit coupling, while in coupled double-dot or triple-dot 'molecules' there is even magnetic exchange and electron transfer. From an experimental point of view, chemistry might be able to offer a new approach to self-assembled quantum dots [51].

Photons in quantum optical cavities also constitute excellent qubit candidates [52]. Resonant coupling of atoms with a single mode of the radiation field was experimentally achieved 25 years ago [53], and eventually the coherent coupling of quantum optical cavities with atoms or (simple) molecules was suggested as a means to achieve stable quantum memories in a hybrid quantum processor [54]. There might be a role to play for molecular spin qubits in this kind of hybrid quantum devices that combine solid-state with 'flying' qubits.

The use as qubits of either flux, charge or phase state of a superconductor with a Josephson junction (e.g. in a SQUID) is experimentally unique because of the macroscopic size of the quantum system [55]. The relation with molecular magnetism is twofold: first, as they share instrumentation, both benefit from technical developments, for example, micro-SQUIDS; second, molecular magnets including lanthanide–lanthanide SIMs could conceivably be placed within SQUIDs or micro-SQUIDS to produce yet another different kind of hybrid quantum processor [56]. Indeed, it has recently been shown, using theoretical methods, that it is possible to achieve strong and ultra-strong coupling regimes between SMM crystals and a superconducting circuit, something that would allow the use of SMMs for storage or manipulation of quantum information, either in the context of quantum computing or of quantum simulations [57]. Analogous schemes have been proposed to couple superconducting flux qubits with the next candidate, NV defects in diamond. Indeed, single spins diluted in a solid-state matrix are very close analogues to molecular spin systems: they function as molecules trapped in a rigid matrix, because the network of covalent bonds does not affect the spin subsystem. The attainment of strong coupling conditions between solid-state spin ensembles and radio-frequency or optical photons might make these systems interesting as quantum memories [58]. Similarly to magnetic molecules, NV centres can be seen effectively as a small number of electronic spins localized in a molecular-sized volume: two carbon atoms are substituted by a nitrogen atom and an extra electron, creating a spin triplet $S = 1$. Thus, NV defects in diamond may serve as a goal one should aim for with molecular spin qubits [59–64].

A related family of solid-state systems is in most regards identical to SIMs: diluted single ions of lanthanides. Indeed, there are several other examples of determinations of coherence times and observations of Rabi oscillations (quantum oscillations resulting from the coherent absorption and emission of photons driven by an electromagnetic wave) [65] in these systems. One should mention the studies on the SIM $Ho_xY_{(1-x)}LiF_4$ [31], a system that has also been studied for its interesting magnetism at different dilutions, where it presents a range of behaviour including spin glass and dipolar Ising ferromagnet. The study by Baibekov *et al.* is also important, as they reported a new family of molecular qubits based on Er^{3+} substituting for Ca^{2+} in a single crystalline matrix of $CaWO_4$ [66].

2.4.3
Molecular Spin Qubits

A wide variety of proof-of-principle systems have been proposed, synthesized and studied in the field of molecular spin qubits. In fact, due to the fast development of the field, several chemical quantum computation reviews using magnetic molecules as spin qubits have been published over the past decade, covering both experimental and theoretical results [67–69]. Only in a minority of experiments implementing non-trivial one- or two-qubit gates has been carried out, so in this aspect this family is clearly not yet competitive with other hardware candidates.[1)] Of course, the main interest of the molecular approach that makes it qualitatively different is that molecules can be chemically engineered to tailor their properties and acquire new functionalities.

Before reviewing existing examples, a very brief explanation on the *mechanisms of decoherence* for molecular spin qubits is necessary; more details are available elsewhere [67]. Broadly speaking, the three decoherence sources for these systems are spin bath decoherence, oscillator bath decoherence and pairwise dipolar decoherence, and can be regulated by a combination of temperature, magnetic field and chemical design of the system [70]. The spin bath mainly consists of nuclear spins, but in general it also includes any localized excitations that can couple to the

1) Here we will focus on electron spin qubits and thus we will not be discussing NMR quantum computing, where molecules played a key role in the early successes of quantum information processing.

qubit state, for example, a crystallographic disorder or the spin of a paramagnetic impurity. The nuclear part can be reduced by working with isotopically purified molecules. Of course, this strategy works better for systems which, like POMs, combine a rich magnetochemistry with elements whose nuclei consist of an even number of protons, such as oxygen, molybdenum and tungsten, as every isotope of every element with an odd number of protons bears a nuclear spin [71]. The oscillator bath, in insulators, is mainly the phonon bath. Pairwise dipolar decoherence is caused by distant qubits being resonant and thus having a probability of spin flip [72]. Although often attacked by dilution [73, 74], this phenomenon is not actually solved by mere dilution, and it requires either low dimensionalities or sophisticated quantum error correction procedures, or both.

Recently reported results show that SMMs functioning as *single qubits* have the potential to attain very high couplings with superconducting circuits [57]. In addition, the great variety of magnetic molecules enables a vast choice of resonance frequencies. However, for many of the best-known SMMs, such as Fe_8 or Mn_{12}, the strong magnetic anisotropy introduces some technical difficulties, that is, the need to work at very high frequencies, above 100 GHz, or the application of strong (above 2 T) and very accurately aligned (within 0.5°) magnetic fields. In this sense, SIMs can be better suited for quantum manipulation by standard EPR equipment by following either of two strategies: (i) using the energy level structure that results from the weak magnetic exchange and/or strong hyperfine coupling that are characteristics of lanthanides or (ii) using Gd^{3+}, where the orbital magnetic moment and thus the CF only arise from an interaction with excited states and therefore are extraordinarily small; although the real power of these two strategies is evidenced in the next section, multi-qubit molecules, for example, if one compares the energy splitting of the ground state in Mn_{12} and in GdW_{30}, one observes that the splitting in the former case is two orders of magnitude higher than in the second case ($100\ cm^{-1}$ compared to $1\ cm^{-1}$) [10]. In addition, SIMs have the advantage of being simpler, their physical response is easier to describe, they offer a greater versatility for the modification of the spin Hamiltonian via the rational design of the local coordination shell surrounding the central magnetic ion and they can be made more robust against decoherence.

It is interesting to note that to actually implement a useful algorithm it is necessary to implement a certain number of quantum operations within the coherence time. Recently, we reported that it was possible to increase the number of coherent rotations by a factor of 10 by matching the Rabi frequency with the frequency of the proton in the polyoxometalate SIM GdW_{30}. Under these conditions, it was possible to perform at least 80 such operations (Figure 2.14) [75].

Research on *multi-qubit molecules* starts with the synthesis and characterization of systems that seem to embody more than one qubit, for example, systems with weakly coupled electron spins. Indeed, many molecular structures include several weakly coupled magnetic ions [76–78]. On a smaller scale, the capability of implementing a Controlled-NOT quantum logic gate using molecular clusters

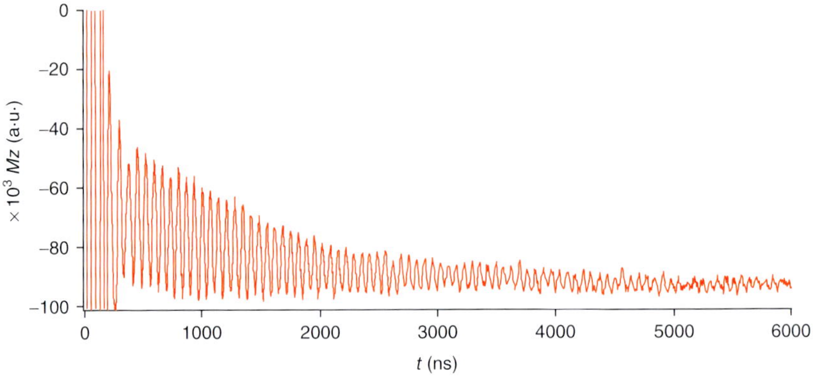

Figure 2.14 Coherent oscillations in $[GdW_{30}P_5O_{110}]^{12-}$.

containing two Tb^{3+} ions was demonstrated. Here, the definition of control and target qubits is based on the strong magnetic anisotropy and the magnetic inequivalence of both ions, which also provides a method to realize a SWAP gate in the same cluster [79]. An introduction to the requirements for a paramagnetic molecule to act as a two-qubit quantum gate is proposed in a tutorial review published by Aromí *et al.* [79], where two synthetic strategies are presented, based on ligand design and inorganic synthesis, preparing dinuclear complexes of anisotropic metal ions that present dissimilar environments and magnetic coupling. This topic is discussed in more detail in the corresponding chapter of the present book [80]. If used as qubits, this richness in the energy level structures can be used to implement elementary quantum gates (CNOT and SWAP) or even, in the case of a complex magnetic chain, act as quantum simulators [81]. The first step is usually the preparation of actual entanglement, that is, of nontrivial superpositions that involve more than one qubit. A preliminary but complete task would be the implementation of a full process of initialization, gating and readout. So far, no example of such a kind of quantum preparation has been reported using lanthanides. With d-transition metal ions, an example using POMs has been reported. In these molecular systems an all-electrical process enabling entanglement and measurement of the quantum state was proposed. This was done using the mixed-valence polyoxometalate $[PMo_{12}O_{40}(VO)_2]^{q-}$, which contains two localized magnetic moments connected through a central redox-active mixed-valence Mo_{12} cluster (Figure 2.15). The electrical control of the number of electrons in the Mo_{12} cluster, for example, by means of a scanning tunnelling microscopic (STM) tip, enables the on/off switching of the magnetic exchange between the localized spins, and thus the application of two-qubit gates. A sequential tunnelling conductance would be used for the read-out process. To finish this part, it is important to notice that to build a useful quantum computer one would

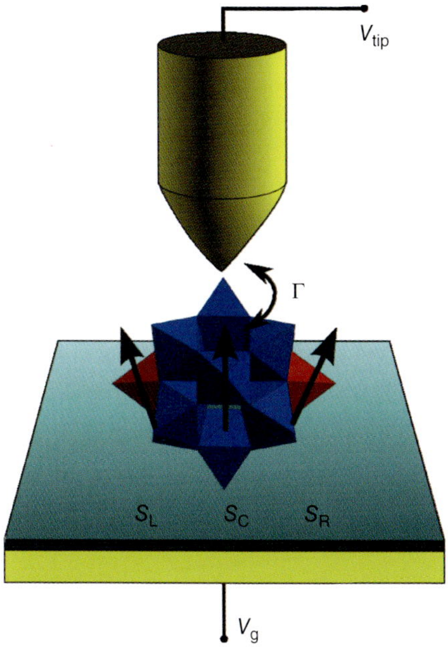

Figure 2.15 Proposed setup to effect two-qubit gates onto Mo12V2, a mixed-valence POM that combines two localized electrons on the lateral vanadyl groups with a variable number of delocalized electrons in the Keggin core.

need to scale up the system indefinitely, so a criterion to evaluate partial achievements is whether one can build upon them to get closer to the ultimate goal.

2.5 Conclusions

The impact of mononuclear lanthanide complexes in molecular magnetism is increasing and evolving towards their applications in nanotechnology. Thus, while these nanomagnets pose complex theoretical problems and tough experimental challenges in molecular magnetism, they open a window for the manipulation of quantum spins at the molecular level, a nanotechnological feat that would be crucial not only in nanomagnetism but also in quantum computing.

We have shown in this chapter that the major electronic features that determine the spin dynamics of SIMs based on lanthanides can be directly correlated with the local coordination environment around the 4f metal ions. By using an effective point-charge model that accounts for covalent effects, we have shown that the splitting of the ground state, J, of the lanthanide into M_J sublevels, caused by the influence of the CF created by the surrounding ligands, is consistent with experimental magnetic susceptibility and, when available, also with spectroscopic data. The power of this approach is that it allows us to predict the energy splitting as well as the nature of the resulting wave functions of a lanthanide in any environment, in terms of the CF parameters. At this point, it is important to insist that, in the case of f electrons, the high-order CF terms arise from the interaction between the electric field and ground state J, whereas for d-transition-metal SMMs, they appear as perturbative corrections. This difference underlines the key influence of these CF parameters on the magnetic properties of lanthanide complexes. Several real lanthanide-based SIMs have been studied with our method, and general magnetostructural correlations have been proposed.

Furthermore, it is interesting to note that the quantum effects in the SIMs are much more pronounced than those in the polynuclear cluster-type SMMs. Hence, this class of nanomagnets provides ideal systems to study quantum phenomena such as tunnelling, relaxation and coherence. In this vein, these simple molecules are better suited for their integration as reliable quantum bits (qubits) for quantum computing than the polynuclear SMMs are. The reason lies in the larger stability of SIMs against decoherence because the robustness of a quantum state decreases with the system's number of degrees of freedom. This idea is further supported by the possibility of chemically controlling the quantum coherence in these systems.

Finally, we have shown that POM chemistry provides excellent examples of mononuclear lanthanide complexes that behave either as SMMs or as qubits with the longest possible coherence time. This is so because the high symmetry of the metal complex and the ideal magnetic insulation imposed by the POM ligand allow us to accommodate the lanthanides in very different symmetries (D_{4d} vs C_5), while offering the possibility of tuning the magnetic anisotropy and keeping them magnetically isolated (the magnetic ordering in these materials only occurs at very low temperatures; typically below 0.01 K). Furthermore, the long coherence times that can be achieved in some of these nanomagnets are a consequence of the possibility offered by POM chemistry to minimize the main sources of quantum decoherence (hyperfine couplings and dipolar spin-spin interactions) by preparing nuclear-spin-free compounds and by diluting the magnetic centres, while conserving the crystallinity.

Acknowledgements

The present work has been funded by the EU (Project ELFOS and ERC Advanced Grant SPINMOL), the Spanish MINECO (grants MAT2011-22785, MAT2007-61584 and the CONSOLIDER project on Molecular Nanoscience), and the Generalidad Valenciana (Prometeo and ISIC Programmes of excellence).

References

1. Leuenberger, M.N. and Loss, D. (2011) Quantum computing in molecular magnets. *Nature*, **410**, 789–793.
2. Sessoli, R., Tsai, H.L., Schake, A.R., Wang, S., Vincent, J.B., Folting, K., Gatteschi, D., Christou, G. and Hendrickson, D.N. (1993) High-spin molecules: [Mn12O12(O2CR)16(H2O)4]. *J. Am. Chem. Soc.*, **115**, 1804–1816.
3. Gatteschi, D. and Sessoli, R. (2003) Quantum tunneling of magnetization and related phenomena in molecular materials. *Angew. Chem. Int. Ed.*, **42**, 268–297.
4. Ishikawa, N., Sugita, M., Ishikawa, T., Koshihara, S.Y. and Kaizu, Y. (2003) Lanthanide double-decker complexes functioning as magnets at the single-molecular level. *J. Am. Chem. Soc.*, **125**, 8694–8695.
5. Ganivet, C.R., Ballesteros, B., de la Torre, G., Clemente Juan, J.M., Coronado, E. and Torres, T. (2012) Influence of peripheral substitution on the magnetic behaviour of single-ion magnets based on homo- and heteroleptic TbIII Bis(phthalocyaninate). *Chem. Eur. J.*, **19**, 1457–1465.
6. AlDamen, M.A., Clemente-Juan, J.M., Coronado, E., Martí-Gastaldo, C. and Gaita-Ariño, A. (2008) Mononuclear lanthanide single-molecule magnets based on polyoxometalates. *J. Am. Chem. Soc.*, **130**, 8874–8875.
7. Aldamen, M.A., Cardona-Serra, S., Clemente-Juan, J.M., Coronado, E., Gaita-Ariño, A., Martí-Gastaldo, C., Luis, F. and Montero, O. (2009) Mononuclear lanthanide single molecule magnets based on the polyoxometalates $[Ln(W_5O_{18})_2]^{9-}$ and $[Ln(\beta_2\text{-}SiW_{11}O_{39})_2]^{13-}$ (Ln^{III}) Tb, Dy, Ho, Er, Tm, and Yb). *Inorg. Chem.*, **48**, 3467–3479.
8. Luis, F., Martínez-Pérez, M., Montero, O., Coronado, E., Cardona-Serra, S., Martí-Gastaldo, C., Clemente-Juan, J.M., Sesé, J., Drung, D. and Schurig, T. (2010) Spin-lattice relaxation via quantum tunneling in an Er^{3+}-polyoxometalate molecular magnet. *Phys. Rev. B*, **82**, 060403.
9. Cardona-Serra, S., Clemente-Juan, J.M., Coronado, E., Gaita-Ariño, A., Camón, A., Evangelisti, M., Luis, F., Martínez-Pérez, M.-J. and Sesé, J. (2012) Lanthanide single-ion magnets based on polyoxometalates with a 5-fold symmetry: the series $[LnP_5W_{30}O_{110}]^{12-}$ (Ln^{3+} = Tb, Dy, Ho, Er, Tm, and Yb). *J. Am. Chem. Soc.*, **134**, 14982–14990.
10. Martínez-Pérez, M.J., Cardona-Serra, S., Schlegel, C., Moro, F., Alonso, P.J., Prima-García, H., Clemente-Juan, J.M., Evangelisti, M., Gaita-Ariño, A., Sesé, J., van Slageren, J., Coronado, E. and Luis, F. (2012) Gd-based single-ion magnets with tunable magnetic anisotropy: molecular design of spin qubits. *Phys. Rev. Lett.*, **108**, 247213.
11. Jiang, S., Wang, B., Sun, H., Wang, Z. and Gao, S. (2011) An organometallic single-ion magnet. *J. Am. Chem. Soc.*, **133**, 4730–4733.
12. Jiang, S., Wang, B., Su, G., Wang, Z. and Gao, S. (2010) A mononuclear dysprosium complex featuring single-molecule-magnet behaviour. *Angew. Chem. Int. Ed.*, **49**, 7448–7451.
13. (a) Car, P.E., Perfetti, M., Mannini, M., Favre, A., Caneschi, R. and Sessoli, R. (2011) Giant field dependence of the low temperature relaxation of the magnetization in a dysprosium(III)–DOTA complex. *Chem. Commun.*, **47**, 3751–3753; (b) Cucinotta, G., Perfetti, M., Luzon, J., Etienne, M., Car, P.E., Caneschi, A., Calvez, G., Bernot, K. and Sessoli, R. (2012) Magnetic anisotropy in a dysprosium/DOTA single-molecule magnet: beyond simple magneto-structural correlations. *Angew. Chem. Int. Ed.*, **51**, 1606–1610.
14. Ishikawa, N., Sugita, M., Okubo, T., Tanaka, H., Iino, T. and Kaizu, Y. (2003) Determination of Ligand-field parameters and f-electronic structures of Double-Decker Bis(phthalocyaninato)lanthanide complexes. *Inorg. Chem.*, **42**, 2440–2446.

15. (a) Buckmaster, H.A. (1962) Tables of matrix elements for the operators *O±12, O±14, O±16, O±52*. *Can. J. Phys.*, **40** (11), 1670–1677; (b) Rudowicz, C. (1985) Transformation relations for the conventional *Okq* and normalised *O'kq* stevens operator equivalents with $k = 1$ to 6 and $-k \leq q \leq k$. *J. Phys. C: Solid State Phys.*, **18**, 1415–1430.
16. (a) Ryabov, I.D. (1999) On the generation of operator equivalents and the calculation of their matrix elements. *J. Magn. Reson.*, **140**, 141; (b) Ryabov, I.D. (2009) On the operator equivalents and the crystal-field and spin hamiltonian parameters. *Appl. Magn. Reson.*, **35**, 481–494.
17. Edvarsson, S. and Klintenberg, M. (1998) Role of the electrostatic model in calculating rare-earth crystal-field parameter. *J. Alloys Compd.*, **275-277**, 230–233.
18. Ishikawa, N. (2003) Simultaneous determination of ligand-field parameters of isostructural lanthanide complexes by multidimensional optimization. *J. Phys. Chem. A*, **107** (30), 5831–5835.
19. Hutchings, M.T. (1964) Point-charge calculations of energy levels of magnetic ions in crystalline electric. *Fields Solid State Phys.*, **16**, 227–273.
20. Figgis, B.N. and Hitchman, M.A. (2000) *Ligand Field Theory and its Applications*, Wiley-VCH Verlag GmbH, New York.
21. Baldoví, J.J., Borrás-Almenar, J.J., Clemente-Juan, J.M., Coronado, E. and Gaita-Arino, A. (2012) Modeling the properties of lanthanide single-ion magnets using an effective point-charge approach. *Dalton Trans.*, **41**, 13705–13710.
22. Baldoví, J.J., Clemente-Juan, J.M., Coronado, E. and Gaita-Arino, A. (2013) Two pyrazolylborate dysprosium(III) and neodymium(III) single ion magnets modeled by a radial effective charge approach. *Polyhedron*, **66**, 39–42.
23. Baldoví, J. J., Clemente-Juan, J. M., Coronado, E., Gaita-Arino, A. and Giménez-Saiz, C., (2014) Construction of a General Library for the Rational Design of Nanomagnets and Spin Qubits Based on Mononuclear f-Block Complexes. The Polyoxometalate Case. *Inorg. Chem.*, **53**, 9976–9980.
24. Baldoví, J.J., Cardona-Serra, S., Borrás-Almenar, J.J., Clemente-Juan, J.M., Coronado, E. and Gaita-Arino, A. (2013) Modeling the properties of uranium-based single ion magnets. *Chem. Sci.*, **4**, 938–946.
25. Baldoví, J.J., Cardona-Serra, S., Clemente-Juan, J.M., Coronado, E., Gaita-Ariño, A. and Palii, A. (2012) Rational design of single ion magnets and spin qubits based on mononuclear lanthanide complexes. *Inorg. Chem.*, **51**, 12565–12574.
26. Barsukova, M., Izarova, N.V., Biboum, R.N., Keita, B., Nadjo, L., Ramachandran, V., Dalal, N.S., Antonova, N.S., Carbó, J.J., Poblet, J.M. and Kortz, U. (2010) Polyoxopalladates encapsulating yttrium and lanthanide ions, $[X^{III}Pd^{II}{}_{12}(AsPh)_8O_{32}]^{5-}$ (X=Y, Pr, Nd, Sm, Eu, Gd, Tb, Dy, Ho, Er, Tm, Yb, Lu). *Chem. Eur. J.*, **16**, 9076–9085.
27. Alvarez, S., Alemany, P., Casanova, D., Cibera, J., Llunell, M. and Avnir, O. (2005) Shape maps and polyhedral interconversion paths in transition metal chemistry. *Coord. Chem. Rev.*, **249**, 1693–1708.
28. Bernal, E.G. (1971) Optical spectrum and magnetic properties of $Er3^{+}$ in CaWO4. *J. Chem. Phys.*, **55**, 2538–2549.
29. Stevens, K.W.H. (1952) Matrix elements and operator equivalents connected with the magnetic properties of rare-earth ions. *Proc. Phys. Soc. London, Sect. A*, **65**, 209–215.
30. Stamp, P.C.E. and Tupitsyn, I.S. (2004) Coherence window in the dynamics of quantum nanomagnets. *Phys. Rev. B*, **69**, 014401.
31. Bertaina, S., Gambarelli, S., Tkachuk, A., Kurkin, I.N., Malkin, B., Stepanov, A. and Barbara, B. (2007) Rare-earth solid-state qubits. *Nat. Nanotechnol.*, **2**, 39–42.
32. Nielsen, M.A. and Chuang, I.L. (2000) *Quantum Computation and Quantum Information*, Cambridge University, Cambridge.

33. Ladd, T.D., Jelezko, F., Laflamme, R., Nakamura, Y., Monroe, C. and O'Brien, J.L. (2010) Quantum computers. *Nature*, **464**, 45–53.
34. Stamp, P.C.E. (2006) The decoherence puzzle. *Stud. Hist. Philos. Mod. Phys.*, **37**, 467–497.
35. DiVincenzo, D.P. (1995) Topics in quantum computers, in *Mesoscopic Electron Transport*, NATO ASI Series E (eds L. Kowenhoven, G. Schoen, and L. Sohn), Kluwer Academic Publisher, Dordrecht, arXiv:cond-mat/9612126 [cond-mat.mes-hall].
36. Lloyd, S. (1993) A potencially realizable quantum computer. *Science*, **261**, 1569–1571.
37. Benjamin, S.C. (2004) Multi-qubit gates in arrays coupled by 'always-on' interactions. *New J. Phys.*, **6**, 61, (1-17).
38. Pérez-Delgado, C.A. and Cheung, D. (2007) Local unitary quantum celular automata. *Phys. Rev. A*, **76**, 032320.
39. Pérez-Delgado, C.A. and Cheung, D. (2007) Local unitary quantum celular automata. *Phys. Rev. A*, **76**, 032320.
40. Baldoví, J.J., Coronado, E., Gaita-Ariño, A., Gamer, C., Giménez-Marqués, M. and Mínguez Espallargas, G. (2014) A SIM-MOF: Three-Dimensional Organisation of Single-Ion Magnets with Anion-Exchange Capabilities. *Chem. Eur. J.*, **20**, 10695–10702.
41. Li, Y., Browne, D.E., Kwek, L.C., Raussendorf, R. and Wei, T.C. (2011) Thermal states as universal resources for quantum computation with always-on interactions. *Phys. Rev. Lett.*, **107**, 060501.
42. Xu, N., Zhu, J., Lu, D., Zhou, X., Peng, X. and Du, J. (2012) Quantum factorization of 143 on a dipolar-coupling nuclear magnetic resonance system. *Phys. Rev. Lett.*, **108**, 130501.
43. (a) Baldoví, J.J., Cardona-Serra, S., Clemente-Juan, J.M., Coronado, E., Gaita-Arino, A. and Palii, A. (2013) SIMPRE: a software package to calculate crystal field parameters, energy levels and magnetic properties on mononuclear lanthanide complexes based on charge distributions. *J. Comput. Chem.*, **34**, 1961–1967; (b) Baldoví, J.J., Clemente-Juan, J.M., Coronado, E., Gaita-Arino, A. Palii, A. (2014) An Updated Version of the Computational Package SIMPRE that Uses the Standard Conventions for Stevens Crystal Field Parameters. *J. Comput. Chem.*, **35**, 1930–1934.
44. Tyryshkin, A.M., Tojo, S., Morton, J.J.L., Riemann, H., Abrosimov, N.V., Becker, P., Pohl, H.-J., Schenkel, T., Thewalt, M.L.W., Itoh, K.M. and Lyon, S.A. (2012) Electron spin coherence exceeding seconds in high-purity silicon. *Nat. Mater.*, **11**, 143–147.
45. Maurer, P.C., Kucsko, G., Latta, C., Jiang, L., Yao, N.Y., Bennett, S.D., Pastawski, F., Hunger, D., Chisholm, N., Markham, M., Twitchen, D.J., Cirac, J.I. and Lukin, M.D. (2012) Room-temperature quantum bit memory exceeding one second. *Science*, **336**, 1283–1286.
46. Kielpinski, D., Monroe, C. and Wineland, D.J. (2002) Architecture for a large-scale ion-trap quantum computer. *Nature*, **417**, 709–711.
47. Friedenauer, A., Schmitz, H., Glueckert, J.T., Porras, D. and Schaetz, T. (2008) Simulating a quantum magnet with trapped ions. *Nat. Phys.*, **4**, 757–761.
48. Simon, J., Bakr, W.S., Ma, R., Tai, M.E., Preiss, P.M. and Greiner, M. (2011) Quantum simulation of antiferromagnetic spin chains in an optical lattice. *Nature*, **472**, 307–312.
49. Loss, D. and DiVincenzo, D.P. (1998) Quantum computation with quantum dots. *Phys. Rev. A*, **57**, 120–126.
50. Awschalom, D.D., Basset, L.C., Dzurak, A.S., Hu, E.L. and Petta, J.R. (2013) Quantum spintronics: engineering and manipulating atom-like spins in semiconductors. *Science*, **339**, 1174–1179.
51. Wang, L., Rastelli, A., Kiravittaya, S., Benyoucef, M. and Schmidt, O.G. (2009) Self-assembled quantum dot molecules. *Adv. Mater.*, **21**, 2601–2618.
52. Maître, X., Hagley, E., Nogues, G., Wunderlich, C., Goy, P., Brune, M., Raimond, J.M. and Haroche, S. (1977)

Quantum memory with a single photon in a cavity. *Phys. Rev. Lett.*, **79**, 769–772.

53. De Martini, F., Innocenti, G., Jacobovitz, G.R. and Mataloni, P. (1987) Anomalous spontaneous emission time in a microscopic optical cavity. *Phys. Rev. Lett.*, **59**, 2955–2958.

54. Rabi, P., DeMille, D., Doyle, J.M., Lukin, M.D., Schoelkopf, R.J. and Zoller, P. (2006) Hybrid quantum processors: molecular ensembles as quantum memory for solid state circuits. *Phys. Rev. Lett.*, **97**, 033003.

55. Nakamura, Y., Pashkin, Y.A. and Tsai, J.S. (1999) Coherent control of macroscopic quantum states in a single-Cooper-pair box. *Nature*, **398**, 786–788.

56. Tejada, J., Chudnovsky, E.M., del Barco, E., Hernandez, J.M. and Spiller, T.P. (2001) Magnetic qubits as hardware for quantum computers. *Nanotechnology*, **12**, 181–186.

57. Jenkins, M., Hümmer, T., Martínez-Pérez, M.J., García-Ripoll, J., Zueco, D. and Luis, F. (2013) Coupling single molecule magnets to quantum circuits. *New J. Phys.*, **15**, 095007.

58. (a) Imamoglu, A. (2009) Cavity QED based on collective magnetic dipole coupling: spin ensembles as hybrid two-level systems. *Phys. Rev. Lett.*, **102**, 083602; (b) Wesenberg, J., Ardavan, A., Briggs, G., Morton, J., Schoelkopf, R., Schuster, D. and Mølmer, K. (2009) Quantum computing with an electron spin ensemble. *Phys. Rev. Lett.*, **103**, 070502; (c) Marcos, D., Wubs, M., Taylor, J.M., Aguado, R., Lukin, M.D. and Sørensen, A.S. (2010) Coupling nitrogen-vacancy centers in diamond to superconducting flux qubits. *Phys. Rev. Lett.*, **105**, 210501.

59. Jelezko, F., Gaebel, T., Popa, I., Gruber, A. and Wrachtrup, A. (2004) Observation of coherent oscillations in a single electron spin. *J. Phys. Rev. Lett.*, **92**, 076401.

60. Hanson, R., Gywat, O. and Awschalom, D.D. (2006) Room-temperature manipulation and decoherence of a single spin in diamond. *Phys. Rev. B*, **74**, 161203.

61. Childress, L., Gurudev Dutt, M.V., Taylor, J.M., Zibrov, A.S., Jelezko, F., Lukin, M.D., Wrachtrup, J. and Hemmer, P.R. (2006) Coherent dynamics of coupled electron and nuclear spin qubits in diamond. *Science*, **314**, 281–285.

62. Gurudev Dutt, M.V., Childress, L., Jiang, L., Togan, E., Maze, J., Jelezko, F., Zibrov, A.S., Hemmer, P.R. and Lukin, M.D. (2007) Coherent dynamics of coupled electron and nuclear spin qubits in diamond. *Science*, **316**, 1312–1329.

63. Bermudez, A., Jelezko, F., Plenio, M.B. and Retzker, A. (2011) Electron-mediated nuclear-spin interactions between distant nitrogen-vacancy centers. *Phys. Rev. Lett.*, **107**, 150503.

64. Kolkowitz, S., Unterreithmeier, Q.P., Bennett, S.D. and Lukin, M.D. (2012) Sensing distant nuclear spins with a single electron spin. *Phys. Rev. Lett.*, **109**, 137601.

65. Rabi, I.I. (1937) Use of rotationg coordinates in magnetic resonance problems. *Phys. Rev.*, **51**, 652–654.

66. Baibekov, E., Kurkin, I., Gafurov, M., Endeward, B., Rakhmatullin, R. and Mamin, G. (2011) Coherence times and Rabi oscillations in CaWO4:Cr5+ crystal. *J. Magn. Reson.*, **1**, 61–68.

67. Stamp, P.C.E. and Gaita-Ariño, A. (2009) Spin-based quantum computers made by chemistry: hows and whys. *J. Mater. Chem.*, **19**, 1718.

68. Troiani, F. and Affronte, M. (2011) Molecular spins for quantum information technologies. *Chem. Soc. Rev.*, **40**, 3119–3129.

69. Ardavan, A. and Blundell, S.J. (2009) Storing quantum information in chemically engineered nanoscale magnets. *J. Mater. Chem.*, **19**, 1754.

70. Takahashi, S., Tupitsyn, I.S., van Tol, J., Beedle, C.C., Hendrickson, D.N. and Stamp, P.C.E. (2011) Decoherence in crystals of quantum molecular magnets. *Nature*, **476**, 76–79.

71. Ardavan, A., Rival, O., Morton, J., Blundell, S., Tyryshkin, A., Timco, G. and Winpenny, R. (2007) Will spin-relaxation times in molecular magnets

permit quantum information processing? *Phys. Rev. Lett.*, **98**, 057201.

72. Morello, A., Stamp, P.C.E. and Tupitsyn, I. (2006) Pairwise decoherence in coupled spin qubit networks. *Phys. Rev. Lett.*, **97**, 207206.

73. Schlegel, C., van Slageren, J., Manoli, M., Brechin, E.K. and Dressel, M. (2008) Direct observation of quantum coherence in single-molecule magnets. *Phys. Rev. Lett.*, **101**, 147203.

74. Bertaina, S., Gambarelli, S., Mitra, T., Tsukerblat, B., Müller, A. and Barbara, B. (2008) Quantum oscillations in a molecular magnet. *Nature*, **453**, 203–206.

75. Baldoví, J.J., Cardona-Serra, S., Clemente-Juan, J.M., Coronado, E., Gaita-Ariño, A. and Prima-García, H. (2013) Coherent manipulation of spin qubits based on polyoxometalates: the case of the single ion magnet $[GdW_{30}P_5O_{110}]^{14-}$. *Chem. Commun.*, **49**, 8922–8924.

76. Timco, G.A., Carretta, S., Troiani, F., Tuna, F., Pritchard, R.J., Muryn, C.A., McInnes, E.J.L., Ghirri, A., Candini, A., Santini, P., Amoretti, G., Affronte, M. and Winpenny, R.E.P. (2009) Engineering the coupling between molecular spin qubits by coordination chemistry. *Nat. Nanotechnol.*, **4**, 173–178.

77. Candini, A., Lorusso, G., Troiani, F., Ghirri, A., Carretta, S., Santini, P., Amoretti, G., Muryn, C., Tuna, F., Timco, G., McInnes, E.J.L., Winpenny, R.E.P., Wernsdorfer, W. and Affronte, M. (2010) Entanglement in supramolecular spin systems of two weakly coupled antiferromagnetic rings (Purple-Cr_7Ni). *Phys. Rev. Lett.*, **104**, 037203.

78. Aromí, G., Aguilà, D., Gamez, P., Luis, F. and Roubeau, O. (2012) Design of magnetic coordination complexes for quantum computing. *Chem. Soc. Rev.*, **41**, 537–546.

79. Luis, F., Repollés, S., Martínez-Pérez, M.J., Aguilà, D., Roubeau, O., Zueco, D., Alonso, P.J., Evangelisti, M., Camón, A., Sesé, J., Barrios, L.A. and Aromí, G. (2011) Molecular prototypes for spin-based CNOT and SWAP quantum gates. *Phys. Rev. Lett.*, **107**, 117203.

80. Aromi, G. *et al.* (2015) *Lanthanides and Actinides in Molecular Magnetism*, Wiley-VCH Verlag GmbH & Co. KGaA.

81. Santini, P., Carretta, S., Troiani, F. and Amoretti, G. (2011) Molecular nanomagnets as quantum simulators. *Phys. Rev. Lett.*, **107**, 230502.

3
Polynuclear Lanthanide Single Molecule Magnets

Jinkui Tang and Peng Zhang

3.1
Introduction

Since the seminal discovery of the single-molecule magnet (SMM) behaviour of thermally excited spin states of a Dy_3 triangle complex with an almost diamagnetic ground state [1–4], interest in polynuclear lanthanide SMMs has been nothing short of astounding [5, 6]. Especially in recent years, an increasing number of scientists working in chemistry, physics and materials area have been rather active in this respect, in the quest for an improved understanding of complex 4f magnetism [2–4] and a possible methodology for their applications [7–10] in information storage, quantum computing and spintronics. This has led to a steady growth of publications on polynuclear lanthanide SMMs every year. To the best of our knowledge, hitherto more than 100 examples of polynuclear lanthanide compounds (Ln_n, $n \geq 2$) in this field have been reported, as evidenced in our recent review [5]. Many outstanding achievements have been reported to date.

The first is the great breakthroughs in the anisotropic barrier and blocking temperature for polynuclear lanthanide SMMs. It is remarkable that, in recent years, the anisotropic barrier records have toppled like dominoes for lanthanide polynuclear SMMs such as Dy_2 network (71 K) [11], Dy_4 defect-dicubane (170 K) [12], coupled Dy_3 triangles (200 K) [13] and Dy_5 pyramids (528 K) [14]. The record of 528 K is six times more than the value of 86 K for Mn_6 [15]. Furthermore, the record blocking temperature of 14 K (0.9 mT s^{-1}) shown by N_2^{3-} radical-bridged Tb_2 SMM [16] also represents a new benchmark when compared to the transition-metal cluster's record of 4.5 K (0.14 T s^{-1}) [15]. Clearly, lanthanide-based polynuclear systems can be seen as a promising avenue to explore SMMs that can achieve the eventual goal, namely, application to devices, especially, in the areas of molecular spintronics and quantum computing [17].

The second is the breakthrough in understanding complex 4f magnetism through the applications of some advanced magnetic measurements and quantum chemistry calculation methods, such as the single-crystal magnetic measurement

Lanthanides and Actinides in Molecular Magnetism, First Edition.
Edited by Richard A. Layfield and Muralee Murugesu.

developed by Sessoli [18–20] and *ab initio* investigation developed by Chibotaru *et al.* [21, 22]. Currently the single-crystal magnetic measurement is mainly applied to lanthanide-based single-ion magnets (SIMs), providing straightforward evidence of the local anisotropy of lanthanide ions [18–20]. However, it has been rarely explored in polynuclear lanthanide SMMs. One successful case is Dy_3 triangles, where the nonmagnetic ground doublet from the noncollinearity of the single-ion easy axes of magnetization of Dy ions has been simultaneously corroborated by single-crystal magnetic studies and *ab initio* methodology [3]. In contrast, *ab initio* investigations have been applied widely to polynuclear lanthanide SMMs to gain insight into complex magneto-structural correlations. For example, the nonmagnetic ground state can be demonstrated perfectly by the toroidal arrangement of magnetic moments calculated by *ab initio* investigation in Dy_3 triangles [2], planar Dy_4 [23] and Dy_6 rings [22]. Importantly, the method has extended from the calculations of anisotropy directions, energy levels and g values to the evaluation of exchange couplings for Dy clusters in a most rigorous manner [21, 22]. Traditionally, it is difficult to separate exchange coupling contributions from ligand-field/Zeeman level thermal depopulation effects in orbitally degenerate lanthanide systems and Chibotaru has pioneered these developments. In addition, density functional theory (DFT) calculations were also carried out in several Dy_2 systems to probe the metal–ligand and metal–metal magnetic interactions [24, 25].

Eventually, although the use of polynuclear lanthanide SMMs to surfaces or spintronics devices is still blank, the above achievements have suggested their great potential for this field. Here, a collaborative research between chemists, physicists and materials scientists is necessary to achieve such an aim [17]. In this way, the assembly of new polynuclear lanthanide SMMs will lay the foundation for further applications. In this chapter, we do not intend to cover the results obtained in the very active and promising field of lanthanide-based SMMs in exhaustive detail; we would rather focus on the synthetic strategies employed to design polynuclear lanthanide systems (Ln_n, $n > 2$) with high barriers along with their recent slow relaxation studies. As such, we hope that it will serve as a valuable resource for both new and current researchers in the field.

3.2 Synthetic Strategies

As pointed out by Sessoli and coworker in their review, a possible approach to increasing the energy barrier of lanthanide-based SMMs is by assembling together several lanthanide ions which interact magnetically either in a zero-dimensional structure (SMMs) or in a one-dimensional structure (Single-Chain Magnets) [26]. The last several years have indeed seen a flurry of results from synthetic chemists in this respect. Especially in our recent review about Dy-based SMMs, polynuclear

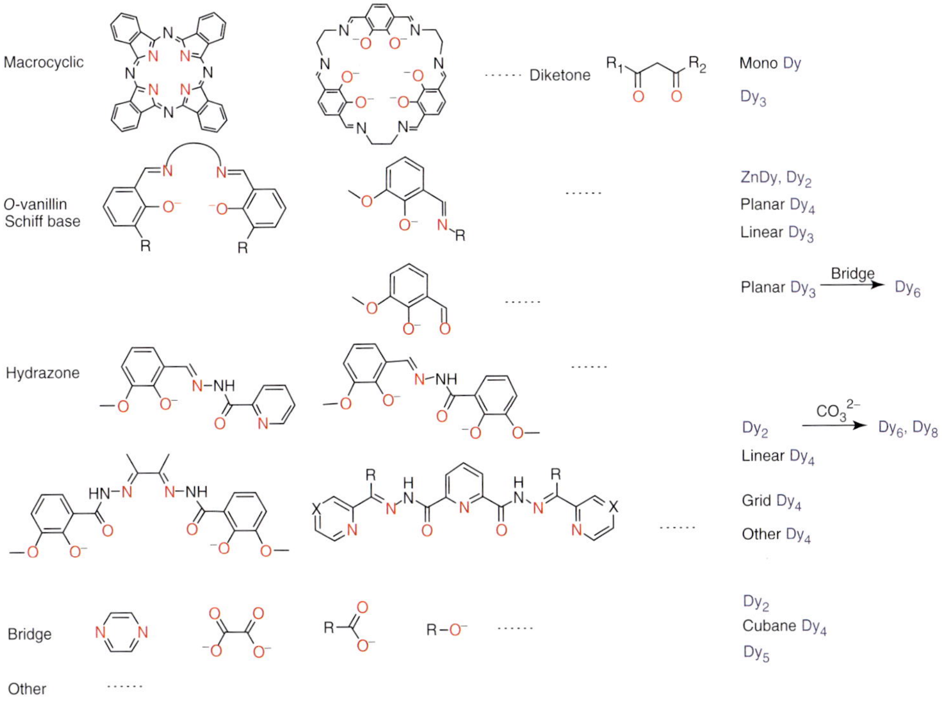

Scheme 3.1 Four classes of ligands used in Ln-SMMs based on the structural motifs. (Reprinted from Ref. [5]. Copyright (2013), with permission from Elsevier.)

Dy-based compounds have been demonstrated widely in view of the different kinds of organic ligands applied (Scheme 3.1) [5]. It has been shown that the weak magnetic interactions due to the internal nature of 4f orbitals in lanthanide ions have been a crucial roadblock for increasing the anisotropic barrier and blocking temperature of SMMs [27]. Although the weak interactions lead to a nonlinear linking of those anisotropic axes of lanthanide ions within the cluster, which thus limits the largest spin ground state reached by a lanthanide cluster [26], high effective barriers surpassing transition metal systems have been achieved and unprecedented magnetic phenomena are observed in many polynuclear lanthanide complexes, such as Dy_3 triangle [1], planar Dy_4 [12] and Dy_5 pyramid [14]. Therefore, crucial structural features favouring a robust SMM performance could be discovered through examining the polynuclear lanthanide SMMs with high energy barriers for the reversal of the magnetization.

First, the Dy_3 triangle reported by Powell *et al.*, showing SMM behaviour of thermally excited spin states, can be regarded as the archetype of the non-collinear Ising antiferromagnetic arrangement in triangular lattices [28], as seen in several coupled Dy_3 (Dy_6) complexes [13, 29, 30]. Second, the excellent

SMM performance of linear Dy complexes indicates that a linear arrangement of metal centres can rationalize the linking of the local anisotropies within a molecule despite the weak magnetic interactions [31, 32]. Third, some planar Dy_4 complexes behave as effective SMMs, especially the Dy_4 reported by Murugesu [12] showing a high effective energy barrier of 170 K and hysteresis loop opening temperature of 7 K. Fourth, the central μ_n-O atom in polynuclear Dy_n complexes might exert a substantial influence in defining a clear unique anisotropy axis, which therefore leads to prominent SMM behaviour [33, 34]. Fifth, the rigid hydrazone ligands appear to be more favourable for the formation of polynuclear lanthanide-based SMMs (Dy_n, $n > 3$) with high barriers in contrast to other Schiff-base ligands [32, 35, 36]. Ultimately, the less-conventional ligands, such as the soft ligands with S donor atoms, can be introduced into lanthanide systems using an organometallic method, which results in the strong organometallic ligand environments and further improves SMM properties [37, 38].

3.2.1
Dy_3 Triangles and Their Derivatives

3.2.1.1 Seminal Dy_3 Triangle

In 2006, the unusual magnetic properties of the triangular $[Dy_3(\mu_3\text{-OH})_2L_3Cl(H_2O)_5]Cl_3$ (**1**, L = ortho-vanillin) complex were reported first by Powell and coworkers [1], which initiated a great interest in the study of Dy_3 triangles and subsequent polynuclear lanthanide complexes. Figure 3.1 shows its structure and the susceptibility of polycrystalline and single crystal samples [1, 3]. The structure (Figure 3.1a) shows the almost C_{3h} symmetry of the molecule, and each of the eight-coordinated dysprosium centres lies in a pentagonal bipyramid where one site in the pentagonal plane is vacant; instead there are two μ_3-hydroxo sites above and below it. The static magnetic behaviour of both polycrystalline and single crystal samples at low temperature reveals the existence of a nonmagnetic ground state (Figure 3.1b). Nevertheless, the strong frequency dependence below 20 K, a signature of SMM behaviour, was still observed, as seen in Figure 3.1c, which should be associated with the thermally populated excited state with a small separation between the two lowest-lying doublets of the Dy_3 molecule.

An effective explanation of this magnetic behaviour is provided by the model of toroidal magnetic moment based on the almost C_{3h} symmetry of the molecule and on the large magnetic anisotropy of the Dy^{3+} ions. These studies combining magnetic measurements, Hamiltonian models and *ab initio* calculations confirm a reliable picture of the electronic structure of the Dy_3 system, revealing the non-collinearity of the single-ion easy axes of magnetization of the Dy ions that lie in the plane of the triangle at 120° from each other (Figure 3.1a) [2–4]. Therefore, a nonmagnetic ground state characterized by a vortex spin-chirality would occur due to the whole antiferromagnetic interaction within the Dy_3 triangle.

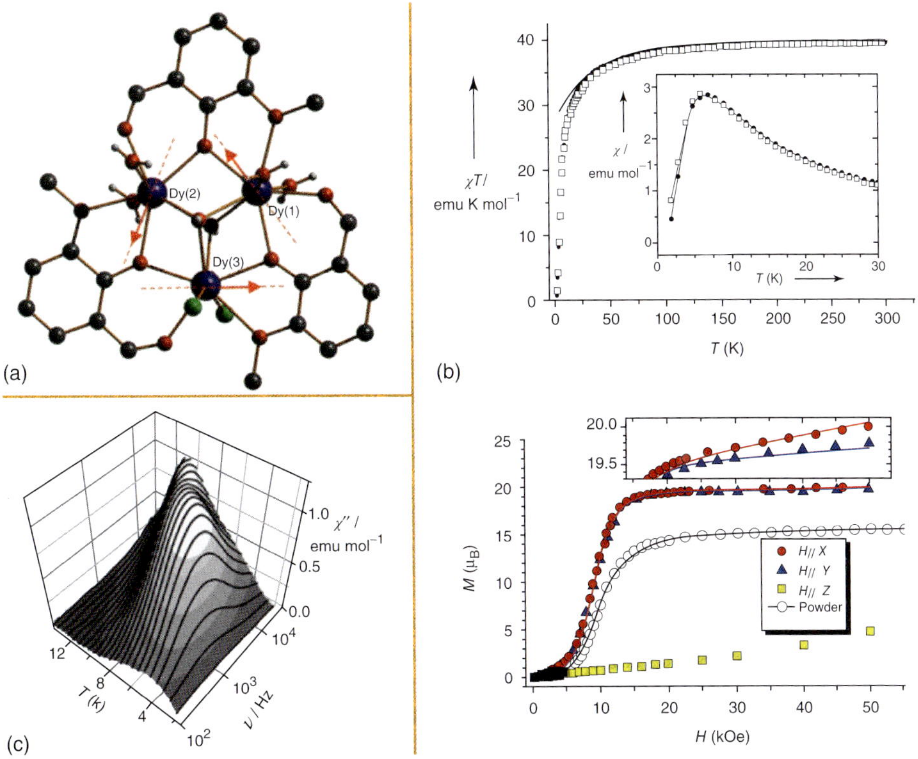

Figure 3.1 (a–c) The structure and magnetic properties of seminal Dy_3 triangle (**1**). (Reprinted with permission from Refs. [1–3]. Copyright 2006 Wiley-VCH, Copyright 2008 Wiley-VCH and Copyright (2008) by The American Physical Society.)

3.2.1.2 Other Triangular Dy_3 Systems

A novel triangular Dy_3 compound (**2**, Figure 3.2a) different from the classical Dy_3 triangle has been reported by Shi and coworkers [39]. Here the three Dy^{3+} ions are doubly bridged by two μ_3-O atoms from the ligands, which resembles the Dy_3 triangle mentioned earlier. However, the difference is that one pair of Dy ions is not bridged by μ_2-O atom, thus leading to a pseudo-isosceles triangle rather than an equilateral triangle. The static susceptibility measurements display an antiferromagnetic interaction at low temperatures (an S-shaped M(H) curve), and the results of *ab initio* calculations favour the toroidal alignment of local magnetic moments of the Dy_3 at the same time. However, a non-vanishing magnetic moment in the Kramers doublet of the Dy_3 can be observed at low temperature, which differs from the previously investigated Dy_3 triangle. This magnetic behaviour should be consistent with the structural features of this Dy_3 compound. The temperature dependent out-of-phase alternating current (ac) signals (χ'') exhibit two distinct peaks within the temperature range of 3–22 K, indicating the

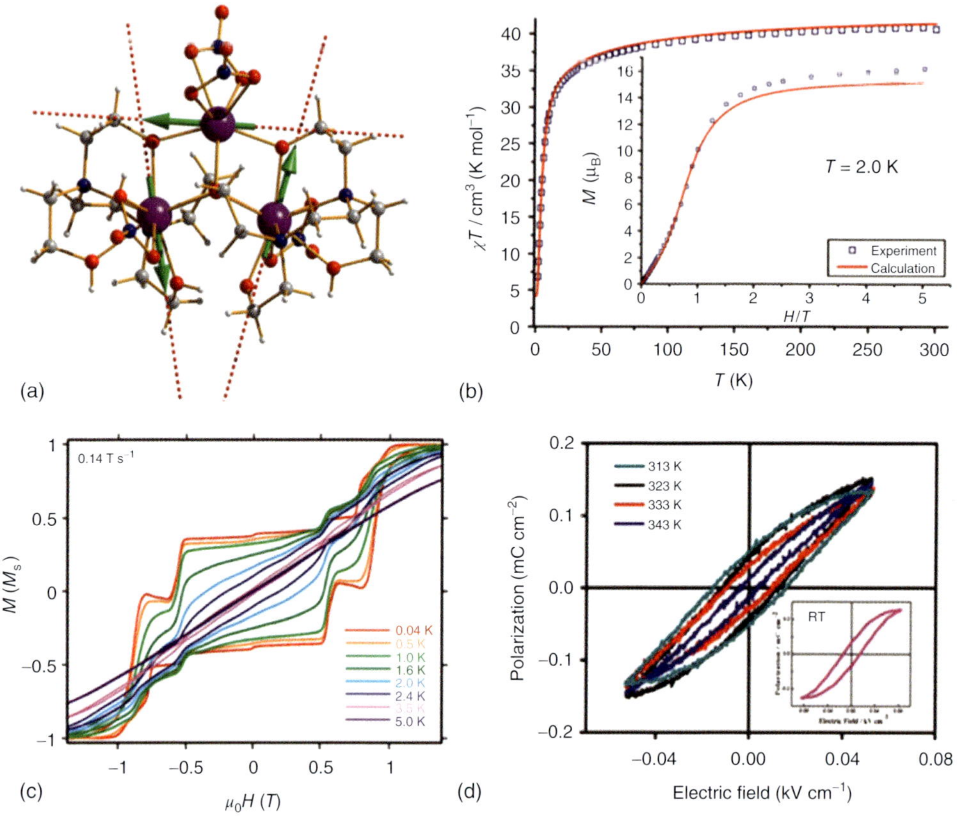

Figure 3.2 (a–d) The structure, magnetism and dielectric properties of compound **2**. (Adapted from Ref. [39]. Reproduced by permission of The Royal Society of Chemistry.)

occurrence of multiple relaxation processes. The open magnetic hysteresis curves (Figure 3.2c) of a single crystal sample show increasing coercivities with decreasing temperature below 5 K, further confirming its SMM properties. Notably, the dielectric study reveals that a paraelectric–ferroelectric phase transition occurs at 470 K for this Dy_3 compound. This is the first reported SMM that also demonstrates a dielectric transition between a paraelectric and a ferroelectric phase.

Two Dy_3 compounds (**3**, **4**) similar to the Dy_3 archetype have been assembled by some of us through grafting bulky hydrazine onto the vanillin group, as seen in Figure 3.3 [40]. The peculiar vortex-spin structure, similar to that of the seminal Dy_3 triangle remains intact as confirmed by direct current (dc) susceptibility measurements (Figure 3.3c). However, the subtle but crucial structural differences between their respective structures give birth to their distinct dynamic magnetic

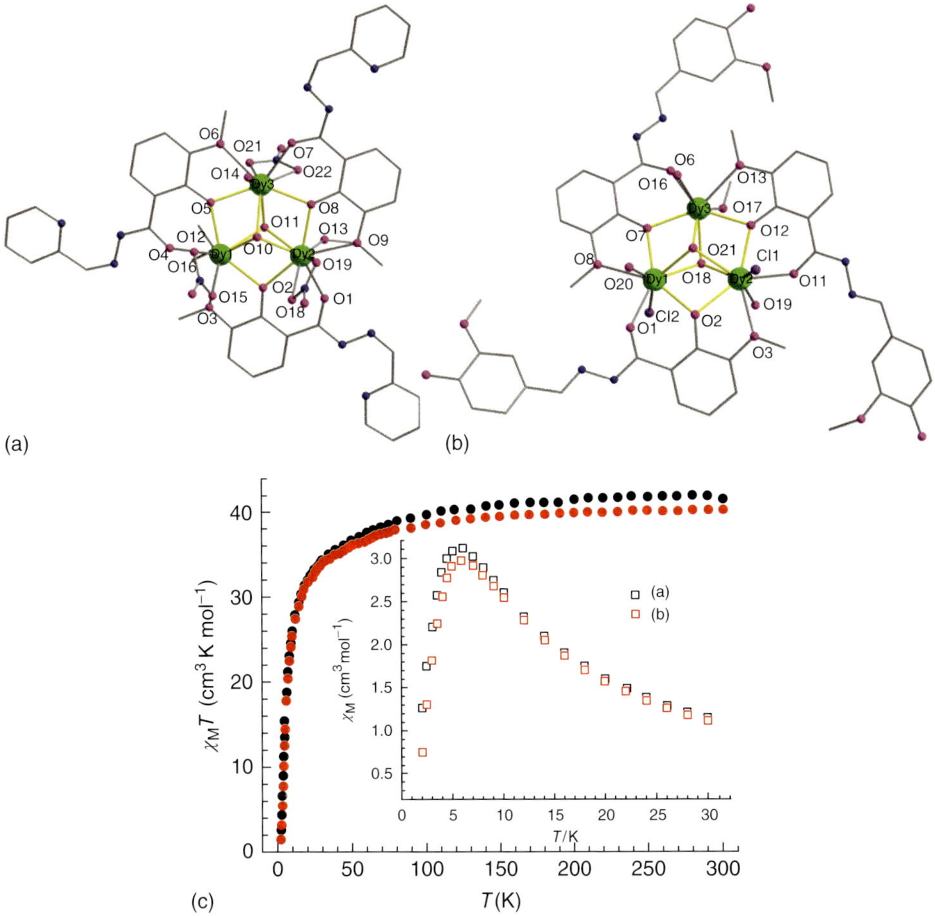

Figure 3.3 (a–c) The structure and static magnetism of compounds **3** and **4**. (Reprinted with permission from Ref. [40]. Copyright (2012) American Chemical Society.)

behaviour revealed by ac susceptibility under zero/nonzero dc field. The almost identical structural features of the compound (**4**) in Figure 3.3b to the Dy_3 archetype in spite of the bulky ligand ensure a similar SMM behaviour for them. In contrast, compound (**3**) in Figure 3.3a exhibits different coordinate geometry at the individual Dy sites due to the terminal η^2-NO_3^- replacing the monodentate Cl^- ions or a solvent molecule. As a result, the out-of-phase ac susceptibility (χ'') curves in zero dc field show the absence of frequency-dependent peaks, observed in other SMMs. The distinctive relaxation dynamics of lanthanide compounds results mainly from the different chemical environment of the metal centre, which governs the large local magnetic anisotropy.

3.2.1.3 The Coupling of Dy_3 Triangles

Meanwhile, the fascinating magnetic properties of the Dy_3 triangle have ignited new interest in the exploitation of coupling several such interesting Dy_3 building blocks to obtain larger Dy^{3+}-based SMMs. Indeed, recent efforts have produced fascinating results with the successful assembly of three typical Dy_6 complexes [13, 29, 30], which not only show structural features similar to Dy_3 triangles in each building block but also retain the functionality of Dy_3 triangles, the toroidal arrangement of single-ion magnetic anisotropy axes. Here, two distinct linking motifs are highlighted by two classical examples in the following text: (i) vertex-to-vertex linkage and (ii) edge-to-edge linkage.

Vertex-to-Vertex Linkage In 2010, Powell and coworkers reported a Dy_6 molecule, $[Dy_6(\mu_3\text{-}OH)_4L_4L'_2(H_2O)_9Cl]Cl_5\cdot 15H_2O$ (**5**, Figure 3.4a; L = *o*-vanillin and H_2L' = 2-hydroxymethyl-6-methoxyphenol), with some exotic magnetic properties, which can be seen as part of their continuing studies on this type of trinuclear system [13]. In the molecule, the resulting alkoxide-O, deriving from the reduction of the aldehyde to an alcohol for one of the three *o*-vanillinato ligands on each triangle, links two Dy_3 units related by an inversion centre in a vertex-to-vertex fashion. Such a linkage destroys the almost C_{3h} symmetry of local Dy_3 triangles, which leads to some differences in its static magnetic

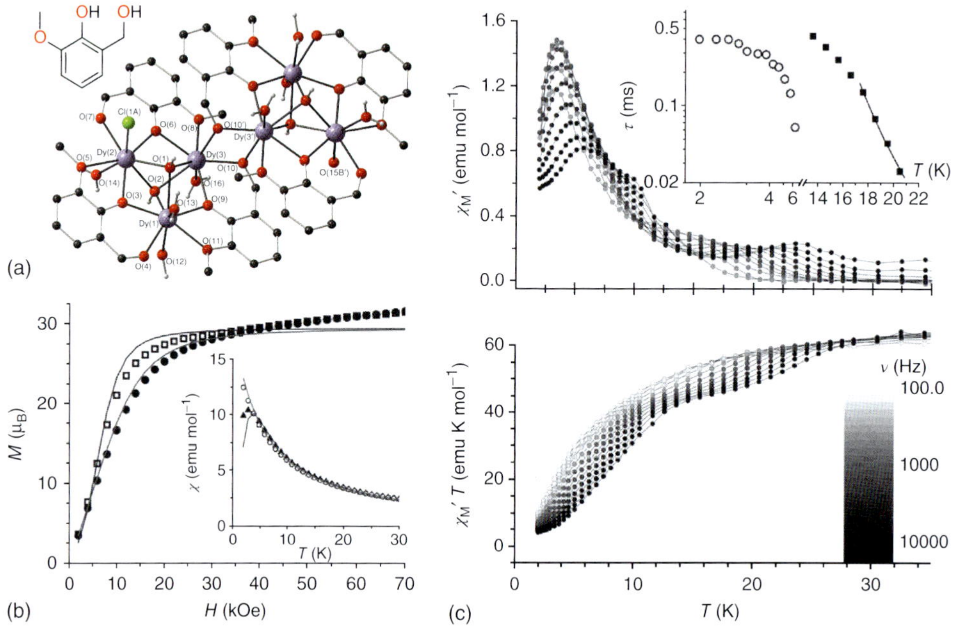

Figure 3.4 (a–c) The structure and magnetic properties of compound **5**. (Reprinted with permission from Ref. [13]. Copyright (2010) Wiley-VCH.)

behaviour from the Dy_3 archetype, such as the absence of the unusual well-defined step in M(H) curves and the lower temperature where a maximum of low temperature susceptibility can be observed (Figure 3.4b). Nevertheless, the nonmagnetic ground states remain intact as revealed by single-crystal magnetic characterization and *ab initio* calculations. The calculations shows the toroidal arrangement of single-ion magnetic anisotropy axes in each Dy_3 triangle, despite the small deviation of the local easy axes of two central Dy ions out of their respective planes of the triangles.

Remarkably, typical SMM behaviour with multiple relaxation processes can be revealed by the ac susceptibility signals (Figure 3.4c), which show a series of frequency-dependent peaks around 25 K, with a second set around 5 K in the out-of-phase component. The application of a dc field has practically no effect on the dynamics, suggesting the negligible tunnelling relaxation in the zero field, which on the contrary was clearly evident in Dy_3 triangle. The Arrhenius analysis for the higher temperature data gives $U_{eff} = 200(10)\,K$ ($\tau_0 = 1.5(5) \times 10^{-9}\,s$). Such a relaxation process can be associated with the antiferromagnetic linking of the two Dy_3 triangles, which favours a simultaneous magnetic flipping on more than one site from/to states high in energy and with high magnetic moments.

Before this, the first example (**6**) coupling two Dy_3 triangles in a vertex-sharing fashion was assembled through an *in situ* aldol condensation reaction between some of the *o*-vanillin molecules and acetone, by Murugesu and coworkers [29]. However, the whole intramolecular ferromagnetic interactions are presented by the increase of the χT product above ~42 K (Figure 3.5b). The temperature-dependent ac susceptibility under zero-dc field displays an obvious out-of-phase (χ'') ac signal, indicating SMM behaviour, below 18 K. More importantly, unusual two-step shaped hysteresis loops were observed with an opening of the loops below 1 K. Such hysteresis loops presumably arise from the toroidal arrangement of the magnetic moments on the Dy^{3+} centres in the triangular units, with the Dy_3 plane being the easy plane of magnetization.

Edge-to-Edge Linkage In continuation of our interest in toroidal spin topologies, by employing a polydentate Schiff-base ligand, we were able to arrange the Dy_3 triangles in a robust edge-to-edge arrangement, whilst perfectly retaining the toroidal arrangement of single-ion magnetic anisotropy axes as revealed by magnetic measurements and *ab initio* calculations [30]. The crystal structure (7) has been exhibited in Figure 3.6a. Such an arrangement and the strong couplings via μ_4-O^{2-} ion stabilize a similar arrangement of toroidal moments in the ground states of each triangle, making the maximum toroidal moment of the complex possible. In the low-temperature susceptibility, a maximum is observed around $T = 7\,K$ under 1 kOe dc field (Figure 3.6b), which is almost identical to that in Dy_3 archetype. The variable-temperature ac susceptibilities show the obvious temperature dependence maximum (Figure 3.6c) and the magnetization hysteresis with a stepped shape (Figure 3.6d) can be observed in

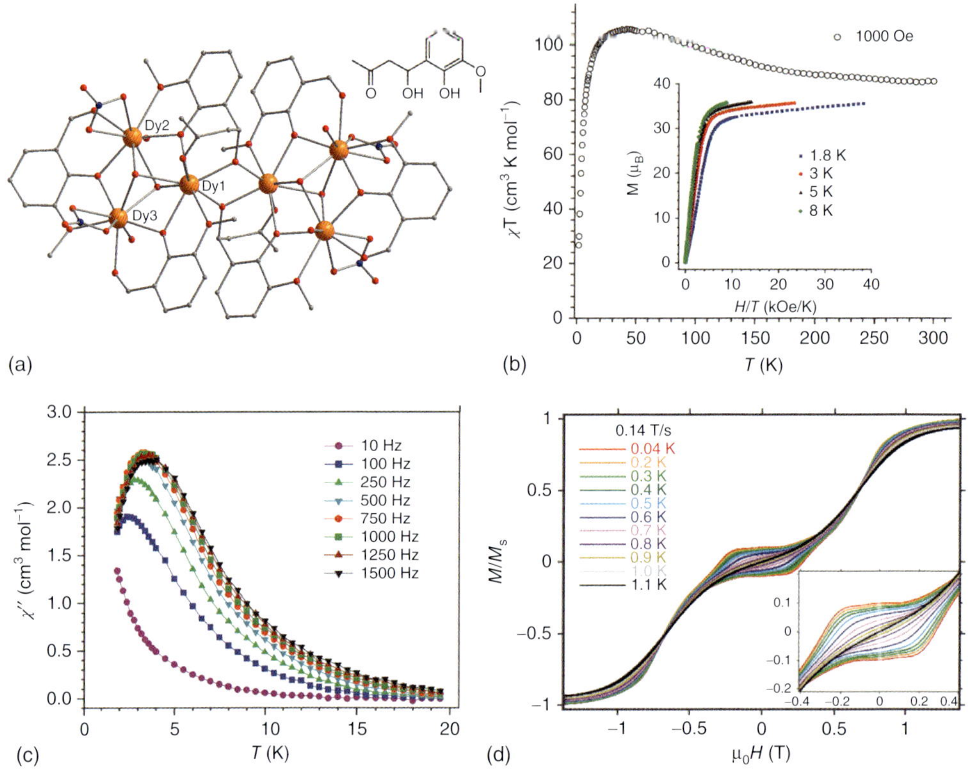

Figure 3.5 (a–d) The structure and magnetic properties of compound **6**. (Adapted from Ref. [29]. Reproduced by permission of The Royal Society of Chemistry.)

a single crystal sample, ensuring the SMM behaviour of this Dy_6. The present results provide a promising strategy for enhancing the toroidal magnetism of polynuclear lanthanide-based compounds via fine-tuning of the arrangements of the lanthanide ions and strengthening couplings between lanthanide ions.

Now we have introduced the typical Dy_3 triangles and their dimers, showing unusual magnetic behaviour. In fact, the investigation of incorporating the Dy_3 triangles into a 3d-4f clusters/chains has also been exploited with one typical example being the heterometallic 1D Cu^{II}/Dy_3 coordination polymers [28]. This complex exhibits an alternation of the trinuclear Dy_3 SMM-building block and a chiral Cu^{2+} bis-valinate complex and, critically, the calculations are indicative of a very strong toroidal magnetization induced by a magnetic field at low temperature. Here, the increasing interest in Dy_3 triangles stems from the presence of their unusual toroidal magnetic moments and their possible further applications in quantum computation, where a nonmagnetic nature of the ground doublet state can reduce decoherence effects from the fluctuation of local magnetic

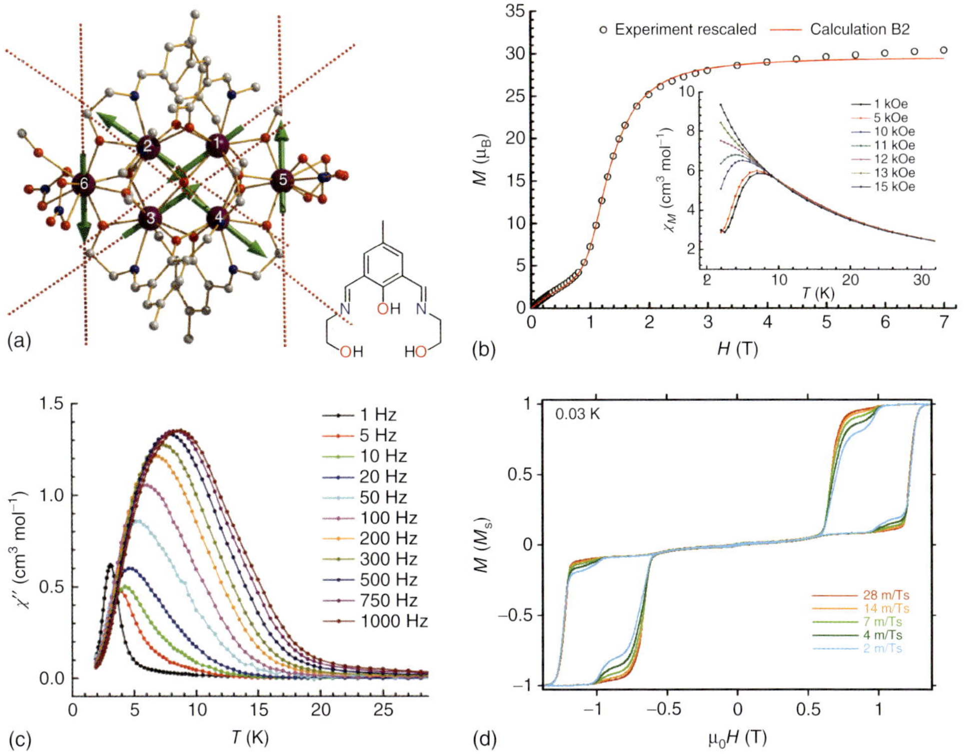

Figure 3.6 (a–d) The structure and magnetic properties of compound **7**. (Reprinted with permission from Ref. [30]. Copyright (2012) Wiley-VCH.)

fields [3]. Further efforts are directed to construct new molecules including such Dy_3 units and to enhance or even discover new properties compared with those of the originally studied examples for future applications [13].

3.2.2
Linear Polynuclear Lanthanide Complexes Showing Robust SMM Behaviour

The review on Dy SMMs from our group has examined the linear polynuclear Dy compounds (Dy_3, Dy_4), most of which show strong SMM behaviour [5]. Especially, the linear Dy_4 from a hydrazone ligand behaves as the best SMM, showing the highest barrier for magnetization reversal, in these linear compounds [32]. In addition, a calculation indicates a nearly collinear arrangement of the easy axes in the linear Dy_3 compound reported by Powell's group [31]. We can draw conclusions that a linear arrangement of metal centres within a molecule can rationalize the linking of the local anisotropies despite the weak magnetic interactions, which is favourable for their SMM behaviour.

3.2.2.1 Linear Dy_3 SMMs

In 2009, Powell *et al.* [31] reported the first linear Dy_3 SMM (**8**), which can be regarded as the opening of a Dy_3 triangle through modification of the original aldehyde ligand with similar coordination geometries of Dy ions relative to those of Dy_3 archetype. The nearly linear molecular structure with a Dy–Dy–Dy angle of 166.29° is shown in Figure 3.7a. The static magnetic susceptibility data exhibits a sharply increasing $\chi_m T$ value with decreasing temperature below 20 K, as a consequence of the intramolecular magnetic interactions. The calculations using a CASPT2/RASSI-SO method indicates the nearly collinear arrangement of the easy axes in the linear Dy_3, which is in agreement with the structural transition from Dy_3 triangle to linear Dy_3. The occurrence of a multiple relaxation process is evident in this compound due to the observations of more than one peak in the χ'' versus T curves (Figure 3.7a). Two barriers (U_{eff}) are calculated to be 28.7 and 69.3 K with $\tau_0 = 6.3 \times 10^{-5}$ and 5.9×10^{-8} s, respectively.

A family of linear Dy_3 and Tb_3 clusters, with almost identical structure to the Dy_3 mentioned in the previous paragraph, was reported by Tong and coworkers in 2011 [42]. The Dy compounds demonstrate similar static and dynamic magnetic

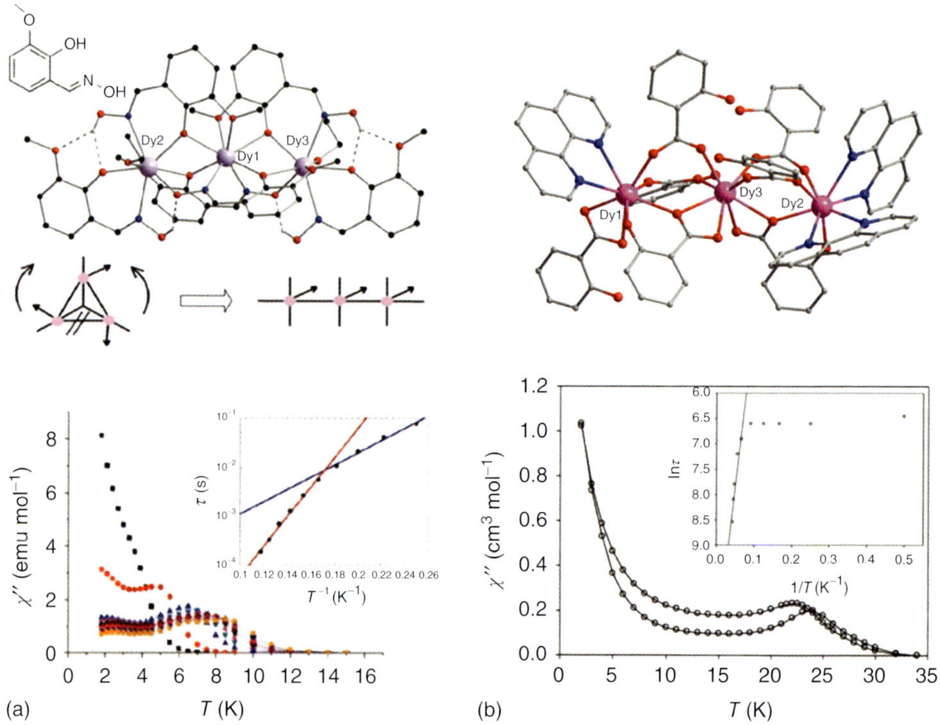

Figure 3.7 (a,b) The structures and dynamic magnetic properties of compounds **8** and **9**. (Adapted from Refs [31, 41]. Reproduced by permission of The Royal Society of Chemistry.)

behaviour and thus expanded the family of linear Dy_3 SMMs based on oxime-type ligands.

Similarly, a linear Dy_3 compound (**9**, Figure 3.7b) with a Dy–Dy–Dy angle of 165.34° was reported by Fang and coworkers [41]. Here the Dy_3 exhibits obvious SMM behaviour as evidenced by well-resolved out-of-phase ac susceptibility maxima between 15 and 35 K.

3.2.2.2 **Linear Dy_4 SMMs**

The Dy_4 compound derived from a rigid hydrazone ligand (Figure 3.8a) shows a nearly linear Dy_4 core with Dy–Dy–Dy angles of 149.99°, where two independent Dy^{3+} centres are present, one being in a distorted bicapped trigonal-prismatic geometry and the other in a nearly perfect mono-capped square-antiprismatic environment [32]. Here more than one peak is observed in the temperature-dependent ac magnetic susceptibility curves (Figure 3.8b), indicating an unusual multiple relaxation mechanism operating in the polynuclear system. Furthermore, in terms of the χ'' versus frequency plots characterized by two clear maxima, two effective barriers can be extracted to be 19.7 and 173 K corresponding to fast and slow relaxation phases (FR and SR), respectively.

$$\chi_{ac}(\omega) = \frac{\chi_2 - \chi_1}{1 + (i\omega\tau_2)^{(1-\alpha_2)}} + \frac{\chi_1 - \chi_0}{1 + (i\omega\tau_1)^{(1-\alpha_1)}} + \chi_0$$

To better understand the nature of the two types of dynamics, the Cole–Cole plots were plotted, and they clearly indicate the evolution from FR to SR with the changing temperature. Strikingly, the plots can be nicely fitted by the sum of two modified Debye functions (Figure 3.8b, inset), which is used to explain such a complex relaxation process. Here, the two separated relaxation processes are most likely associated with distinct anisotropic centres, that is, two Dy ions with

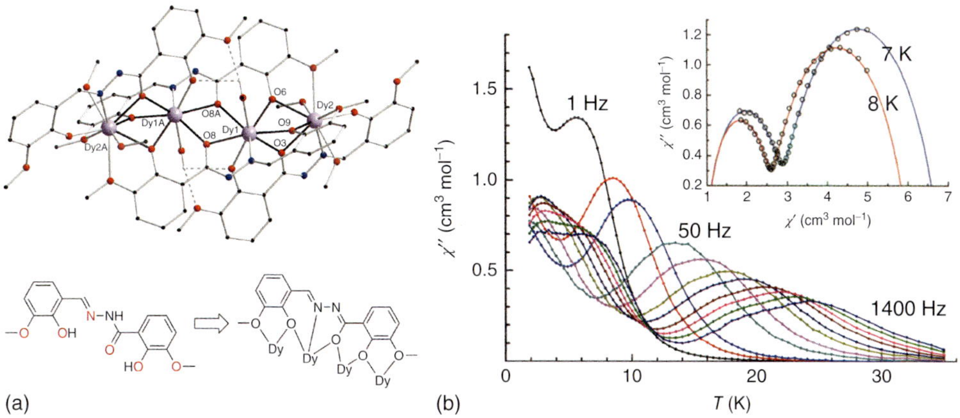

Figure 3.8 (a,b) The structure and dynamic magnetic properties of compound **10**. (Reprinted with permission from Ref. [32] Copyright (2010) American Chemical Society.)

different geometries, in view of the single ion property in polynuclear lanthanide compounds. However, further in-depth studies as well as *ab initio* calculations are required to corroborate this hypothesis.

In fact, two more Dy_4 SMMs with a linear metal core have been reported by our group in 2010 and 2012 (**11** and **12**, Figure 3.9) [43, 44]. In contrast to the Dy_4 mentioned earlier, the Dy_4 compound (**11**) [43] in Figure 3.9a presents a similar structure with two crystallographically independent Dy ions in a nearly perfect square-antiprismatic and a distorted bi-capped trigonal prismatic geometry, respectively. However, a small Dy–Dy–Dy angle of 109.77° is found in this compound. The measured ac susceptibility data shows an obviously temperature-dependent signal with a maximum and a following increase of the χ'' component with decreasing temperature, which is the fingerprint of a crossover from a thermally activated to a direct or tunnelling relaxation process.

Eventually, the Dy_4 compound (**12**) [44] in Figure 3.9b demonstrates the best linear structure characterized by a Dy–Dy–Dy angle of 175.67°. The Dy^{3+} ions are bridged by benzoate ligands in syn–syn μ_2-η^1:η^1 fashion and all show a distorted bicapped trigonal-prismatic geometry. SMM behaviour similar to that of compound **11** can be observed as shown in Figure 3.9.

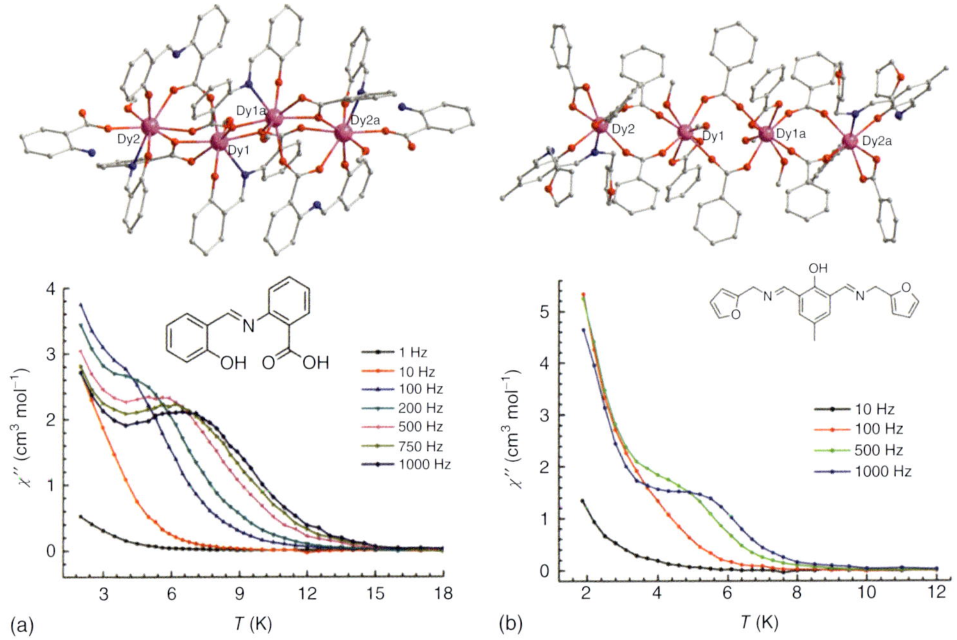

Figure 3.9 (a,b) The structures and dynamic magnetic properties of compounds **11** and **12**. (Adapted from Refs [43, 44]. Reproduced by permission of The Royal Society of Chemistry.)

3.2.3
Planar Dy_4 SMMs

The Dy_4 compound (**13**) presented by Murugesu and coworkers [12] in 2009 exhibited the largest hysteresis loop opening temperature (7 K, Figure 3.10, inset) before the N_2^{3-}-bridged Ln_2 SMMs were discovered [27]. This compound displays a defect-dicubane or planar metal core. The obvious ac signals showing multiple peaks below 40 K are indicative of its SMM behaviour with multiple relaxation modes, which can be associated with two types of Dy ions in the crystal. The almost orthogonal anisotropy axes on the two Dy sites (Dy_1 and Dy_2) were demonstrated by *ab initio* calculations (Figure 3.10, inset), suggesting the weak exchange interactions between the lowest Kramers doublets of those Dy ions. Two effective barriers, 170 and 9.7 K, can be obtained through fitting the ac susceptibility data.

In fact, two planar Dy_4 compounds (**14**-Cl^- and **15**-N_3^-, Figure 3.11a) have been reported by Powell *et al.* [45] before this. These two compounds show

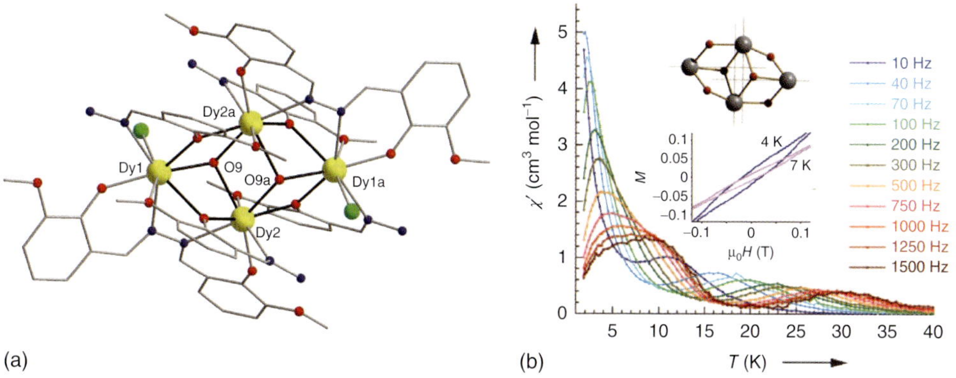

Figure 3.10 (a,b) The structure of compound **14** and dynamic magnetic properties of compound **13**. (Reprinted with permission from Ref. [12]. Copyright (2009) Wiley-VCH.)

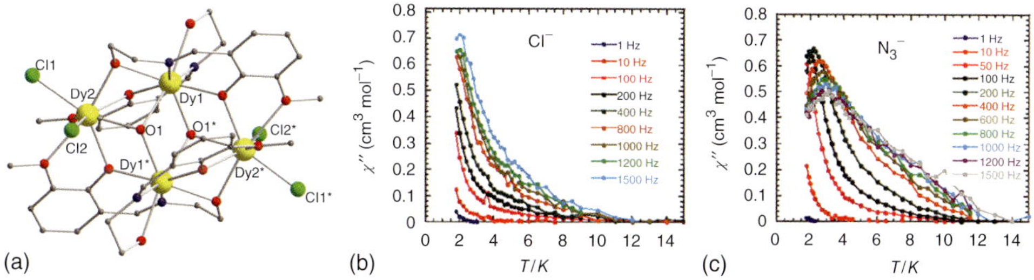

Figure 3.11 (a–c) The structure of compound **14** and dynamic magnetic properties of compounds **14** and **15**. (Reprinted with permission from Ref. [45] Copyright (2008) American Chemical Society.)

isomorphous structures with different coordinating anions, the chlorides in **14** being replaced by azides in **15**. The slow relaxation behaviour was observed below 12 K in both of them, as evidenced in Figure 3.11b,c. However, only compound **15** shows maxima of the out-of-phase (χ'') ac susceptibility signals. An effective barrier can be extracted and is reported to be $U_{eff} = 7$ K. Here, in contrast to compound **13**, the high-temperature relaxation was not observed for complexes **14** and **15**, which could be ascribed to differences in bond lengths and slightly acute bridging angles in the central core as well as different orientations of the anisotropy axes. However, the exact reasons for this are unclear and further comparative calculations are required to test this hypothesis.

Recently, the first Dy_4 SMM with a toroidal magnetic moment in the ground state (**16**), resembling the Dy_3 triangle, was discovered in planar Dy_4 complexes by Tong *et al.* [23]. Notably, the compound was assembled from a rigid multiple nitrogen-ligand with two bidentate chelating sites (Figure 3.12a), which is used to construct polynuclear lanthanide complexes. Here, each Dy ion is surrounded by a distorted square-antiprismatic geometry (Figure 3.12, inset). *Ab initio* calculations show the toroidal arrangement of four anisotropic axes in the ground state and the small separation between the ground state and the first excited state, which could be associated with the SMM behaviour indicated by the ac susceptibility signals (Figure 3.12b). The barrier was fitted to be $U_{eff} = 80$ K ($\tau_0 = 5.75 \times 10^{-6}$ s).

In addition, three other Dy_4 complexes (**17**–**19**; Figure 3.13) showing a planar motif were reported in 2010 and 2011[46–48]. Under zero dc field, compounds **17** (Figure 3.13a) and **18** (Figure 3.13b) show weak SMM behaviour with small effective barriers of 6.2 and 4.5 K, respectively. In contrast, the ac susceptibility signals without maxima were observed under zero dc field in compound **19** (Figure 3.13c). However, the application of a dc field leads to the obvious enhancement of its SMM behaviour, and an effective barrier of 22 K ($\tau_0 = 3.66 \times 10^{-6}$ s) can be obtained.

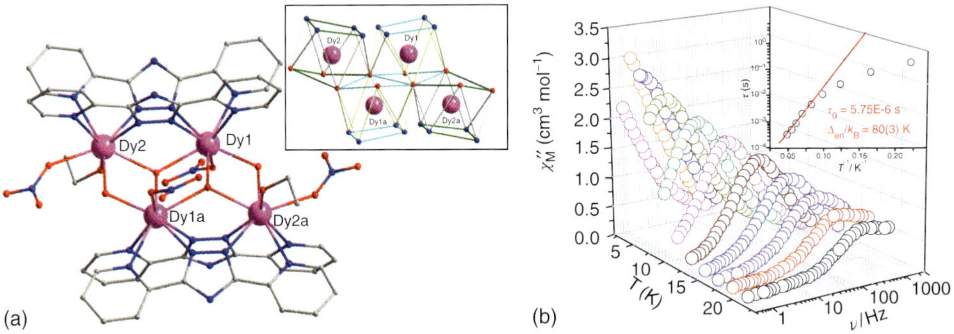

Figure 3.12 (a,b) The structure and dynamic magnetic properties of compound **16**. (Reprinted with permission from Ref. [23]. Copyright (2012) American Chemical Society.)

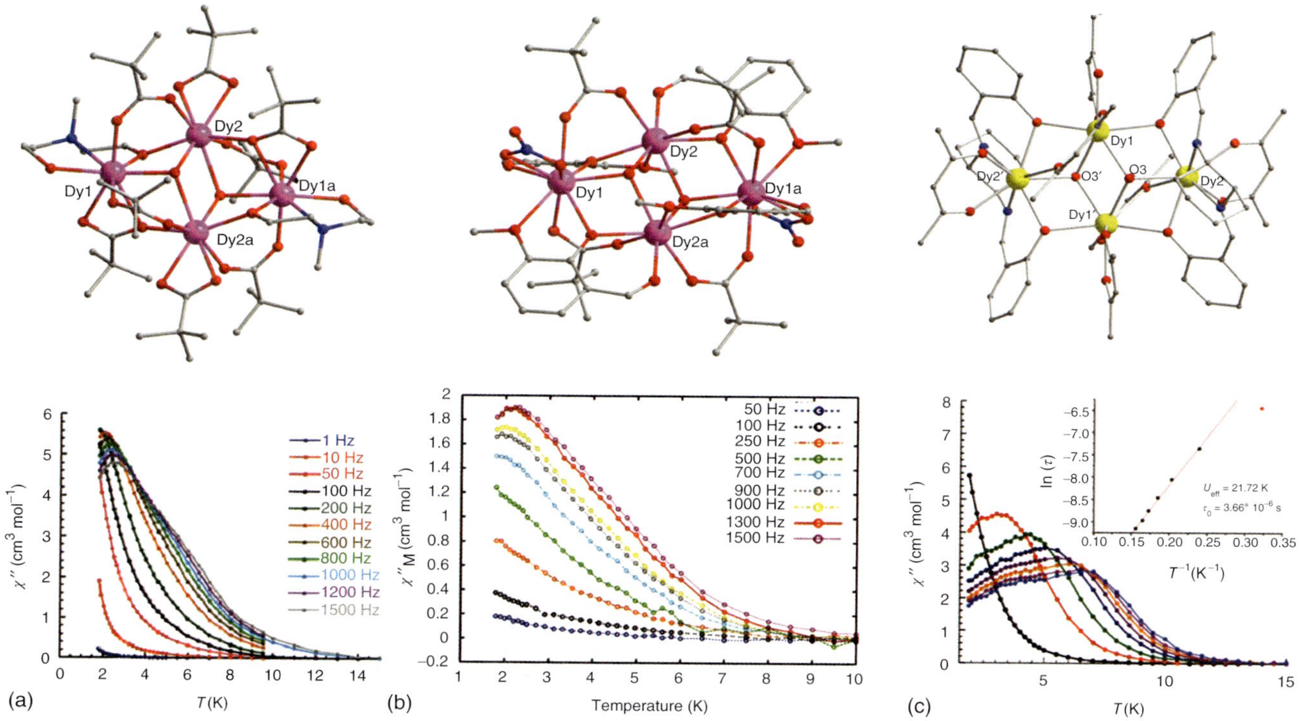

Figure 3.13 (a–c) The structures and dynamic magnetic properties of compounds **17–19**. (Reprinted with permission from Refs [46–48]. Copyright (2010) American Chemical Society, The Royal Society of Chemistry and Copyright (2011) American Chemical Society.)

By comparing these planar Dy_4 compounds (**13–19**), we can discover that only compounds **13** and **16** demonstrate high effective barriers for magnetization reversal, while weak SMM behaviour is observed in the other compounds. The reason should be due to the different local coordination geometry of Dy ions and the combining effects of several magnetic interactions between different Dy ions, but the exact reasons for this are unclear. In this regard, we are investigating a series of planar Dy_4 compounds with the same ligand, showing the enhanced SMM behaviour and a transition from the unclear multiple relaxations to an obvious two-step relaxation, in pursuit of the in-depth exploration of planar Dy_4 SMMs.

3.2.4
Dy_n SMMs Having Multiple μ_n-O ($n > 4$) Bridges

The magnetic investigations of mononuclear or dinuclear lanthanide complexes indicate that some subtle structural details are crucial to the SMM performance of a complex, such as the coordinating H_2O determining the orientation of the magnetic anisotropy axis in Dy/DOTA complex [19], the auxiliary ligands resulting in the enhancement of anisotropy barriers in β-diketone-based Dy SMMs [49] and the spacer modifications of ligands controlling the magnetic anisotropy of three helical Dy_2 SMMs [50]. In contrast, the situation becomes complicated in lanthanide SMMs with larger nuclearity, where the SMM behaviour should arise from a combination of many factors. However, the investigation into a series of Dy_n SMMs fixed by μ_n-O atom reveal that the μ_n-O turns out to be a key factor in defining the clear unique anisotropy axis of each Dy^{3+} ion and creates possibilities for ferromagnetic or antiferromagnetic superexchange paths between metal centres [14, 33].

3.2.4.1 The Dy_4 Grids Fixed by μ_4-O Atom

In 2012, Thompson and coworkers [33] reported two grid-like Dy_4 complexes (**20**, and **21**, Figure 3.14) with the ditopic carbohydrazone ligands, in which an obvious dissimilarity is the absence of μ_4-O in complex **21**. In addition, the μ_2-1, 1-N_3^- groups replacing the μ_2-OH^- in **21** play the bridging roles between two Dy ions in **20**. In complex **20**, Dy–O–Dy angles to the central oxide ion fall in the range 88.9–91.41°, indicating an almost planar $Dy_4(\mu_4\text{-O})$ square. The clear ac susceptibility signals with two maxima below 35 K are indicative of strong SMM behaviour showing two thermally activated relaxations for complex **20**. According to the Arrhenius law, the energy barriers for the reversal of the magnetization were calculated to be 51 K ($\tau_0 = 3.0 \times 10^{-9}$ s) and 91 K ($\tau_0 = 4.5 \times 10^{-7}$ s). Moreover, the application of a dc field, which can reduce the quantum tunnelling of magnetization (QTM), result in an increase of the barrier, $U_{\text{eff}} = 270$ K ($\tau_0 = 4.0 \times 10^{-10}$ s). In contrast, no SMM behaviour was observed in compound **21**.

Furthermore, a similar Dy_4 complex (**22**, Figure 3.15) fixed by μ_4-O atom was reported by our group [34]. Here, all Dy sites are nine-coordinate sites and exhibit

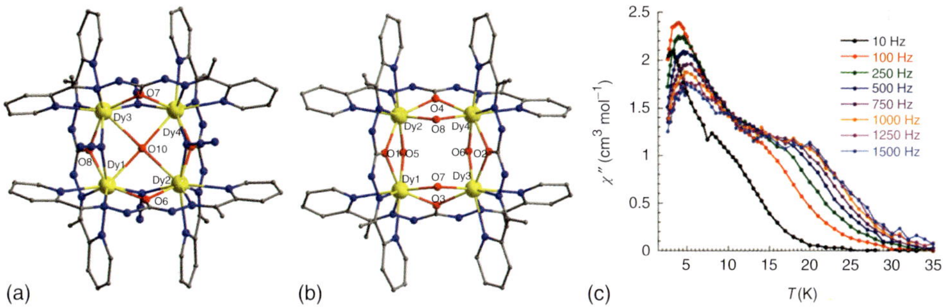

Figure 3.14 (a–c) The structures of compounds **20** and **21** and dynamic magnetic properties of compound **20**. (Adapted from Ref. [33]. Reproduced by permission of The Royal Society of Chemistry.

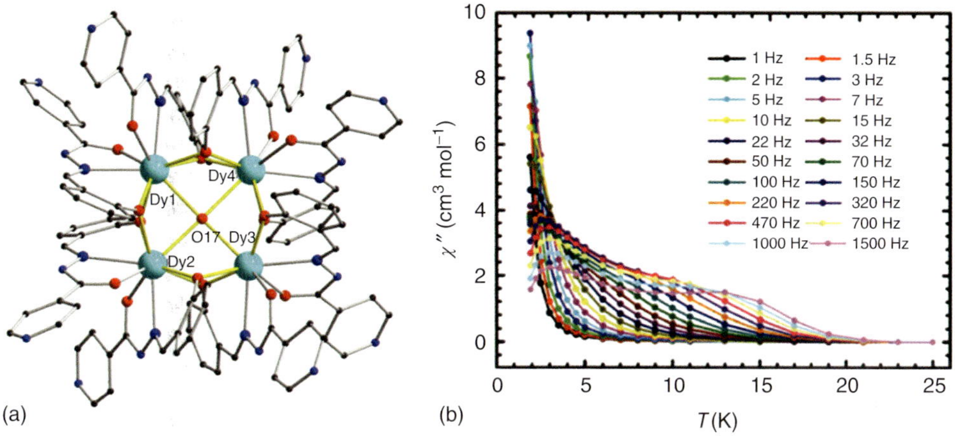

Figure 3.15 (a,b) The structure and dynamic magnetic properties of compound **22**. (Adapted from Ref. [34]. Reproduced by permission of The Royal Society of Chemistry.)

the geometry of a quasi mono-capped square-antiprism. The ac susceptibility measurements reveal similar dynamic behaviour of magnetization to complex **20** under zero dc field. The application of a static field results in a strongly enhanced barrier of $U_{\mathrm{eff}} = 92\,\mathrm{K}$, which is more than triple that of the zero dc field.

Comparing the magnetic properties and structural characteristics of three Dy_4 grid compounds, we can speculate that the μ_4-O atom has a significant influence on their SMM behaviour. If the absence of SMM behaviour in compound **21** is attributable to an effective compensation of each Dy ion's magnetic contributions, the μ_4-O in **20** and **22** might break such a situation and will fine-tune the correlation between the anisotropic axes of Dy ions. Although the exchange interactions via the μ_4-O are weak, the new superexchange paths between Dy ions may advance the magnetic relaxation behaviour effectively at low temperatures, as observed in many Dy^{3+}-based SMMs.

3.2.4.2 The Dy_4 Tetrahedron Fixed by μ_4-O Atom

In 2011, Murugesu *et al.* [51] reported a rare μ_4-O centred Dy_4 tetrahedron (**23**, Figure 3.16) with coordination-induced local chirality, showing SMM behaviour. The local chirality is induced by the coordination of a twisted diazine bridge from the Schiff base ligands. Here four Dy ions occupy the vertexes of a tetrahedron and are fixed by a central μ_4-O atom with a Dy-O distance of 2.9 Å. The *ab initio* calculations indicate four anisotropic axes with different directions for four Dy ions located in different coordinated environments, as seen in Figure 3.16a. The overall antiferromagnetic exchange coupling is presented by the exchange interaction parameter of $J = -0.3\,\text{cm}^{-1}$ obtained via simulations for the experimental dc susceptibility data. The SMM behaviour is indicated by the frequency dependent ac signals with maxima ($U_{\text{eff}} = 23.4\,\text{K}$), and further confirmed by the open hysteresis loops with steps at low temperature (Figure 3.16b).

In 2012, Liao and coworkers [52] reported a Dy_6 octahedron (**24**, Figure 3.17), seen as two edge-sharing $[(\mu_4\text{-O})Dy_4]$ tetrahedra, with two tail-to-tail calixarene ligands. The ac susceptibilities show obvious temperature-dependent both in-phase (χ') and out-of-phase (χ'') signals under a zero dc field, indicating multiple relaxation SMM behaviour.

3.2.4.3 The Dy_5 Pyramid Fixed by μ_5-O Atom

We can compare two Dy_5 compounds (**25**, **26**; Figure 3.18) with the expected square-based pyramidal arrangement, which were reported in 2008 and 2011 by Powell [53] and Winpenny [14], respectively. In compound **25** (Figure 3.18a) [53], the central μ_5-O atom is absent and each Dy ion shows higher coordination numbers than compound **26** (Figure 3.18b) [14]. The ac susceptibility measurements indicate their distinct dynamic magnetization behaviour. Compound **25** displays only ac susceptibility signals below 8 K, while the strong ac susceptibility signals with a blocking temperature as high as 41 K (for $\nu = 1400\,\text{Hz}$) are indicative

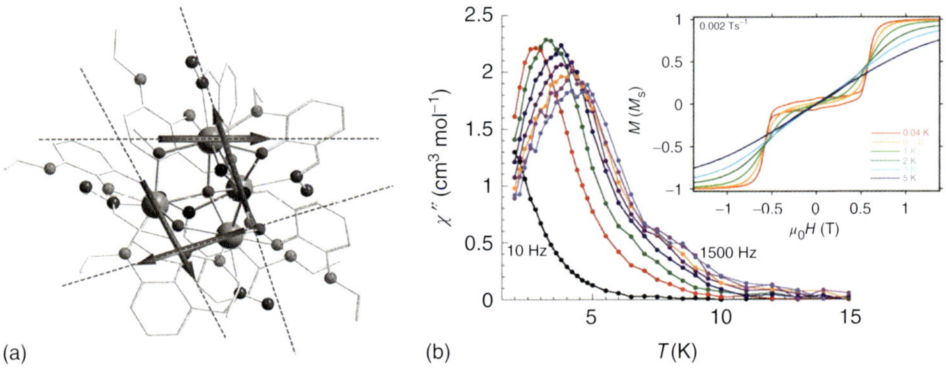

Figure 3.16 (a,b) The structure and magnetic properties of compound **23**. (Reprinted with permission from Ref. [51]. Copyright (2011) Wiley-VCH.)

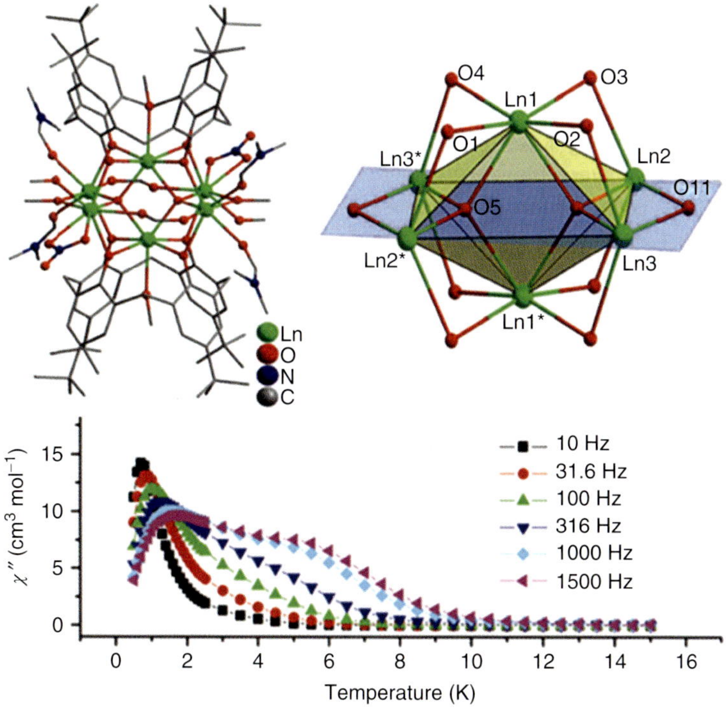

Figure 3.17 The structure and dynamic magnetic properties of compound **24**. (Reprinted from Ref. [52]. Copyright (2012), with permission from Springer.)

of robust SMM behaviour in compound **26**. Therefore, the thermally activated barriers were extracted to be 33 and 528 K for complex **25** and **26**, respectively. As for the disparities of their SMM properties, it should be attributable to the combined factors of μ_5-O, coordinate geometries, symmetries, and so on. However, the central μ_5-O should play a key role in defining the unique anisotropic axis of each Dy ion and modulating the magnetic interactions between Dy ions.

We can see so many lanthanide clusters behave as efficient SMMs in those μ_n-O-fixed complexes. Especially, the energy barrier record (530 K) for polynuclear lanthanide SMMs, held by the μ_5-O-fixed Dy_5 with a square-based pyramid, has not been broken by the end of 2013. Such a structural characteristic should support the isolation of a robust SMM to a great extent, which provides a change to explore new methods for achieving SMMs with high barriers.

In addition, the circular Dy_6 (**27**, Figure 3.19a) [54] and Dy_7 (**28**, Figure 3.19b) [55] complexes were assembled using two different tripodal alcohol ligands by Murray's and Collison's groups, respectively. Complex **28** can be seen as a circular Dy_6 fixed by the central $[Dy(\mu_3\text{-}O)_6]$ group [55], which resembles a μ_6-O atom. Distinct SMM behaviours were observed in these two complexes, complex **27** showing weak ac signals but **28** showing strong SMM behaviour with a high

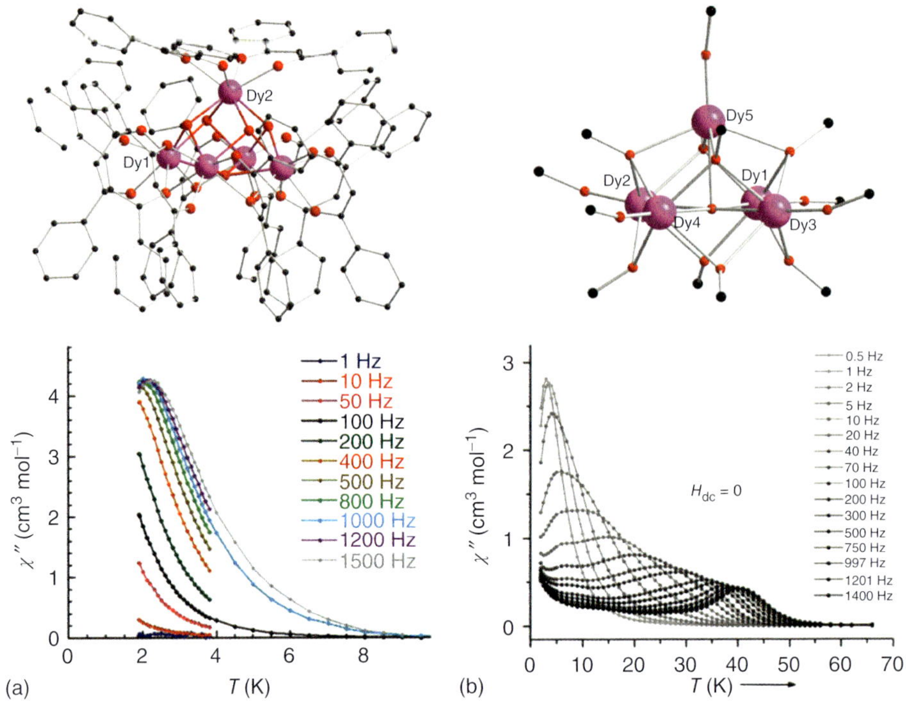

Figure 3.18 (a,b) The structures and dynamic magnetic properties of compounds **25** and **26**. (Reprinted with permission Refs [14, 53]. Copyright (2008) American Chemical Society and Copyright (2011) Wiley-VCH.)

barrier of 140 K below 30 K (Figure 3.19). Recently, *ab initio* calculations were performed for the mononuclear Dy fragments of the Dy_6 compound (**27**) due to its high symmetry [22]. The corresponding results reveal a toroidal moment in the ground state. Considering the different dynamic behaviour of **27** and **28**, the central Dy ion should exert a great influence on the distributions of anisotropic axes of six peripheral Dy ions and the magnetic interactions between them.

3.2.5
Hydrazone-Based Lanthanide SMMs

In a recent review on hydrazone-based dysprosium SMMs published in Current Inorganic Chemistry, we have systematically examined the Dy SMMs with hydrazone ligands from Dy_2 to Dy_8, indicating that the rigid hydrazone ligands appear to be favourable for the isolation of lanthanide-based SMMs with high energy barrier. This is due to their typical structural features, such as a linear arrangement of coordination pockets and versatile coordinating and bridging modes. In particular, several Dy_2 complexes with hula hoop-like geometries

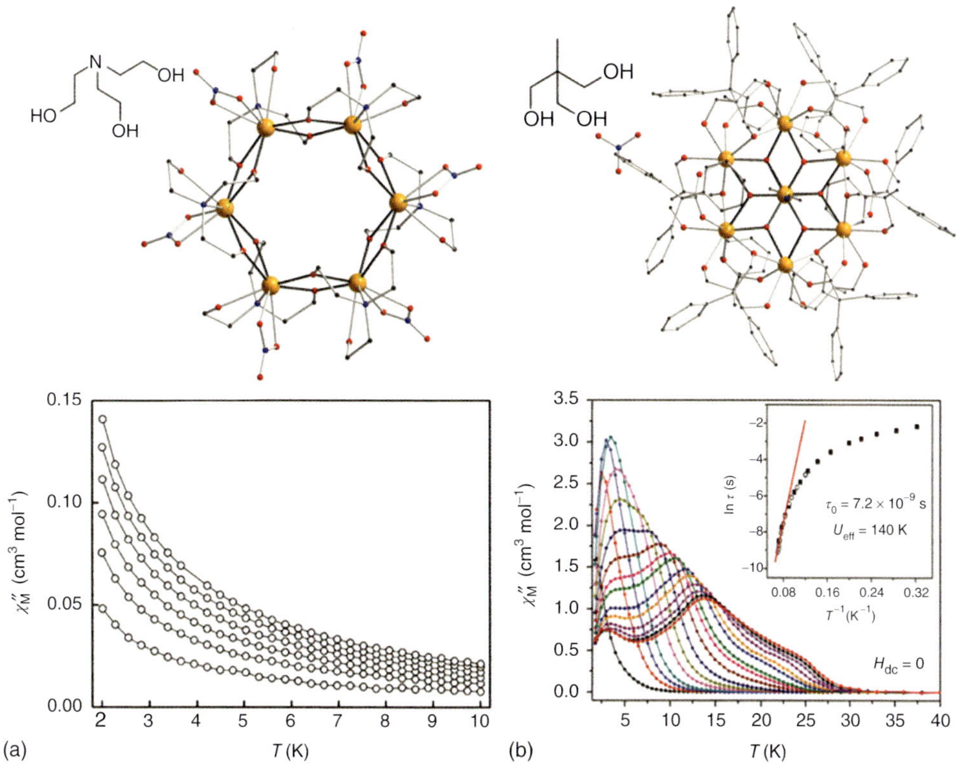

Figure 3.19 (a,b) The structures and dynamic magnetic properties of compounds **27** and **28**. (Adapted from Refs [54, 55]. Reproduced by permission of The Royal Society of Chemistry.)

exhibit strong slow relaxation of the magnetization, which offer a great platform for probing their SMM properties via modulating the local coordinating environments of Dy^{3+} ions [21, 56]. For the larger Dy SMMs with hydrazone ligands, several typical examples will be given to show the favourable features of hydrazone ligands for designing efficient lanthanide SMMs [35, 36]. In fact, two Dy_4 complexes, one being linear (**10**) and the other being grid-like (**22**), have been provided earlier.

3.2.5.1 The Assembly of Dy_6 Triangular Prism with Dy_2 Units

As seen in Figure 3.20, the Dy_6 complex (**29**) [36], showing a triangular prism arrangement, can be considered as the linking of the three Dy_2 skeletons with hula hoop-like geometries, which nicely contribute to the SMM behaviour with a relatively slow tunnelling rate (Figure 3.20). The strongly temperature-dependent ac susceptibilities signals are indicative of typical SMM behaviour, and an effective barrier can be extracted as 76 K with $\tau_0 = 1.2 \times 10^{-6}$ s. In particular, the

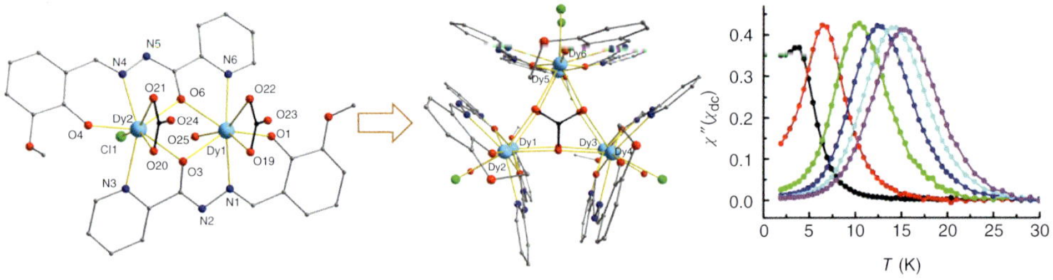

Figure 3.20 The structure and dynamic magnetic properties of compound **29**. (Reprinted with the permission from Ref. [36] Copyright (2012) American Chemical Society.)

absence of increasing out-of-phase (χ'') signals in the low-temperature region signals the 'freezing' of the spins by the anisotropy barrier and the slow quantum tunnelling rate under the zero dc field. This is a successful example of expanding the Dy_2 building units into a larger lanthanide cluster, reserving the efficient SMM behaviour similar to Dy_2 units.

3.2.5.2 A Dy_3 Molecular Cluster Pair (Dy_6)

A Dy_6 compound (**30**, Figure 3.21), derived from the linking of two well-differentiated and symmetry-related groups of Dy_3, was separated by our group through applying a new *o*-vanilloyl hydrazone derivative in 2012 [35]. Eight hydrazone ligands within the compound exhibit four different coordinating modes (Figure 3.21) via facile keto-enol tautomerism, which perfectly demonstrates the coordination versatility of the mixed-donor hydrazone ligands. Ac susceptibility measurements reveal the appearance of an out-of-phase χ'' signal below 15 K. An effective relaxation barrier of $U_{eff} = 29\,K$ can be extracted with $\tau_0 = 1.2 \times 10^{-6}\,s$.

By examining these systems, it can be anticipated that designing new styles of ligands with similar structural features can lead to other complexes with more robust SMM behaviour.

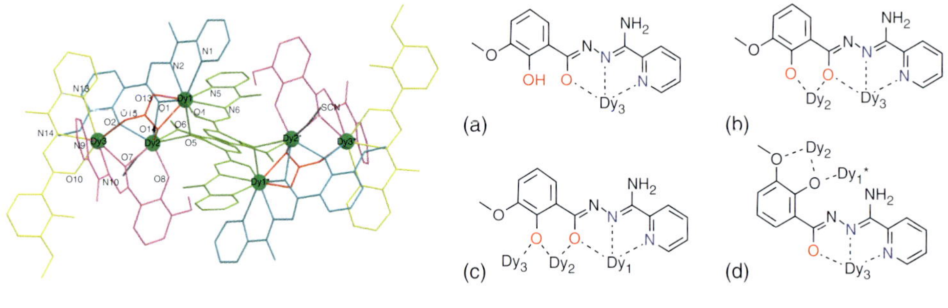

Figure 3.21 The structure of compound **30** and coordination modes of its ligands. (Adapted from Ref. [35]. Reproduced by permission of The Royal Society of Chemistry.)

3.2.6 The Organometallic Synthesis – A New Approach

In general, most ligands, especially bridging ligands, used to construct lanthanide SMMs contain hard O-donors due to the strong oxygen affinity of lanthanide ions. Naturally, the exploration of lanthanide SMMs with new styles of ligands without O-atoms is a promising challenge for a better understanding of complex SMM behaviour and for enhancing their SMM properties [38]. Here the organometallic synthetic approach has shown great potential in designing lanthanide SMMs with high blocking temperatures, as indicated by some remarkable results in mono/dinuclear organometallic lanthanide systems [24, 37, 57–59]. In contrast to common lanthanide systems, a low coordination number, strong ligand field and efficient magnetic interactions could be achieved more easily in those systems. Nevertheless, larger organometallic clusters with more than three lanthanide centres are still very rare within the SMM field. To date, the most typical example is the thiolate-bridged tetralanthanide compounds (**31**, Figure 3.22) published recently in Organometallics [38], where not only Dy_4 but also Tb_4 display SMM behaviour under zero dc field. In fact, such thiolate bridging ligands have been explored in an orgnometallic Dy_2 compound, which displays the special soft ligand environments, strong magnetic interactions and the consequent high magnetization reversal barrier.

The structures and dynamic magnetic behaviours of **31**-Dy_4 and **31**-Tb_4 have been presented in Figure 3.22. Four lanthanide ions are aggregated in the grid-like metal core by a central μ_4-S and eight peripheral μ_2-S atoms from ethanethiol ligands [38]. The individual lanthanide centres occupy distorted six-coordinate $[LnNS_5]$ octahedral coordination environments. Ac susceptibilities measurements reveal pronounced temperature dependence with a series of maxima below 28 K, typical for SMM behaviour, in complex **31**-Dy_4. Furthermore, an

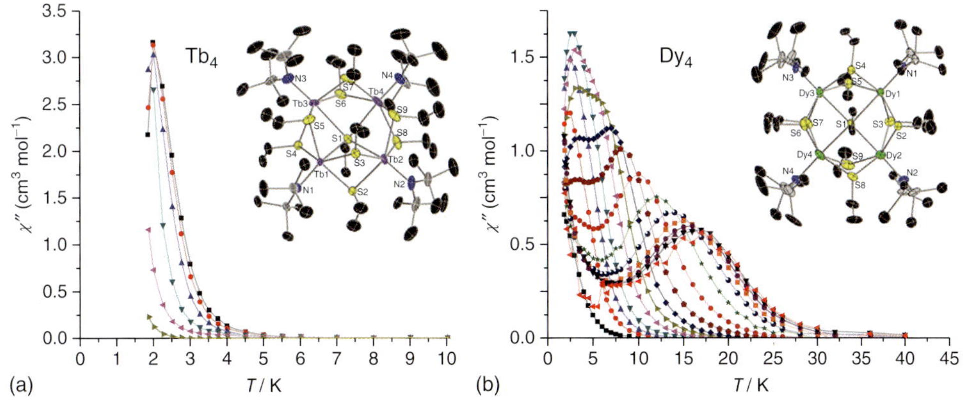

Figure 3.22 (a,b) The structure and dynamic magnetic properties of compound **31**. (Reprinted with permission from Ref. [38]. Copyright (2013) American Chemical Society.)

effective barrier for magnetization reversal can be obtained to be 46 cm^{-1} (66 K) with $\tau_0 = 4.3 \times 10^{-6}$ s. It is noteworthy that the Tb_4 complex shows obvious SMM behaviour as well, despite a small barrier of 4.6 cm^{-1} under a zero applied field, which is rare in polynuclear lanthanide complexes with more than three metal centres. These examples illustrate the potential of organometallic chemistry in synthesizing lanthanide SMMs and the exploration of magnetism under new chemical environments is very helpful for improving the understanding of complex magnetic behaviour.

3.3 Conclusion

Designing a target structure with specific properties and functions represents an eternal aspiration for materials scientists [60]. In the case of SMMs, one can envisage an effective route to design new SMMs with enhanced properties in a controllable manner, with the ultimate goal being to apply these in quantum spintronics and high density storage materials [9, 10]. Even though it is far from this goal, great progress, especially with respect to the high effective barriers and high blocking temperatures achieved in lanthanide-based SMMs compared to transition metal SMMs [14, 16, 27], has recently been made in demonstrating the great potential in this respect.

To date, the most striking developments are concentrated on mono/dinuclear lanthanide SMMs, while the systematic investigation for lanthanide SMMs with larger nuclearity are not enough due to the complexity of the structure, magnetism and their correlations. We have examined here about thirty typical polynuclear lanthanide complexes showing SMM behaviour under zero dc field, and classified them with the expectation that we can learn some lessons in regards to synthesizing polynuclear lanthanide SMMs with high barriers. Some critical structural features potentially leading to the high barriers were highlighted, such as the arrangements of metal centres, μ_n-O fixation, the choice of ligands favouring SMM behaviour and, especially, the novel organometallic synthetic approach, which may lead to some further breakthroughs due to the special organometallic ligand environments and possible strong magnetic interactions. Overall, such a summary could contribute important information towards our understanding of how to improve SMM performance in polynuclear lanthanide complexes for future applications.

References

1. Tang, J., Hewitt, I., Madhu, N.T., Chastanet, G., Wernsdorfer, W., Anson, C.E., Benelli, C., Sessoli, R. and Powell, A.K. (2006) *Angew. Chem. Int. Ed.*, **45**, 1729–1733.
2. Chibotaru, L.F., Ungur, L. and Soncini, A. (2008) *Angew. Chem. Int. Ed.*, **47**, 4126–4129.
3. Luzon, J., Bernot, K., Hewitt, I.J., Anson, C.E., Powell, A.K. and

Sessoli, R. (2008) *Phys. Rev. Lett.*, **100**, 247205.

4. Ungur, L., Van den Heuvel, W. and Chibotaru, L.F. (2009) *New J. Chem.*, **33**, 1224–1230.
5. Zhang, P., Guo, Y.-N. and Tang, J. (2013) *Coord. Chem. Rev.*, **257**, 1728–1763.
6. Woodruff, D.N., Winpenny, R.E.P. and Layfield, R.A. (2013) *Chem. Rev.*, **113**, 5110–5148.
7. Dei, A. and Gatteschi, D. (2011) *Angew. Chem. Int. Ed.*, **50**, 11852–11858.
8. Stamp, P.C.E. and Gaita-Arino, A. (2009) *J. Mater. Chem.*, **19**, 1718–1730.
9. Bogani, L. and Wernsdorfer, W. (2008) *Nat. Mater.*, **7**, 179–186.
10. Mannini, M., Pineider, F., Danieli, C., Totti, F., Sorace, L., Sainctavit, P., Arrio, M.A., Otero, E., Joly, L., Cezar, J.C., Cornia, A. and Sessoli, R. (2010) *Nature*, **468**, 417–421.
11. Lin, P.-H., Burchell, T.J., Clérac, R. and Murugesu, M. (2008) *Angew. Chem. Int. Ed.*, **47**, 8848–8851.
12. Lin, P.-H., Burchell, T.J., Ungur, L., Chibotaru, L.F., Wernsdorfer, W. and Murugesu, M. (2009) *Angew. Chem. Int. Ed.*, **48**, 9489–9492.
13. Hewitt, I.J., Tang, J., Madhu, N.T., Anson, C.E., Lan, Y., Luzon, J., Etienne, M., Sessoli, R. and Powell, A.K. (2010) *Angew. Chem. Int. Ed.*, **49**, 6352–6356.
14. Blagg, R.J., Muryn, C.A., McInnes, E.J.L., Tuna, F. and Winpenny, R.E.P. (2011) *Angew. Chem. Int. Ed.*, **50**, 6530–6533.
15. Milios, C.J., Vinslava, A., Wernsdorfer, W., Moggach, S., Parsons, S., Perlepes, S.P., Christou, G. and Brechin, E.K. (2007) *J. Am. Chem. Soc.*, **129**, 2754–2755.
16. Rinehart, J.D., Fang, M., Evans, W.J. and Long, J.R. (2011) *J. Am. Chem. Soc.*, **133**, 14236–14239.
17. Dunbar, K.R. (2012) *Inorg. Chem.*, **51**, 12055–12058.
18. Bernot, K., Luzon, J., Bogani, L., Etienne, M., Sangregorio, C., Shanmugam, M., Caneschi, A., Sessoli, R. and Gatteschi, D. (2009) *J. Am. Chem. Soc.*, **131**, 5573–5579.
19. Cucinotta, G., Perfetti, M., Luzon, J., Etienne, M., Car, P.-E., Caneschi, A., Calvez, G., Bernot, K. and Sessoli, R. (2012) *Angew. Chem. Int. Ed.*, **51**, 1606–1610.
20. Boulon, M.-E., Cucinotta, G., Luzon, J., Degl'Innocenti, C., Perfetti, M., Bernot, K., Calvez, G., Caneschi, A. and Sessoli, R. (2013) *Angew. Chem. Int. Ed.*, **52**, 350–354.
21. Guo, Y.-N., Xu, G.-F., Wernsdorfer, W., Ungur, L., Guo, Y., Tang, J., Zhang, H.-J., Chibotaru, L.F. and Powell, A.K. (2011) *J. Am. Chem. Soc.*, **133**, 11948–11951.
22. Ungur, L., Langley, S.K., Hooper, T.N., Moubaraki, B., Brechin, E.K., Murray, K.S. and Chibotaru, L.F. (2012) *J. Am. Chem. Soc.*, **134**, 18554–18557.
23. Guo, P.-H., Liu, J.-L., Zhang, Z.-M., Ungur, L., Chibotaru, L.F., Leng, J.-D., Guo, F.-S. and Tong, M.-L. (2012) *Inorg. Chem.*, **51**, 1233–1235.
24. Le Roy, J.J., Jeletic, M., Gorelsky, S.I., Korobkov, I., Ungur, L., Chibotaru, L.F. and Murugesu, M. (2013) *J. Am. Chem. Soc.*, **135**, 3502–3510.
25. Layfield, R.A., McDouall, J.J.W., Sulway, S.A., Tuna, F., Collison, D. and Winpenny, R.E.P. (2010) *Chem. Eur. J.*, **16**, 4442–4446.
26. Luzon, J. and Sessoli, R. (2012) *Dalton Trans.*, **41**, 13556–13567.
27. Rinehart, J.D., Fang, M., Evans, W.J. and Long, J.R. (2011) *Nat. Chem.*, **3**, 538–542.
28. Novitchi, G., Pilet, G., Ungur, L., Moshchalkov, V.V., Wernsdorfer, W., Chibotaru, L.F., Luneau, D. and Powell, A.K. (2012) *Chem. Sci.*, **3**, 1169–1176.
29. Hussain, B., Savard, D., Burchell, T.J., Wernsdorfer, W. and Murugesu, M. (2009) *Chem. Commun.*, 1100–1102.
30. Lin, S.-Y., Wernsdorfer, W., Ungur, L., Powell, A.K., Guo, Y.-N., Tang, J., Zhao, L., Chibotaru, L.F. and Zhang, H.-J. (2012) *Angew. Chem. Int. Ed.*, **51**, 12767–12771.
31. Hewitt, I.J., Lan, Y., Anson, C.E., Luzon, J., Sessoli, R. and Powell, A.K. (2009) *Chem. Commun.*, 6765–6767.
32. Guo, Y.-N., Xu, G.-F., Gamez, P., Zhao, L., Lin, S.-Y., Deng, R., Tang, J. and Zhang, H.-J. (2010) *J. Am. Chem. Soc.*, **132**, 8538–8539.

33. Anwar, M.U., Thompson, L.K., Dawe, L.N., Habib, F. and Murugesu, M. (2012) *Chem. Commun.*, **48**, 4576–4578.
34. Xue, S., Zhao, L., Guo, Y.-N., Chen, X.-H. and Tang, J. (2012) *Chem. Commun.*, **48**, 7031–7033.
35. Xue, S., Zhao, L., Guo, Y.-N., Zhang, P. and Tang, J. (2012) *Chem. Commun.*, **48**, 8946–8948.
36. Guo, Y.-N., Chen, X.-H., Xue, S. and Tang, J. (2012) *Inorg. Chem.*, **51**, 4035–4042.
37. Tuna, F., Smith, C.A., Bodensteiner, M., Ungur, L., Chibotaru, L.F., McInnes, E.J.L., Winpenny, R.E.P., Collison, D. and Layfield, R.A. (2012) *Angew. Chem. Int. Ed.*, **51**, 6976–6980.
38. Woodruff, D.N., Tuna, F., Bodensteiner, M., Winpenny, R.E.P. and Layfield, R.A. (2013) *Organometallics*, **32**, 1224–1229.
39. Wang, Y.-X., Shi, W., Li, H., Song, Y., Fang, L., Lan, Y., Powell, A.K., Wernsdorfer, W., Ungur, L., Chibotaru, L.F., Shen, M. and Cheng, P. (2012) *Chem. Sci.*, **3**, 3366–3370.
40. Xue, S., Chen, X.-H., Zhao, L., Guo, Y.-N. and Tang, J. (2012) *Inorg. Chem.*, **51**, 13264–13270.
41. Liu, C.-S., Du, M., Sanudo, E.C., Echeverria, J., Hu, M., Zhang, Q., Zhou, L.-M. and Fang, S.-M. (2011) *Dalton Trans.*, **40**, 9366–9369.
42. Guo, F.-S., Liu, J.-L., Leng, J.-D., Meng, Z.-S., Lin, Z.-J., Tong, M.-L., Gao, S., Ungur, L. and Chibotaru, L.F. (2011) *Chem. Eur. J.*, **17**, 2458–2466.
43. Ke, H., Xu, G.-F., Guo, Y.-N., Gamez, P., Beavers, C.M., Teat, S.J. and Tang, J. (2010) *Chem. Commun.*, **46**, 6057–6059.
44. Lin, S.-Y., Zhao, L., Ke, H., Guo, Y.-N., Tang, J., Guo, Y. and Dou, J. (2012) *Dalton Trans.*, **41**, 3248–3252.
45. Zheng, Y.-Z., Lan, Y., Anson, C.E. and Powell, A.K. (2008) *Inorg. Chem.*, **47**, 10813–10815.
46. Abbas, G., Lan, Y., Kostakis, G.E., Wernsdorfer, W., Anson, C.E. and Powell, A.K. (2010) *Inorg. Chem.*, **49**, 8067–8072.
47. Langley, S.K., Chilton, N.F., Gass, I.A., Moubaraki, B. and Murray, K.S. (2011) *Dalton Trans.*, **40**, 12656–12659.
48. Yan, P.-F., Lin, P.-H., Habib, F., Aharen, T., Murugesu, M., Deng, Z.-P., Li, G.-M. and Sun, W.-B. (2011) *Inorg. Chem.*, **50**, 7059–7065.
49. Chen, G.-J., Guo, Y.-N., Tian, J.-L., Tang, J., Gu, W., Liu, X., Yan, S.-P., Cheng, P. and Liao, D.-Z. (2012) *Chem. Eur. J.*, **18**, 2484–2487.
50. Habib, F., Long, J., Lin, P.-H., Korobkov, I., Ungur, L., Wernsdorfer, W., Chibotaru, L.F. and Murugesu, M. (2012) *Chem. Sci.*, **3**, 2158–2164.
51. Lin, P.-H., Korobkov, I., Wernsdorfer, W., Ungur, L., Chibotaru, L.F. and Murugesu, M. (2011) *Eur. J. Inorg. Chem.*, **2011**, 1535–1539.
52. Bi, Y., Xu, G., Liao, W., Du, S., Deng, R. and Wang, B. (2012) *Sci. China Chem.*, **55**, 967–972.
53. Gamer, M.T., Lan, Y., Roesky, P.W., Powell, A.K. and Clérac, R. (2008) *Inorg. Chem.*, **47**, 6581–6583.
54. Langley, S.K., Moubaraki, B., Forsyth, C.M., Gass, I.A. and Murray, K.S. (2010) *Dalton Trans.*, **39**, 1705–1708.
55. Sharples, J.W., Zheng, Y.-Z., Tuna, F., McInnes, E.J.L. and Collison, D. (2011) *Chem. Commun.*, **47**, 7650–7652.
56. Guo, Y.-N., Chen, X.-H., Xue, S. and Tang, J. (2011) *Inorg. Chem.*, **50**, 9705–9713.
57. Jeletic, M., Lin, P.-H., Le Roy, J.J., Korobkov, I., Gorelsky, S.I. and Murugesu, M. (2011) *J. Am. Chem. Soc.*, **133**, 19286–19289.
58. Jiang, S.-D., Wang, B.-W., Sun, H.-L., Wang, Z.-M. and Gao, S. (2011) *J. Am. Chem. Soc.*, **133**, 4730–4733.
59. Venugopal, A., Tuna, F., Spaniol, T.P., Ungur, L., Chibotaru, L.F., Okuda, J. and Layfield, R.A. (2013) *Chem. Commun.*, **49**, 901–903.
60. Zhou, H.-C., Long, J.R. and Yaghi, O.M. (2012) *Chem. Rev.*, **112**, 673–674.

4
Lanthanides in Extended Molecular Networks

Roberta Sessoli and Kevin Bernot

4.1
Introduction

In the early stages of molecular magnetism, interest was focussed on the realization of bulk magnets based on molecular networks. The control of the sign and strength of the exchange interactions between the paramagnetic centres was the key issue and extensive investigations led to the magneto-structural correlations [1] exploited even today in this research field. The optimization of the exchange pathways through chemical design revealed to be better feasible in one-dimensional (1D) structures [2], although activity was also focussed on three-dimensional (3D) networks mainly based on cyanide ligands [3, 4].

The main feature of 1D magnetic materials is that long-range magnetic order can only be achieved at absolute zero temperature. This simply results from the thermodynamics of the system. The cost to disorder a chain of aligned (parallel or antiparallel) spins is finite because a finite number of exchange interactions have to be violated to break an infinite chain into two magnetic domains. As the wall between two domains can be created at any site, the gain in entropy is infinite at any temperature different from zero, thus overwhelming the cost of the disordering process [5].

The reasoning in the previous paragraph is only valid in the ideal case of strictly one dimensional materials but in any real crystal, weak interchain interactions are always present leading to magnetic order, although at very low temperatures [6]. If molecular magnets cannot be achieved at room temperature following the 1D approach, the contrary is true: molecular magnetic materials can be purposely designed to strongly reduce interchain interactions by employing bulky and insulating ligands. The exceptionally pronounced one-dimensional character of some 1D molecular systems has been thus exploited in the past to test theories in low dimensions.

Three-positive lanthanide ions are not so promising if the objective is building bulk molecular magnets because of the internal character of 4f orbitals, which

Lanthanides and Actinides in Molecular Magnetism, First Edition.
Edited by Richard A. Layfield and Muralee Murugesu.

precludes strong exchange interactions. For instance Ln_2O_3 shows magnetic order at temperatures close to liquid helium, in contrast to the hundreds of kelvin in the case of 3d metal oxides. Although weak interactions can represent an advantage for some specific applications, for instance magnetic cooling through adiabatic demagnetization at very low temperatures, 4f ions have been often combined with transition metal ions or with organic radicals in order to increase the strength of the magnetic interactions [7]. The latter have shown an unexpected ability in transmitting interactions over very long distances, providing peculiar magnetic behaviours that are presented in this chapter.

The scenario is completely different if magnetic anisotropy is the relevant property of the investigated material, because, in general, lanthanides present an unquenched orbital contribution and much stronger anisotropy than transition metal ions. The key role played by lanthanides in the realization of single-molecule magnets (SMMs) has been extensively described in the earlier chapters. Dealing here with extended systems, we focus on the use of lanthanides to realize another class of materials showing magnetic bistability without long-range magnetic order. This is the case of 1D systems exhibiting the slow dynamics predicted by Glauber [8] for Ising chains. These systems are generally termed *single-chain magnets* (*SCMs*) [9], in analogy with the better known class of SMMs.

Organization of 4f ions into monodimensional molecular edifices is significantly trickier than with 3d ions. The key point of high coordination numbers (from 6 to 12) together with the relatively low symmetry of the coordination polyhedron of lanthanide ions do not favour the creation of regular and symmetric extended edifices [10]. However, chemists have succeeded in obtaining such compounds by using two different strategies: (i) the '*heteroligand strategy*' and (ii) the low connectivity '*tecton strategy*'. In the '*heteroligand strategy*' bridging ligands (that generate the chain) and terminal ligands (that avoid its growth into bi- or tridimensional structures) are associated. The simplest examples are made of a chelating ligand (e.g. carboxylates) and monodentate solvent molecules (e.g. H_2O and mono-alcohols) that complete the coordination sphere of the lanthanide. More stable edifices are obtained when poly-alcohols or poly-pyridines are used as terminal ligands. In the '*tecton strategy*' 4f-based molecular building blocks are made with ligands that constrain the geometry of the tecton and allow few (ideally two) coordination sites free for second-step coordination. These molecular precursors can be neutral or charged depending on the associated co-tecton. Charged 4f-precursors are mainly used with oppositely charged 3d molecular precursors to afford mixed 3d–4f chains. Neutral precursors, the most widely used being β-diketonates, are associated with neutral ligands, in particular organic radicals to build mixed 4f-radical chains.

In organizing the content of this chapter we have made a distinction between isotropic, mainly Gd^{3+}, and anisotropic lanthanide-based molecular networks.

In the first case, 1D materials exhibit a peculiar magnetic behaviour only in a restricted class of compounds comprising nitronyl-nitroxide radicals. Much richer is the literature on anisotropic 4f ions showing slow relaxation of the magnetization due to SCM behaviour.

4.2 Extended Networks Based on Gd^{3+}

4.2.1 Metal-Organic Frameworks

Several examples of chain structures comprising Gd^{3+} as the sole carriers of magnetic moments are available in the literature [11–13]. However, their magnetic features present, at very low temperatures, minor deviations from the Curie behaviour as a consequence of the very weak magnetic interactions. These can be either ferro- or antiferromagnetic and through-space or dipolar contributions, often comparable to through-bond or exchange ones. The situation is somewhat different for regular 3D networks, where the bridging ligands connect the metal centres in a three-dimensional structure [14–16]. This class of compound is also known as metal-organic frameworks, MOFs, and due to the possibility of modifying the chemical nature and dimensions of the voids in the network, they are widely employed in gas storage, sensing, catalysis, and so on [17–19]. In the case of lanthanides, MOFs have also been engineered to enhance their luminescence properties, by using bridging ligands that can act as antenna for energy transfer to the lanthanide centre [20]; they also present interesting magnetic properties, which are briefly discussed in the next two sections.

4.2.1.1 Magneto-Caloric Effect

In recent times Gd-based MOFs have been investigated for application as magnetic refrigerants, exploiting the temperature drop during an adiabatic demagnetization due to the magneto-caloric effect, MCE. When a magnetic system is exposed to a magnetic field that polarizes the magnetic moments thus reducing the degrees of freedom of the system, a drop in the entropy of the system, ΔS_m, is observed. When the field is decreased, that is, the sample is demagnetized, the magnetic contributions to the entropy increase, as shown in Figure 4.1. If the process of demagnetization occurs without any heat flow from the environment to the magnetic system, that is, in adiabatic condition, the change in magnetic entropy of the system has to be compensated by the decrease of the entropy associated with the other degrees of freedom, such as lattice vibrations, and a drop in temperature of the magnetic system, ΔT_{ad}, occurs. In order to understand which are the key parameters to play with in the design of an optimized MCE material it is useful to recall some basic thermodynamic relations [21]:

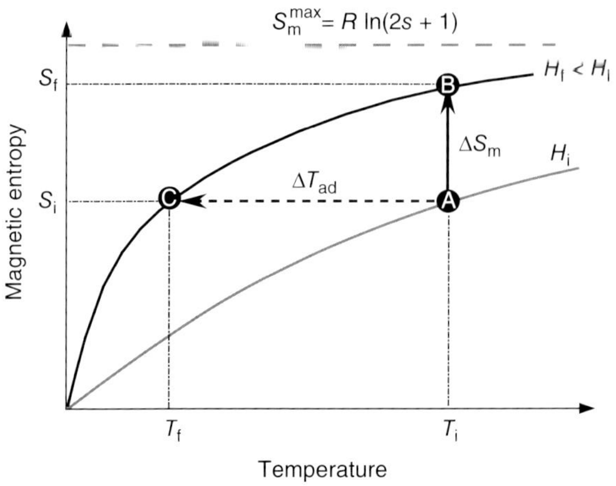

Figure 4.1 Schematic diagram showing the thermodynamics of the demagnetization process. By changing the magnetic field from the initial value H_i to the final one H_f, with $H_f < H_i$ a gain in magnetic entropy, ΔS_m, occurs during the isothermal process from A to B, while a temperature drop, ΔT_{ad}, is observed during the adiabatic process A → C. The maximum magnetic entropy for a system with a spin *s*, $R\ln(2s + 1)$, reached at infinite temperatures, is indicated as a grey broken line.

$$\Delta S_{\text{m}}(T, \Delta H) = \int_{H_1}^{H_2} \left(\frac{\partial M}{\partial T}\right)_H dH \tag{4.1}$$

$$\Delta T_{\text{ad}}(T, \Delta H) = \int_{H_1}^{H_2} \left(\frac{T}{C}\right)_H \left(\frac{\partial M}{\partial T}\right)_H dH \tag{4.2}$$

where H is the applied magnetic field and M and C are the field and temperature dependent magnetization and specific heat, respectively.

The first requirement for large magneto-caloric efficiency is therefore a strong temperature dependence of the magnetization, which for paramagnets is maximized for large spin states and low temperatures. Moreover, the absence of magnetic anisotropy that splits the levels in zero field maximizes the entropy change due to the application of an external field. These conditions are at best satisfied in materials containing Gd^{3+}, and indeed gadolinium gallium garnet, $[Gd_3Ga_5O_{12}]$ (GGG), is the reference material for cryogenic applications [22].

There are many examples of very large spin states in polynuclear 3d and/or 4f metal complexes, and those that are characterized by low magnetic anisotropy have a remarkable change in entropy showing significant MCE properties [23]. Nevertheless, the cooling power is associated with ΔS_m, an extensive property, and what is relevant for applications is the specific value for unit of mass, or unit of volume, of the refrigerant. The gain induced by using magnetic molecules is in part reduced by the lower density of magnetic centres in molecular materials than in paramagnetic salts. This drawback has been partially eliminated in Gd-MOFs, where light and small linkers connect the magnetic centres yielding relatively dense networks.

The first investigated Gd-MOF was $\{Gd_2(fum)_3(H_2O)_4 \cdot 3H_2O\}_n$, where fum^{2-} is the anionic form of the fumaric acid (Figure 4.2) [25]. The structure consists of Gd_2 units bridged by a bis-$\mu_{1,1}$-carboxylate bridge forming molecular ladders. Strong hydrogen bonds and head-to-tail bridging mode of the fumaric acid create a 3D porous network. Exchange interactions are relatively weak and the system orders magnetically at very low temperatures, about 0.2 K. Above the ordering temperature the MCE power was found comparable to that of molecular clusters.

A significant increase in the specific maximum entropic change was achieved by substituting the fumaric acid with the much smaller formic acid, as reported in Table 4.1 where data for a selection of networks based on gadolinium

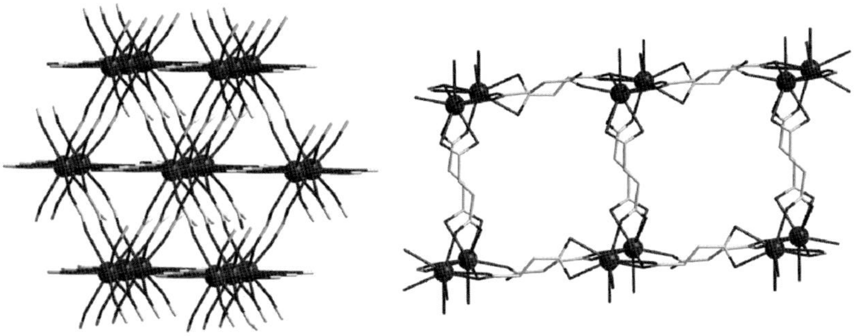

(a) (b)

Figure 4.2 (a) Representation of $\{Gd(HCOO)_3\}_n$ and (b) representation of a molecular ladder of $\{Gd_2(fum)_3(H_2O)_4{\cdot}3H_2O\}_n$. Supramolecular arrangement of these ladders gives rise to a 3D porous network. Gd atoms are represented as dark spheres while H atoms are omitted for clarity. (Redrawn from Refs [24, 25]. Published by Wiley and Elsevier, respectively.)

Table 4.1 Characteristic values of selected gadolinium-based systems exhibiting magnetocaloric effects.

	$-\Delta S_m^{max}$ (mJ K^{-1} cm^{-3})	T_{max} [a)] (K)	ρ (g cm^{-3})	References
$\{Gd_2(fum)_3(H_2O)_4{\cdot}3H_2O\}_n$	45.3	1.0	2.515	[25]
$\{Gd(HCOO)(C_8H_4O_4)\}_n$	68.4	1.9	2.662	[26]
$\{Gd(HCOO)(OAc)_2(H_2O)_2\}_n$	88.9	0.9	2.397	[27]
$\{Gd(HCOO)_3\}_n$	168.5	1.1	3.856	[24]
$[Gd_3Ga_5O_{12}]$ (GGG)	145.0	1.2	7.080	[28]

a) Temperature at which the largest variation in magnetic entropy, ΔS_m^{max}, is observed.

ions are listed. The increase is remarkable for $\{Gd(HCOO)_3\}_n$, given its higher density and the reduced number of nonmagnetic atoms in the lattice [24]. The magneto-caloric efficiency of this MOF overcomes also the reference material, that is, the GGG of formula $[Gd_3Ga_5O_{12}]$. The structure of $\{Gd(HCOO)_3\}_n$ consists of chains of Gd^{3+} bridged by three formate anions in the $\mu_{1,1}$ mode, while the chains are connected to six neighbouring ones by formate anions in the anti–anti $\mu_{1,3}$ mode. The crystal lattice is hexagonal and presents a relatively high Debye temperature, $\Theta_D = 168\,K$. The robustness of the lattice makes $\{Gd(HCOO)_3\}_n$ a good candidate for integration in micro-devices for local refrigeration [29]. It also warrants a better thermal conductivity compared to softer lattices of isolated molecules, thus partially solving one of the drawbacks in the use of molecular magnetic refrigerants.

4.2.1.2 Slow Magnetic Relaxation and Phonon Bottleneck Effects

Some gadolinium-based MOF have also revealed the peculiar dynamics of the magnetization with the appearance of the out-of-phase component of the dynamic magnetization in presence of an applied static magnetic field. This behaviour, which is reminiscent of the field-induced slow relaxation that characterize many lanthanide-based SMMs, is however not related to the magnetic anisotropy, which is vanishingly small in most Gd^{3+} systems.

The first and more deeply investigated material is the $\{Gd_2(fum)_3(H_2O)_4 \cdot 3H_2O\}_n$ MOF, described earlier [25]. Electron paramagnetic spectroscopy and specific heat measurements suggested that the system can be treated as quasi-isolated $S = 7/2$ spins featuring a small negative zero field splitting, $D \approx -0.1\,cm^{-1}$, which, according to the spin Hamiltonian $H_{zfs} = DS_z^2$, induces an overall splitting in the $S = 7/2$ multiplet of about $1.2\,cm^{-1}$. The ac susceptibility measured in the presence of a static field revealed maxima in the out-of-phase component (χ'') versus frequency curves at temperatures as high as 20 K. The extracted relaxation times, when reported in the Arrhenius plot, showed strong deviations from a linear behaviour. Moreover, the activation energy evaluated from the high temperature region resulted in a barrier height of 36 K. This value clearly exceeds the energy separation of the $S = 7/2$ manifold, pointing to a mechanism of relaxation not associated with the magnetic anisotropy.

As already observed for some isotropic polynuclear clusters [30–32], slow relaxation of the magnetization in an external magnetic field can occur because of the inefficient transfer of energy to the environment, for example, the helium bath, and consequent reabsorption of the emitted phonon by the spin system. The phenomenon, also known as *phonon bottleneck* (*PB*), was first introduced by Van Vleck [33]. It is characteristic of low temperatures, where relaxation is dominated by the direct process between closely spaced levels, and results from the low density of phonons with such a long wavelength to match the small energy separation in moderate magnetic fields. Moreover, if the wavelength of the resonant phonons is much larger than the separation between the paramagnetic centres, the excitation can move back and forth between neighbouring spins and the phonons remain locally trapped without reaching the thermal bath. In such a case, the relaxation time of the magnetization depends on the temperature as T^{-2} [34]. Orendáč *et al.* [25] found for $\{Gd_2(fum)_3(H_2O)_4 \cdot 3H_2O\}_n$ a behaviour that is best approximated as $\tau \propto T^{-n}$ with $n = 2.9 \pm 0.1$. It should however be recalled that $Gd^{3+} - Gd^{3+}$ separation, a crucial parameter for the resonant phonon trapping mechanism, is not homogeneous in the crystal lattice, which is formed by weakly interacting dimers.

Another gadolinium MOF, again based on formate, but also entrapping formamidinium (Fmd) cations, of formula $\{(Fmd)Gd(HCCO)_4\}_n$ shows a more regular spacing between the paramagnetic centres [35]. In fact, each Gd^{3+} ion is linked through the anti–anti $\mu_{1,3}$ formate bridge to eight other ions, as shown in Figure 4.3, with distances, evaluated for the isostructural Dy derivative, distributed in the narrow range from 6.68 to 6.76 Å. Ac susceptibility data in a static field of 2 kOe showed maxima up to 15 K for frequencies lower than 10 kHz. The extracted relaxation times are reported in Figure 4.3 and show a significant deviation from linearity in the Arrhenius plot. Interestingly, the power law analysis provided the best fit for $n = 2.01 \pm 0.01$, in remarkable agreement with theoretical expectations for phonon trapping mechanism. The presence of this phenomenon resulting in slow magnetic relaxation should be taken into account when out-of-phase ac susceptibility is observed only in a static magnetic field and there is no experimental evidence of easy axis magnetic anisotropy.

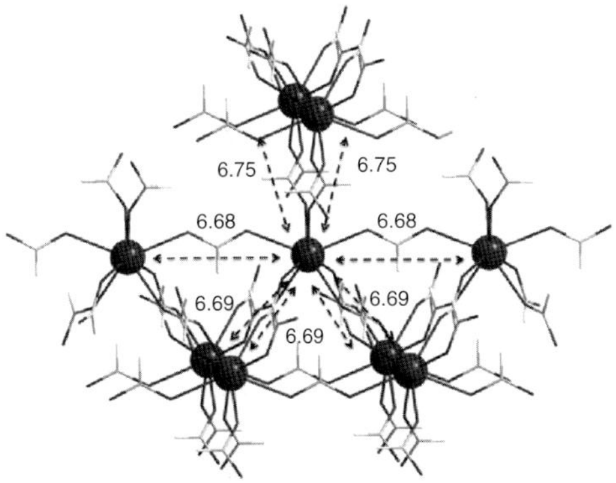

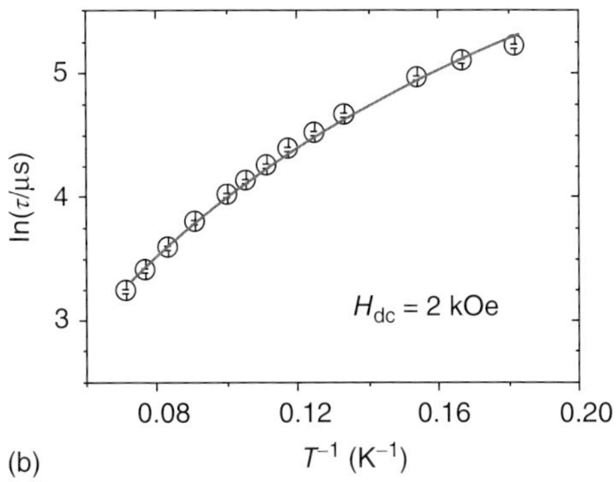

Figure 4.3 (a) Representation of $\{(Fmd)Gd(HCCO)_4\}_n$ with distances between Gd neighbours (in angstrom and calculated on the Dy analogue). Gd atoms as dark spheres, H atoms and Fmd cations omitted for clarity. (Original crystallographic data from Ref. [35].) (b) Temperature dependence of the relaxation time of $\{(Fmd)Gd(HCCO)_4\}_n$ in a static magnetic field of 2 kOe. The grey line corresponds to the best fit assuming that τ is proportional to T^{-n}.

It is interesting to notice that the analogous Dy and Tb derivatives failed to reveal any imaginary component of the ac susceptibility [35], in agreement with the fact that much larger energy splittings characterize these anisotropic anions, with a higher efficiency of energy transfer through the phonon bath.

4.2.2 Magnetic Chains

4.2.2.1 Magnetic Interactions Involving Gd^{3+} Ions

The first investigated molecular chains comprised two different types of 3d paramagnetic centres to exploit the non-compensation of the magnetic moments in presence of antiferromagnetic interaction. The 'ferrimagnetic' approach, introduced by Kahn [36], was successively extended to lanthanide ions, and in particular to Gd^{3+} because the absence of a significant orbital contribution strongly simplified the analysis of the magnetic properties. Investigations on simple oligomeric units revealed an unexpected ferromagnetic interaction, a few wavenumbers in strength, between the Gd^{3+} and Cu^{2+} spins [37]. Similar ferromagnetic interactions were detected when Gd^{3+} was coordinated by organic radicals [38]. The mechanism of this interaction was long debated but it was immediately evident that orbitals of the lanthanide ion, other than the well shielded and too low in energy 4f orbitals, should be involved in the super-exchange mechanism.

The mechanism that was suggested involved a partial charge and spin density transfer from the magnetic orbital of Cu^{2+} to the empty 5d or 6s orbitals of the Gd^{3+} ion. Since these orbitals are orthogonal to the half-filled 4f ones, the latter should experience a spin polarization parallel to that of the copper spin, and therefore a ferromagnetic interaction, as expected according to the Goodenough and Kanamori rules of magnetic exchange [39, 40]. More recent *ab initio* calculations have tried to shed light on this mechanism of interaction, either using self consistent field [41] or density functional [42] approaches. Both converged to evidencing, beyond a small contribution from the Cu(3d) → Gd(5d) charge transfer, a key role played by the delocalization of 4f spin density on the 5d orbitals. A small change in the electron density on the 5d orbitals upon change of the coordination sphere around Gd^{3+} was detected by X-ray absorption spectroscopy performed at the L-edge, although no correlation with the strength of magnetic interactions was found [43]. A less relevant role seems to be played by the empty 6s orbitals. As the resulting magnetic interaction is given by a balance between both ferro- and antiferromagnetic contributions, it is not surprising that exchange interactions of both signs have been observed in Gd-radical complexes [44, 45].

4.2.2.2 Gadolinium-Radical Chains

Although extended 1D systems comprising ferromagnetically coupled gadolinium and copper ions were obtained [46], the exchange interactions were too weak to provide significant effects and the increase of the χT product was observed only at very low temperatures. A much richer scenario was instead presented by 1D structures comprising nitronyl-nitroxide radicals, NIT-R, 2-R-4,4,5,5-tetramethyl-imidazoline-1-oxyl-3-oxide, alternating with Gd^{3+} ions. These radicals, whose general structure is schematized in Figure 4.4, present two equivalent N–O groups on which the unpaired electron is delocalized in an antibonding π^* molecular orbital. They are therefore best suited to transmit magnetic interactions in extended networks [47]. An important drawback is that NIT-R radicals have very poor coordination capability, and thus the Lewis acidity of the metal centre must be enhanced by the use of the very electron-withdrawing ligands such as hexafluoroacetylacetonate ($hfac^-$). A typical chain structure is reported in Figure 4.4 and similar 1D structures have been obtained for several radicals differing in the R group [48, 49].

In Figure 4.5 is reported the temperature dependence of the magnetic susceptibility of four of them, namely R = methyl, ethyl, *i*-propyl and phenyl.

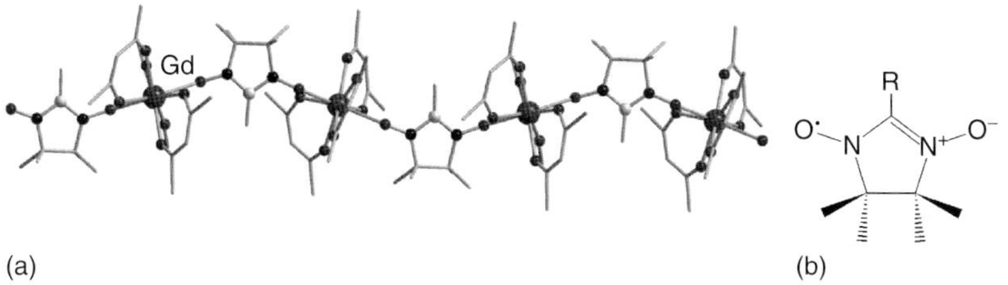

Figure 4.4 (a) Representation of a typical $\{Gd(hfac)_3NIT\text{-}R\}_n$ chain, with $hfac^-$ = hexafluoroacetylacetonate. Gd^{3+} ions are reported as dark spheres, H and F atoms are omitted for clarity. (b) Scheme of 2-R-4,4,5,5-tetramethyl-imidazoline-1-oxyl-3-oxide radical (NIT-R).

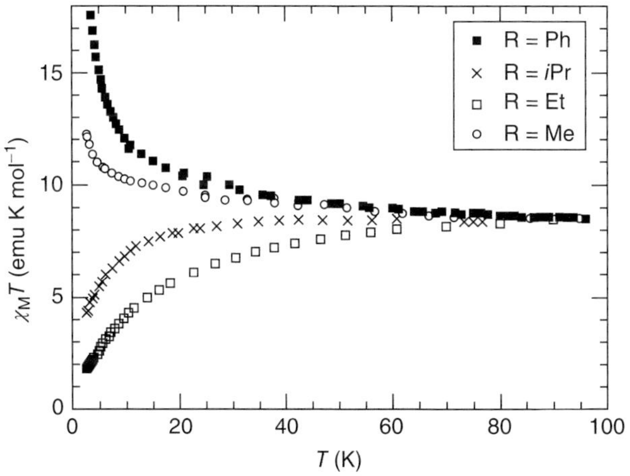

Figure 4.5 Temperature dependence of the χT product of $\{Gd(hfac)_3NIT\text{-}R\}_n$ helical chains with R = methyl, *i*-propyl, phenyl, ethyl. (Original data in Ref. [50].)

Interestingly, with R = ethyl and *i*-propyl the χT product shows a significant decrease at low temperatures, in contrast to what is expected for a chain where spins 1/2 and 7/2 alternate. In fact, even in the case of antiferromagnetic coupling, a divergence at low temperatures can be observed as a consequence of the non-compensation of the magnetic moments. It was immediately clear that introduction of next-nearest neighbour (NNN) antiferromagnetic exchange interactions were necessary to induce a temperature decrease of χT, the chains being too well-shielded to promote interchain interactions comparable to the intrachain ones. Parallel investigations of the Y^{3+} and Eu^{3+} derivatives revealed in both cases a sizeable antiferromagnetic interaction between the radicals, which, with the general spin Hamiltonian written as:

$$H = -J\sum_{i} S_i \cdot S_{i+1} \tag{4.3}$$

was described by J values of about -5 and $-10\,\text{cm}^{-1}$ for the Y^{3+} and Eu^{3+}derivative, respectively [48, 51]. The capability of the nonmagnetic metal ion to transmit magnetic interaction between the radicals is however not sufficient to justify the observed decrease of χT in the Gd^{3+} analogues. In fact, in the presence of dominating antiferromagnetic NNN interactions between the radicals, the Gd^{3+} spins should remain uncorrelated and thus exhibit Curie behaviour. Antiferromagnetic NNN interactions involving the lanthanide ions, despite these being more than 8 Å apart, are therefore active in this interesting class of materials [52].

A quantitative analysis of the magnetic susceptibility was attempted using a more complex spin Hamiltonian:

$$\begin{aligned} H = {} & -J_1\sum_{n=1}^{N/2}(S_{2n-1}\cdot s_{2n} + s_{2n}\cdot S_{2n+1}) - J_2\sum_{n=1}^{N/2}(S_{2n-1}\cdot S_{2n+1}) \\ & - J_2'\sum_{n=1}^{N/2} s_{2n}\cdot s_{2n+2} - g\mu_B H\sum_{n=1}^{N/2}(S_{2n+1} + s_{2n}) \end{aligned} \tag{4.4}$$

where capital S on odd sites represents Gd^{3+} spins while small s on even sites represents the radical spin. $N/2$ is the number of Gd-radical pairs in the chain, J_1 the nearest neighbour (NN) Gd-radical interaction, while J_2 and J_2' represent the $Gd^{3+}-Gd^{3+}$ and radical–radical NNN interactions, respectively. For the Zeeman interaction $g = 2$ can be assumed for both spins, given the negligible orbital contribution. The treatment of such a complex spin Hamiltonian has been simplified by assuming that the spins are classical vectors confined in the plane perpendicular to the chain direction [50].

In case of sufficiently strong antiferromagnetic NNN interactions compared to the NN ones, that is, $2(\delta + \delta') > 1$ with $\delta = |J_2|S^2/J_1Ss$ and $\delta' = |J_2'|s^2/J_1Ss$, the ground state in zero static field has a helical configuration of the spins, as shown in Figure 4.6a, with the angle between adjacent spins, Θ, defined as:

$$\pm\Theta = \pm\cos^{-1}\left[\frac{1}{2(\delta + \delta')}\right] \tag{4.5}$$

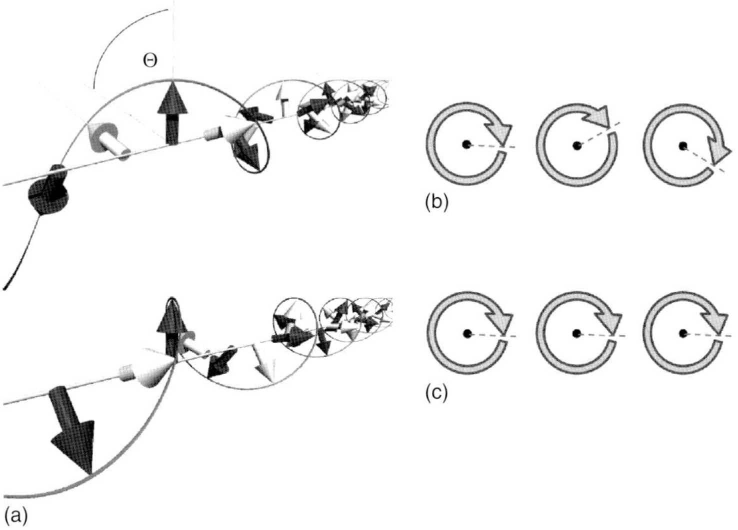

Figure 4.6 (a) Schematic view of the two degenerate helical arrangements of the spins of a 1D frustrated structure with antiferromagnetic next-nearest-neighbour (NNN) interactions. The angle Θ is the one formed between the projections of two adjacent spins, drawn with different colours to represent the alternating spins structure of Gd-radical chains. (b) Schematic representation of the 3D chiral spin liquid phase where all chains have the same chirality but random phases. (c) 3D helical phase where all helices have identical chirality and phase.

where the sign ± corresponds to a clockwise and a counter-clockwise rotation of the spins along the chain, respectively.

The helical pitch was calculated to correspond to $\Theta = \pm 0.397\pi$ and $\pm 0.399\pi$ for the chains with R = Et and *i*-Pr, respectively. The decrease in χT at low temperature is justified by the fact that adjacent Gd^{3+} spins are almost antiparallel in these derivatives. A much smaller pitch, $\Theta = \pm 0.206\pi$ and $\pm 0.175\pi$, was estimated for R = Me and Ph, respectively χ, in agreement with the increase of the χT at low temperature.

A very complex phase diagram was predicted in the seventies by Villain for quasi 1D helimagnets. Configurations with opposite chirality are degenerate and in the paramagnetic phase the chains are composed by domains with opposite chirality separated by a domain wall (DW). On lowering the temperature a 3D chiral spin liquid phase was predicted, which can be schematized as a 3D ensemble of helices all with the same spin chirality but with random phase between different chains, as shown in Figure 4.6b.

At lower temperatures, the standard 3D helical order with all chains having not only the same chirality but also the same phase can be established. The two magnetic phase transitions present very different features. In fact, the transition to the chiral spin liquid phase is not expected to be visible in the magnetic susceptibility but only in the specific heat. The four Gd-radical chains previously discussed were investigated in detail. While the Me and Ph derivatives showed around 0.6 K a typical transition to 3D magnetic order in both the magnetic susceptibility and the specific heat, the Et derivative presented two anomalies in the specific heat, at $T = 2.19 \pm 0.02$ and 1.88 ± 0.02 K, as shown in Figure 4.7. When investigated by either magnetic suscetibility (see Figure 4.7) or muon spin resonance only the anomaly at lowest temperature was visible, thus suggesting that between 1.88 and 2.19 K the material can be actually described as the chiral spin liquid predicted by the Villain conjecture [53].

It is interesting to stress that the spin chirality observed in the gadolinium radical chains differs from the more usual one that characterizes anisotropic materials and is solely due to the significant strength of NNN interactions between lanthanide ions that are very far apart. The mechanism responsible for this interaction remains unclear and the complexity of the system has, up to now, hampered an *ab initio* investigation of the phenomenon.

To conclude this section we would like to recall that the coexistence of structural chirality and magnetism can give rise to a peculiar optical effect, which goes under the name of magneto-chiral dichroism and corresponds to the different absorption of unpolarized light by a chiral material depending on its magnetization state [54, 55]. This effect has been recently investigated in cobalt and manganese radical chiral chains by using hard X-ray photons [56]. Chiral Ln-radical chains have also been more recently reported [57], and it would be very interesting

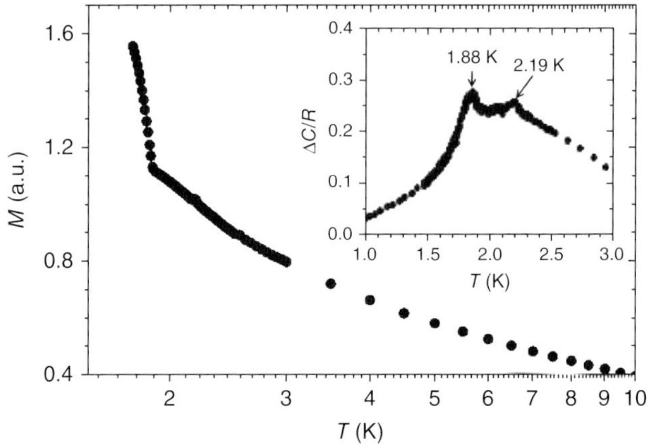

Figure 4.7 Temperature dependence of the magnetization in 5 Oe applied field of a powder sample of $\{Gd(hfac)_3NIT\text{-}Et\}_n$ evidencing a phase transition around 1.8 K. In the inset the magnetic contribution to the specific heat, evaluated by subtracting the lattice contribution to the total specific heat, reveals two distinct phase transitions. (Redrawn from Ref. [53] with data kindly provided by A. Lascialfari. (Published by American physical society).)

to investigate the interplay between structural and spin chirality in the magneto-chiral dichroism.

4.3
Extended Networks Based on Anisotropic Ions

4.3.1
SCM in a Nutshell

At the beginning of this century slow magnetic relaxation in the paramagnetic phase of a 1D system was reported for a Co^{2+} nitronyl-nitroxide chain [58]. The dynamics was under many respects very similar to that of SMMs and these 1D systems were later named SCM to underline the analogies [9]. Since then an intense research activity, though not as spread out as for SMMs, has been devoted to SCMs. The interested reader can find extensive literature on the subject, including exhaustive reviews and a book chapter [59–63]. The aim here is to provide some basic concepts of the phenomenon and to introduce some selected examples of SCMs based on lanthanide ions.

As already stated in the introduction, no long-range order can be achieved at finite temperatures for strictly 1D systems. Therefore the ground state of a spin chain can be schematized as in Figure 4.8 with domains of oppositely oriented

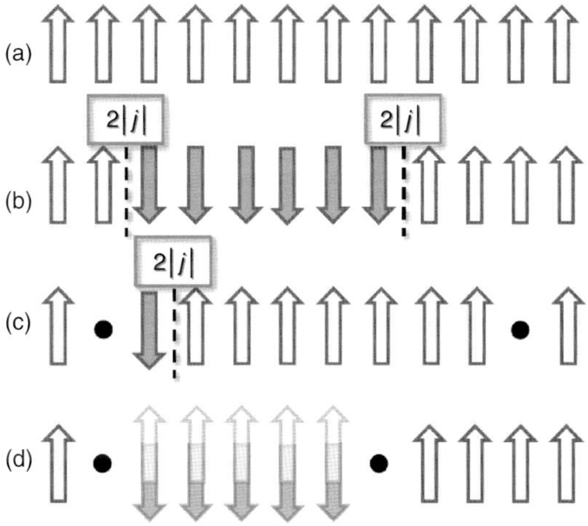

Figure 4.8 (a) Representation of a magnetic chain in its ordered state. (b) The creation of a domain of reversed spins costs four times the exchange interaction, assuming unitary length for the spins. (c) If the reverse domain nucleates on the edge of a finite chain the cost is halved. (d) For short segments, a collective reversal of the spins constitutes a competitive mechanism for magnetic relaxation.

spins separated by DWs. The simplest Hamiltonian to describe a spin chain is the classical Heisenberg model with uniaxial magnetic anisotropy:

$$H = \sum_{i=1}^{\infty} \left[-JS_i \cdot S_{i+1} + DS_i^2 \right] \tag{4.6}$$

where J and D have the same meaning as in the previous sections. In the case of negative D the system presents easy axis anisotropy and Equation 4.6 allows evaluating the cost to create a DW. Two different scenario can be envisaged: (i) the transition from one domain to the next one is abrupt, and the DW is said to be sharp and (ii) the transition is gradual, and the DW is said to be broad [64]. It is rather intuitive that a large $|D|/|J|$ ratio favours sharp DW, with the transition between the two regimes occurring when $|D|/|J| = 2/3$.

The case of $|D|/|J| \gg 1$ corresponds to the classical Ising model, whose dynamics was theoretically investigated in the 1960s by Glauber [8]. The main feature of Glauber dynamics is an exponential divergence of the relaxation time at low temperatures:

$$\tau = \alpha^{-1} \exp\left(\frac{\Delta_G}{k_B T} \right) \tau \tag{4.7}$$

which is very similar to what is observed for SMMs. The activation energy is related to the energy cost to create a segment of reversed spin, which in the Ising model is given by:

$$\Delta_G = 4|J|S^2 \tag{4.8}$$

where the factor 4 is introduced to take into account that two NN interactions are violated to create a domain of reversed spins.

An alternative way to describe the phenomenon is to consider that the ground state of a chain is already divided into domains at any temperatures. In order for the system to follow a small variation of the magnetic field some domains have to reverse their spin orientation. This occurs through a random walk of the DWs, that is, equal probability for the DW to move backward or forward, which implies that the DW needs a time proportional to d^2 to reach the other end of a domain of length d. Given that d scales as the two spins correlation length, ξ, which, for the Ising model, is proportional to $\exp(2J/k_B T)$, for unitary spins, the same exponential relaxation is found

$$\tau = \alpha^{-1}\xi^2 = \alpha^{-1} \exp\left(\frac{4\,|J|\,S^2}{k_B T} \right) \tag{4.9}$$

Several assumptions have been made in this description of the magnetization dynamics of chain systems:

1) DWs have been assumed to be infinitely narrow, that is, just two neighbour spins involved.

2) The chains are considered infinite, or at least of length $L \gg \xi$ at any temperature; this is what is called the *thermodynamic limit*.
3) Small deviations from equilibrium are assumed, which implies small changes in the applied magnetic field as in the case of ac magnetic susceptibility experiments.
4) The pre-exponential factor is assumed to be temperature independent and proportional to the inverse of the attempting frequency, α^{-1}.

In some SCMs the single unit constituting the chain can present on its own a slow dynamics related to its magnetic anisotropy:

$$\tau_{\mathrm{SI}} = \tau_0 \exp\left(\frac{\Delta_{\mathrm{SI}}}{k_{\mathrm{B}}T}\right) = \tau_0 \exp\left(\frac{|D|\, S^2}{k_{\mathrm{B}}T}\right) \tag{4.10}$$

where SI stands for single ion, even if in many SCMs a multi-spin unit constitutes the basic entity.

Substituting τ_{SI} in Equation 4.10 gives:

$$\tau = \tau_0 \exp\left(\frac{|D|\, S^2 + 4|J|S^2}{k_{\mathrm{B}}T}\right) \tag{4.11}$$

Thus the overall barrier Δ for the reversal of the magnetization is constituted by two terms, one related to the SI anisotropy and the second to the intrachain exchange interaction, that is, $\Delta = \Delta_{\mathrm{SI}} + \Delta_\xi$. In this description τ_0 is temperature independent and has the same meaning as in SMMs.

It must be pointed out here that only in a few cases the units constituting the chain show, on their own, slow dynamics of the magnetization, that is, SMM behaviour. In most cases the dynamics of the single unit is dominated by quantum tunnelling, especially in the zero static field and at low temperatures. However, the exchange interaction with the neighbouring units, when these units are connected in the chain structure, acts as an internal field suppressing the tunnel mechanism. Thus the SI slowing dynamics contributes to the barrier for the reversal of the magnetization.

As a final remark it must be mentioned that theoretical and experimental works have been dedicated to investigating the effect of the finite size of the chains [65]. In fact, as ξ grows exponentially, at low temperatures it can become comparable with the distance between two consecutive defects (e.g. impurities and vacancies) which are always present in real systems and hardly separated by more than $10^3 - 10^4$ elementary units. In case of $L < \xi$, the nucleation of the DW is energetically favoured if occurring at the boundaries, because the energy cost is halved. However the probability to have a boundary spin is inversely proportional to L; thus the pre-exponential factor becomes linearly dependent on L, as experimentally found in doped SCMs. As doping occurs at random positions on the chain, a distribution of lengths is observed in a real system. However, as the relaxation time is only linearly dependent on L, a relatively narrow distribution is expected.

In the limit of very short chain segments, for example, at high concentration of diamagnetic dopants, the probability, Γ, of a collective reversal of all the spins of the segment scales as:

$$\Gamma_{\mathrm{coll}} = \left[\alpha \exp\left(-\frac{|D|\, S^2}{k_\mathrm{B} T}\right)\right]^L \tag{4.12}$$

while if the reversal of the single units occurs through quantum tunnelling the probability is simply given by:

$$\Gamma_{\mathrm{coll}} = \alpha_{\mathrm{qt}}^L \tag{4.13}$$

where α_{qt} is the tunnelling frequency for one unit. Given the exponential dependence on L, and the fact that all probabilities are much smaller than 1, the mechanism of collective reversal is competitive only for very large exchange interactions and very short segments. A much wider distribution of relaxation times is expected when collective reversal of spin segments occurs in randomly doped chains, compared to the case of Glauber dynamics, because of the exponential dependence on L in Equation 4.13. Flattening of the Argand plot, that is, the χ'' versus χ' plot, was indeed observed in the case of cobalt-radical chains heavily doped with Zn^{2+} ions [65].

4.3.2 An Overview of Monodimensional Lanthanide Chains Based on Anisotropic Ions

In this sub-chapter, selected examples of lanthanide-based chains are described. We have chosen to comment on only systems with structural characterization and significant dynamic properties. Chains with a sole lanthanide ion as spin carrier are described first. 3d–4f and 3d–3d′–4f heterometallic chains follow. Finally, chains comprising lanthanide and radical ligands conclude this chapter.

4.3.2.1 Chains Based on 4f Ions

Whatever the approach used, '*hetero ligand*' or '*tecton strategy*', chains based only on a 4f ion and neutral ligands are quite hard to obtain. In most cases, chains are a monodimensional arrangement of polynuclear entities (dimers [66, 67] or tetramers [68]) rather than regular chains based on a repetitive monomer [69]. The observable magnetic couplings are then internal to these sub-units and do not concern the whole 1D edifice. In the few rare cases where regular chains are obtained [12, 70, 71] Ln–Ln coupling is weak and poorly transmitted by classical diamagnetic ligands that cannot generate the correlation length necessary for the observation of SCM behaviour. Single molecule magnetic relaxation is then observed, whose features mainly depend on the geometry of Ln^{3+} surroundings. Moreover, although the observed activation energies are significant ($\Delta = 15.8$ K [70], 39.4 K [12], 44 K [71]) slow relaxation is observed only if a static field is applied, a feature not compatible with pure SCM behaviour [72].

A detailed investigation performed by Bartolomé and coworkers [73] on a Dy-furoate chain, named $\{Dy(\alpha\text{-fur})_3\}_n$, highlights that such chains cannot give rise to SCM behaviour. In this compound Dy^{3+} ions are bridged by furoate ligands and water molecules complete the coordination sphere (see the topologic mode (A) in Scheme 4.1). The investigation is severely complicated by the occurrence of two types of chains in the crystal packing, and so two different coordination environments of Dy^{3+} (the influence of noncollinearity of magnetic anisotropy axes is discussed in Section 4.3.3). The resulting $\{Dy(\alpha\text{-fur})_3\}_n$ compound exhibits magnetic slow relaxation below 10 K. Both Dy ions possess Ising anisotropy, as demonstrated by *ab initio* calculations, and are coupled by intrachain ferromagnetic interaction ($J_{intra} = 1.05\,cm^{-1}$, assuming in the Ising Hamiltonian an effective spin of 1/2). $[Dy(\alpha\text{-fur})_3]_n$ should thus be a good candidate for the observation of SCM behaviour. However, the observed thermally activated magnetic relaxation presents an energy barrier of 29 K (chain 1) and 80 K (chain 2). This would require a very strong intrachain interaction, J_{intra}, far from what was observed here. It is clear that the role of the intrachain interaction and its contribution to the energy barrier is practically negligible here. This is confirmed by strong diamagnetic Y^{3+} substitution that is not able to alter significantly the dynamic magnetic parameters. Indeed, as J_{intra} is weak, low temperature heat capacity measurements clearly show a crossover from a SMM regime towards a 3D long-range magnetic order at $T_N = 0.668\,K$. Such ordering is attributed to the occurrence of the interchain antiferromagnetic coupling that is evaluated to be $J_{inter} = -0.188\,cm^{-1}$. The weak ability of furoate ligands to mediate exchange magnetic interaction is underlined by the fact that dipolar coupling contributes to almost 50% of the total interaction energy in $\{Dy(\alpha\text{-fur})_3\}_n$.

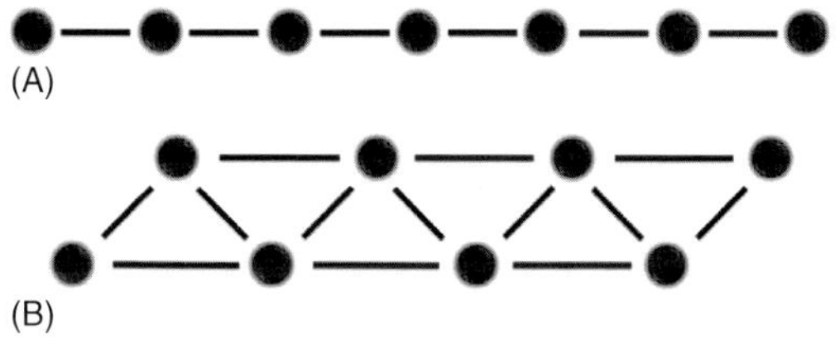

Scheme 4.1 (a,b) Schematic representation of topologic organization of pure 4f-based chains.

$Dy^{3+}-Dy^{3+}$ ferromagnetic coupling has also been observed in a double chain based on the triangular motif (topologic mode (B) in Scheme 4.1) [74]. Slow relaxation is observed ($\Delta = 44.2\,K$, $\tau_0 = 2.4 \times 10^{-8}\,s$) but its 1D origin is not demonstrated, and the onset of 3D ordering is visible at low temperature.

It is thus evident from these selected examples that the exchange interaction between spin carriers constituted solely by anisotropic lanthanide ions has to be drastically enhanced to permit the observation of SCM behaviour at relatively high temperatures.

4.3.2.2 Chains Based on 3d–4f Ions

Super-exchange magnetic interactions can be enhanced by associating 4f and 3d ions. The chemical strategy to build mixed 3d–4f chains is mainly the '*tecton strategy*' as well-documented by Andruh and coworkers [75, 76]. At the early stages of 3d–4f coordination chemistry one-dimensional magnetic edifices were mainly ladder-like or tube-like compounds based on $[Cu^{II}(pba)]^{2-}$ tectons where pba = 1,3-propylenebis(oxamato) ((C) topologic mode in Scheme 4.2) [77]. Chemical variations on the ligands afforded a wide library of compounds that has been used (considering also discrete building blocks) to elaborate Kahn's fundamental theory on the newly created field of Molecular Magnetism [78]. Successively, '*diamagnetic substitution strategy*', that consists in comparing a paramagnetic compound with partially diamagnetic analogues (for example, Zn^{2+} diamagnetic 3d tectons), permitted the extraction of information on the nature of 3d–4f coupling [79].

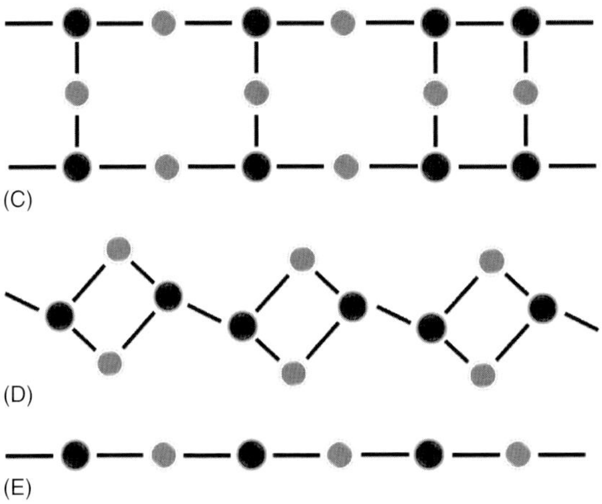

Scheme 4.2 (a–c) Schematic representation of topologic organization of 3d–4f chains. 4f and 3d ions as black and grey spheres, respectively.

A deeper analysis [80] on this Cu–Dy chain showed that NN and NNN interactions are efficient above T_c and that inter spin-ladder ferromagnetic interaction is observed in this family of compounds [79–82]. Attempts were made to use the 3d tectons with other lanthanide precursors, such as hexanuclear lanthanide clusters. Regular chains were obtained with Dy–Cu ferromagnetic interaction but their close packing prevents observation of SCM behaviour [83].

First attempts to obtain 3d–4f-based regular chains were made with Cu–Tb chains that show characteristic features of SCM but without structural characterization [84, 85]. Similarly, Co–Dy based chains were obtained but showed very weak χ''/χ' ratio making the analysis of the magnetic slow relaxation difficult [86]. Lastly, some magnetic interplay between manganese oxide chains and DyO_9 units has been evident on a molecular compound that afforded magnetic hysteresis at 2 K, but clear evidence of its molecular origin is lacking [87].

A ferrimagnetic chain of $\{Dy_2Cu_2\}$ tetramers was obtained by the ingenious use of $Dy(hfac)_3$ and [Cu(dmg)(Hdmg)] as tectons (H_2dmg = dimethylglyoxime) ((D) topologic mode in Scheme 4.2 and Figure 4.9) [88]. Onset of SCM behaviour was observed but at very low temperature. This investigation also represents the first example of the use of high frequency electron paramagnetic resonance to investigate 3d–4f couplings. Its use provided evidence of the presence of two antiferromagnetic Dy–Cu and one ferromagnetic Cu–Cu couplings. This technique, still scarcely used [89], may be useful to rationalize 3d–4f chain behaviours.

The affinity of cyanide groups for lanthanide ions has motivated the use of $[M(CN)_6]^{3-}$ tectons (with M = Cr^{3+}, Mn^{3+}, Fe^{2+}, Fe^{3+}, Co^{3+}) that can give rise to a wide variety of one-dimensional cyanide-bridged structures ((E) topologic mode in Scheme 4.2) [90]. Some noticeable compounds are $\{[Ln(DMF)_4(H_2O)_2Mn(CN)_6]\cdot H_2O\}_n$ chains (DMF, dimethylformamide), where antiferromagnetic coupling was observed between Mn^{3+} tecton and Sm^{3+}, Tb^{3+}, Dy^{3+} and Er^{3+} ions bearing DMF molecules as blocking ligands [91, 92]. In the Sm and Er derivatives, this coupling is strong enough to induce long-range magnetic ordering with critical temperature of $T_c = 18$ and 17.5 K and coercive field of $H_c = 600$ and 980 Oe, respectively [91, 92]. In the Sm derivative, if DMF is changed for DMA (DMA, *N,N'*-dimethylacetamide), T_c diminishes (3.5 K) but H_c is doubled (H_c = 1400 Oe) [93]. Other chemical variations and changes in the 4f and 3d ions afforded similar T_c and H_c values [94–99].

Several experiments have been performed in order to shed some light on magnetic coupling and magnetic anisotropy in this class of compounds. In the particular case of $\{[Ln(pzam)_3(H_2O)Mo(CN)_8]\cdot H_2O\}_n$ (Ln = Nd^{3+}, Sm^{3+}, Tb^{3+}, Er^{3+}), where pzam (pyrazine-2-carboxamide) acts as blocking ligand and the topology corresponds to type (E) in Scheme 4.2 (see also Figure 4.10), different behaviours were observed on varying the lanthanide. For the Tb analogue ferromagnetic Ln–Mo^{5+} interaction ($J = 5\,cm^{-1}$, assuming an Ising model and an effective spin 1/2 for Tb^{3+} ions) and 3D magnetic order (T_c = 1.0 K) have been evidenced from specific heat data [100].

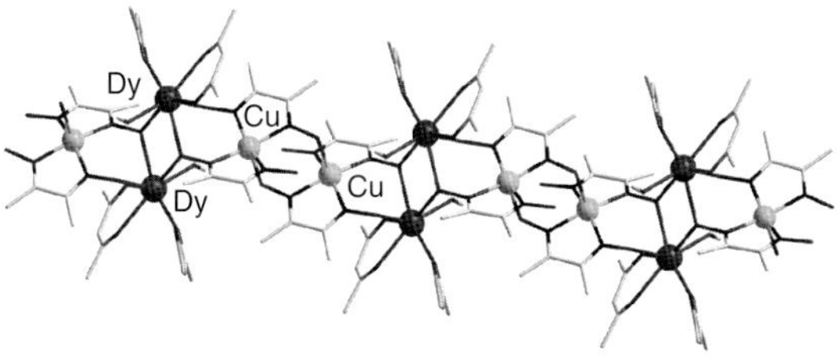

Figure 4.9 Representation of $\{[Dy(hfac)_2(CH_3OH)]_2[Cu(dmg)(Hdmg)]_2\}_n$. Dy and Cu atoms as dark and light spheres, respectively, F and H atoms omitted for clarity. (Redrawn from Ref. [88]. Published by American Chemical Society.)

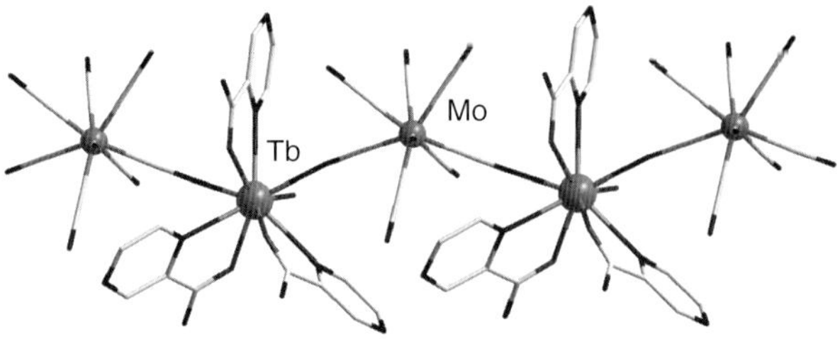

Figure 4.10 Representation of $\{[Tb(pzam)_3(H_2O)Mo(CN)_8]\cdot H_2O\}_n$. Tb and Mo atoms as dark spheres, H atoms omitted for clarity. (From Ref. [100]. Published by Wiley.)

Deeper studies on Nd, Sm and Er analogues showed changes in the lanthanide anisotropy. Antiferromagnetic interactions with XY anisotropy, that is, $J_{xy} \neq 0$ and $J_z = 0$, were observed on the Er derivatives, while intermediate anisotropy, that is, $J_{xy} \neq 0$ and $J_z \neq 0$ with $J_z < J_{xy}$, was observed for Sm and Nd analogues [101, 102]. Further rationalization using *ab initio* calculations showed ligand field influence on the overall properties of the chains and the interplay of magnetic coupling and SI anisotropy on their behaviour [103]. In the Tb analogue, when Mo^{5+} is replaced by W^{5+}, geometric changes in the chain induce a drastic change in the symmetry of the Tb^{3+} ***g*** tensor and 3d–4f magnetic interaction changes from Ising type to an anisotropic XY exchange [104].

In these types of chains, it is therefore required to diminish interchain magnetic interaction in order to observe SCM behaviour. Consequently, keeping the original 3d $[M(CN)_x]^{y-}$ cyanide building block, bulkier pyridine-based ligands have been used to synthesize $\{[Sm(tptz)(H_2O)_4Fe(CN)_6]\cdot 8H_2O\}_n$ (tptz, 2,4,6-tri(2-pyridyl)-1,3,5-triazine) and to isolate the chains in the crystal packing ((E) topologic mode in Scheme 4.2) [105]. Careful use of the '*diamagnetic substitution strategy*' has revealed a ferromagnetic $Sm^{3+}-Fe^{3+}$ interaction. Unfortunately slow relaxation observed below 3.5 K seems to be associated with spin-glass like behaviour.

The problem of the appropriate balance between lanthanide anisotropy and 3d–4f magnetic coupling to afford 3d–4f SCM has been recently addressed. Instead of associating 3d and 4f in distinct tectons, hetero-bimetallic building blocks, where the tecton is made of one lanthanide and one transition metal ion, have been proposed [75, 76, 106]. The advantage of this strategy is that 3d–4f magnetic interaction can be tuned on the discrete tecton by chemical variations. The choice of the ions can be made on these zero-dimensional entities and characterized without the drawback of studying the whole one-dimensional edifice. Once optimized, 3d–4f tectons can be associated with standard ligands or 3d′-based metallo-ligands to afford 3d–4f or 3d–3d′–4f chains, respectively.

An example of such a compound is $\{[Ni_2(H_2O)_2(valpn)_2Dy_2(tfa)_3]\cdot 4CH_3CN\}_n$ (tfa^{2-} = terephthalate and $valpn^{2-}$ = the dianion of the Schiff base resulting from reacting *o*-vanillin with 1,3-propanediamine). Ferromagnetic $Ni^{2+}-Dy^{3+}$ nodes are bridged by terephthalate ligand ((E) topologic mode in Scheme 4.3

and Figure 4.11) [107]. A ladder-like chain is obtained with an arrangement that recalls Kahn's copper based chains. The ferromagnetic interaction within the tecton is retained, organized as a chain, and magnetic slow relaxation is observed ($\Delta = 17.4\,\mathrm{K}$, $\tau_0 = 8 \times 10^{-7}\,\mathrm{s}$) for $H_{dc} = 1000\,\mathrm{Oe}$.

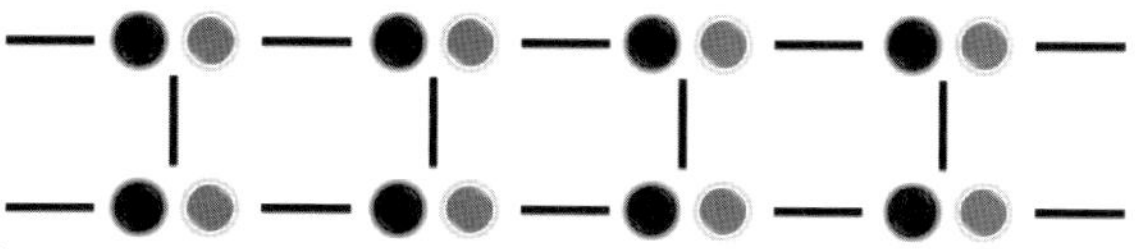

(F)

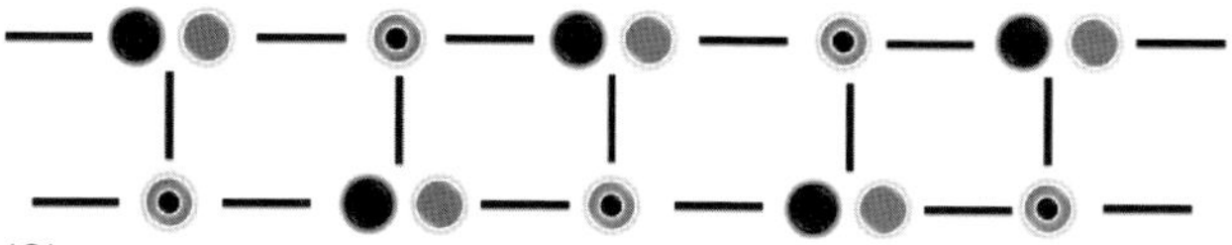

(G)

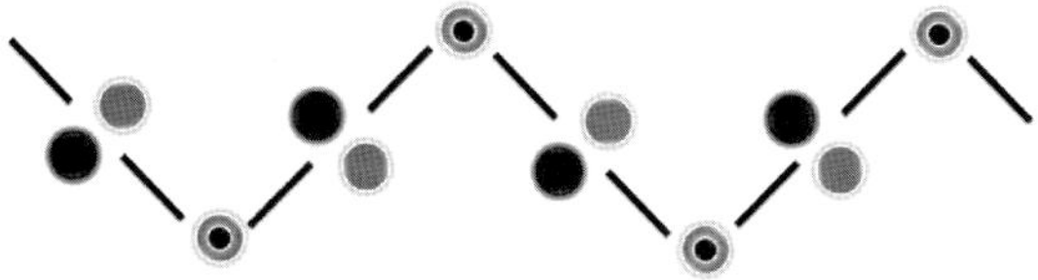

(H)

Scheme 4.3 (a–c) Schematic representation of topologic organization of 3d–4f and 3d–3d′–4f chains. 4f, 3d and 3d′ ions as black, grey and grey-dotted spheres, respectively.

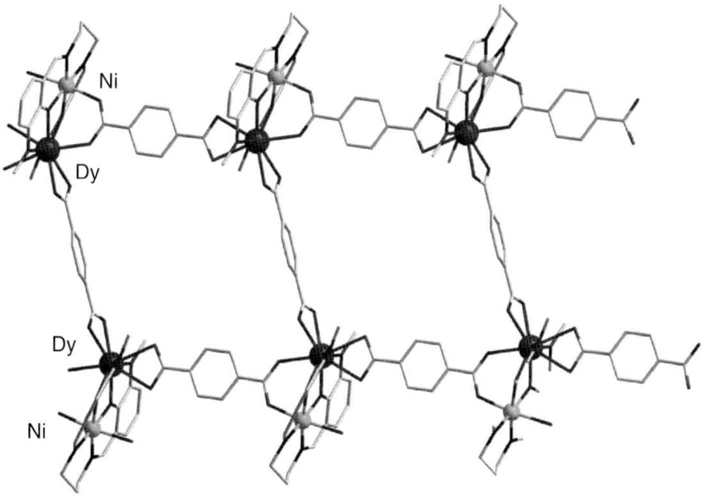

Figure 4.11 Representation of $\{[Ni_2(H_2O)_2(valpn)_2Dy_2(tfa)_3]\cdot 4CH_3CN\}_n$. Dy and Ni atoms as dark and grey spheres, respectively, H atoms omitted for clarity. (Redrawn from Ref. [107]. Published by Royal Society of Chemistry.)

If instead of terephthalate ligand a paramagnetic spacer such as $[M(CN)_x]^{3-}$ with M = Fe^{3+} [108], Mo^{5+} [109] or W^{5+} [106], is used to connect 3d–4f tectons, heterotrimetallic chains are obtained ((G) topologic mode in Scheme 4.3).

For the special case of $\{[(LCu)_2Dy][Mo(CN)_8]\cdot CH_3CN\cdot H_2O\}_n$, where L^{2-} is *N*,*N*′-propylenebis(3-methoxysalicylideneiminato) anion, slow relaxation is observed ($\Delta = 19.1$ K, $\tau_0 = 1.28 \times 10^{-7}$ s) [109] and for $\{[(LCu)_2Tb][Fe(CN)_8]\cdot 4H_2O\}_n$, where $L^{2-}$is the *N*,*N*′-propylene-bis(3-methoxysalicylideneiminato) anion (Figure 4.12), hysteresis has been observed in the millikelvin region using the microSQUID technique [108]. However, there is no definitive evidence that magnetic slow relaxation arises from SCM behaviour and not from SMM tectons.

To increase the intrachain interaction and simultaneously decrease the interchain one, $[Fe(CN)_6]^{3-}$ paramagnetic linkers can be modified to give bulkier $[(Tp^*)Fe(CN)_3]^-$ (Tp* = hydridotris(3,5-dimethylpyrazol-1-yl)borate) [110]. Infinite zig-zag chains of formula $\{[Ni(L)Ln(NO_3)_2(H_2O)Fe(Tp^*)(CN)_3]\cdot 2CH_3CN\cdot CH_3OH\}_n$ with (H_2L = *N*,*N*′-bis(3-methoxysalicylidene)-1,3-diaminopropane and Ln = Nd, Gd, Tb, Dy) were obtained ((H) topologic mode in Scheme 4.3 and Figure 4.13). For the last three derivatives $Ln^{3+}-Ni^{2+}$ ferromagnetic coupling is evident. Moreover, orthogonality of magnetic orbitals of Fe^{3+} and Ni^{2+} gives rise to relatively strong $Fe^{3+}-Ni^{2+}$ ferromagnetic coupling. All derivatives show magnetic slow relaxation with $\Delta = 40.4$ K, $\tau_0 = 4.3 \times 10^{-13}$ s for the Nd derivative, $\Delta = 55.6$ K, $\tau_0 = 4.3 \times 10^{-10}$ s for the Tb derivative and $\Delta = 58.2$ K, $\tau_0 = 4.7 \times 10^{-12}$ s for the Dy derivative. It must also be noticed that the Gd derivative relaxes slowly too, although the temperature dependence of χ'' is more reminiscent of a phase transition around 3 K, as χ'' becomes different from zero at the same temperature for all investigated frequencies.

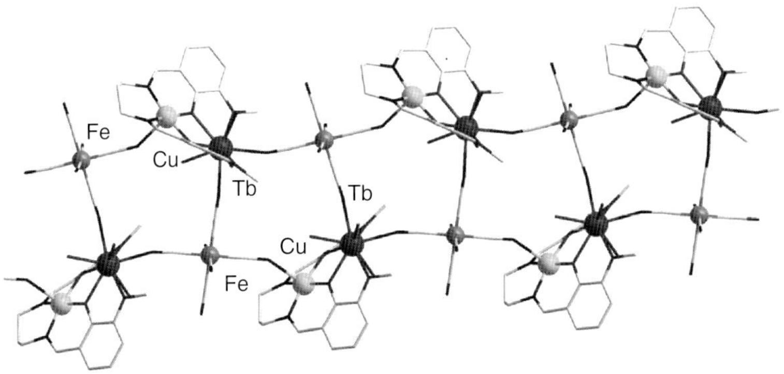

Figure 4.12 Representation of $\{[LCuTb(H_2O)_3][Fe(CN)_6]\cdot 4H_2O\}_n$. Tb, Fe and Cu atoms as black, grey and light grey spheres, respectively, H atoms omitted for clarity. (Redrawn from Ref. [108]. Published by Royal Society of Chemistry.)

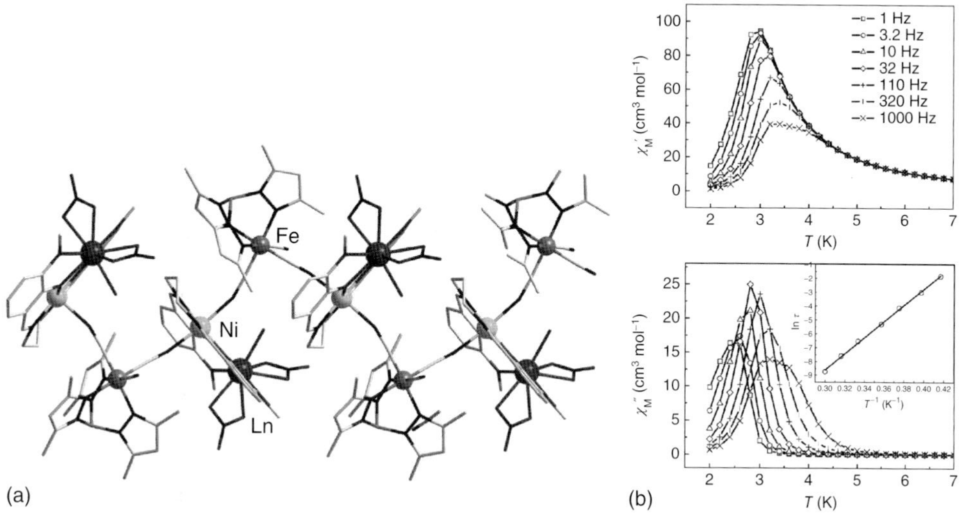

Figure 4.13 (a) Representation of $\{[Ni(L)Ln(NO_3)_2(H_2O)Fe(Tp^*)(CN)_3]\cdot 2CH_3CN\cdot CH_3OH\}_n$·Ln, Fe and Ni atoms as dark, grey and light grey spheres, respectively. H atoms omitted for clarity. (b) Temperature dependence of the in-phase (top) and out-of phase (bottom) susceptibility of the Dy^{3+} derivative with a zero applied dc field; the Arrhenius plot is shown in the inset. (From Ref. [110]. Wiley.)

4.3.2.3 **Chains Based on Radicals and 4f Ions**

A key point for the enhancement of SCM properties is to control and increase magnetic coupling along the chains in order to permit the onset of a significant correlation length at low temperature. Using the experience of Gd-radical chains, where strong 1D magnetic coupling has been demonstrated, synthesis was extended to chains comprising anisotropic lanthanide ions.

From the topological point of view, as the radicals are neutral, they can be associated in different stoichiometry with lanthanide precursors. Hence, the reaction of $Dy(hfac)_3$with NIT-R affords two different families of compounds depending on the Ln/Rad ratio: 1D chains of formula $\{Dy(hfac)_3(NIT\text{-}R)\}_n$ and discrete compounds as $[Dy(hfac)_3(NIT\text{-}R)_2]$. The geometrical environment around the Dy atoms are highly similar in these two series of compounds and consequently $[Dy(hfac)_3(NIT\text{-}R)_2]$, even if bearing two radicals, has been referred to as the *monomer* of the corresponding $\{Dy(hfac)_3(NIT\text{-}R)\}_n$ chain (topology (E) of scheme 4.2). For $R = -CH_2-CH_3$ the chain derivative gives rise to 3D ordering around 4 K [111]. However, if R is changed for bulkier C_6H_4-O-Ph ligand, interchain interactions are dramatically reduced and SCM behaviour is observed [112]. By varying the lanthanide, the anisotropy of the building block is changed and SCM behaviour is enhanced for Tb, with a large hysteresis visible at 1.55 K, and reduced for Ho, Er and Tm when compared to the Dy derivative (see Table 4.2) [113, 118]. A similar behaviour has been observed on closely related chains with magnetic ordering in some cases [114–117].

Another peculiar feature of this family of compounds is the step-like magnetization curves that cannot be explained either by finite size effects or by the intrinsic magnetic behaviour of the chain (ferro or antiferromagnetic chain of alternating spins). To shed some light on this feature, single-crystal magnetic investigations were performed on the corresponding monomer.

A chemical screening among the NIT-R library permitted the obtaining of a $[Dy(hfac)_3(NIT{-}C_6H_4{-}O{-}CH_3CH_2)_2]$ monomer crystallizing in the P-1 space group [119], a necessary condition to perform angular resolved single-crystal magnetometry [120]. Three orthogonal rotations provided a strongly anisotropic susceptibility tensor with $\chi_x T = 0.5\,\mathrm{emu\,K\,mol^{-1}}$, $\chi_y T = 3.2\,\mathrm{emu\,K\,mol^{-1}}$ and $\chi_z T = 33\,\mathrm{emu\,K\,mol^{-1}}(T = 2.5\,\mathrm{K})$. At such a temperature, one can consider that only the ground Kramer doublet of the Dy^{3+} ion is populated and thus can be treated as an effective $S = 1/2$. This affords an experimental $g_z = 18.76$ close to the value of $g = 20$, expected for a pure $M_J = \pm 15/2$, the lowest doublet of the $^6H_{15/2}$ state.

Rotating single-crystal measurements also permitted the extraction of the orientation of the magnetic tensor in the molecular reference frame and the experimental easy axis was found to coincide with the idealized tetragonal axis of the coordination dodecahedron of Dy. Crystal field calculations assuming idealized tetragonal symmetry permitted the reproduction of magnetic susceptibility data for $g_z = 19.9$ and $g_{x,y} \approx 0$ [121]. More elaborated calculations such as *ab initio* post Hartree–Fock CASSCF confirmed this simple analysis [119].

4.3.3
The Key Point of Noncollinearity of Magnetic Anisotropy

Once the anisotropy tensor and its orientation were estimated for the monomer, similar experimental and theoretical treatments were repeated on the 1D $\{Dy(hfac)_3(NIT{-}C_6H_4{-}O{-}Ph)\}_n$ analogue. *Ab initio* calculations afforded a very similar energetic scheme for Dy splitting in both the monomer and the chain. Given the orientation of the coordination dodecahedron in the chain structure, the resulting anisotropy axis is tilted from the chain direction of about 80°. This is indeed experimentally observed in very good agreement with the theoretical description [122]. As the chain structure is generated by a screw axis, adjacent Dy^{3+} ions have their easy-axes not parallel to each other but forming an angle of about 20°, as shown in Figure 4.14. The picture is further complicated as two different symmetry related chains are present in the crystal packing ($P2_12_12_1$ space group).

Table 4.2 Most relevant magnetic parameters of selected examples of lanthanide-based chains giving rise either to slow relaxation or 3D ordering.

Name	Chain type	Ln	Interaction, J (cm^{-1})	Activation energy, Δ (K) (external dc field)	Miscellaneous	References
$\{Dy(4\text{-methyl-3-nitrobenzoate})_3(H_2O)(CH_3OH)\}_n$	Pure 4f	Dy	n.a.	39.4 (5000)	—	[12]
$\{[Dy(HPA)_2(NO_3)_2]NO_3\}_n$	Pure 4f	Dy	n.a.	15.9 (4000)	—	[70]
$\{Dy(2\text{-fur})_3(HOCH_2CH_2OH)\}_n$	Pure 4f	Dy	F + AF	44 (2000)	—	[71]
$\{Dy(\alpha\text{-fur})_3\}_n$	Pure 4f	Dy	$F, J_{intra} = 0.52$ [a)] $AF, J_{inter} = -0.09$ [a)]	$B = 80$ [b)] $A = 29$ $B = 79(4000)$ $A = 38(4000)$	3D ordering at $T_N = 0.66\,K$	[73]
$\{Y_{0.9}Dy_{0.1}(\alpha\text{-fur})_3\}_n$	Pure 4f	Y-Dy	n.a.	$B = 77$ $A = 30$ $B = 80.5(4000)$ $A = 32.4(4000)$	—	[73]
$\{Dy(TDA)_{1.5}(H_2O)_2\}_n$	Pure 4f	Dy	F	44.2	—	[74]
$\{[Dy(hfac)_2(CH_3OH)]_2[Cu(dmg)(Hdmg)]_2\}_n$	3d–4f	Dy–Cu	Dy–Cu, AF, $J = -0.622$ and -0.042, Cu–Cu, $F, \approx 0.7$	—	—	[88]
$\{[Sm(DMF)_4(H_2O)_2Mn(CN)_6]\cdot H_2O\}_n$	3d–4f	Sm–Mn	Sm–Mn, AF	—	$T_c = 18$ K	[91]
$\{[Er(DMF)_4(H_2O)_2Mn(CN)_6]\cdot H_2O\}_n$	3d–4f	Er–Mn	Er–Mn, AF	—	$T_c = 17.5\,K$	[92]
$\{[Sm(DMA)_4(H_2O)_2Mn(CN)_6]\cdot H_2O\}_n$	3d–4f	Sm–Mn	Sm–Mn, AF	—	3D ordering, $T_c = 3.5\,K$	[93]

(*continued overleaf*)

Table 4.2 *(Continued)*

Name	Chain type	Ln	Interaction, J (cm^{-1})	Activation energy, Δ (K) (external dc field)	Miscellaneous	References
$\{[Nd(pzam)_3(H_2O)Mo(CN)_8]\cdot H_2O\}_n$	3d–4f	Nd–Mo	Nd–Mo, F, $J_{perp} = 1.25$[a)]	—	3D ordering	[101]
$\{[Sm(pzam)_3(H_2O)Mo(CN)_8]\cdot H_2O\}_n$	3d–4f	Sm–Mo	Sm–Mo, AF, $J_{par} = -1.81$[a)]	—	3D ordering	[102]
$\{[Tb(pzam)_3(H_2O)Mo(CN)_8]\cdot H_2O\}_n$	3d–4f	Tb–Mo	Tb–Mo, F, $J_{perp} = 2.5$[a)]	—	3D ordering, $T_c = 1.0\,K$	[100]
$\{[Er(pzam)_3(H_2O)Mo(CN)_8]\cdot H_2O\}_n$	3d–4f	Er–Mo	Er–Mo, AF, $J_{perp} = -0.7$[a)]	—	3D ordering	[101]
$\{[Sm(tptz)(H_2O)_4Fe(CN)_6]\cdot 8H_2O\}_n$	3d–4f	Sm–Fe	Sm–Fe, F	—	Spin glass, $T_g = 3.5\,K$	[105]
$\{[L_2Ni_2(H_2O)_2Dy_2(tfa)_3]\cdot 4CH_3CN\}_n$[c)]	3d–4f	Dy–Ni	F, $J_z(Ni - Dy) = 6$[a)] AF, $J_z(Dy - Dy) = -1$[a)]	17.4 (1000)	—	[107]
$\{[(LCu)_2Dy][Mo(CN)_8]\cdot CH_3CN\cdot H_2O\}_n$[c)]	3d–3d′–4f	Cu–Dy–Mo	$J(Mo - Cu1) = -8$ $J(Mo - Cu2) = 11.8$ $J(Dy - Cu) = 15.3$[a)] $J(Dy - Mo) = -8.3$[a)]	19.1	—	[109]
$\{[(LCu)_2\ Tb][Fe(CN)_8]\cdot CH_3CN\cdot H_2O\}_n$[c)]	3d–3d′–4f	Cu–Tb–Fe	Cu–Tb, F	—	Hysteresis at microsquid	[108]
$\{[Ni(L)Nd(NO_3)_2(H_2O)Fe(Tp^*)(CN)_3]\cdot 2CH_3CN\cdot CH_3OH\}_n$[c)]	3d–3d′–4f	Ni–Nd–Fe	Ni–Fe, F	40.4	—	[110]

			Ni–Nd, AF			
{[Ni(L)Gd(NO_3)$_2$ (H_2O)Fe(Tp*)(CN)$_3$]·2CH_3CN·CH_3OH}$_n$ [c)]	3d–3d′–4f	Ni–Gd–Fe	Ni–Fe, F Ni–Gd, F	11.1	—	[110
{[Ni(L)Tb(NO_3)$_2$ (H_2O)Fe(Tp*)(CN)$_3$]·2CH_3CN·CH_3OH}$_n$ [c)]	3d–3d′–4f	Ni–Tb–Fe	Ni–Fe, F Ni–Tb, F (AF inter-chain)	55.6	—	[110]*
{[Ni(L)Dy(NO_3)$_2$ (H_2O)Fe(Tp*)(CN)$_3$]·2CH_3CN·CH_3OH}$_n$ [c)]	3d–3d′–4f	Ni^+–Dy^+–Fe	Ni–Fe, F Ni–Dy, F (AF inter-chain)	58.2	—	[110]*
{|Tb(hfac)$_3$(NIT–C_6H_4–O–Ph)]}$_n$	4f-Rad	Tb	NN ferro, NNN antiferro[d)]	45	—	[113]
{|Dy(hfac)$_3$(NIT–C_6H_4–O–Ph)]}$_n$	4f-Rad	Dy	NN ferro, NNN antiferro[d)] J(Dy – Dy) = −16.7 [a)]	69 (high T) 42 (low T)	—	[112]
{|Ho(hfac)$_3$(NIT–C_6H_4–O–Ph)]}$_n$	4f-Rad	Ho	NN ferro, NNN antiferro[d)]	34 (high T) 18 (low T)	—	[113]
{|Tb(hfac)$_3$(NIT-3Brthien)]}$_n$	4f-Rad	Tb	NN ferro, NNN antiferro[d)]	56.8	3D AF ordering ($T < 4$ K)	[114]

(continued overleaf)

Table 4.2 (*Continued*)

Name	Chain type	Ln	Interaction, *J* (cm^{-1})	Activation energy, Δ (K) (external dc field)	Miscellaneous	References
$\{[Tb_3(hfac)_9(NIT\text{-}2thien)_3]\}_n$	4f-Rad	Tb	NN ferro, NNN antiferro[d)]	77.2	3D AF ordering ($T < 6$ K)	[115]
$\{[Dy(hfac)_3NitPhIm_2]Dy(hfac)_3\}_n$	4f-Rad	Dy	NN ferro, NNN antiferro[d)] .	82.7	—	[116]
$\{Tb(hfac)_3(NIT\text{-}I)\}_n$	4f-Rad	Tb	NN ferro, NNN antiferro[d)]	70.8	3D AF ordering ($T < 4$ K)	[117]

a) Calculated assuming $S_{effective} = 1/2$ and anisotropic *g* factor.
b) Two different chains (A and B) with two different Δ.
c) See cited reference for ligand (L) definition.
d) NN = nearest neighbour interaction, that is, metal radical interaction; NNN = next-nearest neighbour interaction, that is, metal–metal or radical–radical interaction.

HPA = (4-pyridylthio)aceticacid; 2-fur = 2-furoate; α-fur = α-furoate = ($\alpha\text{-}C_4H_3OCOO$); TDA = thiophene-2,5-dicarboxylate; DMF = dimethylformamide; DMA = *N*′-dimethylacetamide; Pzam = pyrazine-2-carboxamide; Tptz = 2,4,6-tri (2-pyridyl)-1,3,5-triazine; Tp* = hydridotris(3,5-dimethylpyrazol-1-yl)borate; $hfac^-$ = hexafluoroacetylacetonate; NIT-R = 2-R-4,4,5,5-tetramethyl imidazoline-3-oxide-1-oxyl; NIT-3Brthien = 2-(3′-bromo-2′-thienyl)-4,4,5,5-tetramethyl-imidazoline-1-oxyl-3-oxide; NIT-2thien = 2-(2′-thienyl)-4,4,5,5-tetramethyl-imidazoline-1-oxyl-3-oxide; NitPhIm = 2-[4-(1-imidazole)-phenyl] nitronylnitroxideradical and NIT-I = 2-iodo-4,4,5,5-tetramethylimidazoline-1-oxyl 3-oxide.

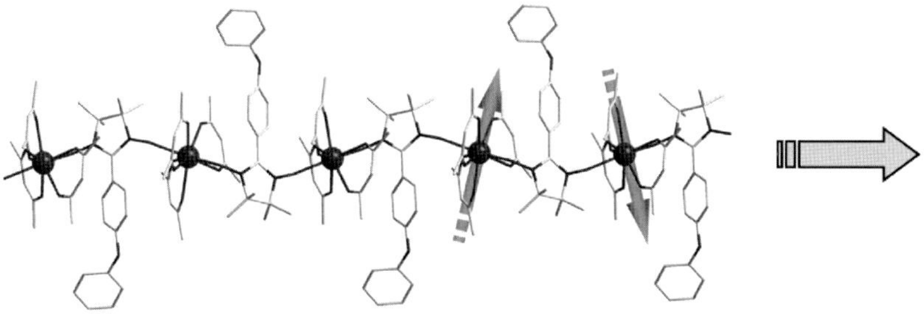
(a)

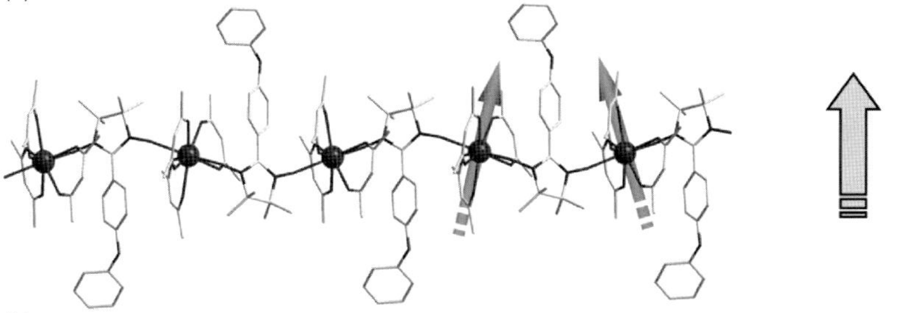
(b)

Figure 4.14 View of the structure of $\{Dy(hfac)_3(NIT\text{–}C_6H_4\text{–}O\text{–}Ph)\}_n$ chain developing along the *b* crystallographic axis. The two possible alignments with antiparallel favoured at low temperature and low magnetic field (a), and parallel favoured at high temperature and strong magnetic field (b), of neighbouring Dy magnetic moments (small grey arrows) give rise to different orientation of the uncompensated magnetic moment (big grey arrow). Dy atoms as spheres, F and H atoms omitted for clarity. (Redrawn from Ref. [122].)

Contrary to what was observed on the monomer, magnetic single-crystal rotations were found to be temperature dependent. Actually, easy- and hard-axes are inverted as the temperature changes. At low temperature the chain axis is the easy-axis but above 6 K it becomes the hard one (Figure 4.15). Such a dramatic change can be explained by considering that SI anisotropy governs the high temperature behaviour whereas the intrachain interaction governs the low temperature one. In fact at high temperature the largest susceptibility is observed in the direction perpendicular to the chain, which is closer to the easy-axis of individual Dy^{3+} ions. However, when at low temperature antiferromagnetic interactions between Dy^{3+} ions fully compensate the components perpendicular to the chain, an uncompensated moment along the chain direction is observed (Figure 4.14). The system thus adopts 1D weak ferromagnetic behaviour. Application of a moderate magnetic field (above 3 T) can overcome the antiferromagnetic interactions, and the direction perpendicular to the chain shows the largest magnetization at high field, as can be seen in the bottom of Figure 4.15. A quantitative analysis based on the transfer matrix method

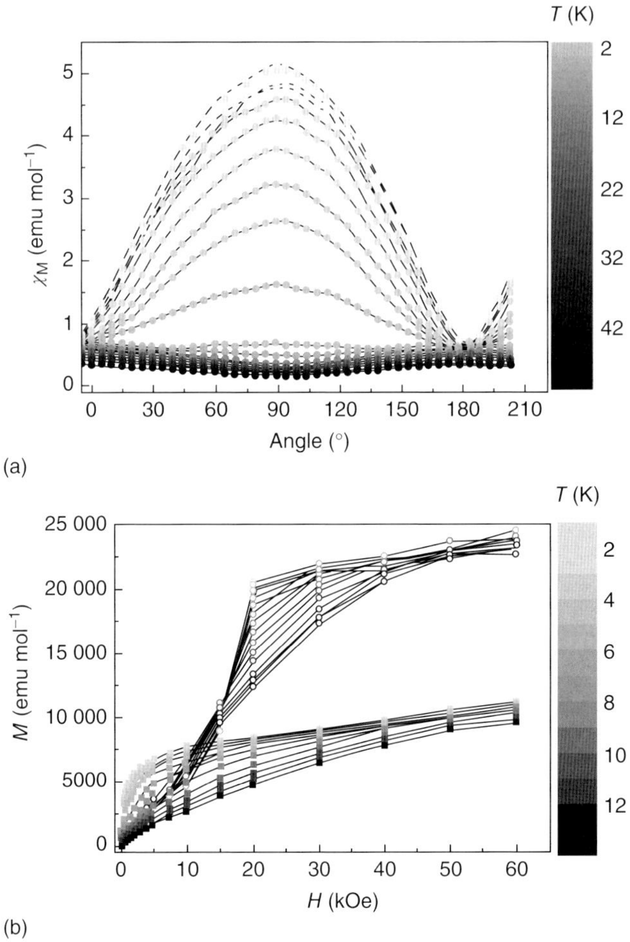

Figure 4.15 (a) Angular dependence of the susceptibility of a single crystal of $\{Dy(hfac)_3(NIT–C_6H_4–O–Ph)\}_n$ measured in an external magnetic field of 1 kOe. 0° and 180° correspond to the direction perpendicular to the chain, whereas, at 90°, the susceptibility is measured along the chain. Temperature colour mapping in the legend. (b) Field dependence of the magnetization along the chain (full squares) and perpendicular to the chain (open circles). Temperature colour mapping in the legend. (Redrawn from Ref. [122].)

permitted the extraction of the $Dy^{3+}-Dy^{3+}$ antiferromagnetic interaction that resulted in a value of $-16.7\ cm^{-1}$ (assuming the usual exchange Spin Hamiltonian and $S_{eff} = 1/2$ for Dy^{3+}) [122].

These results indicate that the $Dy^{3+}-Dy^{3+}$ interaction dominates over the Dy-radical one, despite the fact that the metal centres are separated by more than 8 Å.

This confirms the exceptional efficiency of these radicals in transmitting the magnetic interaction, already evidenced in the case of Gd-radical chains. The strategy based on paramagnetic ligands seems therefore to be the most suited for obtaining 4f based SCM.

4.4 Conclusions

The use of lanthanides in extended magnetic networks is clearly limited by the internal character of 4f orbitals and the consequent poor involvement in the magnetic exchange. This is equally relevant if either bulk molecular magnets are sought for or if the objective is the slow magnetization dynamics in 1D edifices. Nevertheless, the assumption that exchange interactions are vanishingly small is not generally valid, and we hope to have convinced the reader that in some special cases the exchange interaction between 4f ions can be significant even at a relatively large distance when mediated by magnetically non-innocent ligands like nitronyl-nitroxide radicals. However, negligible exchange interactions are not always a drawback, because the significant magnetic anisotropy associated to large magnetic moments can be exploited in the MCE with excellent cooling efficiency in the liquid helium temperature range. It should also be considered that the overwhelming spin-orbit interaction that characterizes the electronic and magnetic properties of these ions make them particularly suitable for investigation of phenomena involving other degrees of freedom, like magneto-optical or magneto-electric effects.

References

1. Willet, R.D. and Gatteschi, D. (1983) *Magneto-Structural Correlations in Exchange Coupled Systems*, Reidel Publishing, Dordrecht.
2. Delhaes, P. and Drillon, M. (1987) *Organic and Inorganic Low-Dimensioanl Crystalline Materials*, Plenum Press.
3. Ferlay, S., Mallah, T., Ouahes, R., Veillet, P. and Verdaguer, M. (1995) *Nature*, **378**, 701.
4. Manriquez, J.M., Yee, G.T., McLean, R.S., Epstein, A.J. and Miller, J.S. (1991) *Science*, **252**, 1415.
5. Steiner, M., Villain, J. and Windsor, C.G. (1976) *Adv. Phys.*, **25**, 87.
6. Villain, J. and Loveluck, J.M. (1977) *J. Phys. Lett.*, **38**, L77.
7. Benelli, C. and Gatteschi, D. (2002) *Chem. Rev.*, **102**, 2369.
8. Glauber, R.J. (1963) *J. Math. Phys.*, **4**, 294.
9. Clerac, R., Miyasaka, H., Yamashita, M. and Coulon, C. (2002) *J. Am. Chem. Soc.*, **124**, 12837.
10. Ribas, J. (2008) *Coordination Chemistry*, Wiley-VCH Verlag GmbH, Weinheim.
11. He, Z., Wang, Z.-M. and Yan, C.-H. (2005) *CrystEngComm*, **7**, 143.
12. Tian, J.M., Li, B., Zhang, X.Y., Li, X.L., Li, X.L. and Zhang, J.P. (2013) *Dalton Trans.*, **42**, 8504.

13. Xu, N., Shi, W., Liao, D.-Z., Yan, S.-P. and Cheng, P. (2008) *Inorg. Chem.*, **47**, 8748.
14. Kitagawa, S. and Uemura, K. (2005) *Chem. Soc. Rev.*, **34**, 109.
15. Fujita, M., Oguro, D., Miyazawa, M., Oka, H., Yamaguchi, K. and Ogura, K. (1995) *Nature*, **378**, 469.
16. Yaghi, O.M., O'Keeffe, M., Ockwig, N.W., Chae, H.K., Eddaoudi, M. and Kim, J. (2003) *Nature*, **423**, 705.
17. Farrusseng, D. (2011) *Metal-Organic Frameworks: Applications from Catalysis to Gas Storage*, Wiley-VCH Verlag GmbH & Co KGaA.
18. MacGillivray, L.R. (2010) *Metal-Organic Frameworks: Design and Application*, John Wiley & Sons, Inc.
19. Rowsell, J.L.C. and Yaghi, O.M. (2005) *Angew. Chem. Int. Ed.*, **44**, 4670.
20. Bunzli, J.C.G. (2006) *Acc. Chem. Res.*, **39**, 53.
21. Pecharsky, V.K. and Gschneidner, K.A. Jr., (1999) *J. Magn. Magn. Mater.*, **200**, 44.
22. McMichael, R.D., Ritter, J.J. and Shull, R.D. (1993) *J. Appl. Phys.*, **73**, 6946.
23. Evangelisti, M. and Brechin, E.K. (2010) *Dalton Trans.*, **39**, 4672.
24. Lorusso, G., Sharples, J.W., Palacios, E., Roubeau, O., Brechin, E.K., Sessoli, R., Rossin, A., Tuna, F., McInnes, E.J.L., Collison, D. and Evangelisti, M. (2013) *Adv. Mater.*, **25**, 4653.
25. Sedláková, L., Hanko, J., Orendáčová, A., Orendáč, M., Zhou, C.L., Zhu, W.H., Wang, B.W., Wang, Z.M. and Gao, S. (2009) *J. Alloys Compd.*, **487**, 425.
26. Sibille, R., Mazet, T., Malaman, B. and François, M. (2012) *Chem. Eur. J.*, **18**, 12970.
27. Lorusso, G., Palacios, M.A., Nichol, G.S., Brechin, E.K., Roubeau, O. and Evangelisti, M. (2012) *Chem. Commun.*, **48**, 7592.
28. Tishin, A.M. and Spichkin, Y.I. (2003) *The Magnetocaloric Effect and its Applications*, Taylor & Francis Group, London.
29. Lorusso, G., Jenkins, M., González-Monje, P., Arauzo, A., Sesé, J., Ruiz-Molina, D., Roubeau, O. and Evangelisti, M. (2013) *Adv. Mater.*, **25**, 2984.
30. Carretta, S., Santini, P., Amoretti, G., Affronte, M., Candini, A., Ghirri, A., Tidmarsh, I.S., Laye, R.H., Shaw, R. and McInnes, E.J.L. (2006) *Phys. Rev. Lett.*, **97**, 207201.
31. Chen, L., Ramsey, C.M., Dalal, N.S., Ren, T., Cotton, F.A., Wernsdorfer, W. and Chiorescu, I. (2006) *Appl. Phys. Lett.*, **89**, 252502.
32. Chiorescu, I., Wernsdorfer, W., Müller, A., Bögge, H. and Barbara, B. (2000) *Phys. Rev. Lett.*, **84**, 3454.
33. Van Vleck, J.H. (1941) *Phys. Rev.*, **59**, 724.
34. Huber, D.L. (1966) *Phys. Rev.*, **142**, 137.
35. Rossin, A., Giambastiani, G., Peruzzini, M. and Sessoli, R. (2012) *Inorg. Chem.*, **51**, 6962.
36. Kahn, O., Pei, Y., Verdaguer, M., Renard, J.P. and Sletten, J. (1988) *J. Am. Chem. Soc.*, **110**, 782.
37. Bencini, A., Benelli, C., Caneschi, A., Carlin, R.L., Dei, A. and Gatteschi, D. (1985) *J. Am. Chem. Soc.*, **107**, 8128.
38. Benelli, C., Caneschi, A., Gatteschi, D., Laugier, J. and Rey, P. (1987) *Angew. Chem. Int. Ed. Engl*, **26**, 913.
39. Goodenough, J.B. (1958) *J. Phys. Chem. Solid.*, **6**, 287.
40. Kanamori, J. (1963) in *Magnetism*, vol. 1 (eds G.T. Rado and H. Suhl), Academic Press, New York.
41. Paulovič, J., Cimpoesu, F., Ferbinteanu, M. and Hirao, K. (2004) *J. Am. Chem. Soc.*, **126**, 3321.
42. Rajaraman, G., Totti, F., Bencini, A., Caneschi, A., Sessoli, R. and Gatteschi, D. (2009) *Dalton Trans.*, 3153.
43. Champion, G., Lalioti, N., Tangoulis, V., Arrio, M.A., Sainctavit, P., Villain, F., Caneschi, A., Gatteschi, D., Giorgetti, C., Baudelet, F., Verdaguer, M. and Moulin, C.C.D. (2003) *J. Am. Chem. Soc.*, **125**, 8371.
44. Lescop, C., Belorizky, E., Luneau, D. and Rey, P. (2002) *Inorg. Chem.*, **41**, 3375.

45. Caneschi, A., Dei, A., Gatteschi, D., Sorace, L. and Vostrikova, K. (2000) *Angew. Chem. Int. Ed.*, **39**, 246.
46. Guillou, O., Bergerat, P., Kahn, O., Bakalbassis, E., Boubekeur, K., Batail, P. and Guillot, M. (1992) *Inorg. Chem.*, **31**, 110.
47. Gatteschi, D. and Rey, P. (1999) *Design, Synthesis, and Properties of Nitroxide Networked Materials in Magnetic Properties of Organic Materials*, (ed. P. M. Lahti), Marcell Dekker, New York.
48. Benelli, C., Caneschi, A., Gatteschi, D., Pardi, L. and Rey, P. (1989) *Inorg. Chem.*, **28**, 275.
49. Benelli, C., Caneschi, A., Gatteschi, D., Pardi, L. and Rey, P. (1990) *Inorg. Chem.*, **29**, 4223.
50. Bartolome, F., Bartolome, J., Benelli, C., Caneschi, A., Gatteschi, D., Paulsen, C., Pini, M.G., Rettori, A., Sessoli, R. and Volokitin, Y. (1996) *Phys. Rev. Lett.*, **77**, 382.
51. Benelli, C., Caneschi, A., Gatteschi, D., Pardi, L. and Rey, P. (1989) *Inorg. Chem.*, **28**, 3230.
52. Benelli, C., Caneschi, A., Gatteschi, D. and Pardi, L. (1990) *J. Appl. Phys.*, **67**, 5613.
53. Cinti, F., Rettori, A., Pini, M.G., Mariani, M., Micotti, E., Lascialfari, A., Papinutto, N., Amato, A., Caneschi, A., Gatteschi, D. and Affronte, M. (2008) *Phys. Rev. Lett.*, **100**, 057203.
54. Barron, L.D. and Vrbancich, J. (1984) *Mol. Phys.*, **51**, 715.
55. Rikken, G.L.J.A. and Raupach, E. (1997) *Nature*, **390**, 493.
56. Sessoli, R., Boulon, M.-E., Caneschi, A., Mannini, M., Poggini, L., Wilhelm, F. and Rogalev, A. (2014), *Nature Phys.*, in press, doi:10.1038/nphys3152.
57. Bogani, L., Cavigli, L., Bernot, K., Sessoli, R., Gurioli, M. and Gatteschi, D. (2006) *J. Mater. Chem.*, **16**, 2587.
58. Caneschi, A., Gatteschi, D., Lalioti, N., Sangregorio, C., Sessoli, R., Venturi, G., Vindigni, A., Rettori, A., Pini, M.G. and Novak, M.A. (2001) *Angew. Chem. Int. Ed.*, **40**, 1760.
59. Coulon, C., Miyasaka, H. and Clérac, R. (2006) in *Single-Molecule Magnets and Related Phenomena*, vol. 122 (ed. R. Winpenny), Springer, Berlin, Heidelberg, p. 163.
60. Lescouëzec, R., Toma, L.M., Vaissermann, J., Verdaguer, M., Delgado, F.S., Ruiz-Pérez, C., Lloret, F. and Julve, M. (2005) *Coord. Chem. Rev.*, **249**, 2691.
61. Sun, H.L., Wang, Z.M. and Gao, S. (2010) *Coord. Chem. Rev.*, **254**, 1081.
62. Zhang, W.-X., Ishikawa, R., Breedlove, B. and Yamashita, M. (2013) *RSC Adv.*, **3**, 3772.
63. Bogani, L., Vindigni, A., Sessoli, R. and Gatteschi, D. (2008) *J. Mater. Chem.*, **18**, 4750.
64. Billoni, O.V., Pianet, V., Pescia, D. and Vindigni, A. (2011) *Phys. Rev. B*, **84**, 064415.
65. Bogani, L., Caneschi, A., Fedi, M., Gatteschi, D., Massi, M., Novak, M.A., Pini, M.G., Rettori, A., Sessoli, R. and Vindigni, A. (2004) *Phys. Rev. Lett.*, **92**, 207204.
66. Luo, Y.-D., Sun, G.-M., Li, D.-M. and Luo, F. (2011) *Inorg. Chem. Commun.*, **14**, 778.
67. Zhu, Y., Luo, F., Song, Y.M., Huang, H.X., Sun, G.M., Tian, X.Z., Yuan, Z.Z., Liao, Z.W., Luo, M.B., Liu, S.J., Xu, W.Y. and Feng, X.F. (2012) *Dalton Trans.*, **41**, 6749.
68. Liu, S.-J., Zhao, J.-P., Song, W.-C., Han, S.-D., Liu, Z.-Y. and Bu, X.-H. (2013) *Inorg. Chem.*, **52**, 2103.
69. Miyasaka, H. and Yamashita, M. (2007) *Dalton Trans.*, 399.
70. Na, B., Wang, Y.-X., Han, T., Shi, W. and Cheng, P. (2013) *Inorg. Chem. Commun.*, **35**, 19.
71. Wang, Y., Li, X.-L., Wang, T.-W., Song, Y. and You, X.-Z. (2009) *Inorg. Chem.*, **49**, 969.
72. Coulon, C., Clérac, R., Wernsdorfer, W., Colin, T., Saitoh, A., Motokawa, N. and Miyasaka, H. (2007) *Phys. Rev. B*, **76**, 214422.
73. Bartolomé, E., Bartolome, J., Melnic, S., Prodius, D., Shova, S., Arauzo, A.,

Luzon, J., Luis, F. and Turta, C. (2013) *Dalton Trans.*, **42**, 10153.
74. Chen, Z., Zhao, B., Cheng, P., Zhao, X.Q., Shi, W. and Song, Y. (2009) *Inorg. Chem.*, **48**, 3493.
75. Andruh, M., Costes, J.-P., Diaz, C. and Gao, S. (2009) *Inorg. Chem.*, **48**, 3342.
76. Andruh, M. (2007) *Chem. Commun.*, 2565.
77. Guillou, O., Kahn, O., Oushoorn, R.L., Boubekeur, K. and Batail, P. (1992) *Inorg. Chim. Acta*, **198–200**, 119.
78. Kahn, O. (1993) *Molecular Magnetism*, Wiley-VCH Verlag GmbH, Weinheim.
79. Kahn, M.L., Mathoniere, C. and Kahn, O. (1999) *Inorg. Chem.*, **38**, 3692.
80. Evangelisti, M., Kahn, M.L., Bartolome, J., de Jongh, L.J., Meyers, C., Leandri, J., Leroyer, Y. and Mathoniere, C. (2003) *Phys. Rev. B*, **68**, 184405.
81. Kahn, M.L., Lecante, P., Verelst, M., Mathoniere, C. and Kahn, O. (2000) *Chem. Mater.*, **12**, 3073.
82. Kahn, M.L., Verelst, M., Lecante, P., Mathoniere, C. and Kahn, O. (1999) *Eur. J. Inorg. Chem.*, **1999**, 527.
83. Calvez, G., Bernot, K., Guillou, O., Daiguebonne, C., Caneschi, A. and Mahé, N. (2008) *Inorg. Chim. Acta*, **361**, 3997.
84. Costes, J.-P., Clemente-Juan, J.M., Dahan, F. and Milon, J. (2004) *Inorg. Chem.*, **43**, 8200.
85. Costes, J.P., Vendier, L. and Wernsdorfer, W. (2010) *Dalton Trans.*, **39**, 4886.
86. Huang, Y.G., Wang, X.T., Jiang, F.L., Gao, S., Wu, M.Y., Gao, Q., Wei, W. and Hong, M.C. (2008) *Chem. Eur. J.*, **14**, 10340.
87. West, J.P., Hwu, S.J. and Queen, W.L. (2009) *Inorg. Chem.*, **48**, 8439.
88. Okazawa, A., Nogami, T., Nojiri, H. and Ishida, T. (2008) *Chem. Mater.*, **20**, 3110.
89. Okazawa, A., Fujiwara, K., Watanabe, R., Kojima, N., Yoshii, S., Nojiri, H. and Ishida, T. (2011) *Polyhedron*, **30**, 3121.
90. Tanase, S. and Reedijk, J. (2006) *Coord. Chem. Rev.*, **250**, 2501.
91. Yan, B., Chen, Z.D., Wang, S.X. and Gao, S. (2001) *Chem. Lett.*, 350.
92. Yan, B., Chen, Z.D. and Wang, S.X. (2001) *Transition Met. Chem.*, **26**, 287.
93. Yan, B. and Chen, Z. (2001) *Helv. Chim. Acta*, **84**, 817.
94. Figuerola, A., Ribas, J., Casanova, D., Maestro, M., Alvarez, S. and Diaz, C. (2005) *Inorg. Chem.*, **44**, 6949.
95. Figuerola, A., Diaz, C., Ribas, J., Tangoulis, V., Sangregorio, C., Gatteschi, D., Maestro, M. and Mahia, J. (2003) *Inorg. Chem.*, **42**, 5274.
96. Kou, H.Z., Gao, S. and Jin, X.L. (2001) *Inorg. Chem.*, **40**, 6295.
97. Yan, B. and Chen, Z.D. (2001) *Prog. Nat. Sci.*, **11**, 401.
98. Estrader, M., Ribas, J., Tangoulis, V., Solans, X., Font-Bardia, M., Maestro, M. and Diaz, C. (2006) *Inorg. Chem.*, **45**, 8239.
99. Guo, Y., Xu, G.F., Wang, C., Cao, T.T., Tang, J.K., Liu, Z.Q., Ma, Y., Yan, S.P., Cheng, P. and Liao, D.Z. (2012) *Dalton Trans.*, **41**, 1624.
100. Prins, F., Pasca, E., de Jongh, L.J., Kooijman, H., Spek, A.L. and Tanase, S. (2007) *Angew. Chem. Int. Ed.*, **46**, 6081.
101. Tanase, S., Evangelisti, M., de Jongh, L.J., Smits, J.M.M. and de Gelder, R. (2008) *Inorg. Chim. Acta*, **361**, 3548.
102. Tanase, S., Evangelisti, M. and de Jongh, L.J. (2011) *Dalton Trans.*, **40**, 8407.
103. Tanase, S., Ferbinteanu, M. and Cimpoesu, F. (2011) *Inorg. Chem.*, **50**, 9678.
104. Pasca, E., Roscilde, T., Evangelisti, M., Burzuri, E., Luis, F., de Jongh, L.J. and Tanase, S. (2012) *Phys. Rev. B*, **85**, 184434.
105. Zhao, H.H., Lopez, N., Prosvirin, A., Chifotides, H.T. and Dunbar, K.R. (2007) *Dalton Trans.*, 878.
106. Andruh, M. (2011) *Chem. Commun.*, **47**, 3025.
107. Pasatoiu, T.D., Etienne, M., Madalan, A.M., Andruh, M. and Sessoli, R. (2010) *Dalton Trans.*, **39**, 4802.
108. Gheorghe, R., Madalan, A.M., Costes, J.P., Wernsdorfer, W. and Andruh, M. (2010) *Dalton Trans.*, **39**, 4734.
109. Visinescu, D., Madalan, A.M., Andruh, M., Duhayon, C., Sutter, J.P., Ungur, L.,

Van den Heuvel, W. and Chibotaru, L.F. (2009) *Chem. Eur. J.*, **15**, 11808.

110. Yao, M.-X., Zheng, Q., Qian, K., Song, Y., Gao, S. and Zuo, J.-L. (2013) *Chem. Eur. J.*, **19**, 294.
111. Benelli, C., Gatteschi, D., Sessoli, R., Rettori, A., Pini, M.G., Bartolomé, F. and Bartolomé, J. (1995) *J. Magn. Magn. Mater.*, **140–144** (Pt. 3), 1649.
112. Bogani, L., Sangregorio, C., Sessoli, R. and Gatteschi, D. (2005) *Angew. Chem. Int. Ed.*, **44**, 5817.
113. Bernot, K., Bogani, L., Caneschi, A., Gatteschi, D. and Sessoli, R. (2006) *J. Am. Chem. Soc.*, **128**, 7947.
114. Liu, R.N., Zhang, C.M., Mei, X.L., Hu, P., Tian, H.X., Li, L.C., Liao, D.Z. and Sutter, J.P. (2012) *New J. Chem.*, **36**, 2088.
115. Liu, R.N., Ma, Y., Yang, P.P., Song, X.Y., Xu, G.F., Tang, J.K., Li, L.C., Liao, D.Z. and Yan, S.P. (2010) *Dalton Trans.*, **39**, 3321.
116. Liu, R.N., Li, L.C., Wang, X.L., Yang, P.P., Wang, C., Liao, D.Z. and Sutter, J.P. (2010) *Chem. Commun.*, **46**, 2566.
117. Tian, H., Wang, X., Mei, X., Liu, R., Zhu, M., Zhang, C., Ma, Y., Li, L. and Liao, D. (2013) *Eur. J. Inorg. Chem.*, **2013**, 1320.
118. Bernot, K., Bogani, L., Sessoli, R. and Gatteschi, D. (2007) *Inorg. Chim. Acta*, **360**, 3807.
119. Bernot, K., Luzon, J., Bogani, L., Etienne, M., Sangregorio, C., Shanmugam, M., Caneschi, A., Sessoli, R. and Gatteschi, D. (2009) *J. Am. Chem. Soc.*, **131**, 5573.
120. Luzon, J. and Sessoli, R. (2012) *Dalton Trans.*, **41**, 13556.
121. Sorace, L., Benelli, C. and Gatteschi, D. (2011) *Chem. Soc. Rev.*, **40**, 3092.
122. Bernot, K., Luzon, J., Caneschi, A., Gatteschi, D., Sessoli, R., Bogani, L., Vindigni, A., Rettori, A. and Pini, M.G. (2009) *Phys. Rev. B*, **79**, 134419.

5
Experimental Aspects of Lanthanide Single-Molecule Magnet Physics

Kasper S. Pedersen, Daniel N. Woodruff, Jesper Bendix, and Rodolphe Clérac

5.1
Introduction

The growing popularity of lanthanides in molecular magnetism has given rise to new difficulties that must be overcome for the successful determination of the relevant physical quantities necessary to describe single-molecule magnet (SMM) behaviour. For transition metal compounds, the magnetization (typically experimentally obtained as a function of temperature and magnetic field) facilitates the estimation of chemically intuitive parameters, for example, magnetic super-exchange coupling constants and magnetic anisotropy parameters. The easy availability of this information has led to several studies of correlations between chemical structures and magnetic properties. In turn, this work initiated fundamental research in tuning, for instance, 3d metal ion-based SMMs. For lanthanide ions, a similar interest surely exists, but several problems arise before a similar level of knowledge can be achieved. First of all, the intrinsic lability of lanthanide ions, combined with the lack of preference for specific coordination geometry and hence the difficulty of accurately predicting coordination geometry, creates a major obstacle for controlling and tuning the synthesis of 4f SMMs [1]. Secondly, and also relevant to this chapter, lanthanide ion magnetism is much harder to model theoretically due to the many ligand (crystal) field parameters typically required in the usual low-symmetry environments adopted by a 4f ion [2]. Therefore, thermodynamically averaged measurements on powdered or polycrystalline samples are not sufficient to provide a detailed understanding of these systems, hence they have to be supplemented by spectroscopic measurements or dedicated single-crystal studies to gain a full insight into the origin of the magnetic anisotropy and SMM behaviour. For 3d ions, the comparable ligand field (LF) and electronic repulsion energies are much stronger than the spin–orbit (SO) coupling energies. For 4f ions, spherical effects (interelectronic repulsion and SO coupling) strongly dominate over LF effects, which, due to the shielded nature of the 4f

Lanthanides and Actinides in Molecular Magnetism, First Edition.
Edited by Richard A. Layfield and Muralee Murugesu.

orbitals, are much smaller than for 3d ions. The result is that lanthanide ions may be treated as free ions in a first approximation, with a small perturbation originating from the nonspherical surroundings experienced in any chemical compound.

Following the seminal discovery of the SMM behaviour in mononuclear $[Ln^{III}(Pc)_2]^-$ complexes (Ln = Tb, Dy, pc = phthalocyaninate dianion) [3], several in-depth studies have been reported to explain their peculiar properties. Subsequently the scientific interest in mono- and poly-nuclear lanthanide SMMs expanded drastically. It should, however, be stressed that the study of the electronic structures of lanthanide ion systems is by no means a new discipline nor confined to SMM systems. In fact, the static and dynamic properties of lanthanide ions have been studied since the early days of electron paramagnetic resonance (EPR) spectroscopy [4]. The fast-growing number of published lanthanide-based SMMs has reinvigorated the interest in determining and understanding the microscopic interactions pertinent to lanthanide complexes. Unfortunately, the complexity of most lanthanide-based SMMs strongly hampers the investigation of the microscopic origin of the slow relaxation of the magnetization hence various theoretical approaches, including *ab initio* calculations [5] and crystal field approaches by using effective charge densities, have recently been developed [6]. Alternatively, methods for magneto-structural correlations considering angular parts of the wavefunctions were put forward [6a]. The performance of these approaches is typically judged by comparison with experimental data collected on powdered samples, for example, the temperature dependence of the magnetization. The most obvious way of comparing the theoretical results with experiment involves *direct* observation of magnetic anisotropy, LF splittings and exchange interactions. The experimental determination of this information is a nontrivial task due to the low-symmetry of most lanthanide systems, coupled with the energy window of about $10-1000\,cm^{-1}$ which is challenging to access by spectroscopic means. In addition, the magnetic exchange interactions in polynuclear systems are typically very weak.

In this chapter, we give an outline of the main available experimental techniques that may be employed to mitigate the gap and provide fundamental experimental information. As this chapter is written by chemists for chemists, we intend to provide an overview of the 'tool box' used to characterize lanthanide-based SMMs instead of an exhaustive review of the vast literature. Although we concentrate mainly on selected SMM examples, most of the techniques are completely general and can be applied to any lanthanide system of interest. Notably, the detailed understanding of the electronic structure is still central for future applications of tailored lanthanide systems where the slow relaxation of the magnetization may be less important, for instance for quantum computing applications. As several, excellent tutorial reviews or monographs are available on the topic of physical characterization of magnetic molecules and lanthanide ions, we have restrained the discussion to encompass more explanatory examples and highlight the lessons learned from the physical characterization. In particular, we focus on the knowledge typically sought by chemists, that is, the determination of the 'anisotropy barrier', from thermodynamic and/or spectroscopic measurements, and the magnetic exchange couplings.

5.2 Manifestation of Single-Molecule Magnet Behaviour

5.2.1 Magnetization and ac Susceptibility Measurements

Let's consider an SMM, which is magnetized to a magnetization, M_0, with a polarizing magnetic field, H. After (fast) removal of the field, the magnetization will decay exponentially with a single characteristic relaxation time (τ):

$$M(t) = M_0 \exp\left(-\frac{t}{\tau}\right) \tag{5.1}$$

In real systems, a distribution in the characteristic time may lead to a stretched exponential decay. In the thermally activated regime where the relaxation of the magnetization is due to the Orbach mechanism, the temperature dependence of the relaxation time may be described by an Arrhenius law of the form:

$$\tau(T) = \tau_0 \exp\left(\frac{\Delta_{\text{eff}}}{k_B T}\right) \tag{5.2}$$

Here, Δ_{eff} is the effective energy barrier for spin-reversal and the preexponential factor (τ_0) takes into account the interactions of the spin with the phonon bath. Commonly, Δ_{eff} and τ_0 are the characteristic parameters reported for an SMM. Despite the simple formulae, it is important to note that the physics behind the magnetization relaxation in lanthanide systems is nontrivial and several barrier-unrelated mechanisms, including Raman and direct processes, may be operative. Most commonly, the dynamic magnetic properties of lanthanide SMMs are characterized in powdered or polycrystalline form using conventional superconducting quantum interference device (SQUID) magnetometers. As the dynamics in most SMMs are typically fast compared to the field ramping and measurement time scales in commercially available magnetometers, the dynamics are more conveniently studied in small oscillating magnetic fields, for instance, ac susceptibility measurements, than by M versus H hysteresis and magnetization decay (*M vs time*) measurements. Application of a static magnetic field during these ac measurements can reduce the probability of quantum tunnelling of the magnetization (QTM) and in cases where no SMM behaviour is observed in the absence of a static field, SMM behaviour can be revealed by the presence of the static field.

In the ac susceptibility experiments, the time dependence of the polarizing field is $H_{ac}\cos(\omega t) + H_{dc}$, where ω is the angular frequency ($\omega = 2\pi\nu$, ν being the ac frequency) and H_{ac} is the amplitude of the ac magnetic field. In the following, we omit the static component, H_{dc}, as most of the measurements are performed in zero dc field. Below a certain temperature and above a certain frequency, the local magnetic moment of the magnetic molecule cannot follow the external ac field and the phase-shift manifests itself in the magnetization as $M(t) = M_{ac}\cos(\omega t - \varphi) = M_{ac}\{\cos(\varphi)\cos(\omega t) + \sin(\varphi)\sin(\omega t)\}$. The time dependent magnetization can be rewritten as $M(t) = H_{ac}(\chi'\cos(\omega t) + \chi''\sin(\omega t))$ introducing the in-phase (χ') and out-of-phase (χ'') components of the total ac susceptibility ($\chi = \chi' + i\chi''$), which can be expressed as $\chi' = (M_{ac}/H_{ac})\cos(\varphi)$ and $\chi'' = (M_{ac}/H_{ac})\sin(\varphi)$ respectively. As the phase shift depends on the applied ac frequency, the ac susceptibility also varies as a function of ω. In the limits, as $\omega \to 0$ and $\omega \to \infty$, the ac susceptibility is called the *isothermal and adiabatic susceptibility* noted χ_T and χ_S, respectively (it should be noted that the isothermal susceptibility is directly comparable to the susceptibility measured in dc fields). Using these two limit susceptibilities and in the case of the spin-lattice relaxation, the ac susceptibility can be rewritten as [7]:

$$\chi = \chi_S + \frac{\chi_T - \chi_S}{1 + i\omega\tau} \tag{5.3}$$

From $\chi = \chi' + i\chi''$, it is straightforward to obtain the expressions of the real and imaginary components of the ac susceptibility: [7]

$$\chi' = \chi_S + \frac{\chi_T - \chi_S}{1 + \omega^2\tau^2} \quad \chi'' = \frac{(\chi_T - \chi_S)\omega\tau}{1 + \omega^2\tau^2} \tag{5.4}$$

For $\omega\tau = 1$, χ' and χ'' display, respectively, an inflection point and maximum reaching values of $(\chi_T + \chi_S)/2$ and $(\chi_T - \chi_S)/2 = \max[\chi'']$. Therefore, the temperature dependence of the relaxation time, $\tau(T)$, is conveniently obtained from ac susceptibility measurements when $\tau = \omega(\max[\chi''])^{-1}$ (in this formula, $\omega(\max[\chi''])$ designates the angular frequency of the χ'' peak maxima). It is worth noting that the $\tau = \omega(\max[\chi''])^{-1}$ relation is always valid when χ'' is measured as a function of the ac frequency at fixed temperatures as χ_T and χ_S are constants. On the other hand, if χ'' is measured at fixed ac frequency as a function of the temperature, the $\tau = \omega(\max[\chi''(T)])^{-1}$ equality is not preserved when χ_T and χ_S are temperature dependent.

In real systems, there will always be a distribution of the relaxation time meaning that Equation 5.3 should more adequately be written as given in Ref. [8] (generalized Debye model):

$$\chi = \chi_S + \frac{\chi_T - \chi_S}{1 + (i\omega\tau)^{1-\alpha}} \tag{5.5}$$

The phenomenological parameter, α, is typically estimated from fitting of Cole–Cole plots (χ'' vs χ' for a fixed temperature) [9]. Small values of α are expected for SMMs having, ideally, one characteristic time. Larger values of α

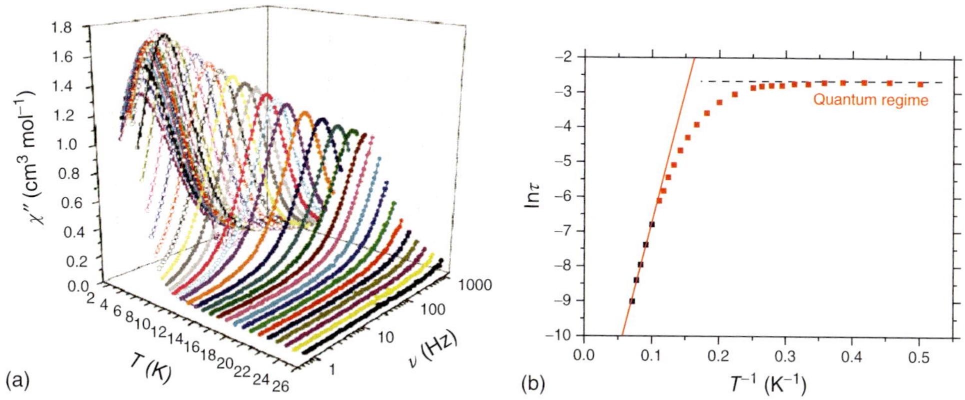

Figure 5.1 (a) Out-of-phase component of the dynamic (ac) susceptibility, χ'', versus frequency, ν_{ac} and temperature, T, for $[Dy_2(valdien)_2(NO_3)_2]$ in zero dc field. (b) Temperature dependence of the magnetization relaxation time plotted as $\ln(\tau)$ versus T^{-1} (Arrhenius plot). (Reprinted with permission from [11]. Copyright 2011, American Chemical Society.)

could be indicative of the presence and/or overlap of a multitude of relaxation processes. For a perspective on the modelling of ac susceptibility data for lanthanide SMMs, the reader is referred to the recent work of Tang and coworkers [10].

A typical thermal variation of the relaxation behaviour is shown in Figure 5.1 for $[Dy_2(valdien)_2(NO_3)_2]$ (H_2valdien = $N1,N3$-bis(3-methoxysalicylidene)diethylenetriamine) [11]. At the highest temperatures, the $\tau(T)$ function is approximately linear as expected for a thermally activated process described by an Arrhenius law (Equation 5.2). At the lower temperatures, the $\ln(\tau)$ vs T^{-1} data deviates from linearity eventually becoming temperature independent suggesting a relaxation processes involving a mixture of the two components of the ground state (i.e. Quantum Tunnelling of the Magnetization, QTM). This observation is common for both non-Kramers and Kramers ions, although, for the latter, QTM is forbidden by the Kramers theorem stating that for half-integer spin systems the degeneracy is always preserved (in the absence of a magnetic field). However, the hyperfine coupling of electronic spin to nuclear spins or dipolar coupling between molecules may be responsible for the apparent pronounced QTM even for Kramers ions.

Moreover, it is not unusual for a lanthanide SMM to have more than one thermally activated relaxation mechanism. This is particularly true for some polynuclear systems that can be considered as consisting of several, weakly interacting, mononuclear SMMs. In a recent communication, Powell and coworkers [12] reported on a $\{Co^{II}{}_2Dy^{III}{}_2\}$ SMM exhibiting interesting relaxation dynamics. The structure of the compound is shown in Figure 5.2a and consists of a di-hydroxide-bridged Co^{II} dinuclear unit decorated with two Dy^{3+} ions. The

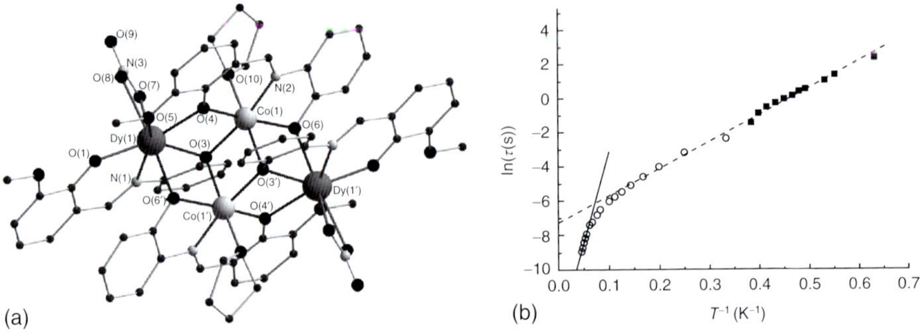

Figure 5.2 Molecular structure (a) and Arrhenius plot (b) of a $\{Co^{II}{}_2Dy^{III}{}_2\}$ SMM. The best-fits of the two Arrhenius regimes are shown as solid and dashed lines. (Reprinted with permission from [12]. Copyright 2012, Wiley.)

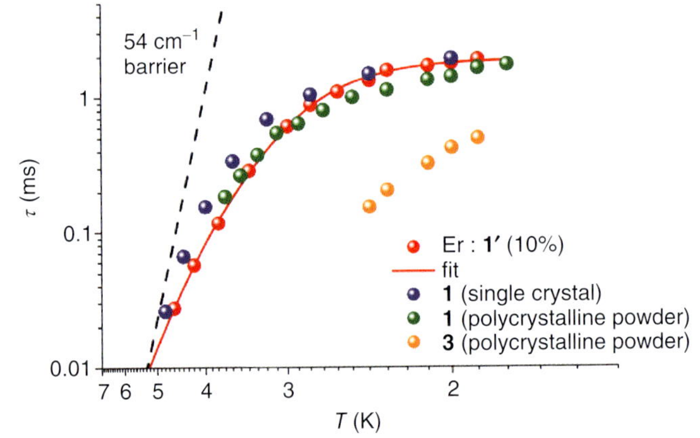

Figure 5.3 Arrhenius plot extracted from ac susceptibility data of Er(trensal) (indicated as '**1**'. '**1**′' designates the isostructural, but diamagnetic analogue Y(trensal)). (Reprinted with permission from [13]. Copyright 2014, Royal Chemical Society.)

Arrhenius plot (cf. Figure 5.2b) features two linear regions, which can be fitted to yield $\Delta_1 = 11.0\ \text{cm}^{-1}$ ($\tau_0 = 7.7 \times 10^{-4}$ s) and $\Delta_2 = 82.1\ \text{cm}^{-1}$ ($\tau_0 = 6.2 \times 10^{-7}$ s) for the temperature regimes 1.6–8 and 18–22 K, respectively. This behaviour has been attributed to distinct Dy^{3+} single-ion and polynuclear $\{Co^{II}{}_2Dy^{III}{}_2\}$ effects where the Δ_1 barrier is believed to originate from the coupled system and the barrier Δ_2 from the localized LF excitations of the Dy^{3+} ions.

Recently, the groups of Sorace, Dreiser and coworkers [13, 14] independently reported the magnetic properties of the Er(trensal) SMM system ($trensalH_3$ = 2,2,2-tris(salicylideneimino)triethylamine) featuring a single Er^{3+} ion in a crystallographically imposed trigonal LF. Notably, the separation between the ground and first excited state is known to be $\Delta = 54\ \text{cm}^{-1}$ from optical and neutron spectroscopic methods (see following text). However, the estimation of the relaxation time from ac susceptibility measurements gives data in the 1.8–5 K temperature regime (Figure 5.3) that are incompatible with a simple Orbach process involving these states. Fitting of the high temperature data yielded an effective energy barrier of ~20 cm^{-1} whereas the Arrhenius prediction $\tau(T) = 9 \times 10^{-12}$ s $\times \exp[\Delta/(k_B T)]$ (dashed line in Figure 5.3; $\Delta = 54$ cm^{-1}) fails to reproduce the experimental data. In fact, this system is a nice example of an SMM in which other mechanisms like Raman processes are relevant. The best fit to $\tau(T)^{-1} = \tau_{QTM}^{-1} + CT^n$, including the temperature-independent QTM term and a CT^n second-order Raman process (with $C = 0.17(3)$ $s^{-1} K^{-n}$, $n = 8.0(1)$ and $\tau_{QTM} = 1.93(5)$ ms) nicely reproduces the shape of $\tau(T)$ in the whole temperature range studied.

The fundamental characteristic of a magnet is the observation of hysteretic behaviour in the field dependence of the magnetization. As described above, lanthanide SMMs often exhibit fast QTM in zero applied field giving rise to 'butterfly' hysteresis loops with small or no remnant magnetization in zero field. A hysteretic process is always dependant on temperature and the magnetic field sweep rate, and thus, in most systems, the lack of magnetization hysteresis appears to be solely a technical issue: that is, the magnetic field sweeping rate of commercial magnetometers is not fast enough. In a few examples reported by Evans, Long and coworkers [15], the coercive field (that is the field required to change the sign of the magnetization) of some lanthanide SMMs was shown to be extremely large, surpassing that of technologically employed hard magnets such as $SmCo_5$. A notable example is shown in Figure 5.4b

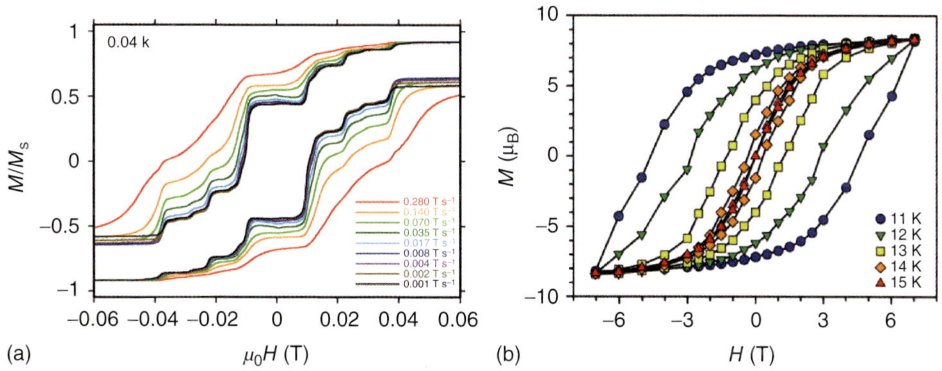

Figure 5.4 (a) Plot of single-crystal magnetization (M) versus dc magnetic field (H) of Tb^{3+} in ($^n Bu_4N$)[$Y(pc)_2$] (2% doping concentration) at $T = 0.04$ K with sweep rates in the range of 0.001–0.280 T s^{-1}. (Reprinted with permission from [16]. Copyright 2005, Wiley.) (b) M versus H hysteresis loops of [K(18-crown-6)$(THF)_2$][{[$(Me_3Si)_2N]_2$(THF)Tb}$_2$(μ-η^2:η^2-N_2)] obtained on a polycrystalline sample with an average sweep rate of 0.9 T s^{-1}. (Reprinted with permission from [15b]. Copyright 2011, American Chemical Society.)

depicting the M versus H hysteresis loops for a polycrystalline sample of $[K(18\text{-crown-6})(THF)_2][\{[(Me_3Si)_2N]_2(THF)Tb\}_2(\mu\text{-}\eta^2:\eta^2\text{-}N_2)]$ incorporating two Tb^{3+} ions linked by the exotic N_2^{3-} ion. Although this system does not constitute a record in terms of the spin-reversal barrier (227.0(4) cm^{-1}, $\tau_0 = 8.2(1) \times 10^{-9}$ s), the coercive field, is truly enormous being 5 T at 11 K.

A number of lanthanide SMMs have been studied by Wernsdorfer using the micro-SQUID technique for which the technical details can be found elsewhere [17]. The advantages of this technique are (i) the possibility of studying very small oriented single-crystals, (ii) the measurements can be performed at millikelvin temperatures (down to 35 mK) in a dilution fridge and (iii) fast variation of the dc field (up to 30 T s^{-1}).

One illustrative example of the usefulness of the micro-SQUID technique is the magnetic hysteresis data (shown in Figure 5.4a) obtained on an oriented single crystal of the diamagnetic $(^nBu_4N)[Y(pc)_2]$ compound doped with 2% of Tb^{3+} [16]. The magnetic field is along the magnetization easy axis of a single-crystal and the magnetic field is swept at different rates. The plot of magnetization versus field showed 'irregularly arranged staircase' hysteresis loops, which are a clear indication of QTM. In this system, the $m_J = \pm 6$ ground doublet state of Tb^{3+} is coupled to the nuclear spin *via* the hyperfine interaction with $A_{hf} = 0.0173$ cm^{-1}. Terbium has a single natural isotope, ^{159}Tb with a nuclear spin of $I = 3/2$. Combining each of the $m_J = +6$ and $m_J = -6$ states with $m_I = \pm 3/2$ and $\pm 1/2$ yields the coupled states $|6\rangle|m_I'\rangle$ and $|6\rangle|m_I\rangle$ which produce avoided level crossings (i.e. removal of the degeneracy) at 13 different magnetic fields each of which correspond to a step in the experimental hysteresis loop.

5.2.2
NMR Spectroscopy

Although nuclear magnetic resonance (NMR) spectroscopy has been used to study a variety of lanthanide based systems like contrast agents for medical imaging applications, only a handful of studies on lanthanide SMMs using NMR have been performed to date. These include the neutral $[Ln(Pc)_2]^0$(Ln = Dy or Tb) [18] and anionic $(^nBu_4N)[Tb(Pc)_2]$ [19] phthalocyaninate complexes. To study the behaviour of lanthanide SMMs, solid state ^{1}H-NMR spectra were recorded on $(^nBu_4N)[Tb(Pc)_2]$ diluted in a nonmagnetic $(^nBu_4N)Br$ matrix in various concentrations ranging from 100% to about 0.7% at temperatures ranging between 2 and 300 K. It was found that useful ^{1}H-NMR spectra of the undiluted $(^nBu_4N)[Tb(pc)_2]$ could only be obtained above 100 K and below 40 K due to fast proton relaxation occurring between 40 and 100 K, which prevented the observation of any NMR signal. This problem was not encountered for the diluted samples, which instead showed dependence of the nuclear spin-lattice relaxation rates on the magnetic and thermal history of each sample. This complication was overcome by annealing the samples at high temperature (about 320 K) followed by a slow cooling ($<1\ \text{K min}^{-1}$) before performing the measurements [19]. The nuclear spin-lattice relaxation rates can be extracted for each sample from the recovery of nuclear magnetization after a saturation recovery pulse sequence. The spin-relaxation time is averaged due to the distribution of relaxation rates of the different ^{1}H nuclei caused by the different hyperfine couplings between ^{1}H nuclei and Tb^{3+} ions. The determination of the relaxation rates allows the investigation of the Tb^{3+} magnetic moment fluctuation caused by transitions between crystal field states within the complex. The lifetime of a LF state ($\approx m_J$) can also be expressed as a function of the probability of a transition between the m_J and $m_J \pm 1$ states. As the lifetimes of each state is dominated by spin-phonon scattering processes, they can be expressed as functions of eigenvalues of the individual crystal field levels. This then allowed the authors to derive the energy difference between the $m_J = \pm 6$ and $m_J = \pm 5$ states. The energy barrier height for both diluted samples is 641 cm^{-1} whilst the undiluted sample exhibits a lower value of 584 cm^{-1} [19]. Intriguingly, these estimations are much larger than those previously reported for (^{n}Bu$_4$N)[Tb(Pc)$_2$] which were obtained via ac susceptibility measurements [20]. Similar measurements were performed on the neutral analogues, [Tb(Pc)$_2$]^0 and [Dy(Pc)$_2$]^0. The spectra recorded below 40 K showed broadening of the signal, indicating the onset of the slowing down of spin dynamics within the [Ln(Pc)$_2$]^0 complexes. As observed with the anionic complex, no spectra between 40 and 140 K were obtained due to the short relaxation times of the ^{1}H nuclei in the sample. To elucidate the dynamic magnetic behaviour of these complexes between 40 and 140 K, muon spin rotation experiments (μ^+SR) were employed [21a,b]. Performing the analogous experiments described for the (^{n}Bu$_4$N)[Tb(Pc)$_2$] complex, energy barriers to magnetization reversal for [Tb(Pc)$_2$]^0 and [Dy(Pc)$_2$]^0 can be extrapolated as 881 and 610 K, respectively [21b]. These values are significantly higher than those determined from magnetometry measurements, highlighting the need for more than one analytical technique to investigate the magnetic properties of lanthanide SMMs.

5.2.3 Muon Spin Rotation

The muon spin relaxation technique uses the implantation and subsequent decay of muons, μ^+, in matter. The muon has a polarized spin of 1/2 [22]. When implanted, the muons interact with the local magnetic field and decay (lifetime = 2.2 μs) by emitting a positron preferentially in the direction of polarization. Adequately positioned detectors are then used to determine the asymmetry of this decay as a function of time, $A(t)$. This function is thus dependant on the distribution of internal magnetic fields within a molecule, which gives essential information about the local (dynamic) magnetic fields [23]. Muon-spin relaxation (μ^+SR) has been predominantly used to study phthalocyaninate complexes [21] and {Dy_3} triangular complexes [24]. In 2006, Powell and coworkers [25] reported a {Dy_3} triangular complex, $[Dy_3(\mu_3\text{-}OH)_2L_3Cl(H_2O)_5]Cl_3 \cdot 4H_2O \cdot 2MeOH \cdot 0.7MeCN$ (where HL = *o*-vanillin), showing SMM behaviour at low temperatures (about < 20 K). However, upon lowering the temperature down to 1.8 K, susceptibility measurements showed that the {Dy_3} complex possesses a non-magnetic $S_T = 0$ ground state despite the odd electron number (see below). To further probe the magnetic dynamic behaviour of the {Dy_3} molecule, μ^+SR spectroscopy was employed [24]. The temperature dependence of the asymmetry function $A(t)$ shows that the muons experience a magnetic field within the sample with two contributions: that is, a static and a dynamic component.

The behaviour of the static component seen in the low temperature curves (about < 35 K) can be attributed either to a slowing down of magnetic moments on each dysprosium ion, or to a magnetic ground state in which the magnetic moments of the dysprosium ions do not cancel each other. Since bulk susceptibility measurements confirm the singlet ground state of the {Dy_3} complex, these μ^+SR results appear to be direct evidence of slow relaxation of the magnetization, that is, SMM behaviour of individual {Dy_3} molecules. The analysis of the asymmetry $A(t)$ in the full temperature range was performed using the Lorentzian Kubo–Toyabe (A_{LKT}) function multiplied by a square-root exponential function, $A(t) = A_{LKT}(t)\exp(-\sqrt{\lambda t})$, thereby allowing access to the temperature dependence of the relaxation rate, λ (cf. Figure 5.5). On decreasing the temperature from 100 K, λ experiences a maximum at about 35 K. This

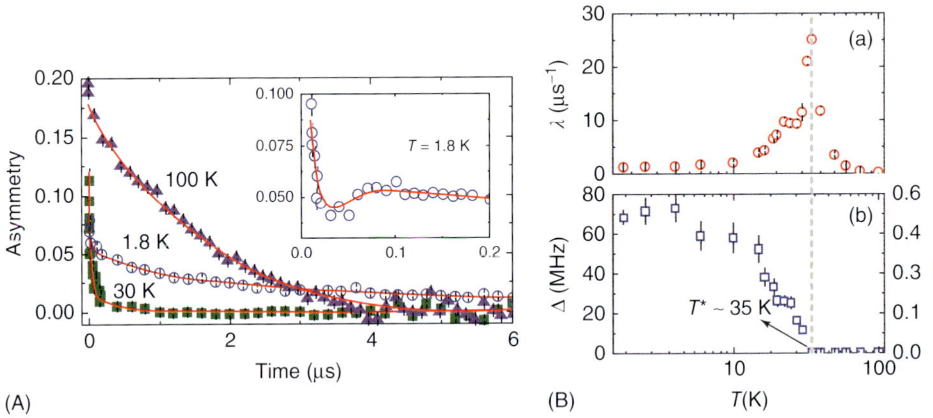

Figure 5.5 (A) Time-dependence of the asymmetry function, $A(t)$. (B) Temperature dependence of the relaxation rate, λ, and the root mean square of the field distribution, Δ, in {Dy_3}. (Reprinted with permission from [24]. Copyright 2010, American Physical Society.)

thermal behaviour is indeed reminiscent of the freezing of the local moments at the muon timescale around 35 K. Due to the absence of any long-range order, this feature has to be of local or molecular origin. From the μ^+SR parameters, the relaxation time was calculated and plotted as a function of inverse temperature allowing an Arrhenius type analysis to be performed. This analysis reveals that the magnetization dynamics are faster than estimated for the bulk samples. For comparison, the Arrhenius parameters, $\Delta_{eff}/k_B = 36(2)$ K and $\tau_0 = 2.5(5) \times 10^{-7}$ s, extracted by ac susceptibility measurements should be compared to $\Delta_{eff}/k_B = 27-39$ K and $\tau_0 = (1.4 - 2.5) \times 10^{-8}$ s estimated from the μ^+SR technique in the 10–25 K temperature regime [24]. The low-temperature saturation value of the fluctuation time is one order of magnitude smaller than observed by ac susceptibility. Two possible explanations were given for this behaviour: (i) μ^+SR spectroscopy measures the relaxation of individual $\{Dy_3\}$ molecules whilst the standard ac susceptibility measurements measure the bulk relaxation dynamics and (ii) due to the μ^+SR sensitivity, a different and unrelated relaxation process is being observed. The lack of definitive conclusion on the observed relaxation by μ^+SR technique highlights again the complexity and the difficulty of understanding in detail the magnetic properties of lanthanide SMMs even when different techniques are used.

5.3 Quantifying the Magnetic Anisotropy

Without any discussion, SMM properties are almost always investigated by dc and ac susceptibility measurements. As will be discussed in the following section, these standard magnetic measurements on polycrystalline or powdered samples are in most cases not sufficient to achieve a complete understanding of these complex magnetic systems, particularly the lanthanide-based complexes. For the magnetic properties of lanthanide systems, the LF, although weak compared to transition elements, remains of primary importance. When polycrystalline samples are available for a series of isostructural SMMs encompassing different lanthanide ions, a simultaneous fitting of all the magnetic susceptibility data can be performed, such as the study reported by Ishikawa [26]. The authors showed that the ligand-field parameters $A_k^q r^k$ of the operators $A_k^q r^k \hat{O}_k^q$ scale linearly to the number of f-electrons (n_f) by $A_k^q r^k = a_k^q + b_k^q(n_f - 10)$. This approach, supported by NMR paramagnetic shifts [27], provides a possible way to circumvent severe over-parameterization of, for example, uncharacteristic $\chi T(T)$ data. This approach was also applied to a series of mononuclear polyoxometallate-based lanthanide SMMs [28].

Magnetic measurements on oriented single crystals are the ideal technique for direct observation of the magnetic anisotropy. Therefore at a first glance, it may appear surprising that relatively few single-crystal studies on lanthanide

SMMs have been reported in literature. However, most lanthanide SMMs exhibit a local low symmetry and crystallize commonly in low-symmetry space groups from which it is difficult to orient single crystals and thus to find the right orientation of the magnetic eigen-axes. In addition, if the crystal could be correctly indexed and magnetic data acquired in different rotation planes, the detailed understanding of the experimental data is strongly hampered by the complexity of the lanthanide LF and also the potential magnetic exchange interactions [29]. The group of Sessoli has extensively employed single-crystal experiments and has interpreted the results using state-of-the-art *ab initio* calculations. One peculiar example is the study of the $Na[Dy(DOTA)(H_2O)] \cdot 4H_2O$ complex ($H_4DOTA = 1,4,7,10$-tetraazacyclododecane-N,N',N'',N'''-tetraacetic acid) [30]. In this system, the Dy^{3+} ion resides in a non-perfect, but approximately tetragonal environment. The deviation arises from a sodium ion placed in the secondary coordination sphere. Importantly, the situation in this system is simplified by the presence of only one molecule orientation in the crystal lattice. The susceptibility tensor can therefore be conveniently extracted from the single-crystal magnetometry. For the measurements, a single-crystal was glued onto a Teflon cube and subsequently indexed relative to the faces of the cube by a single-crystal X-ray diffractometer. Thereafter, the Teflon cube/crystal assembly was mounted on a horizontal rotator in a magnetometer. The χT data at 1.8 K from three orthogonal rotations are shown in Figure 5.6a. The analysis was conducted through the general expressions: $\chi^{\text{rot}\gamma}(\theta) = \chi_{\alpha\alpha}\cos^2(\theta) + \chi_{\beta\beta}\sin^2(\theta) + 2\chi_{\alpha\beta}\sin(\theta)\cos(\theta)$ with cyclic permutation of

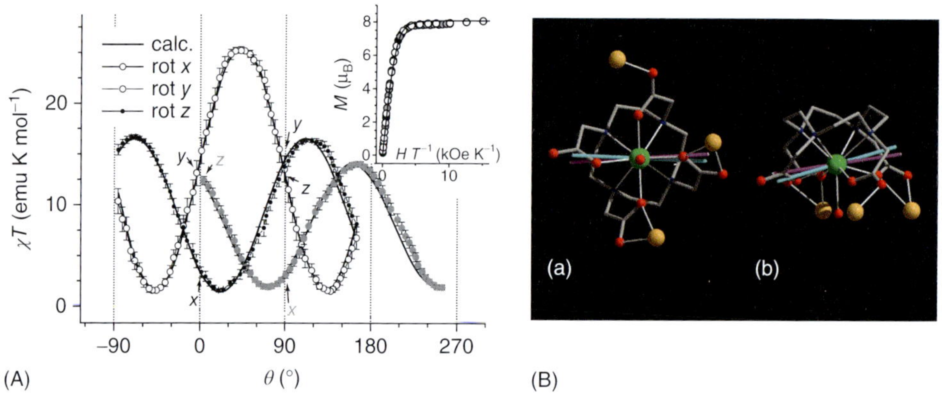

Figure 5.6 (A) Single-crystal magnetization data and best-fits for $Na[Dy(DOTA)(H_2O)] \cdot 4H_2O$ in a dc field of 1 kOe. The inset shows the magnetization at low temperatures plotted against the reduced variable HT^{-1}. The solid lines are the best-fit as described in the text. (B) Experimental (purple) and calculated (cyan) directions of the easy axis of magnetization in $Na[Dy(DOTA)(H_2O)] \cdot 4H_2O$. (Reprinted with permission from [30]. Copyright 2012, Wiley.)

the α, β and γ indices over the xyz reference frame and θ the angle formed by the magnetic field and the α axis. Diagonalizing the susceptibility tensor and treating the Dy^{3+} magnetic centre as an effective $S_{eff} = \frac{1}{2}$ ion at low temperature reveals the principal components of the $\boldsymbol{g}$ tensor: 17.0(1), 4.8(1) and 3.4(1). In the case of a pure $m_J = \pm 15/2$ well-isolated ground state, the effective parallel g-factor would be $g_{\parallel} = 2g_J J = 20$ ($g_J = 4/3$). In Figure 5.6b, the $Na[Dy(DOTA)(H_2O)] \cdot 4H_2O$ SMM is shown along and perpendicular to the approximate fourfold rotation axis. The purple axis is the orientation of the experimentally determined easy axis, which surprisingly is *not* located anywhere near the pseudo-fourfold symmetry axis as would be expected from chemical intuition. This important result, also corroborated by the *ab initio* calculated easy axis (cyan axis in Figure 5.6b), highlights the problems of the high sensitivity of the 4f LF and the obstacles to overcome in order to successfully design lanthanide SMMs.

In the preceding section on muon-spin relaxation, we discussed the case of a $\{Dy_3\}$ complex having a non-magnetic ground state despite the presence of three $J = 15/2$ Dy^{3+} magnetic centres [25]. To further investigate this system, single-crystal magnetometry measurements and CASSCF (Complete Active Space Self-Consistent Field) calculations were employed and revealed that the $\{Dy_3\}$ moiety exhibits a noncollinear Ising spin ground state [31, 32]. The unit cell contains two equivalent $\{Dy_3\}$ molecules, which facilitate the successful extraction of the local magnetization (i) perpendicular to (Z) and (ii) in the $\{Dy_3\}$ plane (X and Y). As shown in Figure 5.7, application of the magnetic field along Z gives rise to a small moment that increases almost linearly with the dc field up to the highest field studied (50 kOe). Conversely, when the dc field is applied along X or Y, the magnetization has an almost identical field behaviour, being very small at weak fields and rapidly increasing around 8 kOe. This behaviour indicates that the Dy^{3+} spins

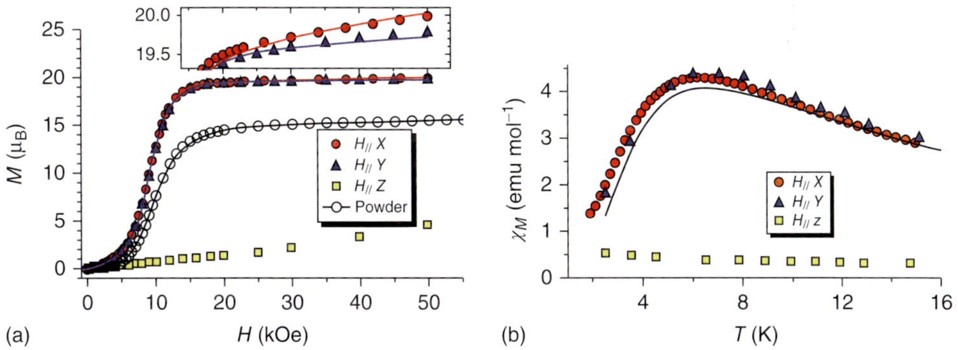

Figure 5.7 Magnetization (a) and susceptibility (b) of the $\{Dy_3\}$ complex obtained on oriented single crystals and powdered samples. The Z direction is orthogonal to the $\{Dy_3\}$ plane. The inset in the left panel shows the difference between the X and Y directions that is only captured by using the excited-state encompassing Hamiltonian described in the text. (Reprinted with permission from [32]. Copyright 2008, American Physical Society.)

lie in the $\{Dy_3\}$ plane and cancel out on rotating the local Dy^{3+} moments by 120° [32]. The low-temperature magnetic powder data could be fitted successfully with Equation 5.6 treating the Dy^{3+} sites as effective spin-1/2 ions with an anisotropic $\boldsymbol{g}$-tensor, using the following parameters: $j_{zz}/k_B = 10.6(4)$ K and $g_z = 20.7(1)$.

$$\hat{H} = \sum_{i=1,2,3}^{i<k} j_{zz}\hat{S}_{z_i}\hat{S}_{z_k} - \mu_B \sum_{i=1,2,3} g_z H_z \hat{S}_{z_i} \tag{5.6}$$

The value of $g_z \approx 20$ corresponds well to the expected value for a pure $m_J = \pm 15/2$ state as described above. A better theory/experiment agreement of the full set of low-temperature data is obtained by introducing the contribution of the first excited state. This improved model, considering only the $m_J = \pm 15/2$ and $m_J = \pm 13/2$ states, was developed by adding $(\delta/14)\sum_i((15/2)^2 - \hat{S}_{z_i}^2)$ to the Zeeman and the exchange coupling terms. This approach leads to a new set of fitting parameters: $j/k_B = -0.092(2)$ K, $g = 1.35(1)$ in good agreement with the Landé g-factor expected for a Dy^{3+} ion ($g_J = 4/3$) and a δ splitting between the ground and first excited state for each of the Dy^{3+} ions of 102(5) K.

In the case of high-symmetry systems, the data acquisition and subsequent data analysis are usually significantly simplified. For the previously mentioned Er(trensal) complex (crystallizing in the trigonal $P\bar{3}c1$ space group), the single crystal susceptibility and magnetization data were recently reported (Figure 5.8) [13]. The anisotropy was found to be of Ising type with an easy axis coinciding with the threefold rotation axis. In fact, Er(trensal) was studied thoroughly by Riley and coworkers [33] in the early 2000s, who unravelled the complete LF

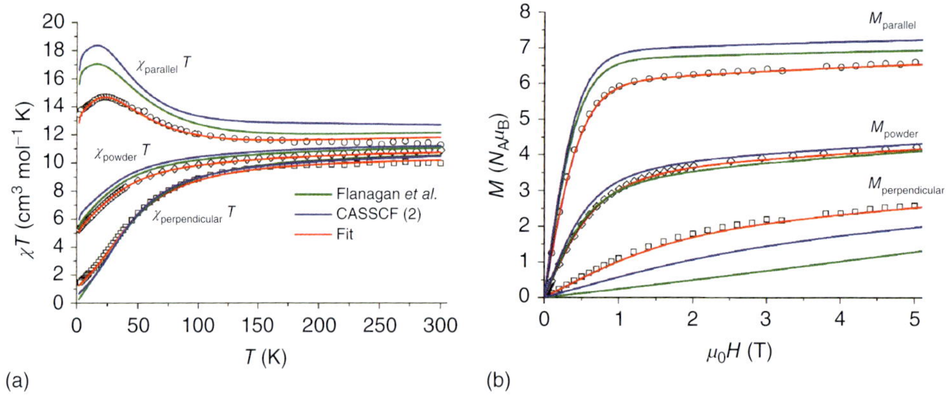

Figure 5.8 Static magnetic data for the Er(trensal) complex in the case of a powdered sample and an oriented single crystal measured in parallel or perpendicular direction to the C_3 rotation axis. (a) Temperature dependence of the magnetic susceptibility shown as the temperature product (H_{dc} = 2000 Oe). (b) Field dependence of the low-temperature magnetization. (Reprinted with permission from [13]. Copyright 2014, Royal Chemical Society.)

picture and extracted all the symmetry-allowed LF parameters. For trigonal symmetry, nine allowed LF parameters remain (in contrast to 27 in the absence of any symmetry) and thus the estimation of all these parameters from the powder magnetization measurements is impossible due to a vast over-parameterization of the fitting procedure. The calculated parallel and perpendicular susceptibilities and magnetizations using Riley's parameters are shown as green lines in Figure 5.8.

It is clearly visible that the spectroscopically determined parameters overestimate the easy axis anisotropy. This observation is further corroborated by X-band EPR data at low temperature, which provide the principal components of the g tensor, $g_{\|} = 11.8(4)$ and $g_{\perp} = 3.6(1)$, for the ground doublet that can be compared to 13 and 1.2, respectively, obtained from the published LF parameters. A similar problem is present in the outcome of the *ab initio* CASSCF calculations (blue lines in Figure 5.8), which overestimate the overall easy axis anisotropy due to small admixture with the first excited state that possesses an easy plane anisotropy. By slightly modifying the zero-field splitting parameters, a unified set of single-crystal and powder magnetic data as well as inelastic neutron scattering (INS) spectra can be successfully reproduced in concert (see following text). This result illustrates the importance of parameterizing the relevant energy window of a given system. Nevertheless, the best set of fitting parameters describing several multiplets of the Er^{3+} ion may not be perfectly suited to describe the splitting of the ground doublet accurately.

5.4 Splitting of the Ground Multiplet

5.4.1 Magnetic Resonance Spectroscopies

EPR spectroscopy applied to lanthanide systems is an old discipline that goes back to early studies on, for example, aqua-complexes of lanthanide ions. For a detailed presentation of this topic, the reader is referred to the classical work of Abragam and Bleaney [4]. For lanthanide SMMs, EPR spectroscopy has been much less employed, which is due partly to the complexity of the resonance spectra due to the high sensitivity of the technique. A useful result obtainable from EPR spectroscopy conducted at low temperature is the determination of the eigenvalues of the $\boldsymbol{g}$ tensor for the ground state or low-lying excited states. Traditionally, two continuous wave modes of EPR are used, the perpendicular ($H_{\mathrm{microwave}} \perp H_0$) and parallel modes ($H_{\mathrm{microwave}} \parallel H_0$), which give rise to the $\Delta m = \pm 1$ and $\Delta m = 0$ selection rules, respectively.

For the approximate D_{4d} symmetry of the mononuclear Er^{3+} SMM, $Na_9[Er(W_5O_{18})_2] \cdot xH_2O$, X-band EPR spectroscopy was performed on a powdered sample [28a]. At around 800 Oe, a prominent resonance is observed that increases on warming from 8 to 25 K indicating a transition between excited states. The temperature dependence of the resonance line intensity suggested the presence of an $m_J = \pm 1/2$ doublet located around $17.5\,cm^{-1}$ above the $m_J = \pm 13/2$ ground state in good agreement with estimations from magnetic measurements. A recent study by Hill and coworkers [34] focussed on the analogous $Na_9[Ho(W_5O_{18})_2] \cdot nH_2O$ system incorporating a Ho^{3+} ion in a close – but non-perfect – tetragonal LF [34]. Interestingly, the high-frequency (50.4 GHz) spectrum at low-temperature shows eight equally spaced lines corresponding to the coupling of the $m_J = \pm 4$ ground doublet to the ^{165}Ho (100%) nuclear spin of $I = 7/2$. In fact, these transitions are $\Delta m_J = \pm 8$ and thus forbidden in D_{4d} symmetry. Therefore, the high sensitivity of the EPR technique gives a direct proof of the non-perfect tetragonal symmetry of the Ho^{3+} site in the low-symmetry space group ($P\bar{1}$) of this system.

5.4.2
Luminescence Spectroscopy

As briefly discussed earlier, the accurate determination of the splitting scheme of the LF states and the nature of the states (i.e. the eigen-values and -vectors of the LF Hamiltonian) remains a heavy and complicated task for most lanthanide SMMs. Unfortunately, the energy span of the ground J multiplet of several hundred reciprocal centimetres is difficult to probe by optical spectroscopic techniques. However, most lanthanide ions luminesce and many of them exhibit high quantum yields. Thus, in principle, luminescence spectroscopy, provided the resolution is sufficiently high, allows easy access to the LF splitting of the ground J multiplet. Thanks to the shielding of the 4f orbitals from the LF by the filled $5s^2$ and $5p^6$ levels, the absorption and emission lines in the visible and near infrared (NIR) areas are extensively narrow. This spectroscopic approach is currently receiving increasing attention as the splittings can be observed often at room temperature and because of the easy availability of the spectrometers. Some representative recent examples are given in Ref. [35]. Most studies in the field of lanthanide SMMs have been focussed on Dy^{3+} complexes and the $^4F_{9/2} \rightarrow {}^6H_{15/2}$ (ground term) emission.

If the spectra are sufficiently resolved, the splitting of the $^6H_{15/2}$ state into a maximum of eight Kramers doublets may be observed. One example of the successful application of this technique to unravel the LF splittings is given for the previously discussed $Na[Dy(DOTA)(H_2O)] \cdot 4H_2O$ system. By assuming that the decay solely originates from the lowest state of the excited $^4F_{9/2}$ multiplet, the observation of eight lines was attributed to the splitting of the $^6H_{15/2}$ state into eight Kramers doublets (cf. Figure 5.9). The separation of $53 \pm 8\,cm^{-1}$ between the lines at the highest energies would therefore provide the energy of the first excited state in the $^6H_{15/2}$ multiplet. Interestingly, this value is relatively close to the value of the spin-reversal barrier extracted from ac susceptibility

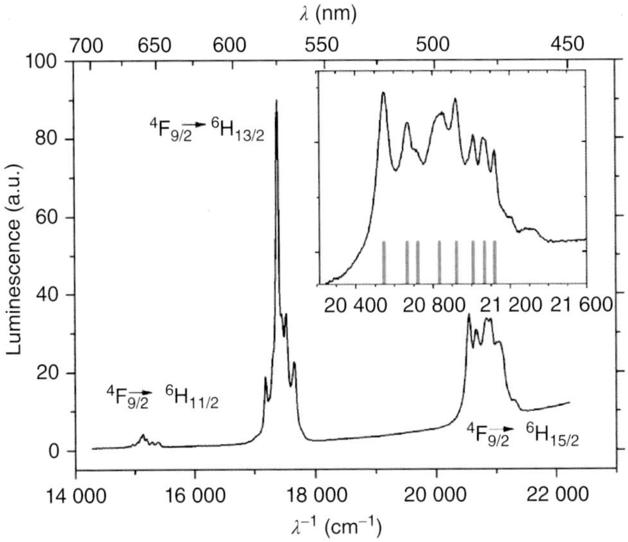

Figure 5.9 Room-temperature luminescence spectrum of $Na[Dy(DOTA)(H_2O)] \cdot 4H_2O$ obtained with an excitation at $\lambda = 350\,nm$. The inset shows the $^4F_{9/2}{\rightarrow}^6H_{15/2}$ emission and the peak positions calculated from CASSCF. (Reprinted with permission from [30]. Copyright 2012, Wiley.)

measurements of about 42 cm^{-1} and the value calculated from the *ab initio* value of 64 cm^{-1} [30]. A particularly nice example of a luminescence study on a molecular system was reported by Riley and coworkers [33] on trigonal Ln(trensal) ($trensalH_3 = 2, 2', 2''$-tris(salicylideneimino)-triethylamine) systems. Taking the Er^{3+} complex as an example, the use of low-temperature spectroscopy facilitates the identification of more than 50 spectroscopic transitions. As previously mentioned, for Er^{3+} ions in a trigonal point group symmetry, nine LF operators are allowed by symmetry. They have been all extracted from the measurements that revealed one of the strongest LFs ever reported for a lanthanide complex.

5.4.3
Inelastic Neutron Scattering

Luminescence spectroscopy provides simple access to the splitting of the ground multiplet but this technique is not always accessible due to nonradiative decay and strong ligand absorptions as encountered, for example, in the $[Ln(Pc)]^{-/0}$ systems. For these reasons, alternative spectroscopic tools should be available for magnetochemists. The use of INS as a spectroscopic probe for molecular magnetic systems has recently been reviewed by Guidi [36], Amoretti *et al.* [37]

and Güdel *et al.* [38]. The neutron has a spin angular momentum of 1/2 that may couple to the magnetic moment of electrons producing the $\Delta S = 0, \pm 1$ and $\Delta m = 0, \pm 1$ selection rules. Despite the promising properties of the $\Delta S = \pm 1$ selection rule and the resulting possibility to directly observe exchange splittings, INS has received very limited attention. Several reasons have hampered the use of INS for lanthanide SMMs. Among them, the complications due to sample background from ^1H scattering, the limited access to the spectrometers, and the requirement for large sample quantities are certainly the most important issues faced by the magnetochemists. However, in particular for lanthanides, INS is a valuable tool to obtain information on ligand-field splittings. Indeed this technique has been employed extensively in solid-state physics while very little use has been reported for molecular lanthanide systems. Probably, the most important limitation of INS for ligand field analysis is the likely observation of only a few LF transitions, which, for low-symmetry compounds, do not provide enough information for a reliable estimation of all LF parameter values. Illustrating the rarity of INS studies on lanthanide SMMs, even for the intensively studied phthalocyaninates, only one study on the non-SMM (^{n}Bu$_4$N)[Tm(Pc)$_2$] has been reported [39]. In this work, the INS spectra were compared to optical data, thus providing a high-resolution description of the first excited states. The first INS investigation of mononuclear lanthanide SMMs was reported only recently by Pedersen *et al.* [13] on the Er(trensal) SMM and two other peripherally substituted analogues. Representative spectra are shown in Figure 5.10 below. The peaks labelled M1, M2 and M3 decrease in intensity on warming from 1.5 to 50 K indicating magnetic ground-state excitations. The energy of peak positions agrees perfectly with those extracted from optical spectroscopy by Riley and coworkers. Confidence in the peak assignment is provided by performing the INS experiment on the isostructural, but diamagnetic, Y(trensal) complex that provides access to the phonon background (PhBG). The intensity of a vibrational excitation scales with the magnitude of the linear momentum transfer $|\boldsymbol{Q}|$ of the scattering process whereas the intensity of a (single-centre) magnetic excitation follows the magnetic form factor that is a decaying function of $|\boldsymbol{Q}|$. The assignment is further corroborated by measurements at higher temperatures, which can be used to estimate the PhBG in the spectra (orange trace in Figure 5.10) according to the $[1-\exp(-\hbar\omega/(k_B T))]^{-1}$ Bose factor ($\hbar\omega$ is the neutron energy loss). Following this term, it appears that the intensity of the phonon scattering is increasing with increasing temperature, offering a way to discriminate features from phonons or magnetic excitations.

Simulating the INS spectra using the Stevens parameters extracted by Riley and coworkers provides a nice reproduction of the INS spectra. However, as previously discussed, it was necessary to adjust these parameters in order to reproduce the static magnetic data (see above, Figure 5.8). Fitting the linear region of the relaxation time for the Er(trensal) SMM yields an energy barrier of 20(1) cm^{-1}

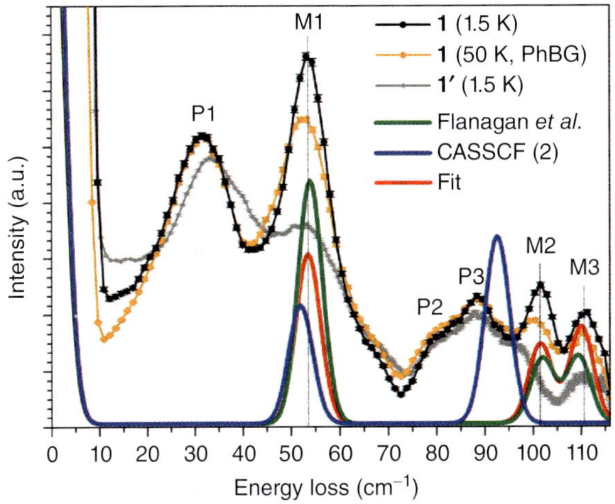

Figure 5.10 INS spectra of the Er(trensal) SMM with $\lambda_i = 2.2$ Å at $T = 1.5$ and 50 K. 'PhBG' refers to the phonon background extracted from downscaling the high-temperature data as described in the main text. (Reprinted with permission from [13]. Copyright 2014, Royal Chemical Society.)

(in a dc field of 900 Oe, Figure 5.3). However, the presence of the first excited state at 54 cm^{-1} shown by optical and INS spectroscopy suggests that relaxation processes other than the thermally activated Orbach process are important in the studied temperature domain between 1.9 and 5 K. Interestingly, it was found that similar Er^{3+} complexes with chemically substituted benzene rings of the trensal^{3-} ligand have quite different magnetization dynamics. The structural overlays of the Er(trensal) moiety with the substituted analogues are shown in Figure 5.11 together with the energy level diagrams established by INS spectroscopy. The INS transitions for these three related systems are very intense and therefore easily assignable as described for Er(trensal). Therefore, despite very similar first coordination spheres, these mononuclear Er^{3+} SMMs display pronouncedly different eigen-values and -vectors of the lowest-lying LF states. This result strongly underlines the complications chemists face to design lanthanide based SMMs without a rigid and controllable coordination environment.

For exchange-coupled lanthanide SMMs, the first example to be studied by INS was the $[Cu^{II}_2Tb^{III}_2L_2(hfac)_4]$ complex (noted $\{Tb^{III}_2Cu^{II}_2\}$; H_3L = 1-(2-hydroxybenzamido)-2-(2-hydroxy-3-methoxybenzylideneamino)-ethane; hfacH = 1,1,1,5,5,5-hexafluoroacetylacetone) [40]. This compound behaves as an SMM (with $\Delta_{eff}/k_B = 21$ K and $\tau_0 = 2.7 \times 10^{-8}$ s), but, interestingly, the closely related $\{Tb^{III}_2Ni^{II}_2\}$ system incorporating square-planar diamagnetic Ni^{2+} does not. This observation suggests that the slow relaxation of the magnetization in this system is not solely the result of the strong anisotropy of the Tb^{3+}

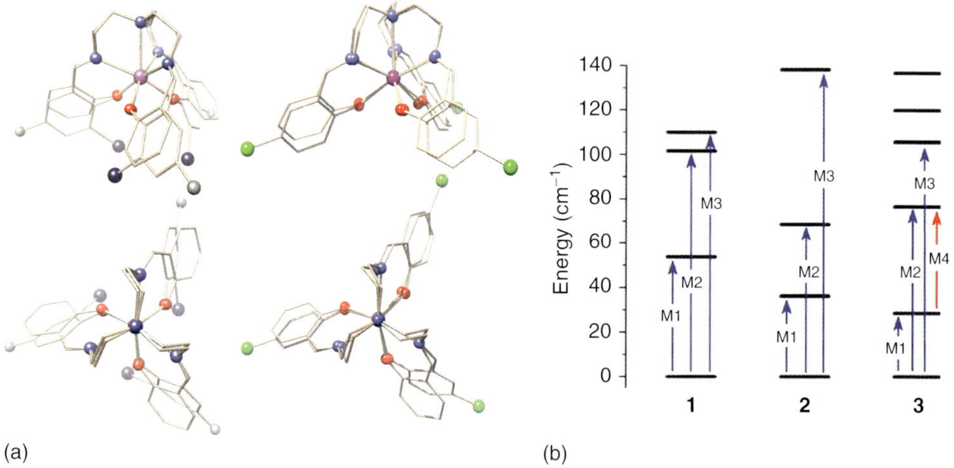

Figure 5.11 (a) Structural overlays of the Er(trensal) molecular structure with two substituted analogues; Er(3-I,5-Me-trensal) (**2**), Er(5-Cltrensal) (**3**). (b) Comparison of the lower-lying energy states of Er(trensal) (**1**), and peripherally substituted analogues (**2** and **3**). (Reprinted with permission from [13]. Copyright 2014, Royal Chemical Society.)

ions but that the exchange interactions are also essential. The INS spectra are shown in Figure 5.12a–c.

In the low-temperature spectrum, one prominent peak (denoted 'I', Figure 5.12a–c) is present at the neutron energy loss of 19.1 cm^{-1}. At higher temperatures, the intensity of I decreases indicating a 'cold' magnetic excitation, that is, a transition originating from the ground state. Simultaneously, a peak (α) at lower energy loss (about 12.5 cm^{-1}) gains intensity, suggesting an energy loss transition from an excited state. A shoulder at about 25 cm^{-1} is almost temperature independent and can therefore be attributed to an excitation of magnetic origin. For the Ni^{2+} analogue, no magnetic peaks were observed in good agreement with the calculated first excited LF state of the Tb^{3+} ion at 110 cm^{-1} that is not in the energy window accessible with the employed setup. This observation also suggests that the magnetic excitations in the Cu^{2+} system are indeed excitations from the whole complex rather than localized excitations.

In a recent report by Kofu *et al.* [41], a LF excitation was also observed for a fully deuterated dinuclear $\{Cu^{II}Tb^{III}\}$ complex. This system shows two prominent ground state magnetic excitations at 1.7 and 12.3 meV as evidenced by the temperature and $|\boldsymbol{Q}|$ dependence (Figure 5.12d,e). An analysis of the available data through a spin-Hamiltonian approach suggested that the energy associated with the first peak (i) (Figure 5.12d,e) originates from the Tb^{3+} – Cu^{2+} interactions whereas the peak at 12.3 meV is the signature of a LF excitation intrinsic

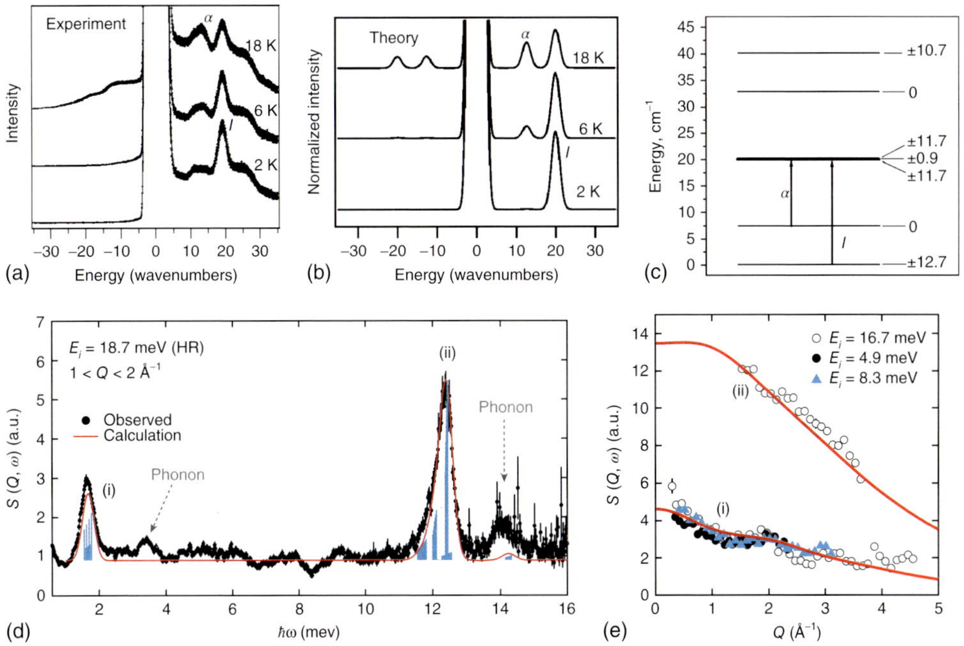

Figure 5.12 (a–c) Experimental (a) and simulated (b) INS spectra of the $\{Cu^{II}{}_2Tb^{III}{}_2\}$ complex. The calculated energy level scheme of this complex is shown in (c) and the states labelled according to the expectation value of the complex $\hat{J}_z$ operator. (Reprinted with permission from [40]. Copyright 2013, American Chemical Society.) (d,e) Inelastic neutron scattering spectra of dinuclear $\{Tb^{III}Cu^{II}\}$ complex at $T = 3.5$ K (d) and $|\boldsymbol{Q}|$-dependence of the transitions (i) and (ii) (e). Solid lines are best fits to Equation 5.6 as described in the main text. (Reprinted with permission from [41]. Copyright 2013, American Physical Society.)

to the Tb^{3+} ion. This conclusion is supported by the $\boldsymbol{Q}$-dependence of the intensity given by the Equation 5.7.

$$I(|\boldsymbol{Q}|) \propto Af_{\mathrm{Tb}}^2 + Bf_{\mathrm{Cu}}^2 + Cf_{\mathrm{Tb}}f_{\mathrm{Cu}}\frac{\sin(|\boldsymbol{Q}|d)}{|\boldsymbol{Q}|d} \tag{5.7}$$

where A, B and C are numerical coefficients, f_{Tb} and f_{Cu} the magnetic form factors of Tb^{3+} and Cu^{2+}, respectively, and d is the $Tb^{3+}-Cu^{2+}$ distance. The lack of oscillatory motion in the intensity of the peak at 12.3 meV with $|\boldsymbol{Q}|$ suggests a localized magnetic excitation whereas the clear oscillations of the intensity for (i) with $|\boldsymbol{Q}|$ demonstrates that the involved wavefunctions contain contributions from both Tb^{3+} and Cu^{2+} metal ions. This result is in perfect agreement with the previous results [40] on the more complicated $\{Tb^{III}{}_2Cu^{II}{}_2\}$ complex. Furthermore, the energy of the first excited states agrees well with the effective spin-relaxation barrier of 1.5 meV suggesting that the relaxation of magnetization takes place by a thermally assisted QTM through the first excited state.

5.5 Observation of the Signatures of Exchange Coupling

The fact that the magnetic coupling constant in lanthanide systems is often on the order of 1 cm^{-1} or less makes the accurate estimation of this parameter a rather complicated task. In practice, the interaction values determined solely from fitting of magnetic data obtained on bulk, polycrystalline or powdered samples is usually associated with a high degree of uncertainty. On top of this problem comes the discussion on how to treat magnetic interactions between orbitally degenerate metal ions [42]. From bulk magnetometry, notable results have been obtained by Noyiri and coworkers [43] treating the lanthanide ions as Ising spins and evaluating exchange interactions from steps in the field dependence of the magnetization. The most fundamental question for chemists is usually dealing with the nature of interaction being ferromagnetic or antiferromagnetic. As the energy splittings due to the LF are much larger than the energy involved in the magnetic exchange interactions, the thermal variation of, for instance, the χT product of the polynuclear complex is mainly governed by depopulation of LF levels. In this situation, the estimation of the exchange energies becomes highly dubious. Various methodologies have been reported to gain insight into the exchange coupling in lanthanide systems such as the studies of isostructural diamagnetically substituted analogues or more recently using X-ray magnetic circular dichroism (XMCD) spectroscopy.

5.5.1 Chemical Substitution

The idea of obtaining qualitative information on magnetic exchange by diamagnetic substitution of the magnetic lanthanide ion was proposed in 1999 by Kahn *et al.* [44]. In this work, the substitution of Cu^{2+} by Zn^{2+} in a series of $Cu^{II} - 4f$ coordination polymers allows the deconvolution of the effects from the ligand-field and the exchange interactions. Due to the strong similarities of the first coordination sphere for the lanthanide ion in the two analogous compounds, the LF perturbation can be considered conserved as a first approximation. Thus, for the Zn^{2+}-incorporating compound, the decrease of the χT product on lowering the temperature can be subtracted directly from the χT versus T data of the Cu^{2+} system in which the magnetic exchange is active. When the magnitude of the interaction parameter is too small to be detectable, the $\chi T(4f - Cu^{II}) - \chi T(4f - Zn^{II})$ difference should be temperature independent and approach the Curie constant of the Cu^{II} metal ion. Deviations from this behaviour are thus related to either ferro- or antiferromagnetic $Cu^{II} - 4f$ coupling evidenced by an increase or a decrease, respectively, of the $\chi T(4f - Cu^{II}) - \chi T(4f - Zn^{II})$ difference on lowering temperature. Although this provides no direct information about the magnitude of the interaction, it does provide an interesting fingerprint of the interaction that is somewhat hidden. This approach has been used quite extensively since the pioneering work of Kahn, where the magnetic lanthanide ion is substituted by diamagnetic analogues such as Lu^{III}, La^{III} or even Y^{III} ions.

5.5.2 X-Ray Magnetic Circular Dichroism

XMCD is a relatively new technique that provides convenient access to element specific magnetizations by means of synchrotron radiation. The basis for the technique relies on the difference in absorption of right- and left-circularly polarized X-ray photons by magnetized systems, for instance, paramagnetic ions in magnetic fields. The XMCD signal is hence obtained as $I_{XMCD}(E) = I_+(E) - I_-(E)$ and the integral of the XMCD signal is proportional to the specific element magnetization. Furthermore, the employment of the so-called sum rules allows for the determination of absolute values of spin and orbital magnetic moments [45]. For SMMs, the first XMCD study was performed by Hamamatsu *et al.* [46] who qualitatively confirmed the ferromagnetic nature of the intra-complex interactions for the aforementioned $\{Cu^{II}_2Tb^{III}_2\}$ complex and the analogous $\{Cu^{II}_2Dy^{III}_2\}$ system deduced from bulk magnetization measurements. The first report on the use of XMCD to determine the magnitude of magnetic exchange interactions in a 3d–4f complex was reported by Dreiser *et al.* [47] for a trinuclear

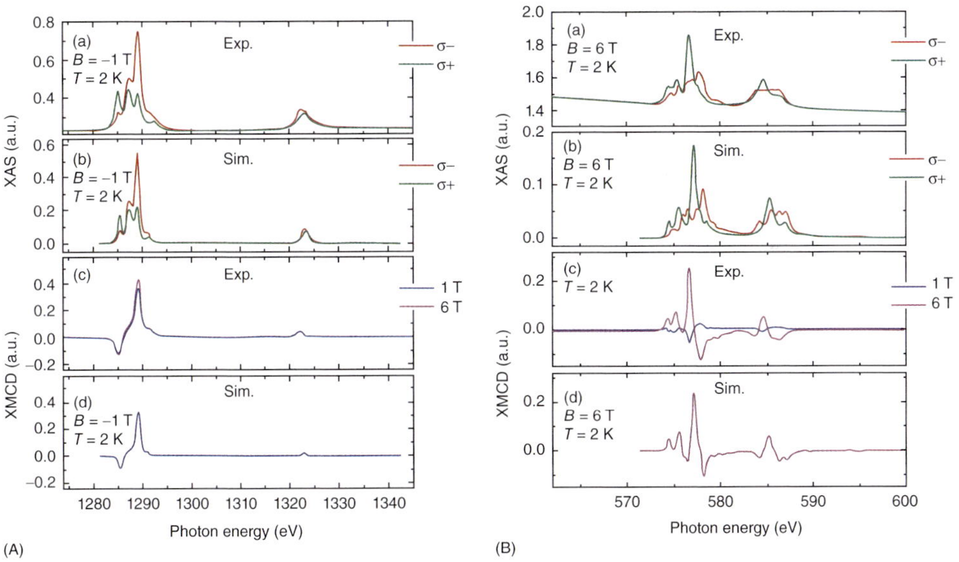

Figure 5.13 X-ray Absorption Spectroscopy (XAS) spectra (top) and XMCD spectra (bottom) for Dy^{3+} at the $M_{4,5}$ edges (A) and Cr^{3+} at the $L_{2,3}$ edges (B) in the {DyCrDy} complex. The calculated spectra were obtained from ligand-field multiplet calculations. (Reprinted with permission from [47]. Copyright 2012, Royal Chemical Society.)

$Dy^{III} - Cr^{III} - Dy^{III}$ complex. Both Cr $L_{2,3}$ and Dy $M_{4,5}$ absorption edges were investigated (Figure 5.13).

In a small field of 1 T, the dichroic signal from the Dy^{3+} ion is almost saturated whereas the Cr signal has a much more dramatic field dependence (cf. Figure 5.13). At low field, the Cr magnetization estimated from the XMCD spectra is of opposite sign to the Dy magnetization that is aligned parallel to the external field due to the larger moment. This simple observation indicates without ambiguity the antiferromagnetic nature of the Dy–Cr interaction. At a critical field, as evidenced by the change of sign of the dichroic signal, the 'all-aligned moment' excited state becomes lower in energy as the magnetic field overcomes the intra-molecular antiferromagnetic interactions. The Cr^{3+} and Dy^{3+} magnetizations extracted from the XMCD signals are shown in Figure 5.14. The magnetization curve for the Cr^{3+} ion exhibits a 'wiggle-shape' characteristic of the moment flip induced by the applied magnetic field, whereas the corresponding Dy^{3+} is more featureless. The data have been modelled at different levels of complexity. In the simplest case, the Dy^{3+} centre is treated as an effective $S = 1/2$ ion and the Cr^{3+} site as a simple $S = 3/2$ spin with a small axial anisotropy. The best-fits of the experimental data are shown in Figure 5.14. Notably, the magnetic parameters deduced from XMCD data are in good agreement with magnetometry analysis. This system was also analysed similarly to the previously described $\{Dy_3\}$ triangle [31, 32] by taking into consideration not only the ground state but also, for one of the two non-equivalent Dy^{3+} ions, the first excited state approximately placed at 53 cm^{-1} above the ground state (as supported by INS spectroscopy) [47].

One of the few advantages of the weak exchange interactions present in most lanthanide systems is that the Zeeman energy associated with typical laboratory

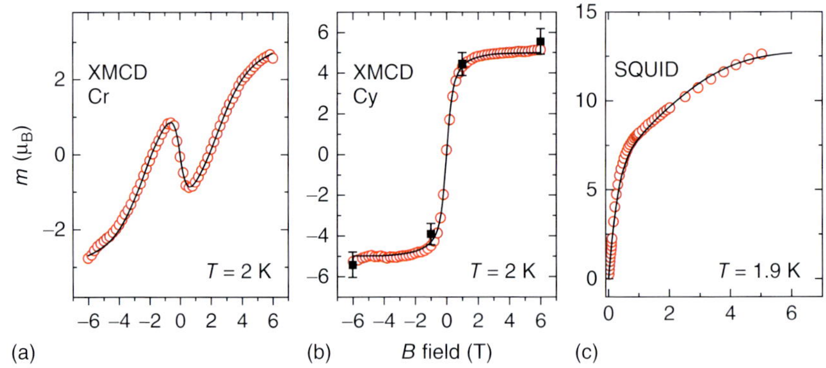

Figure 5.14 XMCD derived magnetizations for Cr^{3+} (a), Dy^{3+} (b) and the complex magnetization obtained by bulk magnetometry (c) for the {DyCrDy} complex. (Reprinted with permission from [47]. Copyright 2012, Royal Chemical Society.)

fields easily exceeds the energy of the exchange interaction. Therefore the characteristic 'wiggle-shape' curve should be experimentally observable in many antiferromagnetically coupled 4f systems. For ferromagnetically coupled systems, the response will be less clear; however, the field dependence of the magnetization is, in this case, expected to increase more steeply than a simple Brillouin-type paramagnetic function. Although the critical field in antiferromagnetically coupled systems can easily be observed, the subsequent estimation of the interaction parameter is troublesome. In the simplest cases, the lanthanide can be treated as an effective $S = 1/2$ system with an anisotropic $\boldsymbol{g}$-tensor, as previously described, and then the magnetic interactions can be estimated. But in most cases, the wavefunctions of the lowest-lying doublet are a priori unknown, which hampers the possibility of estimating and thus comparing magnetic exchange interactions from different complexes.

5.6 Concluding Remarks and Perspectives

The magnetic and optical properties of lanthanide ions are fascinating and are at the forefront of novel and revolutionary technological applications. Conclusively, the interest of chemists in lanthanide-based SMMs and derived magnetic materials is still growing and numerous papers appear in scientific journals every week. Unfortunately for the vast majority of reported systems, the understanding of the underlying physics remains at a rather phenomenological level. This is indeed mainly due to the overwhelming physical complexity of magnetic systems with a significant orbital momentum such as those based on lanthanide ions. However the development of possible applications, for example, molecular spintronic devices, using SMMs and in particular lanthanide-based SMMs necessitates a thorough understanding of the microscopic properties of these systems. As outlined and discussed in this chapter, this information is not easily accessible. Scientists are thus forced to use a plethora of complementary experimental techniques combined with state-of-the-art computational approaches to be able to access the desired knowledge. The large number of new complexes synthesized by chemists requires an easily available screening-type set of experiments to probe the potentially interesting properties of these molecular species. One of the conclusions of this chapter for these chemists is that the availability and interplay between single-crystal magnetometry, luminescence spectroscopy and computational analyses by widespread software codes are a promising and efficient combination of techniques to characterize lanthanide-based SMMs, even if more specialized experiments (INS, EPR, μ^+SR, XMCD, NMR, and so on) will always be necessary in the case of systems with remarkable properties.

References

1. Sessoli, R. and Powell, A.K. (2009) *Coord. Chem. Rev.*, **253**, 2328.
2. Sorace, L., Benelli, C. and Gatteschi, D. (2011) *Chem. Soc. Rev.*, **40**, 3092.
3. Ishikawa, N., Sugita, M., Ishikawa, T., Koshihara, S. and Kaizu, Y. (2003) *J. Am. Chem. Soc.*, **125**, 8694.
4. Abragam, A. and Bleaney, B. (1986) *Electron Paramagnetic Resonance of Transition Ions*, Dover.
5. Chibotaru, L.F. (2013) *Advances in Chemical Physics*, John Wiley & Sons, Inc., Hoboken, NJ, p. 397.
6. (a) Rinehart, J.D. and Long, J.R. (2011) *Chem. Sci.*, **2**, 2078; (b) Baldoví, J.J., Cardona-Serra, S., Clemente-Juan, J.M., Coronado, E., Gaita-Ariño, A. and Palii, A. (2012) *Inorg. Chem.*, **51**, 12565; (c) Chilton, N.F., Collison, D., McInnes, E.J.L., Winpenny, R.E.P. and Soncini, A. (2013) *Nat. Commun.*, **4**, 2551 (d) Reu, O.S., Palii, A.V., Ostrovsky, S.M., Tregenna-Piggott, P.L.W. and Klokishner, S.I. (2012) *Inorg. Chem.*, **51**, 10955.
7. Casimir, H.B.G. and Du Pré, F.K. (1938) *Physica V*, **6**, 507.
8. Cole, K.S. and Cole, R.H. (1941) *J. Chem. Phys.*, **9**, 341.
9. Gatteschi, D., Sessoli, R. and Villain, J. (2006) *Molecular Nanomagnets*, Oxford University Press, New York, p. 72.
10. Guo, Y.-N., Xu, G.-F., Guo, Y. and Tang, J. (2011) *Dalton Trans.*, **40**, 9953.
11. Long, J., Habib, F., Lin, P.-H., Korobkov, I., Enright, G., Ungur, L., Wernsdorfer, W., Chibotaru, L.F. and Murugesu, M. (2011) *J. Am. Chem. Soc.*, **133**, 5319.
12. Mondal, K.C., Sundt, A., Lan, Y., Kostakis, G.E., Waldmann, O., Ungur, L., Chibotaru, L.F., Anson, C.E. and Powell, A.K. (2012) *Angew. Chem. Int. Ed.*, **51**, 7550.
13. Pedersen, K.S., Ungur, L., Sigrist, M., Sundt, A., Schau-Magnussen, M., Vieru, V., Mutka, H., Rols, S., Weihe, H., Waldmann, O., Chibotaru, L.F., Bendix, J. and Dreiser, J. (2014) *Chem. Sci.*, **5**, 1650.
14. Lucaccini, E., Sorace, L., Perfetti, M., Costes, J.-P. and Sessoli, R. (2014) *Chem. Commun.*, **50**, 1648
15. (a) Rinehart, J.D., Fang, M., Evans, W.J. and Long, J.R. (2011) *Nat. Chem.*, **3**, 538; (b) Rinehart, J.D., Fang, M., Evans, W.J. and Long, J.R. (2011) *J. Am. Chem. Soc.*, **133**, 14236; (c) Meihaus, K.R. and Long, J.R. (2013) *J. Am. Chem. Soc.*, **135**, 17952.
16. Ishikawa, N., Sugita, M. and Wernsdorfer, W. (2005) *Angew. Chem. Int. Ed.*, **44**, 2931.
17. Wernsdorfer, W. (2001) *Adv. Chem. Phys.*, **118**, 99.
18. Branzoli, F., Filibian, M., Carretta, P., Klyatskaya, S. and Ruben, M. (2009) *Phys. Rev. B*, 220404.
19. Branzoli, F., Carretta, P., Filibian, M., Zoppellaro, G., Graf, M.J., Galan-Mascaros, J.R., Fuhr, O., Brink, S. and Ruben, M. (2009) *J. Am. Chem. Soc.*, **131**, 4387.
20. (a) Ishikawa, N., Sugita, M., Okubo, T., Tanaka, N., Iino, T. and Kaizu, Y. (2003) *Inorg. Chem.*, **42**, 2440; (b) Ishikawa, N., Sugita, M., Ishikawa, T., Koshihara, S. and Kaizu, Y. (2004) *J. Phys. Chem. B*, **108**, 11265.
21. (a) Branzoli, F., Carretta, P., Filibian, M., Graf, M.J., Klyatskaya, S., Ruben, M., Coneri, F. and Dhakal, P. (2010) *Phys. Rev. B*, **82**, 134401; (b) Branzoli, F., Filibian, M., Carretta, P., Klyatskaya, S. and Ruben, M. (2009) *Phys. Rev. B*, **79**, 220404; (c) Hofmann, A., Salman, Z., Mannini, M., Amato, A., Malavolti, L., Morenzoni, E., Prokscha, T., Sessoli, R. and Suter, A. (2012) *ACS Nano*, **6**, 8390.
22. For an introduction see for instance: Blundell, S.J. (1999) *Contemp. Phys.*, **40**, 175 and Schenck, A. (1986) Muon spin rotation, *Principles and Applications in Solid State Physics*, Hilger, Bristol.
23. Yaouanc, A. and de Réotier, P.D. (2010) *Muon Spin Rotation, Relaxation, and Resonance: Applications to Condensed Matter*, Oxford University Press, Oxford.
24. Salman, Z., Giblin, S.R., Lan, Y., Powell, A.K., Scheuermann, R., Tingle, R. and

Sessoli, R. (2010) *Phys. Rev. B*, **82**, 174427.

25. Tang, J., Hewitt, I., Madhu, N.T., Chastanet, G., Wernsdorfer, W., Anson, C.E., Benelli, C., Sessoli, R. and Powell, A.K. (2006) *Angew. Chem. Int. Ed.*, **45**, 1729.
26. Ishikawa, N., Iino, T. and Kaizu, Y. (2002) *J. Phys. Chem. A*, **106**, 9543.
27. The paramagnetic shift is dipolar in origin and the contributions can be written as $\Delta\delta = (3\cos^2\theta - 1)(\chi_{zz} - \overline{\chi})/(2R^3)$; Bleaney, B. (1972) *J. Magn. Reson.*, **8**, 91.
28. (a) AlDamen, M.A., Cardona-Serra, S., Clemente-Juan, J.M., Coronado, E., Gaita-Arino, A., Martí-Gastaldo, C., Luis, F. and Montero, O. (2009) *Inorg. Chem.*, **48**, 3467; (b) Cardona-Serra, S., Clemente-Juan, J.M., Coronado, E., Gaita-Ariño, A., Camoń, A., Evangelisti, M., Luis, F., Martínez-Peŕez, M.J. and Sese, J. (2012) *J. Am. Chem. Soc.*, **134**, 14982.
29. Bernot, K., Luzon, J., Bogani, L., Etienne, M., Sangregorio, C., Shanmugam, M., Caneschi, A., Sessoli, R. and Gatteschi, D. (2009) *J. Am. Chem. Soc.*, **131**, 5573.
30. Cucinotta, G., Perfetti, M., Luzon, J., Etienne, M., Car, P.-E., Caneschi, A., Calvez, G., Bernot, K. and Sessoli, R. (2012) *Angew. Chem. Int. Ed.*, **51**, 1606.
31. Chibotaru, L.F., Ungur, L. and Soncini, A. (2008) *Angew. Chem. Int. Ed.*, **120**, 4194.
32. Luzon, J., Bernot, K., Hewitt, I.J., Anson, C.E., Powell, A.K. and Sessoli, R. (2008) *Phys. Rev. Lett.*, **100**, 247205.
33. Flanagan, B.M., Bernhardt, P.V., Krausz, E.R., Lüthi, S.R. and Riley, M.J. (2001) *Inorg. Chem.*, **40**, 5401.
34. Ghosh, S., Datta, S., Friend, L., Cardona-Serra, S., Gaita-Ariño, A., Coronado, E. and Hill, S. (2012) *Dalton Trans.*, **41**, 13697.
35. (a) Pointillart, F., Guennic, B.L., Golhen, S., Cador, O., Maury, O. and Ouahab, L. (2013) *Chem. Commun.*, **49**, 615; (b) Boulon, M.-E., Cucinotta, G., Luzon, J., Degl'Innocenti, C., Perfetti, M., Bernot, K., Calvez, G., Caneschi, A. and Sessoli, R. (2013) *Angew. Chem. Int. Ed.*, **52**, 350; (c) Pointillart, F., Guennic, B.L., Cauchy, T., Golhen, S., Cador, O., Maury, O. and Ouahab, L. (2013) *Inorg. Chem.*, **52**, 5978; (d) Li, Q.-W., Liu, J.-L., Jia, J.-H., Leng, J.-D., Lin, W.-Q., Chen, Y.-C. and Tong, M.-L. (2013) *Dalton Trans.*, **42**, 11262.
36. Guidi, T. (2012) in *Neutron Spectroscopy of Molecular Nanomagnets in Molecular Cluster Magnets* (ed R.E.P. Winpenny), World Scientific.
37. Amoretti, G., Caciuffo, R., Carretta, S., Guidi, T., Magnani, N. and Santini, P. (2008) *Inorg. Chim. Acta*, **361**, 3771.
38. Basler, R., Boskovic, C., Chaboussant, G., Güdel, H.U., Murrie, M., Ochsenbein, S.T. and Sieber, A. (2003) *ChemPhysChem*, **4**, 910.
39. Magnani, N., Caciuffo, R., Colineau, E., Wastin, F., Baraldi, A., Buffagni, E., Capelletti, R., Carretta, S., Mazzera, M., Adroja, D.T., Watanabe, M. and Nakamura, A. (2009) *Phys. Rev. B*, **79**, 104407.
40. Klokishner, S.I., Ostrovsky, S.M., Reu, O.S., Palii, A.V., Tregenna-Piggott, P.L.W., Brock-Nannestad, T., Bendix, J. and Mutka, H. (2009) *J. Phys. Chem. C*, **113**, 8573.
41. Kofu, M., Yamamuro, O., Kajiwara, T., Yoshimura, Y., Nakano, M., Nakajima, K., Ohira-Kawamura, S., Kikuchi, T. and Inamura, Y. (2013) *Phys. Rev. B*, **88**, 064405.
42. For a recent review see: Palii, A.V., Tsukerblat, B., Klokishner, S., Dunbar, K.R., Clemente-Juan, J.M. and Coronado, E. (2011) *Chem. Soc. Rev.*, **40**, 3130.
43. Mori, F., Nyui, T., Ishida, T., Nogami, T., Choi, K.-Y. and Noyiri, H. (2006) *J. Am. Chem. Soc.*, **128**, 1440.
44. Kahn, M.L., Mathonière, C. and Kahn, O. (1999) *Inorg. Chem.*, **38**, 3692.
45. (a) Carra, P., Thole, B.T., Altarelli, M. and Wang, X. (1993) *Phys. Rev. Lett.*, **70**, 694; (b) Thole, B.T., Carra, P., Sette, F. and van der Laan, G. (1992) *Phys. Rev. Lett.*, **68**, 1943.
46. Hamamatsu, T., Yabe, K., Towatari, M., Osa, S., Matsumoto, N., Re, N., Pochaba, A., Mrozinski, J., Gallani, J.-L., Barla, A., Imperia, P., Paulsen, C. and Kappler, J.-P. (2007) *Inorg. Chem.*, **46**, 4458.

47. Dreiser, J., Pedersen, K.S., Piamonteze, C., Rusponi, S., Salman, Z., Ali, M.E., Schau-Magnussen, M., Thuesen, C.A., Piligkos, S., Weihe, H., Mutka, H., Waldmann, O., Oppeneer, P., Bendix, J., Nolting, F. and Brune, H. (2012) *Chem. Sci.*, **3**, 1024.

6
Computational Modelling of the Magnetic Properties of Lanthanide Compounds

Liviu Ungur and Liviu F. Chibotaru

6.1
Introduction

Lanthanide ions are at the core of the most promising magnetic materials obtained to date. After the discovery of Ishikawa's terbium double decker $[Tb(Pc)_2]^-$ [1], the importance of lanthanide ions in the field of molecular magnetism grew rapidly [2, 3]. Significant effort has been made to rationalize the experimental results using traditional theoretical approaches, such as the phenomenological crystal field (CF) theory [4]. The growing amount of detailed experimental data on lanthanide compounds called for higher level theoretical tools for the description of their electronic and magnetic properties. In the recent years, explicitly correlated *ab initio* calculations, based on the complete active space self-consistent field (CASSCF) approach proved to be such a tool. Given its success in describing the magnetic properties of lanthanide compounds, we give here a thorough description of this *ab initio* methodology. The present chapter also revises several findings and insights in the field of lanthanide molecular magnetism obtained with the *ab initio* approach.

6.2
Ab Initio Description of Lanthanides and its Relation to Other Methods

A straightforward definition of the term '*ab initio*' is 'from the beginning', 'from the first principles'. *Ab initio* calculation of some property implies not using any experimental data or input, but relating the calculated property only to the fundamental constants entering the applied quantum-physical laws.

According to the basic principle of quantum mechanics, any measurable property can be computed *ab initio* if the total *wave function* ψ describing the quantum eigenstate of the system is known, since it contains the complete

Lanthanides and Actinides in Molecular Magnetism, First Edition.
Edited by Richard A. Layfield and Muralee Murugesu.

information about the system. It is obtained as a solution of the Schrödinger equation:

$$\widehat{\mathbf{H}}\psi = E\psi \tag{6.1}$$

where $\widehat{\mathbf{H}}$ is the energy operator of the system (the Hamiltonian), E is the energy eigenvalue and ψ is the eigenfunction. In the most general case, the Hamiltonian should include the kinetic energy of all particles, attraction between particles with opposite charges and repulsion between particles with identical charges, and account for relativistic effects (relativistic movement of the particles, the influence on their spin, orbital and magnetic moments, the interaction between the spin and orbital moments, etc.). Fully accounting for these interactions is an impossible task nowadays; therefore, several approximations are made in order to make the calculation feasible:

- Separation of the wave functions describing the motion of nuclei and the wave function describing the electrons (Born–Oppenheimer approximation). This approximation is based on the fact that the nuclear particles are much heavier than the electrons, and therefore much slower than the latter. In such a situation, the electronic wave function can be found at fixed positions of the nuclei.
- Representation of each molecular orbital as a linear combination of atomic orbitals (atomic basis sets). Atomic basis sets are usually represented as Slater type orbitals or as combinations of Gaussian functions. The latter is very popular, due to a very fast algorithm for the computation of bielectronic integrals.
- Treatment of the relativistic effects within the Douglass–Kroll–Hess (DKH) formalism [5–7], which is a reliable approximation for the much more complex Dirac equation for most atoms.
- Mean-field approximation of quasi-free electrons (the Hartree–Fock approximation). The total wave function is described, in this case, by a single Slater determinant.
- Accounting for electron correlation in a second step, via the mixing of a limited number of Slater determinants in the total wave function. Electron correlation is very important for correct treatment of interelectronic interactions and for a quantitative description of covalence effects and of the structure of multielectronic states. Accounting completely for the total electronic correlation is computationally extremely difficult, and is only possible for very small molecules, within a limited basis set. Formally, electron correlation can be divided into *static*, when all Slater determinants corresponding to all possible electron populations of frontier orbitals are considered, and *dynamic* correlation, which takes into account the effects of dynamical screening of interelectron interaction.

Further, we give a short review of the key points of the computational scheme suitable for the description of electronic structure in lanthanides.

6.2.1
Ab Initio Approach for the Electronic Structure of Lanthanides

The wave functions of the ground and excited states of lanthanides have a truly *multiconfigurational* character.[1] Therefore, computational description of both the ground state and the low-lying excited states, which are important for magnetic behaviour, is only possible by a multiconfigurational *ab initio* method. In this respect, the CASSCF method proved to be a reliable tool for the description of electronic properties of lanthanide complexes.

6.2.1.1 Accounting for *Static* Electron Correlation within CASSCF

CASSCF is one of the most widely used *ab initio* multiconfigurational approaches [8–10]. It is based on an efficient partitioning of molecular orbitals into three main blocks: inactive (double-occupied), active and virtual (empty). The inactive orbitals are kept doubly occupied while the virtual orbitals are kept empty in all configurations employed to build the CASSCF wave function. The remaining electrons (active) span the explicitly correlated orbitals. The CASSCF wave function is a linear combination of all possible configurations (Slater determinants) which are built by following this partitioning scheme. Energy minimization is done self-consistently with respect to the coefficients of the Slater determinants (configuration interaction (CI) coefficients) and the molecular orbitals. A major limitation of the CASSCF calculation is the number of Slater determinants which are generated by a specific active space. In practice, the calculation becomes almost impossible if the wave function has more than 5–10 million configuration state functions (CSF, or Slater determinants), which is easily achieved for an active space consisting of 15–16 electrons spanning the same number of active orbitals. Various implementations aimed at lowering the number of the CSFs have been performed, the most promising one being the restricted active space self-consistent field (RASSCF) method [11, 12]. The success of the CASSCF method is due to its ability to describe near-degeneracies, which are often met for various problems, in particular, for the description of the near-degenerate ground states of transition metal (TM) and lanthanide compounds. Efficient implementations of the CASSCF methodology are present in MOLCAS [13], Gaussian [14], Orca [15], MOLPRO [16] and many other packages. In particular, it is the ability of the CASSCF method to describe near-degeneracies and excited states, which made it particularly suitable for the description of ligand-field states in lanthanide compounds.

6.2.1.2 Accounting for Dynamical Electron Correlation: An Important Step Towards Accurate Predictions

CASSCF wave function includes only the *static* correlation: only a small number of electrons spanning frontier orbitals are correlated between them, while

1) Except Gd^{3+}, which is a single-determinant due to its f^7 electronic configuration: $\Psi = |\uparrow\uparrow\uparrow\uparrow\uparrow\uparrow\uparrow|$

the correlation including inactive electrons also (which are usually much more numerous!), that is, the *dynamical* electron correlation, has to be accounted for in order to achieve accurate results. Currently there are two approaches for (partial) account of dynamical correlation to a multiconfigurational CASSCF wave function: (i) perturbative and (ii) CI methods.

Perturbative methods (CASPT2 [17], NEVPT2 [18]) add the dynamical correlation in an effective way, using multiconfigurational second-order perturbation theory on the CASSCF input states. These methods have proved to be suitable for studying problems in spectroscopy, photochemistry, and so on [19, 20].

CI methods [21] add a certain number of excited Slater determinants, usually selected by the excitation type (e.g. single, double, triple excitations), which were initially not present in the CASSCF wave function, and treat them in a non-perturbative way. Inclusion of additional configurations allows for more degrees of freedom in the total wave function, thus improving its overall description. These methods are extremely costly and therefore, are only applicable to small systems. Among this class of methods, DDCI (difference-dedicated configuration interaction) [22] and CISD (single- and double excitations) [21] are the most popular.

6.2.1.3 Accounting for Relativistic Effects within the Douglas–Kroll–Hess Theory

Accounting for relativistic effects exactly is only possible by solving the Dirac equation exactly. The original Dirac equation is of four components, which automatically means it is much more difficult to be solved than the Schrödinger equation (6.1). Transformations employed by Douglas and Kroll [23] together with the so-called 'decoupling scheme' employed by Hess [24, 25] allowed for a much simpler and practical way to account for relativistic effects without the need to solve the complete system of Dirac equations. In this treatment, the relativistic effects are divided into two parts: the scalar contribution and the spin–orbit coupling. Scalar relativistic corrections are accounted for in the basis set generation, when effective-core potentials (ECPs) or atomic natural orbitals with relativistic correction (e.g. ANO-RCC [26, 27] or ANO-DK3 [28]) are employed for the description of individual atoms. Accuracy in the treatment of relativistic effects can be achieved by gradually increasing the order of the DKH expansion [7]. However, for most chemical elements, treatment of relativistic effects in the second-order DKH expansion is sufficient. Practical implementations of the second-order DKH are present in almost all quantum chemical packages. For the computation of lanthanides, an important aspect is the accuracy of the spin–orbit coupling treatment. For complexes with the ground orbital state well separated from excited states, when the energy separation is several times larger than the spin–orbit coupling, magnetic properties are reasonably well described by treating spin–orbit coupling perturbatively. In the opposite case, when the spin–orbit coupling is larger than the energies of few excited orbital states, for example, in the near-degenerate cases (almost all lanthanide compounds), perturbative treatment of the spin–orbit coupling fails to describe correctly the electronic and magnetic properties. In the MOLCAS program package [13], spin–orbit coupling is treated non-perturbatively within the mean-field theory in the restricted active space state interaction (RASSI) method [29]. In this method, the spin–orbit state-interaction matrix is built on the basis of the input states optimized at the CASSCF level. Diagonal energies of the state interaction matrix can be those of the initial CASSCF calculation or taken from a more advanced CASPT2 calculation. Diagonalization of the spin–orbit matrix leads to spin–orbit multiplets.

It is important to mention that the RASSI method treats all wave functions as 'frozen', that is, the CASSCF wave functions do not change during the calculation. Allowing the relaxation of the CASSCF wave functions during the RASSI step, would, probably, improve the results. There are several attempts to consider spin–orbit coupling in a variational way [30–32]. The main drawback of the methods involving variational treatment of spin–orbit coupling is that the dynamical correlation becomes effectively impossible to be treated, either perturbatively or by CI. The reason for this is the large number of spin orbitals, which is doubled compared to the number of orbitals in the non-relativistic methods.

6.2.1.4 Spin–Orbit Multiplets of Free Lanthanide Ions: Relativistic CASSCF/RASSI Method in Work

Owing to the fact that a correct description of the spin–orbit coupling for lanthanides requires a large number of spin terms to be included in the calculation, it is computationally very difficult to include dynamical correlation for them in the case of real lanthanide compounds, irrespective of the approach. On the other hand, since most of the excited states of lanthanide compounds, important for the magnetism, are of ligand-field type (various arrangements of n electrons within the 4f shell) the dynamical correlation is expected to be of similar importance for the entire ground atomic term, leading to minor relative energy changes for the low-lying J-multiplets compared to a CASSCF/RASSI calculation. In fact, as we show in Table 6.1, minimal relativistic CASSCF/RASSI calculations give already quite accurate energies of low-lying multiplets which are responsible for magnetism, in all lanthanides. This agreement, however, is not as good for excited terms as it is for the ground one presented in the Table 6.1. This is seen, for example, in the comparison between experimental and calculated J-multiplets originating from the excited ^{6}P atomic term of the Gd^{3+} ion. For an accurate description of higher excited J-multiplets, originating from excited atomic terms, dynamical correlation is very important and should not be neglected.

In lanthanide complexes, the ligand field created by the surrounding ligands will split the atomic J-multiplets into several components. The latter are doubly degenerate (Kramers doublets (KDs)) for systems with odd number of electrons and non-degenerate (in the absence of symmetry) for systems with even number of electrons.

Table 6.1 Energies of the low-lying *J*-multiplets predicted within the CASSCF/RASSI approach for the free Ln^{3+} ions. ANO-RCC of single-zeta quality were employed. The active space of the CASSCF method included only *n* electrons spanning the 4f shell.

	Term	*J*-multiplet	Experiment [33]	Calculation		*J*-multiplet	Experiment [33]	Calculation
Ce^{3+}, f^1	^2F	5/2	0	0	Yb^{3+}, f^{13}	7/2	0	0
		7/2	2 253	2 415		5/2	10 214	9 771
Pr^{3+}, f^2	^3H	4	0	0	Tm^{3+}, f^{12}	6	0	0
		5	2 152	2 120		5	8 090	7 952
		6	4 389	4 359		4	12 500	12 724
Nd^{3+}, f^3	^4I	9/2	0	0	Er^{3+}, f^{11}	15/2	0	0
		11/2	1 880	1 841		13/2	6 480	6 290
		13/2	3 860	3 835		11/2	10 110	10 138
		15/2	5 910	5 939		9/2	12 350	12 807
Pm^{3+}, f^4	^5I	4	0	0	Ho^{3+}, f^4	8	0	0
		5	1 490	1 442		7	5 050	4 887
		6	3 110	3 045		6	8 550	8 412
		7	4 820	4 763		5	11 110	11 048
		8	6 580	6 559		4	13 180	13 109
Sm^{3+}, f^5	^6H	5/2	0	0	Dy^{3+}, f^5	15/2	0	0
		7/2	1 080	1 011		13/2	3 460	3 349
		9/2	2 290	2 213		11/2	5 780	5 734
		11/2	3 610	3 557		9/2	—	7 579
		13/2	4 990	5 004		7/2	9 060	9 007
		15/2	6 470	6 518		5/2	10 100	10 070
Eu^{3+}, f^6	^7F	0	0	0	Tb^{3+}, f^6	6	0	0
		1	370	353		5	2 052	1 982
		2	1 040	989		4	3 314	3 263
		3	1 890	1 828		3	4 292	4 242
		4	2 860	2 806		2	4 977	4 940
		5	3 910	3 874		1	5 431	5 400
		6	4 940	4 998		0	5 653	5 626
Gd^{3+}, f^7	^8S	7/2	0	0				
	^6P	7/2	32 120	39 406				
		5/2	32 720	39 927				
		3/2	33 290	40 474				

6.2.2
Ab Initio Versus Two-Component DFT

Two-component DFT methods aim at solving the reduced form of the four-component Dirac equation using DFT formalism. The full relativistic treatment allows describing, for instance, a noncollinear magnetic density in the ground state of complexes [34]. Although relativistic effects are accurately described by these methods, an adequate solution is only obtained for wave functions which could be represented by a single Slater determinant. In this respect, two-component DFT methods are not applicable for the description of electronic properties of lanthanides, since the wave function for most of them has an intrinsic multideterminantal character. In the case of lanthanides (f^n system) we have the following situations:

- For $n < 7$, the total spin of the system S couples antiferromagnetically with the orbital momentum L into the total momentum in the ground state: $J = |L - S|$ [35]. The individual $|J, M_J\rangle$ multiplets cannot be, in principle, described by a single Slater determinant. This is a consequence of the general rules of coupling of two angular momenta: eigenfunctions of all projections of the total angular momentum $|J, M_J\rangle$, which is lower than the maximal possible $(L + S)$, are represented as linear combinations of several determinants [36].
- For $n = 7$, the case of Gd^{3+}, the ground term is orbitally non-degenerate ($L = 0$). Then, $J = S$ and the ground state wave function is adequately described by a single Slater determinant, that is, DFT methods may work.
- For $n > 7$, the spin–orbit interaction stabilizes the maximal total momentum in the ground state, $J = L + S$ [35]. Depending on the character of the splitting of the ground J-multiplet, the following situations may occur:
  - $M_J < M_{\mathrm{Jmax}}$ – the wave function is multideterminantal.
  - $M_J = M_{\mathrm{Jmax}}$ – this is the only case, which can, in principle, lead to a single-determinant wave function. This situation may be encountered in some limited highly axial systems, such as $Dy\text{-}O^+$ [37]. However, usual compounds do not have any site symmetry, which means that several projections of the J-manifold will be mixed by the low-symmetry components of the CF, leading to a multideterminantal character of the wave function.

In conclusion, the DFT approach, even in a two-component scheme, is generally not applicable to treat lanthanides.

6.2.3
Ab Initio Versus Phenomenological Crystal Field Theory for Lanthanides

CF (ligand field) theory was one of the first theoretical tools employed for rationalization of experimental data for lanthanide compounds. There are two different CF approaches employed for the description of lanthanide complexes: (i) the phenomenological treatment of the lowest J-multiplet involving effective CF parameters specific for this group of states [38, 39] and (ii) the microscopic CF model, which considers all single- and bielectronic interactions in the f^n shell [4, 40, 41]. However, both these approaches can give only a limited insight into the nature of ligand field states due to the following reasons:

- The relative contributions of covalent, electrostatic and other interactions to the CF are not known a priori. The existent models include exclusively either covalent (angular overlap model [42–44] or electrostatic [40, 41, 45] interactions, which seems unjustified.)
- The 'effective' CF models, intended to include covalence effects via effective charges and shielding parameters [46] (superposition model [47], effective charge model [48], simple overlap model [49, 50]), keep the radial ($M-L$ distance) dependence of the CF parameters as in the simple (point charge) electrostatic model. Dedicated studies have shown, however, that the radial dependence of these parameters deviates strongly from the latter for the whole series of lanthanide ions [51, 52].
- The existent CF models neglect or describe very approximately the effect of multiatomic structure, conjugation and multidentate character of the ligand on the CF parameters. The effect of second and more distant coordination spheres is described as electrostatic, in some models, while it was proved by *ab initio* calculations that it is mainly covalent by its nature [53, 54].
- The bottleneck of phenomenological models is the large number of independent parameters (27 in low-symmetric complexes) required for the description of the CF, which cannot be reliably extracted from experiment in a unique manner. As a rule, these models are confronted with the description of a limited amount of experimental data, while it is not possible in principle to provide the entire set of CF parameters. The latter strongly depend on fitted experiments and, therefore, are not reliable (an example is described below).

6.3
Ab Initio Calculation of Anisotropic Magnetic Properties of Mononuclear Complexes

The increased power of *ab initio* methods underlay an important progress manifested in two aspects. First, they allow nowadays treating large complexes (with more than a hundred of atoms) fully *ab initio*. Second, these methods provide accurate solutions for multielectronic eigenstates, which often allows making predictions of different properties of complexes with an experimental accuracy. Representative quantum chemistry software is the MOLCAS package [13]. As was mentioned before in Section 6.2.1, the DKH treatment of relativistic effects in combination with explicitly correlated multiconfigurational description of the multielectronic wave functions provides very good solutions for molecular multiplets of the complexes. The latter can be used in principle for the calculation of magnetic properties of different complexes. Indeed, since these properties are simply a response to a Zeeman interaction induced by an external magnetic field **B**, in order to evaluate them, we only need to know the matrix elements of the operator of total magnetic moment of the complex $\hat{\mu}$ calculated on the basis of multiplet eigenstates.

$$\hat{\mathbf{H}}_{\mathrm{Zee}} = -\hat{\mu}\mathbf{B}, \ \hat{\mu} = -\mu_{\mathrm{B}} \sum_{i=1}^{N_{\mathrm{el}}} g_{\mathrm{e}}\hat{\mathbf{s}}_i + \hat{\mathbf{l}}_i \tag{6.2}$$

6.3.1
Implementation of *Ab Initio* Methodology: SINGLE_ANISO Program

The SINGLE_ANISO program is a routine which allows for non-perturbative calculation of effective spin (pseudospin) Hamiltonians and static magnetic properties of mononuclear complexes and fragments completely *ab initio*, including the spin–orbit interaction. As a starting point, it uses the results of a CASSCF/(CASPT2)/RASSI calculation for the ground and several excited spin–orbit multiplets. A description of the computational methodology of pseudospin Hamiltonians is given in [55, 56] while the derivation of (macroscopic) magnetic properties is given in [57].

The SINGLE_ANISO program is able to calculate the following:

- Parameters of pseudospin magnetic Hamiltonians:
  1. First rank (linear after pseudospin) Zeeman splitting tensor $g_{\alpha\beta}$ (i.e., the conventional g tensor), its main values, including the sign of the product $g_X g_Y g_Z$, and the main magnetic axes (X_m, Y_m, Z_m) in the initial coordinate frame.
  2. Second rank (bilinear after pseudospin) zero-field splitting tensor $D_{\alpha\beta}$ (i.e. the conventional D tensor), its main values and the main anisotropy axes (X_a, Y_a, Z_a).
  3. Higher rank ZFS tensors (D^4, D^6, ... , etc.) and Zeeman splitting tensors (G^3, G^5, ... , etc.) for complexes with moderate and strong spin–orbit coupling.
  4. Angular moments along the main magnetic axes.

  Magnetic Hamiltonians are defined for a desired group of N electronic states obtained in the *ab initio* calculation, to which a pseudospin $\tilde{S}$ (it reduces to a true spin S in the absence of spin–orbit coupling) is subscribed according to the relation $N = 2\tilde{S} + 1$. For instance, the two wave functions of a KD correspond to $\tilde{S} = 1/2$. The implementation is done for *any dimension* of the pseudospin.
- A complete decomposition of the *ab initio* computed CF matrix in irreducible tensor operators (ITOs) and in extended Stevens operators. The parameters of the multiplet-specific CF acting on the ground atomic multiplet of lanthanides, and the decomposition of the CASSCF/RASSI wave functions into functions with definite projections of the total angular momentum on the quantization axis are provided.

- Static magnetic properties:
  1. Van Vleck susceptibility tensor $\chi_{\alpha\beta}(T)$
  2. Powder magnetic susceptibility function $\chi(T)$
  3. Magnetization vector $\vec{\mathbf{M}}(\vec{\mathbf{B}})$ for specified directions of external magnetic field $\vec{\mathbf{B}}$
  4. Powder magnetization $M_{\text{mol}}(B)$

The calculation of magnetic properties takes into account the contribution of excited states (the ligand-field and charge transfer states of the complex or mononuclear fragment included in the RASSI calculation) via their thermal population and Zeeman admixture. The intermolecular exchange interaction between magnetic molecules in a crystal can be taken into account during the simulation of magnetic properties by a phenomenological parameter zJ' specified in the input.

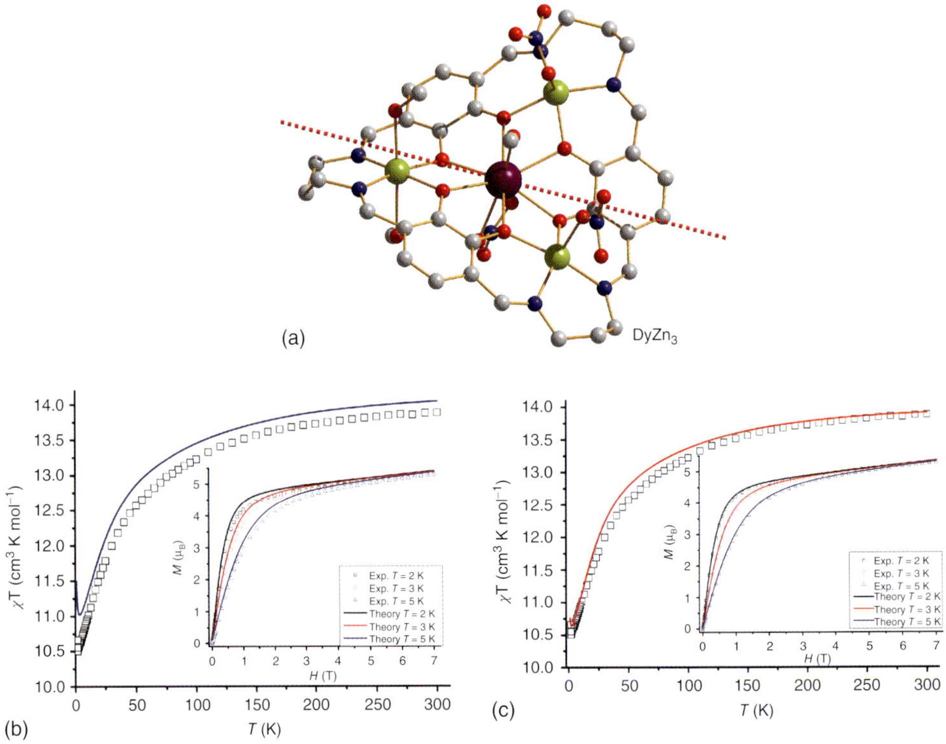

Figure 6.1 (a) Molecular structure of the macrocyclic $DyZn_3$ compound [58]. Red dashed line shows the main magnetic axis in the ground Kramers doublet. Colour scheme: Dy, violet; Zn, yellowgreen; O, red; N, blue; C, grey; H, white. (b) *Ab initio* calculated magnetism for ^6H,^6F and ^6P terms mixed by the spin–orbit coupling. (c) the same as (b) for all sextets, 128 quartets and 130 doublets originating from the $4f^9$ electronic configuration of Dy(III).

6.3.2
Temperature-Dependent Magnetic Susceptibility and Field-Dependent Magnetization

An example of the application of relativistic CASSCF/RASSI/SINGLE_ANISO calculations for the description of magnetism of lanthanide complexes is the recently studied $DyZn_3$ compound (Figure 6.1) [58]. All three Zn^{2+} ions have a $3d^{10}4s^0$ electronic configuration in the ground state, which means that they are non-magnetic ($S = 0$). This makes the $DyZn_3$ complex magnetically equivalent to a mononuclear Dy complex, since the magnetism of this compound is solely due to the presence of the Dy^{3+} ion, which has a $4f^9 5d^0 6s^0$ open-shell electronic configuration. Table 6.2 shows the results of the *ab initio* CASSCF/RASSI calculations. These calculations have been performed using ANO-RCC basis sets of the following contractions: [8s7p5d4f2g1h] for Dy, [5s4p2d1f] for Zn, [4s3p1d] for O and N from the first coordination sphere, [3s2p] for C, and all remaining N and O, and [2s] for H; the active space of the CASSCF method included nine electrons spanning seven 4f orbitals, CAS (9 in 7). The number of states included in the spin–orbit mixing was gradually increased from the minimal one (^{6}H, ^{6}F and ^{6}P, 21 spin states) towards the maximal possible one (279 spin states, limited by the hardware). We see that a minimal description of the spin–orbit coupling gives already a reasonable description of the magnetism, while the mixing of a large number of spin terms leads to a much better description of magnetism.

Table 6.2 Energies of the lowest spin-free states originating from the ^{6}H multiplet and the energies of the low-lying Kramers doublets of the $DyZn_3$ complex.

Atomic multiplet	Spin free states	*J* multiplet	Kramers Doublets
	0.0		0.0
	29.1		41.7
	88.3		75.4
	95.6		217.7
	173.8	$\frac{15}{2}$	273.6
^{6}H	192.9		298.9
	365.4		343.1
	387.1		391.9
	497.3		
	538.6		
	581.7		
Main values of the g tensor in the ground Kramers doublet			
g_X			0.105
g_Y			0.255
g_Z			18.148
Sign of the $g_X g_Y g_Z$			+

6.3.3 Magnetic Anisotropy in Low-Lying Doublets

The ground atomic J-multiplet of a lanthanide ion f^n splits in low-symmetry complexes into $2J + 1$ non-degenerate for even n (usually grouped in doublets), and in $(2J + 1)/2$ degenerate KDs for odd n. The *ab initio* calculations described above should be able to describe adequately the magnetic anisotropy of these doublet states. This was indeed found to be the case for individual Dy centres in Dy_3 triangles [59], for which *ab initio* calculations predicted the directions of local anisotropy axes [60] very close to the experimental results [61]. An example of a mononuclear Ln complex for which an ample comparison of experimentally measured single-ion magnetic anisotropy with *ab initio* predictions has been performed is the recently investigated Cp*ErCOT single-molecule magnet (SMM) [62]. Figure 6.2 shows the polar plot of angular dependence of single-crystal magnetization in the $a - b$ plane for each of the two molecules in the unit cell of the crystal. The relative shift of the theoretical plots from the experimental ones is due to the deviation of the calculated main anisotropy axes on the two molecules from the experimental ones by a few degrees [62]. Comparable accuracy of the *ab initio* calculated directions of local anisotropy axes has been demonstrated for other mononuclear lanthanide complexes as well by direct comparisons with the results of single-crystal investigations [54, 63, 64]. An indirect indication of the high accuracy of the predictions of local anisotropy axes in *ab initio* calculations of mononuclear Ln fragments (see the next section) can be inferred from good reproduction of magnetic properties of polynuclear lanthanide complexes, which crucially depend on relative orientations of these axes [65–75].

An important feature of the lanthanides is the *high axiality* of ground doublet state often achieved even in low-symmetry Ln complexes and fragments. This means a much larger g factor along the easy axis compared to hard axes (transversal g factors) in the case of Ln ions with odd number of electrons, and small splitting of the ground doublet in the case of Ln ions with even number of electrons. The high axiality of the ground doublet state explains the suppression of quantum tunnelling of magnetization (QTM) and the SMM behaviour of most of the lanthanide complexes, even when they completely lack any symmetry [37]. As an example, Table 6.3 shows that the transversal g factors (g_X, g_Y), for a low symmetry Dy(III) fragment of a Dy_3 triangle [37] are very small for the ground state KD1, that is, it is strongly axial. The table shows that for higher doublets the axiality strongly diminishes, becoming the lowest in the middle of the spectrum, while starting to increase again for the highest states. Concomitantly, the main anisotropy axes of the doublet state gradually rotate with respect to each other [37]. A similar behaviour is observed for Ln ions with even number of electrons in low-symmetry environment. Table 6.4 shows the results of *ab initio* calculations for a Tb(III) centre in the Tb_3 complex [76]. We see again that the lowest and the highest Ising doublets exhibit the smallest splitting (i.e. show the highest axiality),

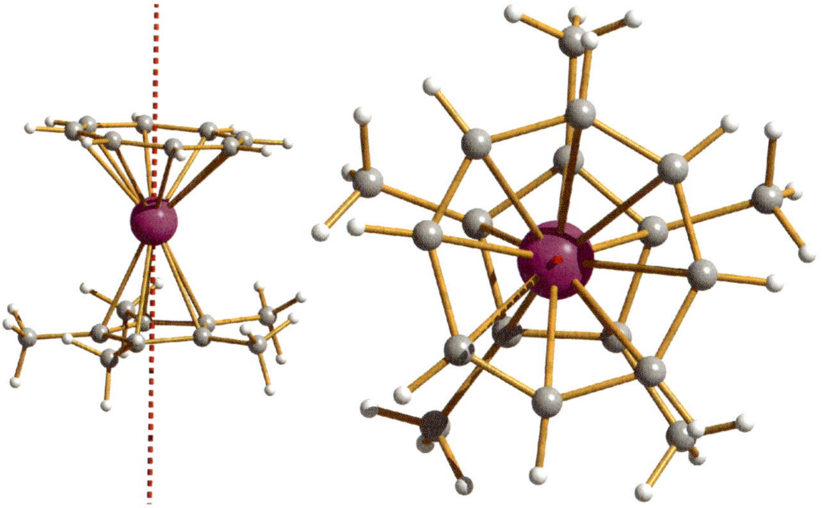

(a)

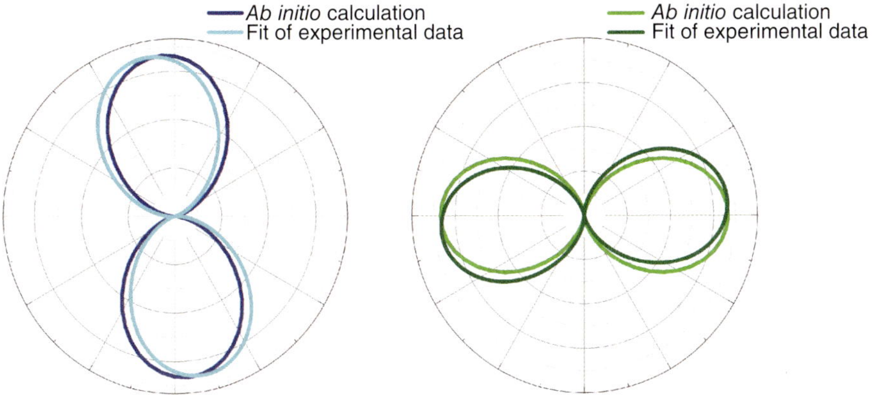

(b)

Figure 6.2 (a) Orientation of the main magnetic axis in the ground Kramers doublet of [Cp*ErCOT], [62]. Colour scheme: Er, purple; C, yellow; H, white. (b) Polar plot of the angular dependence of the single-crystal magnetic susceptibility recorded in the *a–b* plane, comparison between the results of the *ab initio* calculation for two molecules in the unit cell (left and right plot, respectively) and the contributions extracted from the fit of the experimental data (colour code in insets). Copyright © 2013 WILEY VCH Verlag GmbH & Co. KGaA, Weinheim.

while the main magnetic axes rotate gradually when passing from one doublet to the other [37]. We stress that the almost perfect axiality of the ground doublet state of low-symmetry lanthanide centres is an important feature enabling them to function as SMM. This is especially important in polynuclear complexes where Ln centres do not have, as a rule, any site symmetry.

Table 6.3 Energies (per centimetre) and main values of the g tensors of the lowest Kramers doublets at the Dy_3 site of the dysprosium triangle (from [37]).

Kramers doublet	Energy	g_X	g_Y	g_Z
1	0.000	0.008	0.034	19.667
2	103.476	0.398	0.645	17.750
3	141.038	3.495	3.670	11.366
4	171.100	2.214	4.969	11.748
5	208.691	0.069	3.223	14.098
6	228.586	1.215	3.047	16.450
7	287.051	0.104	0.281	18.743
8	395.074	0.002	0.008	19.556

Source: From Ref. [37].

Table 6.4 Energies (per centimetre) and the g_Z values[a)] of the lowest Ising doublets at the Tb_3 site of the Tb_3 complex

Ising doublet	Energy	Δ_{tun}	g_Z
1	0.000	0.090	17.837
	0.090		
2	106.225	1.271	14.711
	107.496		
3	171.446	6.940	11.853
	178.386		
4	214.780	—	—
5	235.191	7.797	9.904
	242.988		
6	275.819	1.693	14.113
	277.512		
7	485.772	0.046	17.685
	485.818		

a) $g_X = g_Y = 0$ according to the Griffith's theorem [77].
Source: From Ref. [37].

6.3.4 *Ab Initio* Crystal Field

As was mentioned in Section 6.2.3, phenomenological CF models have been extensively used in the past to rationalize the electronic structure of lanthanide complexes [4, 40–50], and they are still in use nowadays [38, 39, 78, 79]. The *ab initio* approach described above is, in principle, able to provide accurate energies and wave functions of multiplets in lanthanide complexes. Nevertheless, even if *ab initio* calculations appear to be successful, CF description is still of great interest because it includes in a concentrated way all basic interactions responsible for magnetic and optical (luminescence) properties of lanthanides. In this sense, the parameters of CF Hamiltonians play the role of physical characteristics of the complexes, much similar to the parameters of the effective zero-field splitting and how Zeeman Hamiltonians do for their magnetic properties. For the latter, the most relevant is the CF splitting of the ground atomic J-multiplet, that is, the multiplet-specific CF Hamiltonian:

$$\hat{\mathbf{H}}^{J}_{CF} = \sum_{n,m} (B^m_n O^m_n \left(\frac{\hat{\mathbf{J}}}{J}\right) + C^m_n \Omega^m_n \left(\frac{\hat{\mathbf{J}}}{J}\right) \tag{6.3}$$

where O^m_n and Ω^m_n are Stevens operators [35] defined in the space of angular momentum operator $\hat{\mathbf{J}}$ of the ground atomic multiplet. We have developed recently an approach based on ITO projection, which allows a unique determination of all 27 parameters B^m_n and C^m_n entering equation 6.3 from the *ab initio* calculations described above [80]. We stress that this is the only way to derive the set of *true* CF parameters, which would not be the case if a fitting procedure would have been applied. Another advantage of this procedure is that it permits to project out the CF contribution (6.3) from the total energy matrix and to assess the importance of higher rank ($n > 6$) non-CF contributions. This methodology is currently implemented in the SINGLE_ANISO module [81] of the MOLCAS program package [13] and can be used routinely.

As an example, we consider the Ishikawa's terbium double decker exhibiting an SMM behaviour, $[Tb(Pc)_2]^-$, where Pc is phtalocyanine [38]. Figure 6.3 shows the calculated spectrum of levels originating from the ground $J = 6$ multiplet of Tb(III) for the experimental geometry. The corresponding CF parameters calculated with SINGLE_ANISO routine as described above are shown in Table 6.5 [80]. This CF analysis also shows that the ground doublet state is very close to $M_J = \pm 6$ origin, while the first excited doublet is predominantly of $M_J = \pm 5$ type. For comparison, in Figure 6.3 is shown the spectrum obtained by phenomenological CF model involving only three axial parameters (Table 6.5) extracted from the fitting of magnetic susceptibility and of the chemical shift [1]. Although the nature of the ground and first excited doublet obtained is the same as in *ab initio* calculations, the energy separation is quite different, being closer to the experimental value (260 cm^{-1}) [1] in the latter case. The discrepancy between the two spectra becomes larger for the excited states. Another important observation is the very strong dependence of the CF spectrum on the deviation from the ideal symmetry. Although $[Tb(Pc)_2]^-$ is considered as a prototype of a high-symmetry Ln complex showing deviations from an ideal D_{4d} geometry of hundredths of angstroms in the Ln–N bond lengths, its symmetrization results in strong reconstruction of the *ab initio* spectrum (Figure 6.3). This is also seen in the strong difference between the CF parameters corresponding to the real and symmetrized geometries (Table 6.5). Obviously, such sensitivity of the CF at Ln ions on the geometry of their environment can hardly be simulated

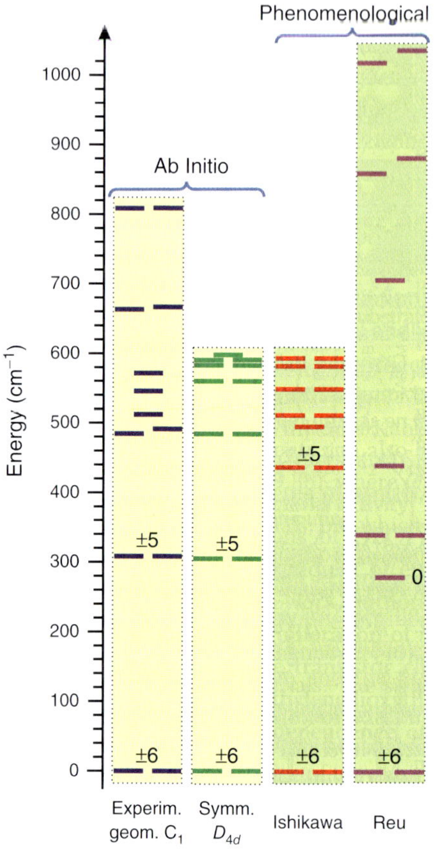

Figure 6.3 Comparison between ligand field states of the $[Tb(Pc)_2]^-$ obtained *ab initio* for the real geometry, for the symmetrized geometry up to the D_{4d} point group symmetry, and extracted from the fitting of the experimental data within CF approaches of Ishikawa [1] and Reu [78], assuming that the $[Tb(Pc)_2]^-$ compound has exact D_{4d} and C_4 symmetry, respectively.

realistically by phenomenological models. As an example, we show in Figure 6.3 the calculated CF spectrum for symmetrized $[Tb(Pc)_2]^-$ within the exchange charge model of the CF [78]. Despite the fact that the parameters in this model have been fitted to the same experimental data as in the previous CF model [1], the CF spectrum of the two models differ drastically. In particular, the first excited state predicted by the exchange charge model corresponds to the $M_J = 0$ [78] and not to the $M_J = \pm 5$ predicted by the previous CF model (and the *ab initio* calculations). This comparison shows that phenomenological CF models can encounter serious problems in the prediction of realistic spectrum so that the

Table 6.5 Crystal-field parameters for Ishikawa's $[Tb(Pc)_2]^-$ complex obtained *ab initio* for the experimental geometry [80], for the symmetrized geometry [80] and those obtained from the fitting of the experimental data [1, 78].

Rank	Projection	Real geometry		D_{4d}- geometry		Fitting by Ishikawa [1]		Fitting by Reu[78]	
		B_n^m	C_n^m	B_n^m	C_n^m	B_n^m	C_n^m	B_n^m	C_n^m
2	0	608.8	–	530.4	–	414	–	173	–
	1	−19.1	4.6	0	0	x	x	*x*	x
	2	19.2	317.9	0	0	x	x	*x*	x
4	0	−98.5	–	−112.1	–	−228	–	−631.4	–
	1	−11	0.5	0	0	x	x	*x*	x
	2	3.5	59.8	0	0	x	x	*x*	x
	3	0.6	−2.2	0	0	x	x	*x*	x
	4	−15	5	0	0	x	x	397.4	953
6	0	−13.9	–	−7.5	–	33	–	79.7	–
	1	−0.2	−0.9	0	0	x	x	*x*	x
	2	−0.7	24.1	0	0	x	x	*x*	x
	3	1.3	−1.2	0	0	x	x	*x*	x
	4	−5.3	3	0	0	x	x	256.4	667
	5	3.4	−4.1	0	0	x	x	*x*	x
	6	12.1	17.9	0	0	x	x	*x*	x

x −parameters that were ignored in the fitting of experimental data [1, 78];

ab initio approach becomes indispensable, especially, when all 27 CF parameters are non-negligible.

6.4 *Ab Initio* Calculation of Anisotropic Magnetic Properties of Polynuclear Complexes

The *ab initio* calculations of the exchange spectrum and the derivation of exchange parameters for polynuclear lanthanide complexes is a much more difficult task than the calculation of CF splitting and single-ion magnetism. This is the reason why such calculations have not yet been done. DFT calculations of exchange coupling parameters have been done only for Gd-TM and Gd-radical (R) complexes [82], but cannot be applied to other lanthanide ions due to reasons mentioned in Section 6.2.2. Fortunately, the exchange interactions between two strongly axial lanthanide ions (Ln1−Ln2) or between a strongly axial lanthanide ion and an isotropic spin, such as Ln-TM, or Ln-R exchange coupled pairs, reduces the exchange interaction of the Ising type for the ground doublet states. The latter can be efficiently described within a semiempirical two-step approach reviewed in subsequent sections.

6.4.1 Two-Step Approach for the Calculation of Electronic Structure of Polynuclear Lanthanide Complexes

Localization of the magnetic orbitals observed in TM complexes and networks [83] is even more pronounced in lanthanide-containing mono- and polynuclear compounds. This fact allows treating the electronic and magnetic structure of polynuclear compounds in a two-step procedure. In the first step, reasonable fragmentation of the cluster into mononuclear fragments is performed and reliable *ab initio* calculations for each fragment are done. In the second step, the magnetic interaction between the fragments is introduced in an effective way. The exchange interaction between magnetic centres is considered within the Lines model [84], while accounting for the dipole–dipole magnetic coupling is treated exactly. The Lines model [84] is an approximation that allows describing the anisotropic exchange interaction between magnetic sites via a single parameter. To this end, an isotropic Heisenberg model involving the true spins of the two centres is introduced with an effective parameter (the Lines exchange parameter). In the second step, the matrix of this model is constructed on the basis of products of localized lowest states of the two centres obtained in fragment *ab initio* calculations of corresponding centres with spin–orbit coupling included. The exchange matrix obtained describes exactly the anisotropic exchange interaction in two limiting cases:

1) Of two strongly axial doublets on the two sites (the case of extreme magnetic anisotropy)
2) Of isotropic spins on the two sites (lack of magnetic anisotropy).

The Lines approximation is expected to be quite accurate for the description of the exchange interaction between a strongly axial doublet and an arbitrary isotropic spin. For all other cases, the Lines model [84] is a reasonable approximation. Efficient implementation of the Lines model was done in the program POLY_ANISO.

6.4.2 Key Rules for Cluster Fragmentation

One important point in this two-step approach is the cluster fragmentation into mononuclear fragments containing unpaired electrons (spins) only on one magnetic centre. By fragmentation, we try to achieve the preservation of the electronic and magnetic structure of each of the magnetic centres as close as possible to their form in the initial polynuclear cluster (if one makes abstraction of relatively weak exchange and dipolar interaction between the sites).

From previous experience, one can formulate some rules for building reliable fragments:

- If the computer power is sufficient, it is not recommended to remove any atom from the initial polynuclear molecule. To calculate the magnetic states on one centre, neighbouring metal ions have to be described by their closest diamagnetic equivalents. For example, Fe^{2+}, Co^{2+}, Ni^{2+} and Cu^{2+} could be reasonably well described by diamagnetic Zn^{2+}, while Ti^{2+}, V^{2+}, Cr^{2+} and Mn^{2+} are probably better described by diamagnetic Ca^{2+}. In this respect, the charge of the neighbouring ions should also be taken into account correctly: Ti^{3+}, V^{3+}, Cr^{3+} and Mn^{3+} can be simulated by diamagnetic Sc^{3+}. Metals which are known to have a very stable closed-shell (i.e. diamagnetic) configuration in the ground state (e.g. octahedral Co^{3+}) are recommended to be kept as they are in the *ab initio* calculation.
- If the cluster is too large to be treated as described above, then some distant groups of atoms have to be removed or replaced. It is always recommended to start the removal of atomic groups from distant regions of the considered magnetic centre. At this step chemical intuition is of great help. For example, $-CH_3$ group is chemically quite similar to the H atom; therefore, such a substitution at a significant distance from the magnetic centre is not supposed to change the electronic and magnetic structure on the metal site much. *t*-Butyl groups could be efficiently modelled by a $-CH_3$ group. However, it is not accurate, in our opinion, to replace conjugated rings (e.g. phenyl) by H atoms, as they are quite large, and may induce $\pi-\pi$ stacking effects, which influence intra- and intermolecular magnetic interactions. Such a replacement is acceptable only for a crude computational model. In general, it is better to avoid breaking of any conjugated part of the molecule, or to cut the double or triple bond.
- The position of the hydrogen atoms which are bonded directly to the first-coordinated ligand atoms (usually O, N) is very important, as they effectively orient the orbitals of the first-coordinated ligand atoms, influencing directly the ligand field acting on the magnetic centre. In general, the effect of the second coordination sphere is of great importance for the correct description of the ligand field, as has been previously shown [53, 85].

6.4.3 Implementation of *Ab Initio* Methodology: POLY_ANISO Program

The POLY_ANISO program is a computational tool which allows treatment of exchange interaction within the Lines model, dipole–dipole magnetic coupling, calculation of effective spin (pseudospin) Hamiltonians and static magnetic properties of polynuclear complexes using the *ab initio* calculated wave functions and spectrum of individual magnetic centres (fragments). The computational methodology for pseudospin Hamiltonians is similar to mononuclear fragments [55, 56]. The derivation of macroscopic magnetic properties is given in [57].

The current version of the POLY_ANISO program is able to calculate the following quantities:

- Anisotropic exchange interaction within the Lines model described above. Dipole–dipole magnetic interaction between centres is calculated fully *ab initio* and is added to the exchange matrix.
- Parameters of pseudospin magnetic Hamiltonians (see the description in Section 6.3.1, points 1–4). It is important to note that the multiplicity of the pseudospin can become quite large in the case of polynuclear compounds with weakly interacting magnetic sites.
- Static magnetic properties (see Section 6.3.1, points 1–4): The calculation of magnetic properties takes into account the contribution of all exchange states and also excited states on individual magnetic centres (the ligand field and charge transfer states of the complex or mononuclear fragment included in the RASSI calculation) via their thermal population and Zeeman admixture. The intermolecular exchange interaction between magnetic molecules in a crystal can be taken into account during the simulation of magnetic properties by a phenomenological parameter zJ' specified in the input.

6.4.4
Noncollinear Magnetic Structure of Ln_n Complexes

A prominent feature of polynuclear lanthanide complexes, which distinguishes them from TM complexes, is their noncollinear magnetic structure. TM complexes are usually found in the strong exchange limit [86], when the exchange splitting in the complex is much larger than the zero-field splitting on the metal sites. In such systems, the ionic and exchange anisotropy leads to zero-field splitting of exchange multiplets, characterized by a total spin S into individual levels or into exchange doublets $|S, \pm M\rangle$ (or their combinations) when the axial anisotropy of the entire complex is predominant, as in Mn_{12}acac and Fe_8 complexes [87]. In the state of saturated magnetization $+M$ or $-M$ of a given doublet, the local magnetizations on the metal sites (and radicals, if any) are parallel or antiparallel to its total magnetization, because the corresponding wave functions $|S, +M\rangle$ or $|S, -M\rangle$, are eigenstates of isotropic (Heisenberg) exchange interaction. Contrary to TM compounds, lanthanide complexes are found in the weak exchange limit because the CF splitting of the ground atomic multiplet of Ln ions ($> 500\text{cm}^{-1}$) exceeds considerably the exchange interaction between them (few per centimetres). As a result, in the case of lanthanides, which are in addition strongly axial ($g_Z \gg g_X, g_Y$) in low-symmetric environment, the local magnetizations in magnetically saturated states will point along the main magnetic axes (Z) of corresponding Ln ions, which are generally not parallel to each other.

As an example of such a noncollinear magnetic structure revealed by *ab initio* calculations, Figure 6.4 shows the calculated local anisotropy axes on three Dy sites of the Dy_3 triangle and the relative arrangement of the local magnetic moments in the ground state [60]. The three magnetic moments form an almost toroidal arrangement (Figure 6.4) and sum up into a total magnetic moment $<0.5\mu_B$ (calculated local magnetic moments on Dy sites are $\approx 10\mu_B$), in full

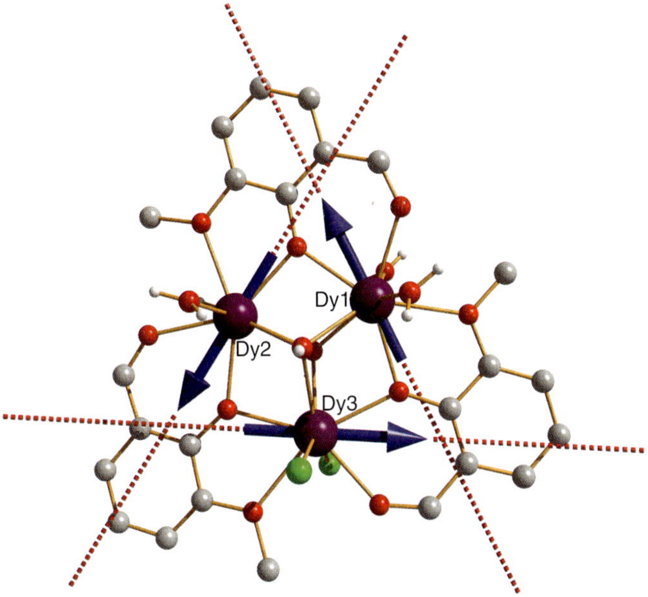

Figure 6.4 Structure of the Dy_3 triangular unit [60, 67]. Colour scheme: Dy^{III}, purple; Cl, green; O, red; C, grey; H, white. Dashed lines show the calculated anisotropy axes on the dysprosium sites and the arrows show the ordering of local magnetizations in the ground state of the complex.

accord with the experimental slope of $M(H)$ and χT at low temperature [59]. It is worth noting that besides a good description of magnetism, the applied *ab initio* methodology proved to be surprisingly accurate in the prediction of anisotropic properties of individual Ln ions. Thus, the Dy_3 triangle is the first molecular system where an almost perfect toroidal magnetic moment was detected. Indeed, the experimentally derived directions of local anisotropy axes [61] show deviations from the calculated ones by only a few degrees [67].

Another important insight offered by the *ab initio* calculations is the structure and the anisotropy of exchange multiplets of Dy_3 complex, arising from anisotropic exchange and dipolar interaction between Dy centres (Section 6.4.1). Figure 6.5 shows the spectrum of lowest exchange multiplets arising from the interaction between the ground KDs at Dy sites [67]. The three excited exchange multiplets correspond to reversal of one or two local magnetic moments with respect to the ground magnetic configuration (Figure 6.5c). As a result, the complex acquires in these states a magnetic moment of about 20 μ_B, which mostly lies in the Dy_3 plane as the Zeeman splitting shows (Figure 6.5a). The almost perfect cancellation of the sum of three magnetic moments in the ground state of dysprosium triangle is due to its nearly trigonal symmetry and to the fact that the local magnetic axes lie mostly in the Dy_3 plane (Figure 6.4). Deviation

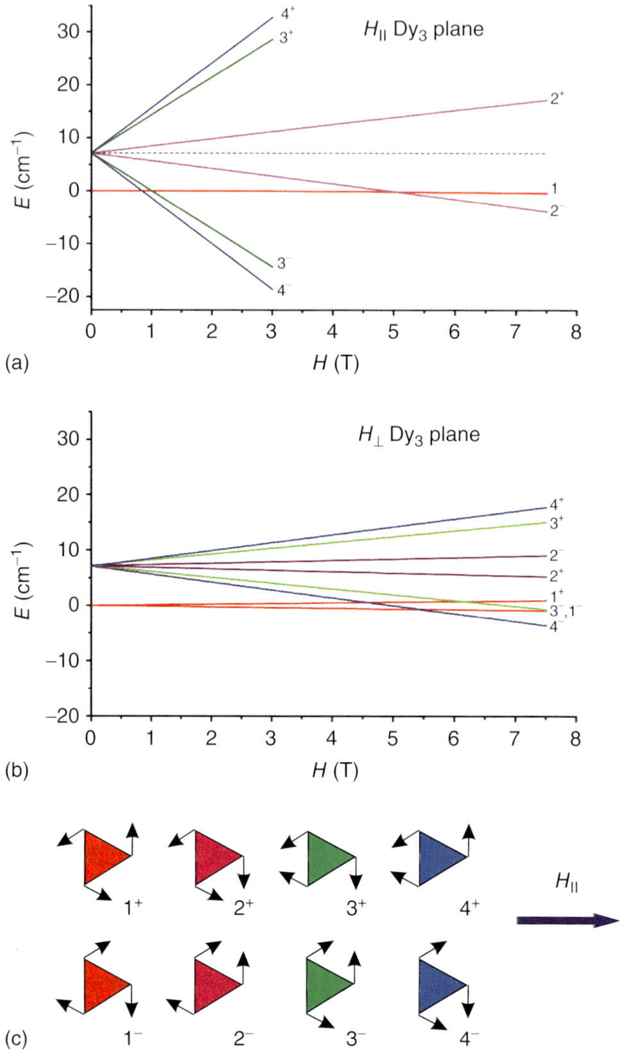

Figure 6.5 Evolution of the lowest magnetic states of Dy_3 complex with applied magnetic field, simulated for $J = -0.6\ cm^{-1}$: (a) H is applied in the Dy_3 plane. Reproduced from Ref. [67] with permission from the Centre National de la Recherche Scientifique (CNRS) and The Royal Society of Chemistry; (b) H is perpendicular to the Dy_3 plane; (c) schematic noncollinear magnetic structure of the ground (1) and three excited (2–4) Kramers doublets and the direction of the in-plane applied field in (a). The two signs in the superscripts denote the two time reversal components of the Kramers doublets.

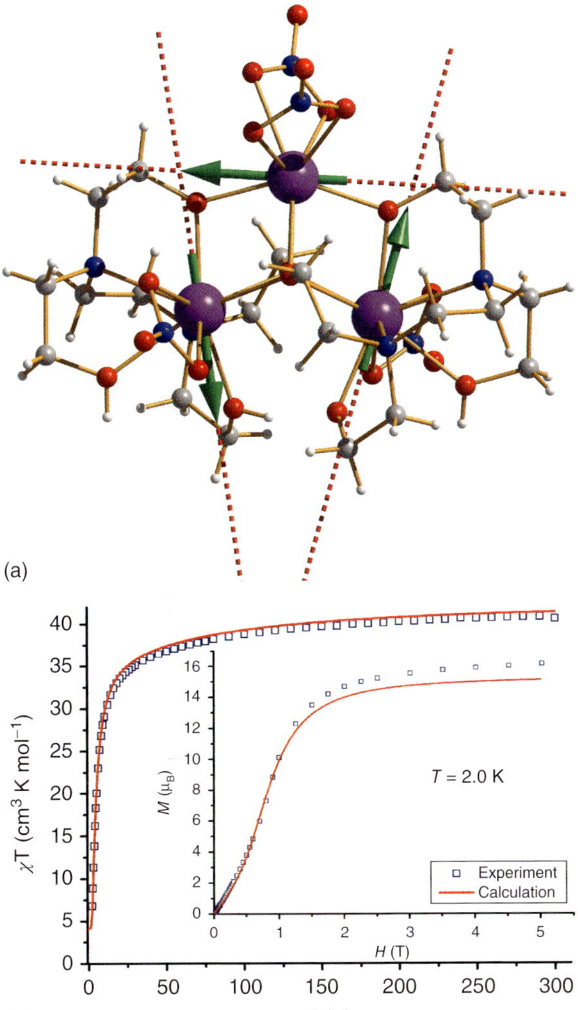

Figure 6.6 (a) Calculated local anisotropy axes on Dy(III) sites (dashed lines) and the local magnetic moments in the ground state (arrows). (b) Temperature dependence of the χT product for Dy_3-isoscelles complex. The red line is the *ab initio* calculated curve. Inset: molar magnetization at 2.0 K. (Reproduced from Ref. [72] with permission from The Royal Society of Chemistry.)

from the geometry of an equilateral triangle will destroy this cancellation resulting in a magnetic ground state of the complex. As an example, Figure 6.6 shows a complex where one oxygen bridge connecting two of three Dy centres is removed [72]. The calculated main magnetic axes in the ground KD of Dy(III) ions still lie in the Dy_3 plane; however, the local magnetic moments in the ground exchange multiplet do not compensate each other anymore, resulting in a non-negligible total magnetic moment (Figure 6.6a). The latter manifests itself in a high slope of magnetization $M(H)$ and χT curves at low temperatures (Figure 6.6b). We also note that the *ab initio* methodology gives quite an accurate description of excited exchange multiplets, which is testified by close reproduction of experimental magnetization steps [72]. This approach was found successful also for the rationalization of anisotropic magnetic properties and the elucidation of non-collinear magnetic structure in a number of Ln_n complexes of higher nuclearity such as Dy_4 parallelogram [88] and Dy_6 wheel [71].

6.4.5 Mixed Lanthanide-Transition Metal Compounds

One of the intensively explored directions in the lanthanide research is the design of mixed lanthanide-TM compounds. The main motivation behind this research is the much stronger exchange interaction (by about one order of magnitude) in Ln–TM pairs compared to Ln–Ln pairs, which in combination with strong anisotropy on Ln sites, makes the mixed compounds more promising candidates for the design of efficient SMMs. The magnetic structure in the ground exchange multiplets of mixed compounds is richer than in pure lanthanide complexes and depends on the nature of TM ions and on the relative orientations of the main magnetic axes on Ln ions. Since the latter are strongly axial as a rule, the exchange interaction between them and isotropic TM ions (or radicals) is of collinear Ising type. Then, given that the Ln ions contained in the complex have all parallel main magnetic axes, the arrangements of all magnetic moments in the ground exchange doublet of mixed complex will be parallel, that is, fully collinear, with all local magnetic moments pointing along these axes. As an example, Figure 6.7a shows a $Cr_2{}^{III}Dy_2{}^{III}$ tetranuclear complex, where *ab initio* calculations have revealed such a magnetic structure [75]. If the local magnetic axes on Ln sites are not parallel to each other, then the isotropic spin of a TM, which interacts with two (or more) of them will align along an intermediate direction, to provide the maximal gain of exchange energy in the ground state. For example, in $Cr^{III}_4Dy^{III}_4$ square (wheel) complex (Figure 6.7b) the isotropic spin on each Cr ion lies in the plane bisecting the angle between the directions of the main magnetic axes of two neighbouring Dy sites [89]. Finally, when the TM ions are not isotropic, their magnetic moments in the ground exchange multiplet of the complex will not align completely to the magnetic axes of the lanthanide ions with which they interact, but will occupy an intermediate direction between the latter and their own anisotropy axes. As an example, Figure 6.7c shows the arrangement of local magnetic moments in the ground state of $Co^{II}_2Dy^{III}_2$ complex [70], from which we can see that the magnetic moments on Co ions are not parallel to the magnetic moments on Dy ions, albeit the latter have parallel magnetic axes.

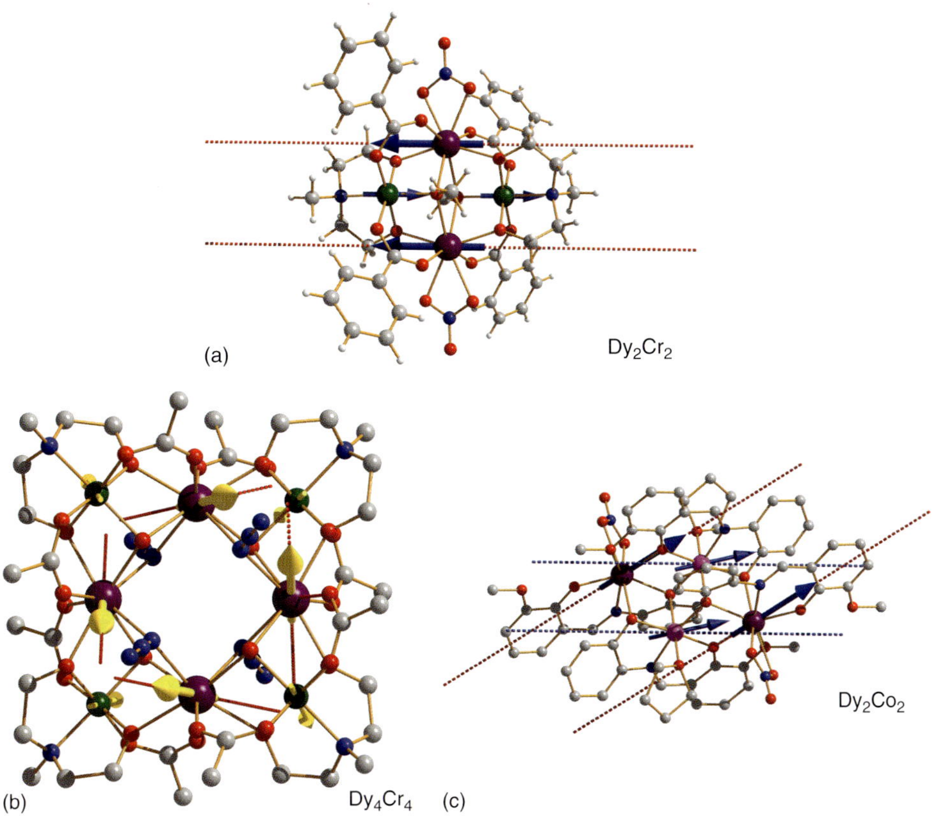

Figure 6.7 (a) Molecular structure of the $Dy^{III}_2Cr^{III}_2$ complex [75]. (b) Molecular structure of the $Dy^{III}_4Cr^{III}_4$ complex [89]. Copyright © 2010 WILEY VCH Verlag GmbH & Co. KGaA, Weinheim. (c) Molecular structure of the $Dy^{III}_2Co^{II}_2$ complex [70]. In (a), magnetic moments on Cr ions are exactly antiparallel to the moments on Dy^{III} ions due to inversion symmetry of the entire complex (i.e. main magnetic axes on Dy are parallel), strong Dy–Cr antiferromagnetic coupling and lack of intraionic anisotropy on Cr^{III} ions. In (b), the magnetic moments on Cr^{III} ions are oriented in between the magnetic moments of neighbouring Dy^{III} ions. In (c), there is a competition between exchange interaction, which tends to align all magnetic moments along a certain axis, and local magnetic anisotropy, which orients the magnetic moments on metal sites along their intrinsic anisotropy axes.

Another distinct property of mixed complexes which we mention here is the possibility to observe two activation regimes of relaxation of magnetization. One of them corresponds to reversal of magnetization of individual ions, at higher temperatures, and the other is related to climbing over the barrier built from the exchange multiplets of the complex. The coexistence of these two relaxation regimes has been recently revealed in the $Co^{II}_2Dy^{III}_2$ complex in a combined study including ac susceptibility and recovery magnetization measurements and the *ab initio* calculations [70]. On the other hand, two activation barriers are not seen concomitantly in pure Ln_n complexes due to a weaker exchange/dipolar interaction and, as a consequence, a much lower exchange barrier. Besides providing relatively strong exchange interaction, the largest being recently attained in some $Ln_2(N_2)^{3-}$ lanthanide-radical systems [90, 91], the mixed complexes can involve isotropic magnetic centres with large spin, which is an additional factor contributing to the design of efficient barriers for magnetization blocking [73].

6.4.6 Lanthanide-Containing Magnetic Chains

It is well known that the magnetic properties of long (infinite) chains of exchange-coupled magnetic centres are generally not described exactly, even at a model level of theory. Two practical tools for the treatment of chains of isotropically (slightly anisotropically) coupled spins are the classical treatment in the case of large spins ($S \gg 1/2$) [92] or via Monte Carlo simulations [93]. Contrary to isotropic spin chains, the chains containing lanthanide ions allow for a surprising simplification of their theoretical description and, as a consequence, to an exact treatment of their magnetic properties in some important cases. These are situations when the chain can be divided into units delimited by strongly axial pseudospins described by the ground multiplets of Ln ions. It is supposed that these pseudospins have Ising exchange interactions with left and right (multicentre) units, while the latter do not interact directly. In this case, the thermodynamical properties of the chains can be calculated exactly by the transfer matrix method [94]. In combination with *ab initio* treatment of anisotropic magnetic properties of individual metal centres described above, this approach allows for the full treatment of magnetism in many Ln-containing magnetic chains. The approach described has been applied for the *ab initio* investigation of $\{[(CuL)_2Dy][Mo(CN)_8]\}_\infty$, the first studied chain with the 3d–4f–5d motif [66]. Figure 6.8a shows that the chain can be conceived as Cu1–Mo–Cu2 magnetic units linked via Dy(III) ions. The *ab initio* calculations have shown that the ground KD at each Dy(III) site is strongly axial [66]. Then, applying the transfer matrix method the magnetic susceptibility and field-dependent magnetization have been simulated (Figure 6.8b,c). The relative arrangement of magnetic moments on Dy sites in the ground state of the chain is shown in Figure 6.8a.

Another series of lanthanide-containing chains which have been treated *ab initio* are the chiral $\{Dy_3Cu\}_\infty$ 1D polymers [69]. These polymers represent Dy_3 units, with structure similar to the dysprosium triangles discussed above (Figure 6.4), linked via Cu(II) ions (Figure 6.9a). The latter are connected to Dy_2 ion in the right unit and to Dy_1 in the left one. The *ab initio* calculations show that all Dy ions in the units are strongly axial, with relative orientations of main magnetic axes very close to the dysprosium triangle (Figure 6.4). As in

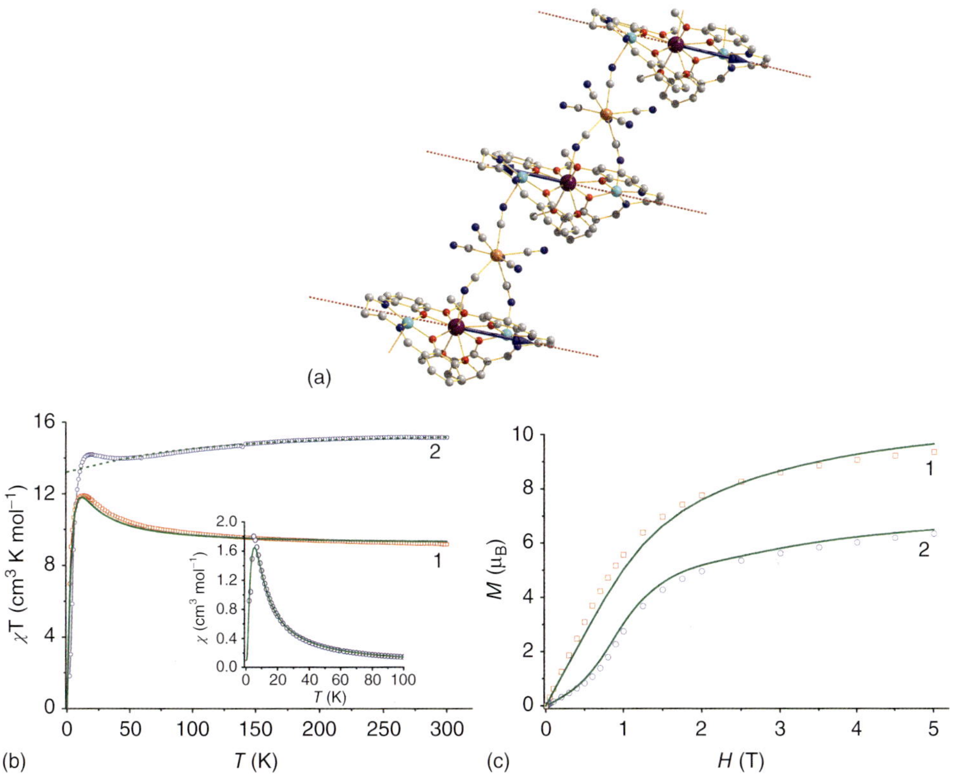

Figure 6.8 (a) Direction of local anisotropy axes (dashed lines) and the low-temperature alignment of magnetic moments (blue arrows) on dysprosium ions along the chain in crystal $\{[(CuL)_2Dy][Mo(CN)_8]\}_\infty$ [66]. (b) Temperature dependence of $\chi_M(T)$ for $\{[(CuL)_2Gd][Mo(CN)_8]\}_\infty$ (**1**) and $\{[(CuL)_2Dy][Mo(CN)_8]\}_\infty$ (**2**) (experimental in squares (**1**) and circles (**2**), and simulated in solid line), the dashed curve corresponds to *ab initio* calculation of $\chi_M(T)$ for non-interacting magnetic centres in **2**; the inset shows the experimental (circles) and simulated (line) susceptibility χ_M of (**2**) taking into account the exchange interaction between metal centres [66]. (c) Field dependence of the magnetization recorded for 2.0 K **1** (squares) and **2** (dots), the continuous curves are the *ab initio* based simulations for **1** and **2** when exchange interactions between magnetic ions is taken into account.

the latter case, the exchange/dipolar interaction between Dy centres stabilizes the toroidal arrangement of local magnetic moments in each Dy_3 unit. The calculated magnetic properties of the chain are shown in Figure 6.9b. The exchange spectrum and anisotropic magnetic properties of individual Dy_3 units is similar to those shown in Figure 6.5 for the dysprosium triangle. The spectrum of low-lying excitations in the chain corresponds to Ising exchange interaction between the ground state exchange KD of Dy_3 units, describing the two values

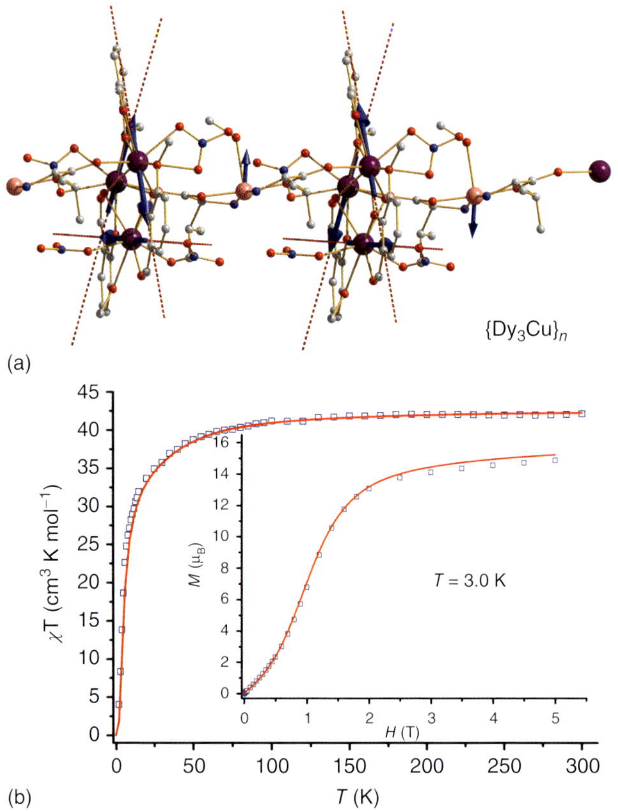

Figure 6.9 (a) Calculated anisotropy axes for the lowest Kramers doublets on Dy ions in the chiral $\{Dy_3Cu\}_\infty$ 1D polymers [69]. Blue arrows show the orientation of the local magnetic moments in the ground state on Dy and Cu ions. (b) A comparison between measured (empty squares) and calculated (red line) magnetic susceptibility for the $\{Dy_3Cu\}_\infty$ chain. Inset: a comparison of the measured and calculated molar magnetization at 3.0 K. (Reproduced from Ref. [69] with permission from The Royal Society of Chemistry.)

of the toroidal moment ($\tau = \pm 1/2$) and the projections of magnetic moment of Cu(II) ion ($s_z = 1/2$) along their direction in the ground state. (Figure 6.9a).

6.5 Conclusions

Theoretical description of molecular magnets has a history of several decades. The most investigated systems in the past were isotropic or weakly anisotropic magnetic molecules, for which the phenomenological modelling of their magnetic properties was sufficient [86, 92]. The strongly anisotropic magnetic complexes, in particular, the lanthanide compounds, represent more complex magnetic properties requiring much more involved models for their rationalization. To be adequate, these models should involve tens of parameters, which can hardly be extracted from experiment in a reliable way. At this stage, the *ab initio* calculations based on explicitly correlated wave functions of the complex with non-perturbative (exact) treatment of spin–orbit coupling can become an indispensable research tool. We have shown here that the *ab initio* calculations based on CASSCF/RASSI approach implemented in MOLCAS package [13] provide the description of magnetic properties with experimental accuracy, while the pseudospin magnetic Hamiltonians describing these properties can be routinely derived from the results of these calculations. The *ab initio* approach is supplemented by phenomenological description of anisotropic exchange interactions within the Lines model, proved to be also highly successful for the description of many polynuclear compounds involving lanthanides. Given the constantly increasing computer power and predictive capabilities of modern quantum chemistry, the *ab initio* methods are expected to become a standard research tool for lanthanide-based magnetic materials.

References

1. Ishikawa, N., Sugita, M., Ishikawa, T., Koshihara, S. and Kaizu, Y. (2004) *J. Phys. Chem. B*, **108**, 11,265–11,271.
2. Woodruff, D.N., Winpenny, R.E.P. and Layfield, R.A. (2013) *Chem. Rev.*, **113**, 5110–5148.
3. Sessoli, R. and Powell, A.K. (2009) *Coord. Chem. Rev.*, **253**, 2328–2341.
4. Wybourne, B.G. (1965) *Spectroscopic Properties of Rare Earth Ions*, Interscience, New York.
5. Douglas, M. and Kroll, N.M. (1974) *Ann. Phys.*, **82**, 89–155.
6. Hess, B.A. (1986) *Phys. Rev. A*, **33**, 3742–3748.
7. Reiher, M. (2012) *WIREs Comput. Mol. Sci.*, **2**, 139–149.
8. Siegbahn, P.E.M., Heiberg, A., Roos, B.O. and Levy, B. (1980) *Phys. Scr.*, **21**, 323–327.
9. Roos, B.O., Taylor, P.R. and Siegbahn, P.E.M. (1980) *Chem. Phys.*, **48**, 157–173.
10. Siegbahn, P.E.M., Almlöf, J., Heiberg, A. and Roos, B.O. (1981) *J. Chem. Phys.*, **74**, 2384–2396.
11. Olsen, J., Roos, B.O., Jørgensen, P. and Jensen, H.J.A. (1988) *J. Chem. Phys.*, **89**, 2185.
12. Malmqvist, P.A.A., Rendall, A. and Roos, B.O. (1990) *J. Phys. Chem.*, **94**, 5477–5482.
13. Aquilante, F., Vico, L.D., Ferré, N., Ghigo, G., Malmqvist, P.A.A., Neogrády, P., Pedersen, T.B., Pitoňák, M., Reiher, M., Roos, B.O., Serrano-Andrés, L., Urban, M., Veryazov, V. and Lindh, R. (2010) *J. Comput. Chem.*, **31**, 224.
14. Frisch, M.J., Trucks, G.W., Schlegel, H.B., Scuseria, G.E., Robb, M.A. *et al.* (2009) *Gaussian 09 Revision A.1*, Gaussian Inc., Wallingford, CT.
15. Neese, F. (2010) *ORCA - An Ab Initio, Density Functional and Semiempirical Program Package, Version 2.9.0.*, Universitat Bonn, Bonn.
16. Werner, H.-J., Knowles, P.J., Manby, F.R., Schutz, M. *et al.* (2009) *MOLPRO, Version 2009.1, A Package of ab Initio Programs*, Universitat Stuttgart, Stuttgart, *http://www.molpro.net* (accessed 09 September 2014).

17. Andersson, K., Malmqvist, P.A.A. and Roos, B.O. (1992) *J. Chem. Phys.*, **96**, 1218.
18. Angeli, C., Cimiraglia, R., Evangelisti, S., Leininger, T. and Malrieu, J.P. (2001) *J. Chem. Phys.*, **114**, 10 252.
19. Roos, B.O., Andersson, K., Fulscher, M.P., Malmqvist, P.A.A., Serrano-Andres, L., Pierloot, K. and Merchan, M. (1996) *Adv. Chem. Phys.*, **93**, 219–331.
20. Schapiro, I., Sivalingam, K. and Neese, F. (2013) *J. Chem. Theory Comput.*, **9**, 3567-3580.
21. Helgaker, T., Olsen, J. and Jørgensen, P. (2000) *Molecular Electronic-Structure Theory*, John Wiley & Sons, Ltd.
22. Garcia, V.M., Castell, O., Caballol, R. and Malrieu, J.P. (1995) *Chem. Phys. Lett.*, **238**, 222-229.
23. Douglas, N. and Kroll, N.M. (1974) *Ann. Phys.*, **82**, 89.
24. Hess, B.A. (1986) *Phys. Rev. A*, **33**, 742.
25. Hess, B.A., Marian, C., Wahlgren, U. and Gropen, O. (1996) *Chem. Phys. Lett.*, **251**, 365.
26. Roos, B.O., Lindh, R., Malmqvist, P.A.A., Veryazov, V. and Widmark, P.O. (2004) *J. Phys. Chem. A*, **108**, 2851–2858.
27. Roos, B.O., Lindh, R., Malmqvist, P.A.A., Veryazov, V. and Widmark, P.O. (2005) *J. Phys. Chem. A*, **109**, 6575–6579.
28. Tsuchiya, T., Abe, M., Nakajima, T. and Hirao, K. (2001) *J. Chem. Phys.*, **115**, 4463.
29. Malmqvist, P.A.A., Roos, B.O. and Schimmelpfennig, B. (2002) *Chem. Phys. Lett.*, **357**, 230.
30. Ganyushin, D. and Neese, F. (2013) *J. Chem. Phys.*, **138**, 104 113.
31. Vallet, V., Maron, L., Teichteil, C. and Flament, J.P. (2000) *J. Chem. Phys.*, **113**, 1391–1402.
32. Verma, P. and Autschbach, J. (2013) *J. Chem. Theory Comput.*, **9**, 1052–1067.
33. Zalubas, W.C.M.R. and Hagan, L. (1978) *Atomic Energy Levels - The Rare-Earth Elements*, US Government Printing Office, Washington, DC.
34. Sandratskii, L.M. (1998) *Adv. Phys.*, **47**, 91–160.
35. Abragam, A. and Bleaney, B. (1970) *Electron Paramagnetic Resonance of Transition Ions*, Oxford University Press, Oxford.
36. Wigner, E.P. (1959) *Group Theory: And its Application to the Quantum Mechanics of Atomic Spectra*, Academic Press, London.
37. Ungur, L. and Chibotaru, L.F. (2011) *Phys. Chem. Chem. Phys.*, **13**, 20 086–20 090.
38. Ishikawa, N., Sugita, M., Ishikawa, T., Koshihara, S. and Kaizu, Y. (2003) *J. Am. Chem. Soc.*, **125**, 8694-8695.
39. AlDamen, M.A., Clemente-Juan, J.M., Coronado, E., Martí-Gastaldo, C. and Gaita-Ariño, A. (2008) *J. Am. Chem. Soc.*, **130**, 8874-8875.
40. Judd, B.R. (1963) *Operators Techiques in Atomic Spectroscopy*, McGraw-Hill.
41. Morrison, C.A. and Leavitt, R.P. (1982) *Handbook on the Physics and Chemistry of Rare Earths*, vol. 5, North-Holland, Amsterdam, pp. 461–692.
42. Jørgensen, C.K. (1963) *Adv. Chem. Phys.*, **5**, 33–146.
43. Gerloch, M. and Wooley, R.G. (1983) *Prog. Inorg. Chem.*, **31**, 371–446.
44. Figgis, B.N. and Hitchman, M.A. (1999) *Ligand Field Theory and its Applications*, Wiley-VCH Verlag GmbH.
45. Görller-Walrand, C. and Binnemans, K. (1996) *Handbook on the Physics and Chemistry of Rare Earths*, vol. 23, North-Holland, Amsterdam, pp. 121–283.
46. Erdos, P. and Razafimandimby, H.A. (1979) *J. Phys. (Paris)*, **40**, 171–172.
47. Newman, D.J. (1971) *Adv. Phys.*, **20**, 197–446.
48. Morrison, C.A. (1982) Lectures on Crystal Field Theory. Tech. Rep. HDL-SR-82-2, Harry Diamonds Laboratory.
49. Malta, O.L. (1982) *Chem. Phys. Lett.*, **87**, 27-29.
50. Malta, O.L. (1982) *Chem. Phys. Lett.*, **88**, 353–356.
51. Nekvasil, V. (1978) *Phys. Status Solidi B*, **87**, 317–323.
52. Nekvasil, V. (1979) *Czech. J. Phys.*, **88**, 785–796.
53. Petit, S., Pillet, G., Luneau, D., Chibotaru, L.F. and Ungur, L. (2007) *Dalton Trans.*, **40**, 4582–4588.

54. Boulon, M.E., Cucinotta, G., Luzon, J., DeglInnocenti, C., Perfetti, M., Bernot, K., Calvez, G., Caneschi, A. and Sessoli, R. (2013) *Angew. Chem. Int. Ed.*, **52**, 350–354.
55. Chibotaru, L.F. and Ungur, L. (2012) *J. Chem. Phys.*, **137**, 064 112.
56. Chibotaru, L.F. (2013) *Adv. Chem. Phys.*, **153**, 397–519.
57. Ungur, L. (2010) Ab initio methodology for the investigation of magnetism in strongly anisotropic complexes. PhD thesis, KU Leuven.
58. Feltham, H.L.C., Lan, Y., Klöwer, F., Ungur, L., Chibotaru, L.F., Powell, A.K. and Brooker, S. (2011) *Chem. Eur. J.*, **17**, 4362–4365.
59. Tang, J.K., Hewitt, I., Madhu, N.T., Chastanet, G., Wernsdorfer, W., Anson, C.E., Benelli, C., Sessoli, R. and Powell, A.K. (2006) *Angew. Chem. Int. Ed.*, **45**, 1729–1733.
60. Chibotaru, L.F., Ungur, L. and Soncini, A. (2008) *Angew. Chem. Int. Ed.*, **47**, 4126–4129.
61. Luzon, J., Bernot, K., Hewitt, I.J., Anson, C.E., Powell, A.K. and Sessoli, R. (2008) *Phys. Rev. Lett.*, **100**, 247 205.
62. Boulon, M.E., Cucinotta, G., Liu, S.S., Jiang, S.D., Ungur, L., Chibotaru, L.F., Gao, S. and Sessoli, R. (2013) *Chem. Eur. J.*, **19**, 13 726–13 731.
63. Bernot, K., Luzon, J., Bogani, L., Etienne, M., Sangregorio, C., Shanmugam, M., Caneschi, A., Sessoli, R. and Gatteschi, D. (2009) *J. Am. Chem. Soc.*, **131**, 5573–5579.
64. da Cunha, T.T., Jung, J., Boulon, M.E., Campo, G., Pointillart, F., Pereira, C.L.M., Guennic, B.L., Cador, O., Bernot, K., Pineider, F., Golhen, S. and Ouahab, L. (2013) *J. Am. Chem. Soc.*, **135**, 16 332–16 335.
65. Blagg, R.J., Ungur, L., Tuna, F., Speak, J., Comar, P., Collison, D., Wernsdorfer, W., McInnes, E.J.L., Chibotaru, L.F. and Winpenny, R.E.P. (2013) *Nat. Chem.*, **5**, 673-678.
66. Visinescu, D., Madalan, A.M., Andruh, M., Duhayon, C., Sutter, J.P., Ungur, L., den Heuvel, W.V., and Chibotaru, L.F. (2009) *Chem. Eur. J.*, **15**, 11 808–11 814.
67. Ungur, L., den Heuvel, W.V. and Chibotaru, L.F. (2009) *New J. Chem.*, **33**, 1224-1230.
68. Guo, Y.N., Xu, G.F., Wernsdorfer, W., Ungur, L., Guo, Y., Tang, J., Zhang, H.J., Chibotaru, L.F. and Powell, A.K. (2011) *J. Am. Chem. Soc.*, **133**, 11 948–11 951.
69. Novitchi, G., Pilet, G., Ungur, L., Moshchalkov, V.V., Wernsdorfer, W., Chibotaru, L., Luneau, D. and Powell, A.K. (2012) *Chem. Sci.*, **3**, 1169–1176.
70. Mondal, K.C., Sundt, A., Lan, Y., Kostakis, G.E., Waldmann, O., Ungur, L., Chibotaru, L.F., Anson, C.E. and Powell, A.K. (2012) *Angew. Chem. Int. Ed.*, **51**, 7550-7554.
71. Ungur, L., Langley, S.K., Hooper, T.N., Moubaraki, B., Brechin, E.K., Murray, K.S. and Chibotaru, L.F. (2012) *J. Am. Chem. Soc.*, **134**, 18 554–18 557.
72. Wang, Y.X., Shi, W., Li, H., Song, Y., Fang, L., Lan, Y.H., Powell, A.K., Wernsdorfer, W., Ungur, L., Chibotaru, L.F., Shen, M.R. and Cheng, P. (2012) *Chem. Sci.*, **3**, 3366–3370.
73. Ungur, L., Thewissen, M., Costes, J.P., Wernsdorfer, W. and Chibotaru, L.F. (2013) *Inorg. Chem.*, **52**, 6328-6337.
74. Vieru, V., Ungur, L. and Chibotaru, L.F. (2013) *J. Phys. Chem. Lett.*, **4**, 3565-3569.
75. Langley, S.K., Wielechowski, D.P., Vieru, V., Chilton, N.F., Moubaraki, B., Abrahams, B.F., Chibotaru, L.F. and Murray, K.S. (2013) *Angew. Chem. Int. Ed.*, **52**, 12 014–12 019.
76. Guo, F.S., Leng, J.D., Lin, Z., Tong, M.L., Song, G., Ungur, L. and Chibotaru, L.F. (2011) *Chem. Eur. J.*, **17**, 2458–2466.
77. Griffith, J.S. (1963) *Phys. Rev.*, **132**, 316.
78. Reu, O.S., Palii, A.V., Ostrovsky, S.M., Tregenna-Piggott, P.L.W. and Klokishner, S.I. (2012) *Inorg. Chem.*, **51**, 10 955-10 965.
79. Chilton, N.F., Collison, D., McInnes, E.J.L., Winpenny, R.E.P. and Soncini, A. (2013) *Nat. Commun.*, **4**, 2551.
80. Ungur, L. and Chibotaru, L.F. (2014) *Chem. Sci.*, submitted.
81. Ungur, L. and Chibotaru, L.F. (2006-2013) *SINGLE_ANISO Program*, KU Leuven, Belgium, *http://www.molcas.org* (accessed 09 September 2014).

82. Rajamaran, G., Totti, F., Bencini, A., Caneschi, A., Sessoli, R. and Gatteschi, D. (2009) *J. Comput. Chem.*, **17**, 3153–3161.
83. Anderson, P.W. (1959) *Phys. Rev.*, **115**, 2–13.
84. Lines, M. (1971) *J. Chem. Phys.*, **55**, 2977.
85. Cucinotta, G., Perfetti, M., Luzon, J., Etienne, M., Car, P.E., Caneschi, A., Calvez, G., Bernot, K. and Sessoli, R. (2012) *Angew. Chem. Int. Ed.*, **51**, 1606–1610.
86. Bencini, A. and Gatteschi, D. (1990) *Electron Paramagnetic Resonance of Exchange Coupled Systems*, Springer.
87. Gatteschi, D., Sessoli, R. and Villain, J. (2006) *Molecular Nanomagnets*, Oxford University Press.
88. Guo, P.H., Liu, J.L., Zhang, Z.M., Ungur, L., Chibotaru, L.F., Leng, J.D., Guo, F.S. and Tong, M.L. (2012) *Inorg. Chem.*, **51**, 1233–1235.
89. Rinck, J., Novitchi, G., van den Heuvel, W., Ungur, L., Lan, Y., Wernsdorfer, W., Anson, C.E., Chibotaru, L.F. and Powell, A.K. (2010) *Angew. Chem. Int. Ed.*, **49**, 7583–7587.
90. Rinehart, J.D., Fang, M., Evans, W.J. and Long, J.R. (2011) *Nat. Chem.*, **3**, 538–542.
91. Rinehart, J.D., Fang, M., Evans, W.J. and Long, J.R. (2011) *J. Am. Chem. Soc.*, **133**, 14 236-14 239.
92. Kahn, O. (1993) *Molecular Magnetism*, VCH Publishers, New York.
93. Cano, J. and Journaux, Y. (2005) *Magnetism: Molecules to Materials*, vol. 6, Wiley-VCH Verlag GmbH, pp. 189–222.
94. Van den Heuvel, W. and Chibotaru, L.F. (2010) *Phys. Rev. B*, **82**, 174,436.

7
Lanthanide Complexes as Realizations of Qubits and Qugates for Quantum Computing

Guillem Aromí, Fernando Luis, and Olivier Roubeau

7.1
Introduction to Quantum Computation

7.1.1
General Introduction

The diverse facets of quantum information processing (QIP), be it communication, simulation or computation, rely on two well-accepted, albeit still counter-intuitive, concepts of quantum mechanics: superposition and entanglement [1–3]. Quantum computation (QC) is based on the idea that quantum physical laws can be used to implement logic operations [4, 5] and the certainty that this new paradigm will simplify some computational tasks and allow access to currently untreatable problems. Originally, it was the difficulty of simulating quantum systems by conventional (or 'classical') computers that made Feynman and Deutsch [6–8] contemplate exploiting quantum degrees of freedom to encode and process information. Indeed, the size of the computer register required to simulate an N-component quantum physical system grows exponentially with N in classical, Boolean logics, while only linearly in a quantum machine [4]. The computational gain arises naturally from the use of quantum information units or qubits, instead of classical ones. The main difference between these two resides in the fact that each qubit can not only be in states '0' or '1', as a classical bit, but also in any arbitrary quantum superposition of these two, that is, any quantum state of the form $|\varphi\rangle = \alpha|0\rangle + \beta|1\rangle$. These coherent superpositions can thus naturally and efficiently encode the huge Hilbert space of a general quantum state. Specific quantum algorithms were then developed to take advantage of this potential. These include the Deutsch–Josza algorithm for finding if a function f is either constant or balanced [9], the Shor algorithm for prime numbers factorization [10] and the Grover algorithm for searching in large databases [11]. The latter two are particularly important due to their applications in cryptography and data management, respectively. Yet, the applications of QIP go beyond computation

Lanthanides and Actinides in Molecular Magnetism, First Edition.
Edited by Richard A. Layfield and Muralee Murugesu.

and include the development of secure quantum communication channels as well as of quantum simulators for complex physical problems. The latter are in fact expected to provide the first practical applications of QIP, as they require a lower, thus more realistic, threshold of qubits [12].

Before these expectations become a reality, a number of issues, ranging from fundamental physics to engineering, need to be addressed. A first, necessary ingredient is the development of suitable physical realizations of quantum bits and gates [5, 13]. Proposals for the quantum hardware include ion traps [14, 15], photons [16], nuclear spins [17], superconducting circuits [18, 19] or spin-based systems, such as electrons in quantum dots [20, 21], atomic impurities in solids [22, 23] or organic and coordination molecules [24–27], thus spanning a wide spectrum of materials. Nevertheless, it is still unclear which (if any) will prove successful, even though much experimental progress has been made in the last decade. Crucial challenges, common to most of the above-mentioned proposals, are the efficient protection against errors caused by the interaction with the environment [28] and the implementation of scalable QC architectures. The latter must include efficient methods to initialize, read-out and coherently control and interconnect the elementary hardware units [13], and can be specific to each qubit implementation.

In this chapter, we focus on the use of lanthanides as spin-based hardware for QC. The remainder of this introductory section provides some essential concepts and definitions and then it succinctly describes some of the existing proposals for QC. The second section provides a brief overview of results obtained with spin-based systems other than lanthanides. The following two sections review experiments made on qubits and quantum gates, respectively, based on lanthanides, highlighting their specific properties and advantages for QC applications.

7.1.2
Definition of Qubits, Qugates, Timescales and Essential Requirements

The basic element of a quantum computer is the *quantum bit or qubit*. It is the QC counterpart of the Boolean bit, a classical physical system with two well-defined states. A material realization of a qubit is a quantum two-level system, with energy eigenstates, $|0\rangle$ and $|1\rangle$, and an energy gap ΔE, which can be in any arbitrary superposition $|\varphi\rangle = \cos(\theta/2)|0\rangle + \exp(i\phi)\sin(\theta/2)|1\rangle$. These 'pure' superposition states can be visualized by using a Bloch sphere representation (see Figure 7.1).

A quantum algorithm can be seen as the controlled time evolution of a physical system obeying the laws of quantum mechanics. It is therefore of utmost importance that each qubit may be coherently manipulated, between arbitrary superpositions, via the application of external stimuli. Furthermore, all these manipulations must take place well before its quantum wave function, thus the information it encodes, is corrupted by the interaction with external perturbations. The need to properly isolate qubits but, at the same time, to rapidly

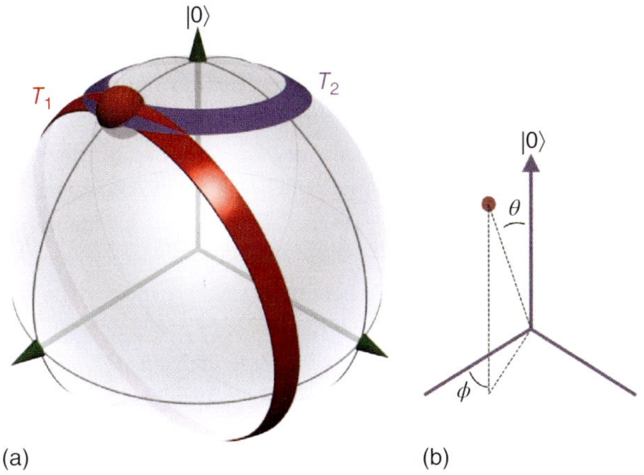

Figure 7.1 (a,b) Schematic representation of a qubit state. Each point on the surface of the Bloch sphere describes a quantum superposition of states $|0\rangle$ and $|1\rangle$ of the form $|\varphi\rangle = \cos(\theta/2)|0\rangle + \exp(i\phi)\sin(\theta/2)|1\rangle$. In a coherent manipulation of the qubit, both the amplitude θ and the phase ϕ have to be controlled. The randomization of θ and ϕ, indicated in red and purple, correspond respectively to the relaxation and decoherence timescales T_1 and T_2.

manipulate and read them represents one of the most difficult dilemmas faced by any QC proposal. In order to describe this question in a more quantitative manner, it is useful to define three relevant timescales. First, the relaxation time T_1 describes transitions from, say, $|1\rangle$ to $|0\rangle$ and thus measures a randomization of the amplitude θ, which eventually brings the system to thermal equilibrium. Second, the decoherence time T_2 accounts for changes in the phase of the qubit ϕ. Finally, we must consider the period of coherent Rabi oscillations $1/\Omega_R$, induced by an external stimulus, between different qubit states (Figure 7.2). For a qubit to become of any practical use, T_1 and T_2 must be sufficiently long as compared with $1/\Omega_R$. Often, $T_1 \gg T_2$, thus it is T_2 that effectively determines the damping of Rabi oscillations (Figure 7.2). The number of qubit operations that can be made without loss of coherence is actually used to evaluate the quality of a quantum system. This 'figure of merit' Q_M is therefore simply $\Omega_R T_2$. Eventually, averaging over an ensemble of qubits may be experimentally necessary. Because slight differences in the qubit oscillation frequencies lead to destructive interferences and thus also to an additional damping of coherent oscillations, a so-called dephasing time T_2^* is then used while often $T_2^* < T_2$.

In both quantum and classical computation science, any complex logic operation can be decomposed into a set of few elementary operations, known as *universal gates*. A basic criterion for building a quantum computer is therefore the proper definition of *universal quantum gates or qugates* [4]. It has been shown that the implementation of single qubit rotations together with one type of entangling

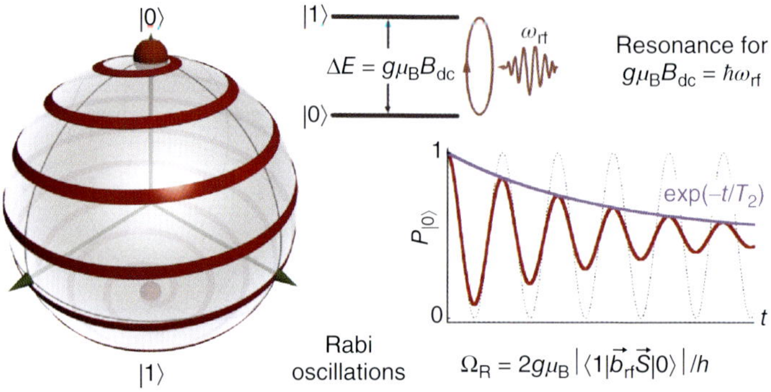

Figure 7.2 Schematic representation of coherent oscillations between states |0⟩ and |1⟩ of a qubit. For an electron spin placed in a *dc* magnetic field, oscillations can be induced via the application of an external radio frequency (rf) pulse resonant with the Zeeman energy. The amplitude θ of the quantum superposition changes periodically at the Rabi frequency Ω_R while the phase ϕ accumulates due to precession around the dc field. Phase kicks cause the damping of these oscillations and thus the loss of coherence, as measured through the decoherence time T_2.

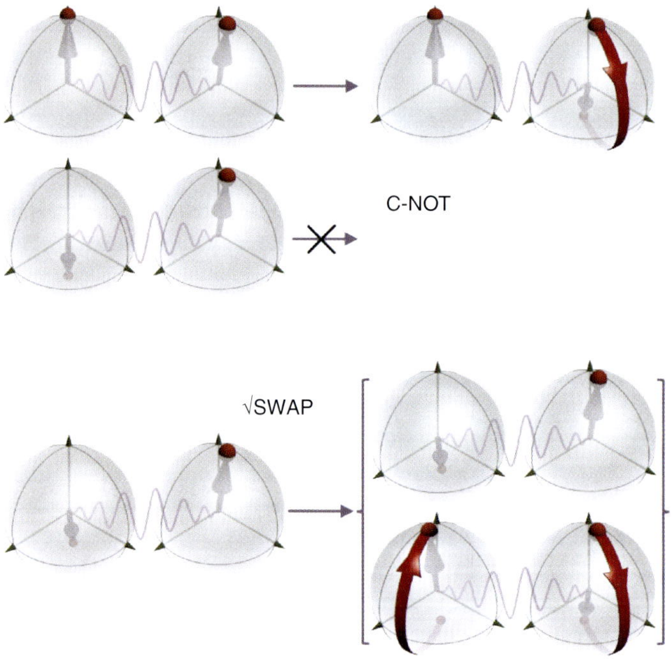

Figure 7.3 Schematic representation of the operations of some quantum logic gates acting on two qubits. In quantum computation, single qubit rotations (Figure 7.2) and CNOT (controlled-NOT) or $\sqrt{}$SWAP quantum gates are universal.

two-qubit gates, such as the CNOT or $\sqrt{\text{SWAP}}$ qugates shown in Figure 7.3, is sufficient for universal QC [4, 13, 29]. By analogy with a classical CNOT gate, that switches a target bit if and only if a control bit is in a particular and defined state (indistinctly defined as either '0' or '1'), the operation of the quantum CNOT gate (for simplicity on a target in state $|0\rangle$ and defining here '0' as the control state that changes the state of the target) gives the expression in Equation 7.1.

$$(\alpha|0\rangle_C + \beta|1\rangle_C) \otimes |0\rangle_T \rightarrow_{\text{CNOT}} (\alpha|0\rangle_C \otimes |1\rangle_T + \beta|1\rangle_C \otimes |0\rangle_T) \tag{7.1}$$

In this equation, C and T refer to control and target qubits, respectively. The resulting state (output of the qugate) is said to be an 'entangled state' of the two qubits, that is, a state that cannot be written as a product of states for each qubit [30]. The occurrence of such entangled states is another characteristic trait of QC, at the basis of secure quantum communication or cryptography. It also implies that, as opposed to what happens with a classical bit, an arbitrary quantum bit cannot be copied (the COPY classical operation is, in fact, based on the application of a succession of classical CNOT gates) [4].

The time required for the operation of a qugate is usually referred to as *gate time* or τ_{gate}, and it should also be much shorter than the decoherence time of each qubit [31]. In fact, the total computation time of a quantum algorithm should also be shorter than T_2 of each of the qubits involved. This very stringent condition seemed to put a fundamental limit to the progress of QC until quantum error correction schemes were developed [32, 33]. These codes replace individual qubits by 'logical qubits', made of several physical qubits, thereby ensuring a redundant codification of quantum information and providing means to recover from some errors.

Another practical tool is dynamical decoupling, a technique that uses sequences of fast qubit rotations to mitigate the effects of decoherence. The pulse sequences are designed such that the interactions of each qubit with its environment tend to average out [34, 35]. While still a major concern, decoherence may thus not be the strong impediment it originally seemed to represent for the advent of QC.

7.1.3
Current Proposals for the QC Hardware

7.1.3.1 Trapped Ions

Single atomic ions confined in radio frequency traps and cooled by laser beams (Figure 7.4a) formed the basis for the first proposal of a CNOT quantum gate with an explicit physical system [14]. The first experimental realization of a CNOT quantum gate was in fact demonstrated on a system inspired by this scheme [37]. In this proposal, two internal electronic states of alkaline-earth or transition metal ions (e.g. Ba^{2+} or Yb^{3+}) define the qubit basis. These states have excellent coherence properties, with T_1 and T_2 in the range of seconds [15]. Each qubit can be

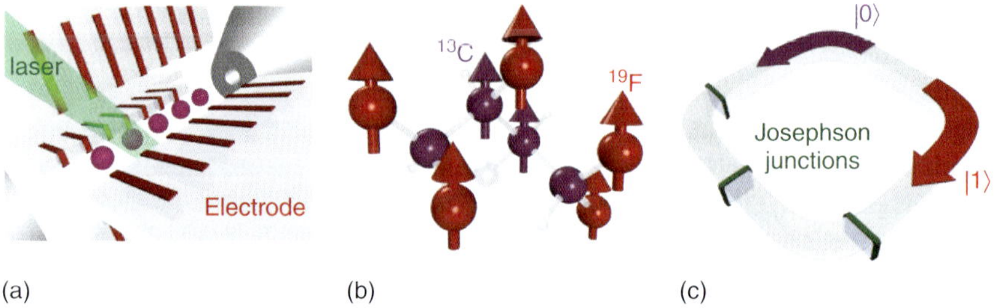

Figure 7.4 Proposals of hardware for QC. (a) Ions trapped in a radiofrequency cavity and cooled by lasers; (b) molecule used to implement Shor's algorithm with NMR using seven nuclear spin qubits [36] and (c) two states of electrical current flow in a superconducting loop with three Josephson junctions (flux qubit).

prepared and addressed with high efficiency via the application of laser pulses focused on each ion. The coupling of two or more qubits through Coulomb interactions is controlled via excitations of the centre of mass motion of these ions. Up to about 10 qubits can be efficiently entangled and addressed within linear arrays of trapped ions. Scaling beyond this limit remains a challenge, although some strategies, involving either the shuttling of ions between different traps or the use of photons to communicate different registers, are being explored [15, 38, 39]. The advent of microfabricated chips integrating photonic components could soon provide access to a true quantum simulator.

7.1.3.2 Nuclear Spins

Another early approach to QC used nuclear magnetic resonance (NMR) techniques [17]. The qubit states are defined by different projections of the nuclear spin along an external magnetic field. The nuclear gyromagnetic ratios and Larmor frequency depend on the element, resulting in different Zeeman splittings. Each qubit can then be addressed and controlled by the application of electromagnetic pulses of the appropriate frequency. Two-qubit interactions arise from indirect coupling through molecular electrons. Another attractive feature is that nuclear spins in molecules maintain their spin orientation thanks to rapid molecular motion in solution, resulting in T_2 that reach (many) seconds, thus similar to those of trapped ions. This led, in 1998, to the experimental realization of the Deutsch quantum algorithm in a solution of chloroform ($CHCl_3$) [40], using the ^1H and ^{13}C nuclear spins as the input and output qubits of the algorithm, respectively. Thanks to advances in radio frequency pulse techniques, the manipulation of up to a dozen qubits [41], the implementation of Shor's algorithm in a molecule with seven qubits (Figure 7.4b) [36] and some quantum error correction protocols have been achieved with liquid-state NMR. However, several major hurdles may prove difficult to jump. The already long T_2 times remain shorter than initialization and measurement (made on ensembles of qubits) timescales, while the pseudo-pure-state techniques of initialization used in these early studies are not scalable.

7.1.3.3 Superconducting Qubits

A macroscopic physical realization of a qubit can be obtained with superconducting quantum circuits [18, 19]. There are three types of superconducting qubits, so-called charge, flux (Figure 7.4c) and phase qubits, depending on the characteristics of inductance, capacitance and tunnel element of the circuit. The performance and, in particular, the coherence time scales of all types have constantly improved, T_1 and T_2 currently reaching values of several microseconds, about 10–100 times more than those required for initialization and readout. The engineering of superconducting circuits with the ability to couple two such qubits, either capacitively or inductively, is rather simple. This led in 2007 to the demonstration of CNOT gate operations on two coupled flux qubits [42]. Eliminating further decoherence remains an important challenge, especially for readout and initialization. Although superconducting qubits have the advantage of using established microfabrication techniques developed for silicon technology, scaling-up this technology might face some technological difficulties due to very stringent conditions on the accuracies required for the fabrication of identical qubits.

7.1.3.4 Spin Qubits

The spin carried by a single electron is the prototype of a two-level quantum system. Indeed, the two quantized orientations or eigenstates of the spin provide a basis to encode the qubit states [5]. By extension, any spin-based system with pseudo-spin-1/2 state can, in principle, serve as a qubit [24–27, 43–46]. Many embodiments of qubits based on electronic spins have been proposed, the most studied being single electron spins confined in semiconducting nanostructures (quantum dots, see Figure 7.5a) [20], impurities such as phosphorus atoms implanted in silicon and nitrogen-vacancy (NV) centres in diamond (Figure 7.5b) [22], or lanthanide ions in crystalline hosts [47]. Organic radicals and more recently, paramagnetic coordination complexes with one or more spin carriers are also actively studied in this field [27]. All these are discussed in more detail in the following sections.

7.1.3.5 Photons

A photonic realization of qubit can be obtained through the polarization state of a photon or using the continuous phase and amplitude of a many-photon laser beam [5, 48]. At first, the difficulty in achieving significant photon–photon interactions necessary for multi-qubit operations can be seen as a drawback of this proposal. However, it was demonstrated that scalable QC is possible using only linear optical circuits and single-photon sources and detectors [16]. The method (known as the *KLM scheme* for Knill, Laflamme and Milburn) [49] uses quantum interference with auxiliary photons at a beam splitter as the source of interactions, and has

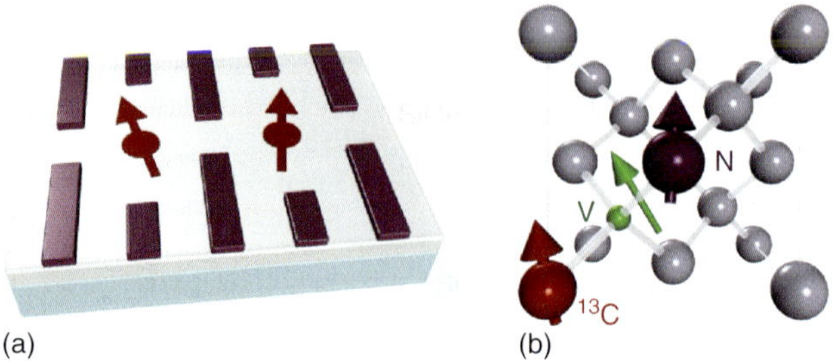

Figure 7.5 (a) Artificial quantum dot architecture showing the confined electron spins. (b) A diamond unit cell showing a NV centre – a nitrogen defect and a carbon vacancy – with an $S = 1$ electronic spin (green arrow) that coupled to a nuclear spin (that of a neighbouring ^{13}C, in red or that of ^{14}N, in purple) can be used as a hybrid quantum register.

allowed the realization of Shor's algorithm [50, 51]. Many efforts in improving detectors, sources and waveguides are being made [5, 52], but the loss of photons remains a major hurdle, that can be compared with decoherence in matter qubits.

7.1.3.6 Hybrid Proposals and Quantum Circuits

Each of the above proposals shows some strong aspects but also its drawbacks. Combining two different quantum technologies offers a clear advantage, provided one can retain the best of both worlds. This idea received a big boost when it was found that radio frequency photons stored in on-chip superconducting resonators can coherently couple to superconducting qubits [53]. Photons can then be used to read-out and manipulate individual qubits [54, 55] and they provide an efficient communication channel or 'quantum bus', between them [56]. In the same spirit, we also find proposals to couple such quantum circuits to ensembles of either electric (e.g. polar molecules) [57] or magnetic (e.g. NV-centres in a diamond crystal or molecular spins) qubits [58–60]. The resulting hybrid devices combine the advantageous properties of the ensembles as single-photon memories with the fast processing capabilities provided by the quantum circuit.

7.2 Quantum Computing with Electron Spin Qubits

7.2.1 Electronic Spins in Semiconductors: QDs and Dopants

Single (or multiple) electrons or holes can be bound locally to a small semiconductor nanostructure or to a single impurity in a solid. The resulting discrete energy levels can be used to define a spin qubit. Coherent control and read-out of individual spins have been demonstrated in several such materials, and some devices with multiple entangled qubits have been built [22]. The following are a few illustrative examples.

7.2.1.1 Quantum Dots

The confinement of electrons (or holes) in quantum dots (QDs) is achieved in different manners and with diverse materials. For example, electrostatic confinement can be achieved by the application of controlled voltages on metallic gates defined lithographically (Figure 7.5a). Also, the random growth of some group III/V semiconductors naturally provides the adequate potentials, for example, in islands of InAs on a GaAs surface. The latter are called *self-assembled QDs*, typically operate at temperatures above 4 K and can be controlled optically, while the former are controlled electrically at lower temperatures. Considerable experimental progress has been made since the early QC proposal [21]. Achievements include the demonstration of single-spin measurements [61] and the control of multiple qubits through exchange interactions. Recent reviews provide very complete overviews to which the reader is directed [20, 22]. Perhaps the major critical issue in QDs with group III/V semiconductors resides in the decoherence induced by the nuclear spins of the semiconductor host. This results in decoherence times T_2 in the range of microseconds and T_2^* of nanoseconds. The use of nuclear-spin-free group IV semiconductors (silicon in particular, with a natural abundance of only 5% of nuclear spin-bearing ^{29}Si) helps to suppress this source of decoherence [22, 62].

7.2.1.2 Dopants and Defects

As an alternative to QDs, silicon can be doped with single atom impurities, in particular phosphorus, which acts as an electron donor. Donors can be implanted individually with a precision of about 10 nm. Either the ^{31}P nuclear spin or the unpaired electron can be used as qubits [63, 64]. An advantage of silicon is its widespread use in current electronics, meaning that QC might profit from methods and technologies already developed for their classical 'cousins'. Also, spins in silicon can attain extremely high coherence times: experiments on ^{28}Si-enriched silicon show spin coherence times T_2 exceeding 10 s [65]. The read-out and coherent manipulation of individual spin qubits in silicon have been recently achieved [66].

Another group IV solid material of interest is diamond. NV centres (Figure 7.5b) are naturally formed by substitution of a lattice site by a nitrogen atom (N) adjacent to a carbon vacancy (V) [23]. In its negatively-charged ground-state, the NV centre is an electron spin triplet with $S = 1$ that can be efficiently initialized in a singlet spin-state through sets of optical transitions. Coherent manipulation can be done with resonant microwave fields, while spin-dependent fluorescence allows fast read-out of individual qubits with an optical microscope. Moreover, thanks to the unique properties of diamond (weak spin–orbit interactions, high Debye temperature, large band-gap), decoherence is reduced to a minimum, even at high temperatures. With purified and ^{12}C-enriched diamond, a decoherence time T_2 of several ms has been reached at room temperature, while T_1 is expected to be of the order of seconds [22].

An additional asset of both donors in silicon and NV centres is that nuclear spins, of ^{31}P in the former and of N and neighbouring ^{13}C in the latter, can be employed as long-term storage quantum memories [63], or even to build multiple qubit registers [22, 67].

7.2.2 Electronic Spins in Molecules: Organic Radicals and Transition Metal Complexes

Chemistry offers a very versatile way for realizing QC using electronic spins, through the synthetic engineering of molecules containing one or more open shell moieties (either p, d or f), disposed in the optimal environment and organization to function as qubits or qugates. A differentiating factor with respect to other technologies is that single molecules can adequately organize weakly coupled qubits [27], thus allowing to encode multiple qubit states or even to perform quantum algorithms and simulations [68]. Of course, this relies on a true synthetic design, for example of specific polynucleating ligands, as opposed to the serendipitous self-assembly, vastly applied in the molecular magnetism community. A very attractive feature of chemical routes with respect to scalability is that huge numbers of *truly identical* molecules are obtained.

7.2.2.1 Organic Radicals

A spin-based qubit only requires a pair of spin states amenable to coherent initialization and manipulation with electromagnetic radiation pulses, with long enough spin coherence times. The simplest approximation to this is an organic radical with its net spin $S = 1/2$. Existing proposals contemplate qubits based on stable radicals as well as qugates in which quantum operations may be implemented by turning on and off, in a controlled manner, a Heisenberg exchange coupling between the spins of two radicals [69]. Although some systems may appear adequate, such as a switchable pair of nitronyl nitroxide radicals [70], experimental studies oriented towards QC are very limited, for example on malonyl radicals for which entanglement and spin manipulation was achieved [24]. So far, research efforts have mainly focused on the realization of so-called molecular spintronic devices [71], where molecular spin carriers would encode classical information. On the other hand, coupling of stable organic radicals with a nuclear spin 1/2 qubit, such as N@C_{60} fullerenes has been used via coherent control of the latter through the electronic qubit [72].

7.2.2.2 Transition Metal Complexes

Single paramagnetic metal ions clearly represent a vast choice of electron spin carriers with true or effective spin 1/2, and thus potential qubits. An advantage over organic-based radicals is the possibility of tuning the energy levels, spin dynamics and hyperfine couplings through the choice of the transition metal (3d, 4d, 5d, 4f, 5f) and the modification of its environment, in particular the local symmetry. Coordination chemistry also provides the means to organize several qubits within molecular architectures [27]. An illustrative and simple example is found in strings of weakly coupled qubits synthesized by disposing paramagnetic Mn(II) ions as hexacoordinated centres along the axes of triple stranded oligoimidazole based coordination helicates [73]. The change in the principal axes of the gyromagnetic g tensor from one metal to the next, caused by the molecular helical twist, provides a means for discriminating the different qubits of the array, thus making them addressable individually by electron paramagnetic resonance (EPR).

A cluster, that is, an array of strongly coupled spins with a true or effective overall spin 1/2, could also be used as a single qubit. Actually, calculations indicate that an antiferromagnetic array would perform better as a qubit than a single electron spin when implementing a CNOT qugate [44]. A first prototype of such paramagnetic coordination assembly is a $[V_{15}]$ cluster, $[V_{15}As_6O_{42}(H_2O)]^{6-}$, of 15 V^{4+} ions (each with $S = 1/2$) coupled antiferromagnetically (Figure 7.6a). The energy level structure consists of two lowest-lying doublets, split by only 80 mK, with a first excited $S = 3/2$ multiplet lying 3.8 K above. Rabi oscillations of the molecular spin were observed and a coherence time T_2 of the order of hundreds of nanoseconds was found at 4 K [74].

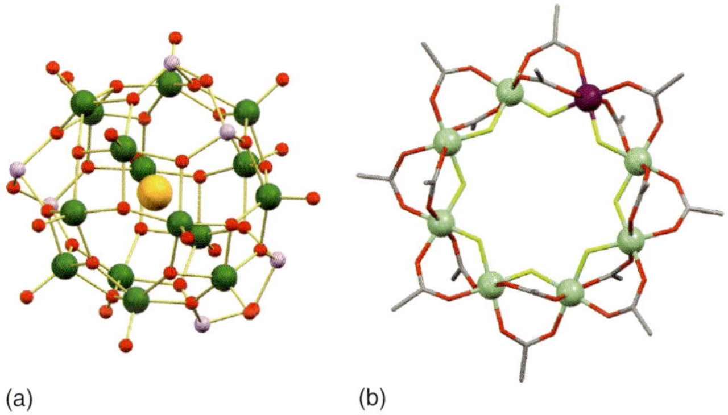

Figure 7.6 Molecular structures of the cation $[V_{15}As_6O_{42}(H_2O)]^{6-}$ (a) and of the anionic molecular ring $[Cr_7NiF_8(piv)_{16}]^-$ (b). Hydrogen atoms and methyl groups omitted for clarity. Colour code: Dark green, V(IV); light green, Cr(III); dark purple, Ni(II); light purple, As(III); red, O; grey, C; yellow, F and orange, O from central H_2O.

Another example is a family of clusters with a magnetic core consisting of a $[Cr_7Ni]$ ring (Figure 7.6b) [75]. Metals in these clusters are held together by a combination of bidentate carboxylate ligands and monoatomic bridges (typically fluoride or alkoxide groups). Antiferromagnetic interactions result in an $S = 1/2$ ring ground state, with a first excited $S = 3/2$ level lying 13 K above [76]. At sufficiently low temperatures, a proper definition of qubit states $|0\rangle$ and $|1\rangle$ is thus provided by the two spin orientations $m_S = \pm 1/2$, whose energies can be split via the application of a magnetic field. Standard pulsed EPR experiments performed on solutions of these molecules give a maximum decoherence time $T_2 = 15\,\mu s$ [77, 78]. In sufficiently dilute solutions, decoherence seems to be dominated by hyperfine interactions. These experiments also show that chemical modifications provide a method for optimizing quantum coherence. In particular, coherence can be enhanced by replacing hydrogen (nuclear spin $I = 1/2$) by deuterium ($I = 1$) and by choosing ligands in order to avoid fluctuations of the local hyperfine fields associated with molecular motions [78]. Chemical design has allowed also the organization of various types of these heterometallic 'wheels' into supramolecular ensembles [75, 79]. Importantly, the inter-ring magnetic interactions could be varied and probed [80], and the existence of entangled spin states was deduced from magnetic susceptibility in some of these supramolecular aggregates [81].

An early example of a similar assembly of two magnetic clusters, albeit with higher spin ground states, is a dimer of two $S = 9/2[Mn_4]$ single-molecule magnets (SMMs) with strong uniaxial magnetic anisotropy. The inter-cluster magnetic coupling results in entanglement, the magnetic states of the dimer being quantum superpositions of the wave functions for each constituent [82].

The deliberate disposition of two or more well-defined and weakly interacting metal aggregates within one molecular assembly is a delicate task, especially considering that the magnetic properties of each aggregate need to be predefined and maintained. Some extended multidentate ligands may facilitate both the aggregation of metals as strongly coupled clusters and the assembly of two or more such entities within stable, covalent molecules. A few relevant examples from literature are (i) a molecular pair of $[Mn_7]$ clusters, each showing SMM properties, linked by bis-hydroxyphenylpyrazolyl ligands that also cement the metals within each cluster [83], (ii) four $[Mn_3]$ SMMs contained in one molecule [84] and (iii) two well separated SMM [TbCu] units linked into one molecule by double dinucleating Schiff base moieties [85].

One of our initiatives towards the preparation of molecules containing metal cluster qubits, each of them with total or *effective* $S = 1/2$ ground states, uses a family of extended bis-β-diketone ligands [86] to favour the assembly of molecular pairs of clusters [27]. These may be of three classes, depending on the structural role played by the H_nLx ligand: (i) the bis-β-diketone ensures the aggregation of metals as clusters, also keeping the latter within one molecule, as in $[Cu_2Ni_2(L4)_2(py)_6]$ [87], which features two weakly interacting $S = 1/2$ moieties, (ii) H_nLx only links metal aggregates that have been formed by other auxiliary ligands

[88, 89] and (iii) the extended ligand is responsible for gathering metals as clusters that are then linked into molecular pairs by auxiliary ligands [90]. Even if some of these systems could satisfy some of the requirements of qubits and qugates, the symmetric nature of all these molecular cluster pairs, as well as most of those present in the literature, does not allow distinguishing between both qubits. As this is necessary for example to realize a universal CNOT gate (see following text), this represents a remaining synthetic challenge.

Lanthanide ions have emerged as a very promising category of chemically accessible realizations of spin-based qubits. Their suitability for this task, which results from their physical, chemical and quantum mechanical properties, is discussed in the following sections.

7.3
Single Lanthanide Ions as Spin Qubits

In lanthanide Ln^{3+} ions, the magnetic properties arise from the strong coupling between the orbital L and spin S angular momenta, which results in a total angular momentum of $J = L + S$. Given the energy differences between levels of different J, the magnetic behaviour can often be accurately described by considering only the ground state multiplet. The $2J + 1$ magnetic levels of this ground multiplet are split by the magnetic anisotropy, arising from the interaction with the crystal field, and by the Zeeman interaction with magnetic fields. Within this multiplet, the spin Hamiltonian can be written, in terms of Steven's operators, as follows [91]:

$$\hat{H} = \sum_{n,m} B_n^m \hat{O}_n^m - g_J \mu_B (H_x J_x + H_y J_y + H_z J_z) \tag{7.2}$$

In Equation 7.2, B_n^m, with n even and $m \leq n$, are numerical coefficients (or anisotropy constants), $\hat{O}_n^m$ are effective spin operators containing powers of J_x, J_y and J_z and g_J is the lanthanide gyromagnetic ratio. The anisotropy terms that do actually appear in the spin Hamiltonian (Equation 7.2) are determined by the local coordination, while their sign and magnitude also vary depending on each particular lanthanide ion. In many cases, this Hamiltonian gives rise to a doublet of lowest lying electronic levels, well separated energetically from all excited levels. Under these conditions, a lanthanide ion can be pictured as an *effective* $S = 1/2$, thus as a potential qubit. In contrast with 'real' $S = 1/2$ systems, the energies and wave functions, thus also, ΔE and Ω_R, depend on the orientation of the external magnetic fields with respect to the crystal axes.

Some properties of these ions make them particularly appealing as solid state spin qubits. The fact that they can be diluted into diamagnetic crystals offers a simple method to optimize their quantum coherence. Similarly to trapped ions, they are simple: each qubit is embodied by a single atom. Yet, their immediate environment brings additional possibilities. In this respect, the use of chemistry offers fascinating perspectives for the design of spin qubits, via the control of the local coordination that ultimately determines the electronic levels and wave functions. In taking this idea a step further, interactions between two or more ions within simple molecular complexes can be engineered in order to realize basic quantum gates. The following sections illustrate these possibilities and examine some proposals of qubits and qugates based on lanthanide ions.

7.3.1
Quantum Coherence of Lanthanide Ions Doped into Crystalline Solids

As mentioned before, a major criterion for evaluating the goodness of a quantum system as a qubit is the figure of merit, $Q_M = \Omega_R T_2$, which gives the number of qubit operations that can be performed before quantum information is lost to the environment. In the case of spin qubits, the main sources of decoherence arise from interactions with nuclear spins and with other electronic spins. An attractive aspect of lanthanide ions is that they can be diluted, at very low concentrations, into diamagnetic solid matrices that, when chosen conveniently, might also have a low concentration of nuclear spins. In this way, hyperfine and dipolar interactions can be made arbitrarily weak, thus maximizing the electron spin coherence times.

A recent and very promising example of this strategy is provided by the study of Rabi oscillations in a rare earth system made of Er(III) ions ($J = 15/2,\ g_J = 6/5$) doped into the structure of the scheelite-like salt $CaWO_4$ [47]. The most abundant stable isotopes of calcium, tungsten and oxygen carry no nuclear spin, meaning that interactions to the 'nuclear spin-bath' are reduced to a minimum in this material. Each Er(III) sits in a distorted cubic environment with S_4 symmetry [92]. The electronic ground state consists of a Kramers doublet. In ions of the ^{167}Er isotope, with a natural abundance of about 23%, the hyperfine interaction with the nuclear $I = 7/2$ spin gives rise to a manifold of electronuclear states. These additional states can be used to encode multiple qubits in a single ion.

A suitable method for measuring the quantum coherence and the figure of merit of a spin-based qubit is by using pulsed EPR techniques [77, 78, 93]. Transitions between pure crystal field states (when $I = 0$) or between electro-nuclear states (when $I = 7/2$) can be selected by tuning the magnitude and orientation of an external magnetic field. Conventional spin-echo experiments performed on a crystal with 10^{-5} atomic erbium concentration give intrinsic decoherence times $T_2 \approx 7\ \mu s$ at $T = 3.5\ K$ for the $I = 0$ isotopes. These experiments also show that T_2 strongly depends on temperature and spin concentration. Extrapolations to 1.5 K allow estimations of coherence times of the order of 100 μs. The resulting Q_M could reach values of the order of 10^4, which is considered sufficient for QIP with error correction codes.

A different sequence of electromagnetic pulses can be used to directly monitor Rabi oscillations between the two spin qubit basis states (Figure 7.7). As expected, the Rabi oscillation frequency Ω_R depends on the orientation of the microwave field, as a result of the crystal field anisotropy, and it is proportional to its amplitude. Rabi oscillations decay exponentially, with a characteristic timescale τ_R. A striking observation is that τ_R strongly depends on the microwave power, which controls Ω_R. As it is shown in the bottom panel of Figure 7.7, τ_R approaches T_2 only in the limit of very low power but remains much shorter otherwise. The fast decay of Rabi oscillations seems to be associated with inhomogeneities in the crystal field parameters, which lead to slightly different Ω_R for different crystal sites. These results reveal the existence of additional sources of decoherence when working with spin ensembles. They also underline the crucial importance, common to most proposals and especially challenging for scaling up QC to architectures involving more than a few qubits, of dealing with exact qubit replicas.

7.3.2 Control of the Magnetic Anisotropy of Lanthanide Ions: Chemical Design of Spin Qubits

7.3.2.1 Mononuclear Single Molecule Magnets

The synthesis of mononuclear molecular complexes, in which a single ion is wrapped by a shell of organic ligands, provides an alternative method for creating arrays of nearly isolated lanthanide spin qubits. The study of these materials was boosted by Ishikawa and coworkers' [94] discovery of magnetic hysteresis in

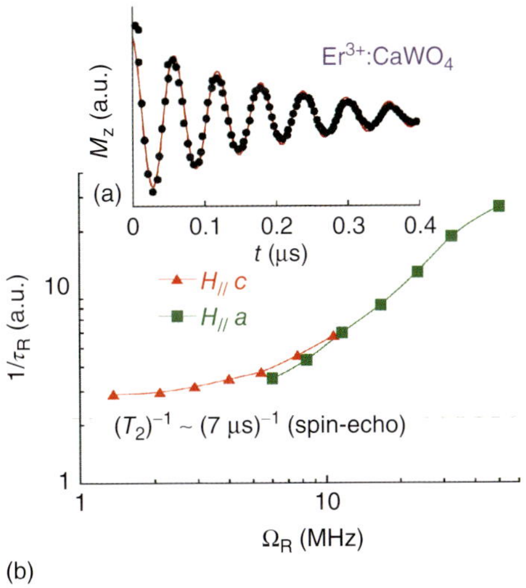

Figure 7.7 (a) Rabi oscillations of the magnetic moments of Er(III) ions diluted in a single crystal of $CaWO_4$, measured under a magnetic field $\mu_0 H = 0.522$ T applied along the c crystallographic axis and at $T = 3.5$ K. (b) Damping rate of Rabi oscillations as a function of the Rabi frequency Ω_R, which is proportional to the amplitude h_0 of the rf magnetic field.

crystals of double decker $[LnPc_2]^-$ molecules (Pc = phthalocyanine; Ln = Tb, Dy). This groundbreaking result underscored the potential of such SMMs, that is, as classical magnetic bits. Slow relaxation was later observed in members of several other families of mononuclear Ln(III) complexes [95–105]. In particular, polyoxometalate (POM) cluster moieties may act as complex ligands encapsulating 4f metal ions in precise and robust coordination environments.

Some of the physical ingredients required for a good SMM, in particular the existence of a strong magnetic anisotropy and a well-defined magnetic ground state doublet, are also useful for the proper definition of qubit states. In the case of lanthanide ions intercalated into inorganic crystals, the magnetic anisotropy is, to a large extent, inherited from the solid host. By contrast, the local coordination in artificial molecular complexes is subject to chemical modifications and, within some limitations, to 'intelligent' design. Molecular chemistry offers the possibility of combining the best of two worlds: lanthanide qubits, with their low decoherence and well defined qubit states, and SMMs, characterized by a nearly perfect uniformity and a practically unbound freedom for design.

7.3.2.2 Gadolinium(III) POMs as Spin Qubits

In order to illustrate these ideas, we first consider a particularly simple situation. As a free ion, Gd(III) exhibits no orbital angular momentum ($L = 0$). Its magnetic anisotropy then arises solely from the mixing, induced by the crystal field interaction, of states from the $L = 0$ multiplet with states from excited multiplets having $L \neq 0$. This ion is therefore an ideal crystal-field probe: its electronic level structure can be easily modified by introducing changes to its local coordination. A recent study of this possibility focused on two families of gadolinium POMs, namely $[Gd(W_5O_{18})_2]^{9-}$ and $[GdP_5W_{30}O_{110}]^{12-}$ (hereafter referred to as GdW_{10} and GdW_{30}, respectively) [106]. The molecular structures of these two materials are shown in Figure 7.8. In the elongated GdW_{10} molecule, two anionic moieties coordinate to the central Gd(III). Each moiety is twisted by an angle $\alpha = 44.2°$ with respect to the other, leading to a distorted square antiprism coordination (ideal D_{4d} symmetry). By contrast, in the 'donught'-shaped GdW_{30}, the first coordination shell of Gd(III) has a close to C_{5v} symmetry.

The simultaneous fit of heat capacity, magnetic susceptibility and EPR spectra enables the determination of the anisotropy constants B_n^m. In the case of GdW_{10}, a good fit was obtained with $B_2^0 = -0.041\ \mathrm{cm^{-1}}$ and $B_4^4 = 2.8 \times 10^{-4}\,\mathrm{cm^{-1}}$, both of which are allowed by the ideal D_{4d} local symmetry. For GdW_{30}, the first coordination shell is compatible with a $B_6^5\hat{O}_6^5$ term. However, this term alone cannot account for EPR data collected at varying temperatures and frequencies. Reasonably good fits were achieved introducing $B_2^0 = 0.013\ \mathrm{cm^{-1}}$ and a second-order off-diagonal term $B_2^2\hat{O}_2^2$ with $B_2^2 = 0.013\ \mathrm{cm^{-1}}$. The presence of this term, which would in principle be forbidden by the local symmetry, has been attributed to the influence of ions located beyond the first coordination sphere and to the departure of the whole GdW_{30} from a C_{5v} symmetry [106].

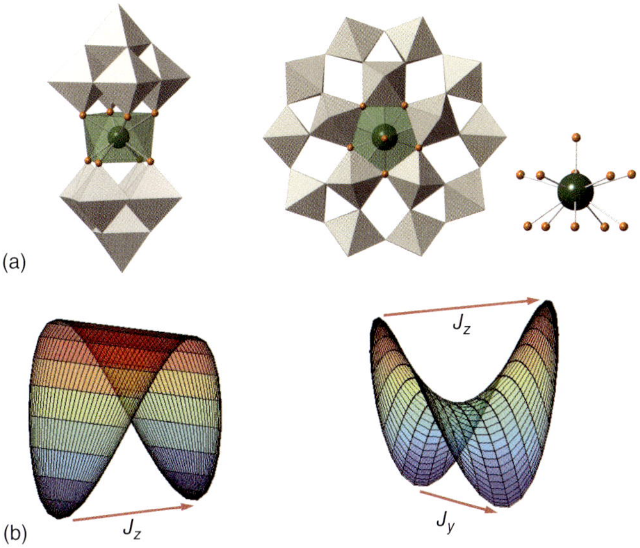

Figure 7.8 (a) Molecular structures of two families of SMMs based on polyoxometalate moieties encapsulating a central lanthanide ion. Left, LnW_{10} [103, 104]; right, LnW_{30} [107]. (b) Three-dimensional plots of the magnetic anisotropy energies of GdW_{10} (left) and GdW_{30} (right).

It follows that the magnetic anisotropies of Gd(III) in GdW_{10} and GdW_{30} differ quantitatively, and even qualitatively, from each other. These differences are illustrated by the bottom panels in Figure 7.8, which show classical potential energy landscapes for both molecules. While the former corresponds to an easy axis magnetic anisotropy along the long molecular axis (z), the latter favours the magnetization to lie within the molecular disk (xy plane), with a preferred axis y determined by the in-plane anisotropy. The existence, in both cases, of energy barriers preventing the magnetic moment reversal gives rise to slow magnetic relaxation, that is, to SMM behaviour, although at very low temperatures on account of the weak magnetic anisotropy [106]. These results evidence the existence of a close correspondence between the molecular symmetry and the electronic level structure and spin dynamics of Gd(III) single ion magnets.

Let us now consider the use of these molecules as spin qubits. The weak magnetic anisotropy of Gd(III) might enable the use of zero-field split levels (instead of the customary ground doublet) for the definition of the qubit basis $|0\rangle$ and $|1\rangle$. Indeed, the two lowest lying level doublets are separated by $\Delta E/h = 22\,\text{GHz}$ in GdW_{10} and by $\Delta E/h = 6.3\,\text{GHz}$ in GdW_{30}, thus within the reach of widespread microwave technologies and, more importantly, of superconducting quantum circuits, such as on chip resonators and tunable-gap superconducting quantum interference devices (SQUIDs). A further advantage is that the overlap between the wave-functions of such states, which determines the Rabi frequency Ω_R of

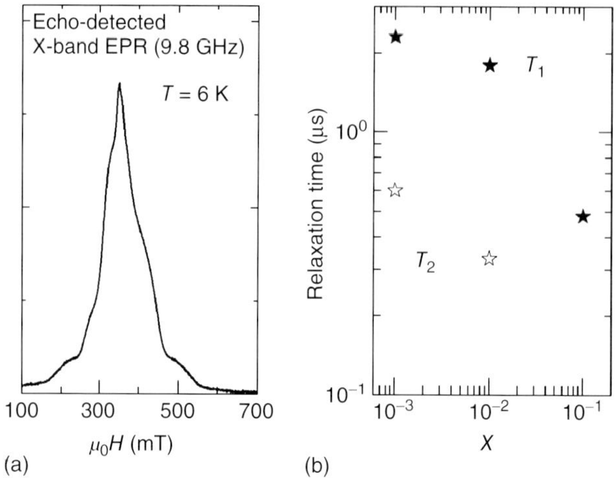

Figure 7.9 (a) Echo-detected X-band EPR spectrum of a powder sample of $Gd_{0.01}Y_{0.99}W_{30}$ at $T = 6$ K. (b) Decoherence times T_1 (solid symbols) and T_2 (open symbols) measured at $\mu_0 H = 0.347$ T, as a function of the atomic concentration of Gd ions.

coherent oscillations induced by the microwave radiation, is relatively large and depends only weakly on the magnetic field. This choice of basis avoids the need for applying strong magnetic fields to 'tune' the qubit to the circuit while, at the same time, making the qubit operation more robust with respect to the intensity and orientation of the magnetic field. These properties might therefore considerably simplify the integration of molecular qubits into hybrid QC architectures, as was recently proposed [59].

The relaxation time T_1 and dephasing time T_2 of GdW_{30} molecules diluted into a crystalline matrix of a diamagnetic analogue (YW_{30}) have been determined by means of pulsed EPR experiments [106]. Some representative results are shown in Figure 7.9. As usual, $T_1 \gg T_2$, thus quantum coherence is mainly limited by the latter. Not surprisingly, T_2 is found to rapidly decrease with increasing concentration and temperature. The best values, obtained for a sample with a 0.1% concentration of GdW_{30} at a temperature of $T = 6\,K$, give $T_2 \approx 410\,ns$, yielding a figure of merit of $Q_M \equiv 2\Omega_R T_2 \geq 50$ for $h_{rf} = 1$ mT. Coherent Rabi oscillations have indeed been observed on diluted samples of GdW_{30} [108]. These values are of the same order of magnitude as those found, under equivalent conditions, for Er(III) ions diluted in $CaWO_4$, which have been described in the previous paragraph.

7.3.2.3 Mononuclear SMMs of Ln(III) Ions with Nonzero Orbital Moment

For lanthanides other than Gd(III), the magnetic anisotropy still depends on the the local coordination environment: it determines which crystal field anisotropy terms in Equation 7.2 are allowed by symmetry, the rest simply vanish. However, the magnitudes and signs of these parameters also depend on the electronic structure of each particular ion. A qualitative model was recently proposed that relates crystal field symmetry with the nature of the ground doublet [109]. This model takes into account the shapes of the charge clouds representing $4f^n$ states, which depend on m_J. For a metal with an oblate electronic distribution, such as Ce(III), Pr(III), Tb(III) or Dy(III), strong axial ligands minimize the crystal field energy of the maximum projection $m_J = \pm J$. By contrast, for metals with a prolate electron cloud, such as Sm(III), Tm(III) and Er(III), the largest m_J levels are stabilized by ligands that furnish a strongly equatorial crystal field. Because of the localized character of 4f electrons, some information on the magnetic anisotropy can be obtained within a point-charge approximation, that is, by calculating the electric field gradients generated by point-like charges located at the equilibrium positions of the ions surrounding the central one. This approximation was employed to describe the anisotropy and magnetic properties of the $[Ln(W_5O_{18})_2]^{9-}$ POM family of complexes, which exhibit axially *compressed* square antiprismatic coordination geometry (idealized D_{4d} symmetry; see earlier text) [110]. It was found that the axial compression of the O donors around the metal gives rise to a negative B_2^0 for Er(III), that is, to a strongly uniaxial anisotropy that leads to a ground state with $m_J = \pm 15/2$. On the other hand, the same coordination makes B_2^0 positive for Tb(III), Dy(III) and Ho(III), which accounts for the fact that the most stable levels do not correspond to states with the largest m_J.

The previous discussion refers mainly to diagonal anisotropy terms $B_n^0 \widehat{O}_n^0$, which only contain powers of J_z and are often dominant. Deviations from pure axial symmetry lead to non-zero off-diagonal terms $B_n^m \widehat{O}_n^m$ with $m \neq 0$. These terms mix states m_J and m'_J, such that $|m_J - m'_J|$ is a multiple of m, and split the otherwise degenerate m_J doublets by a quantum tunnel splitting Δ. An interesting situation arises for non-Kramers ions (for which J is integer) when one of the off-diagonal terms allowed by the local symmetry is on 'good speaking terms' with the ground state doublet $\pm m_J$ determined by the diagonal terms, that is, when $2m_J$ is multiple of m. Examples are HoW_{10} (ground state $m_J = \pm 4$, D_{4d} symmetry allowing a $B_4^4 \widehat{O}_4^4$ term) [103] and TbW_{30} (ground state $m_J = \pm 5$, C_{5v} symmetry allowing a $B_6^5 \widehat{O}_6^5$ term) [107]. The result is a two-level quantum spin system characterized by a large zero-field tunnel splitting Δ and therefore a high Rabi frequency Ω_R. More importantly, Ω_R is very robust towards the action of external magnetic fields and of dipolar and hyperfine interactions, thus making these molecules excellent candidates to act as qubits or quantum memories in hybrid computation devices [59, 111].

An additional and very attractive aspect of molecular qubits is the fact that they are stable in solution, and that the ligand shell can be functionalized with specific chemical groups. In recent years, this has enabled depositing molecular clusters onto different substrates and grafting them to nanostructures or devices, such as carbon nanotube single electron transistors or point contacts [112]. These devices are sufficiently sensitive to detect the spin flip of individual lanthanide phthalocyanine complexes and provide methods to read-out their nuclear spin state [113].

7.4
Lanthanide Molecules as Prototypes of Two-Qubit Quantum Gates

7.4.1
A Family of Asymmetric [Ln_2] Complexes with Weak Magnetic Coupling

The good promise of Ln(III) as potential realizations of qubits, described in the earlier text, immediately opens the important question as to how to integrate these with the appropriate technology in order to approach the goal of performing quantum computing operations [114]. This prospect will require a major multidisciplinary effort and remains still quite distant. Nevertheless, another necessary target and fundamental step forward would be the realization of universal quantum gates using molecules, with the individual qubits embodied by the spin of single lanthanide ions. This may be achieved through the chemical design of appropriate discrete coordination complexes containing two lanthanide ions. In fact, the chemical literature contains hundreds of examples, some of which could be regarded as potential candidates. However, the requirements for one such species to fulfil this function are quite stringent. Indeed, as mentioned in Section 1.2, any algorithm in quantum computing could be realized by combining single qubits with CNOT (two qubits) or $\sqrt{\text{SWAP}}$ quantum gates, which necessitate two coupled qubits. From the definition of the CNOT quantum gate (see earlier text), it follows that if a molecule is to perform this operation based on the spins of lanthanide ions, it must contain two such metal centres, *magnetically different* from each other, and only weakly coupled. The last condition is perhaps the most accessible as the magnetic super-exchange between lanthanide ions within molecular species is well known to be weak [115] (in comparison with the coupling usually seen between the metals of the d block). However, the requirement that both metals of the dinuclear system be different is much more challenging, because chemistry tends to produce symmetric entities. Indeed, a few exceptions exist [116–125], which could be seriously considered as potential 2qubit quantum gates. However, no experimental studies have yet been aimed at verifying such potential.

Quite recently, a strategy was implemented successfully for preparing a large family of dinuclear [Ln_2] complexes featuring systematically two chemically and magnetically non-equivalent lanthanide ions. This was achieved by reacting LnX_3 salts with the ligand H_3L (6-[3-oxo-3-(2-hydroxyphenyl)-propionyl]pyridine-2-carboxylic acid, Figure 7.10) [126]. This ligand exhibits seven different donor atoms and, at least, three potentially chelating pockets, all of which are disposed in a completely asymmetric manner. In the absence of a strong base,

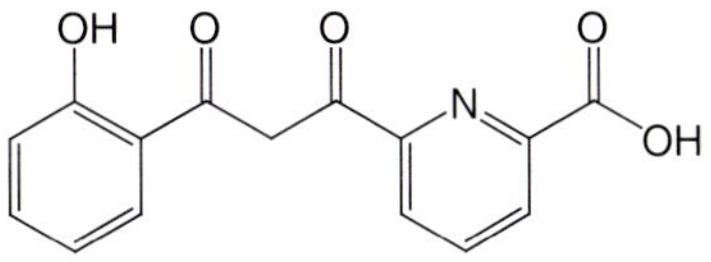

Figure 7.10 Schematic representation of the ligand H_3L used for the synthesis of a series of unsymmetric [Ln_2] and heterometallic [LnLn′] complexes.

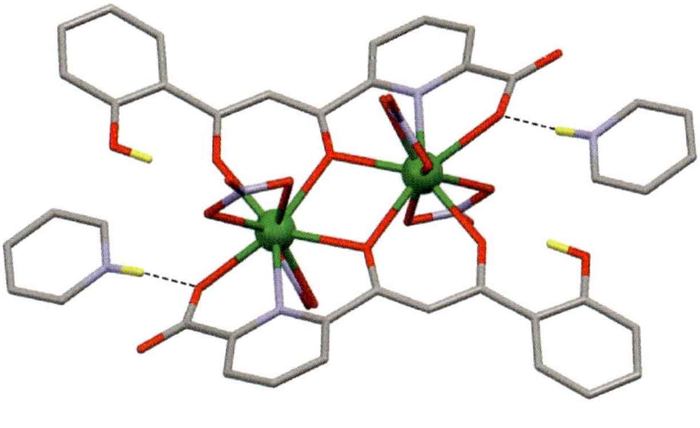

Figure 7.11 Molecular structure of the symmetric complex $(Hpy)_2[Lu_2(HL)_2(NO_3)_4]$. Colour code: Dark green, Lu(III); light purple, N; red, O and grey, C. Only hydrogen atoms on heteroatoms are shown (in yellow).

the phenol group of H_3L remains protonated, thus encouraging the ligand to perform a dinucleating function, by chelating two metals at the β-diketonate and dipicolinate-like coordination domains, respectively. This does not ensure the formation of asymmetric molecules, as demonstrated for example with the synthesis of complexes like $(Hpy)_2[Lu_2(HL)_2(NO_3)_4]$ (py = pyridine; Figure 7.11), (G. Aromí and O. Roubeau, unpublished work) which exhibits two HL^{2-} moieties in a head-to-tail conformation linking 2 equiv Ln(III) centres in a μ:η^4 coordination mode.

Nevertheless, in the appropriate experimental conditions, the versatility in coordination geometries of these ions has allowed the stabilization of dinuclear molecules with two metals bridged by deprotonated forms of three H_3L ligands, thus necessarily lying in different chemical environments. The three ligands are also in the μ:η^4 mode, occupying in total 15 coordination sites (Figure 7.12).

Three (or four, see following text) additional coordination bonds are established with two solvent molecules (pyridine, H_2O) and the anion X^- (Cl^- or NO_3^-). The good behaviour of this reaction system led to the preparation and crystallographic characterization of a comprehensive *quasi*-isostructural series of

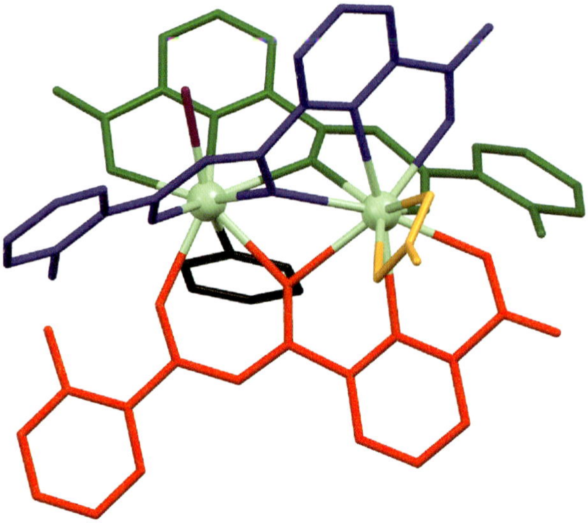

Figure 7.12 Molecular structure of the asymmetric complex $[Ln_2(HL)_2(H_2L)(NO_3)(py)(H_2O)]$ with the three deprotonated ligands resulting in two lanthanide sites with differing chemical environments. Each deprotonated form of H_3L is in a different colour (red, blue, green), pyridine is in black, water in purple, nitrate in orange and La(III) in light green. Hydrogen atoms are not shown.

asymmetric dimers with formula $[Ln_2(HL)_2(H_2L)(NO_3)(py)(H_2O)]$ (Ln = Y and all lanthanides except Pm; Figure 7.12) [127]. The lutetium analogue is the only member of the list that could not be characterized by single crystal X-ray diffraction. Structural access to such a complete series allowed a study of the *lanthanide contraction* [128]. The metric parameters that are proportional to the metal ionic radius were found to follow a quadratic decay with the number of f electrons, as recently unveiled for the elements of the 4f series (Figure 7.13) [129]. Another observation was that, as a result of the contraction, the coordination mode of NO_3^- changes from η^2 for the larger elements (La to Eu) yielding coordination number (CN) 10, to η^1 for the smaller metals (Eu to Lu; CN = 9). The frontier between both regimes is found here on europium, which was found as of both of the isomeric $[Eu_2]$ complexes, displaying CN of either 9 or 10. These two isomers must thus coexist in equilibrium. The most interesting tendency observed in this group of complexes is that one of the metal sites (*Site* 1) offers a cavity that is systematically larger than the other one (*Site* 2). This can be visualized by comparing the Ln–O distances of both sites (Figure 7.13). There is a difference in the average distances of both sites of 0.03–0.05 Å. This could be exploited for facing one of the most ambitious challenges in synthetic coordination chemistry; the controlled preparation of heterometallic 4f–4f′

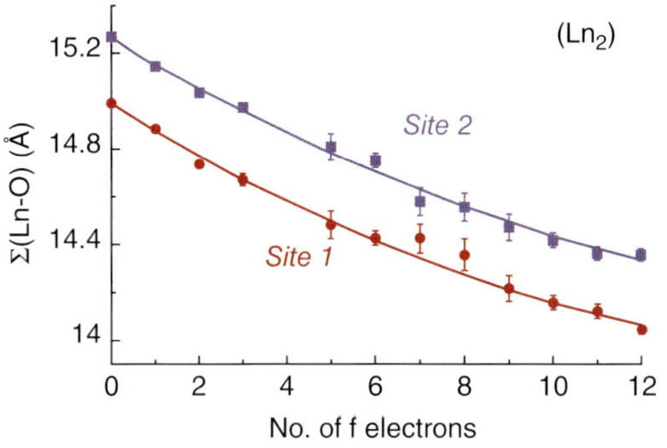

Figure 7.13 The evolution along the series of lanthanides of the sum of Ln–O bond lengths in homometallic [Ln_2] complexes for each crystallographic site. The solid lines are best fit to the quadratic function $\Sigma(n) = a + bn + cn^2$ with n being the number of f electrons of the Ln(III) ion.

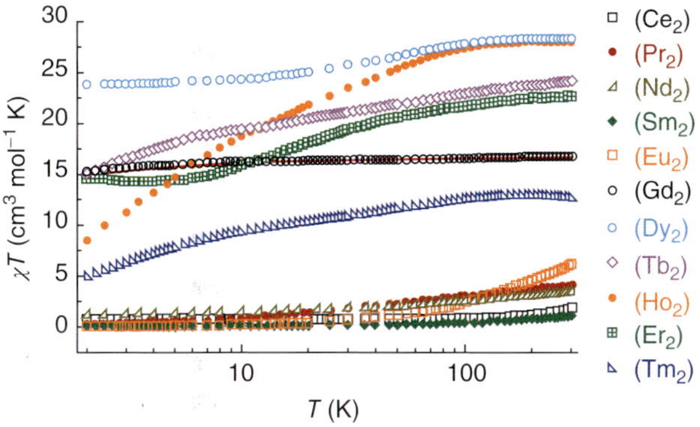

Figure 7.14 $\chi_M T$ versus T plots of the series of homometallic [$Ln_2(HL)_2(H_2L)(NO_3)(py)(H_2O)$] complexes.

complexes (see following text). As expected, the coupling between the Ln(III) magnetic moments within the molecules was found to be weak. It could be quantified from bulk magnetic susceptibility data (Figure 7.14) for the complexes of Gd; [$Gd_2(HL)_2(H_2L)X(py)(H_2O)$] ($J = 0.03$ and $0.04\,cm^{-1}$ for $X^- = Cl^-$ and NO_3^-, respectively). Similar values could be estimated for some of the complexes exhibiting spin–orbit coupling by using Ising models and very low temperature measurements (see following text).

From the above, it is apparent that most members of the extensive [Ln_2] family described fulfil the basic requirements to behave as 2qubit CNOT or $\sqrt{}$ SWAP quantum gates. As part of the process to analyse this possibility, the qubit properties of the individual Ln(III) ions needed to be evaluated.

7.4.2
Heterometallic [LnLn′] Complexes: A Fabric of Chemical Asymmetry and Individual Qubits

In order to evaluate the individual properties as qubit of each Ln(III) ion within the pertinent [Ln_2] complexes, it is necessary to study each ion without the interference of the other. This could be done if it were possible to prepare dinuclear complexes containing the Ln(III) to be studied in the desired position accompanied by a diamagnetic Ln(III) centre (or analogue) at the other site. This was possible thanks to the exceptional structural characteristics of the [Ln_2] family.

As mentioned earlier, the particular disposition of ligands HL^{2-} and H_2L^- around the metals in the dinuclear complexes discussed here causes one of the metal sites (*Site* 1) to exhibit longer Ln–O bond distances than the other (*Site* 2), this occurring systematically for the whole [Ln_2] series, independently from the effects of the lanthanide contraction. While the difference is only subtle (between 1% and 2% of the total bond distance) the added effect on all the Ln–O bonds was exploited for the deliberate preparation of heterometallic [$LnLn'(HL)_2(H_2L)(NO_3)(py)(H_2O)$] complexes. Indeed, very few crystal structures of 4f–4f′ heterometallic complexes had been published before, demonstrating the difficulty of their preparation [130–135]. The reason is the high similarity in chemical properties shown by the metals of the 4f series [136], which makes their recognition and differentiation a very challenging task. However, the systematic decrease in atomic (or ionic) radius along the period had already been used as an attempt to discriminate different 4f metals within chemical systems. Previous strategies are the use of macrocyclic ligands [137], unsymmetrical ditopic ligands [137] or the preparation of Ln 'complexes as ligands' followed by the assembly of a second Ln metal [131, 135, 138].

Here the desired [LnLn′] complexes could be obtained as pure single crystals from one pot reactions, by dissolving stoichiometric $LnX_3 + Ln'X_3$ mixtures with H_3L ($X^- = Cl^-$ or NO_3^-) [139]. For sufficiently well separated metals, single crystal X-ray crystallography was unambiguous in establishing the identity and purity of the heterobimetallic phases. Several other techniques have also provided supporting evidence for some relevant [LnLn′] examples, such as induced coupled plasma metal analysis, EPR, bulk magnetization measurements or specific heat studies. The most revealing technique is however, positive-ion electro-spray ionization (ESI) mass spectrometry experiments. This analysis was suited for unveiling with precision the core of, both the homo- and heterometallic pure complexes, and most importantly, the strict absence of [Ln_2] or [Ln'_2] residues when analysing the purity of the corresponding [LnLn′] analogue (Figure 7.15). While a comprehensive study of the 105 different combinations (excluding Pm from the study) is not yet available (121 if considering Y also), ESI-MS has already confirmed the purity of all the combinations that have been important

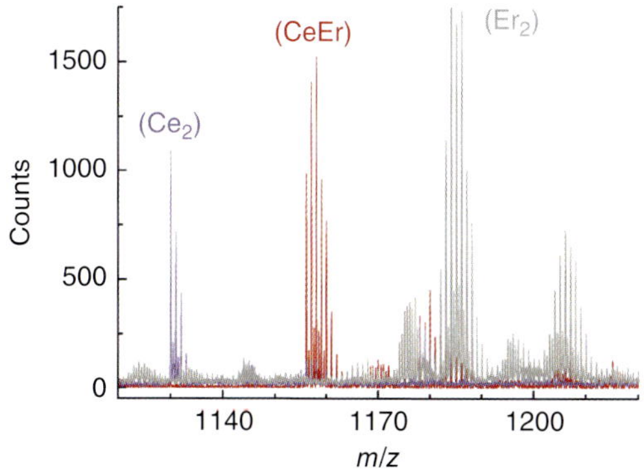

Figure 7.15 Superposition of the ESI mass spectrograms of the [Ce_2], [Er_2] and [CeEr] complexes, emphasizing the absence of the characteristic peaks from any of the homometallic analogues on the graph from the heterometallic species.

for the study of potential 2qubit quantum gates (see following text). However, this technique has revealed already that the reaction system is *not* capable of discriminating pairs of Ln atoms adjacent in the periodic table. Indeed, because the key property for the separation is the difference in ionic size, pairs of adjacent metals are too similar for this method to be effective. Besides this limitation, a large number of pure heterometallic [LnLn′] complexes have been prepared and effectively exploited. We focus below on [LaTb] and [TbLu] that were prepared to allow the comparative study of the isolated qubits involved in the [Tb_2] complex, proposed as a prototype of CNOT and SWAP gates. Another example is the preparation of [$CeEr(HL)_2(H_2L)(NO_3)(py)(H_2O)$] (Figure 7.16), a chemically asymmetric candidate of CNOT quantum gate. This heteronuclear combination was chosen for its minimal amount of nuclear spins (i.e. none for the Ce(III) ion and only 22% of the ^{167}Er isotope with $I = 7/2$ for the Er(III) ion), a priori ideal to minimize decoherence. The complex [CeEr] has been exhaustively characterized and its purity unambiguously established [139]. In this assembly, the larger cation Ce(III) occupies the cavity of *Site* 1, while the smaller metal Er(III) is located at *Site* 2.

7.4.3 Evaluating Qubit Properties

The capacity of preparing pure [LnLn′] complexes with compositions almost at will allows the study of individual Ln ions within such assemblies without the magnetic influence of the other one, just by making the analogue with the position of the second metal occupied by a diamagnetic ion. The homodimetallic complex [$Tb_2(HL)_2(H_2L)Cl(py)(H_2O)$] was investigated as potential CNOT and

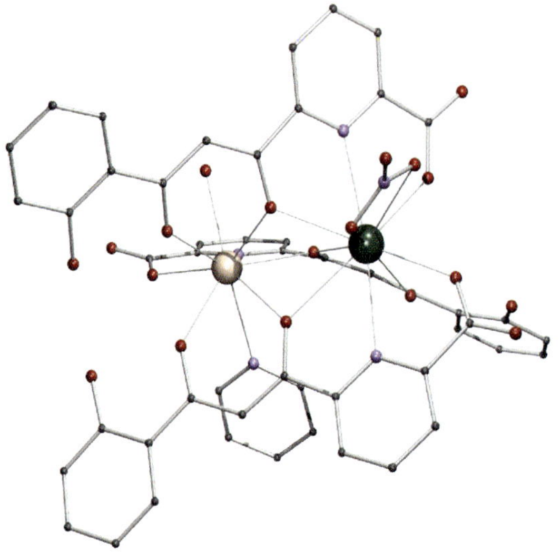

Figure 7.16 View of the molecular structure of the heterometallic complex [CeEr]. Colour code: red, O; purple, N; grey, C; green, Ce and brown, Er. Hydrogen atoms not shown.

$\sqrt{}$ SWAP 2qubit quantum gates (see following text). In this case, the metals that would embody both qubits are of the same element; however, they lie in different coordination environments, which ensures their magnetic non-equivalence. The controlled synthesis of complexes [TbLu(HL)$_2$(H$_2$L)Cl(py)(H$_2$O)] ([TbLu]) and [LaTb(HL)$_2$(H$_2$L)Cl(py)(H$_2$O)] ([LaTb]), containing Tb(III) on *Site 1* and 2, respectively, was performed by combining Tb(III) with either a smaller (Lu(III)) or a larger (La(III)) diamagnetic ion, occupying respectively *Sites 2* and *1* in the assembly.

A combination of low-temperature magnetic susceptibility, specific heat and EPR measurements was used for characterizing the magnetic ground state of Tb(III) within the complexes [LaTb] and [TbLu], respectively. The ground state of these Tb(III) ions is $^7F_6$ ground state multiplet with $g_J = 3/2$. At room temperature, this ground state is sufficiently isolated from excited levels; therefore, determining the magnetic properties of such ions exclusively. The data were thus simulated using the Hamiltonian of Equation 7.2. The $\chi_M T$ versus T plot of the complex [LaTb] (Figure 7.17a) is consistent with that for a free $^7F_6$ ion at room temperature (close to the expected value of 11.825 cm^3 K mol^{-1}), reaching a plateau near 10.2 cm^3 K mol^{-1} below 20 K, which indicates that only the ground state Kramers doublet is populated at these temperatures. The experimental data can be reproduced using Equation 7.2 with only one parameter, $B_2^0 = 35.4$ cm^{-1}. The $\chi_M T$ value of the plateau can be ascribed to $m_J = \pm 6$ doublets, likely made of quantum superpositions of pure m_J sub-levels. The assignment is consistent

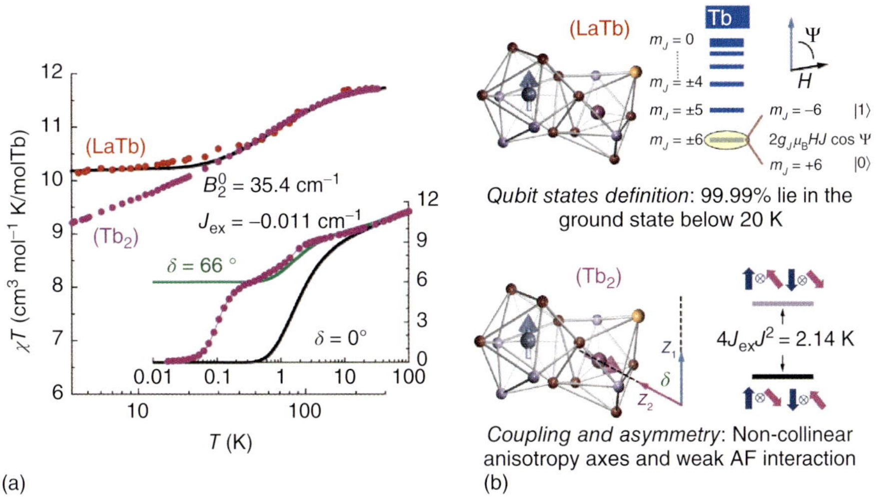

Figure 7.17 (a) Magnetic properties of [LaTb] and [Tb_2] in the form of χT versus T plot per mole of Tb(III). (b) Schematic representation of the qubit definition, weak coupling and asymmetry, as derived from magnetic and heat capacity data.

with specific heat, c_p, measurements, which served to tune the fit parameters defining the crystal field and the gyromagnetic ratios. Above 10 K, the specific heat is dominated by the contributions from lattice vibrations (c_{latt}). These could be determined experimentally, again, by exploiting the excellent versatility of the present chemical system, which provides the possibility of fabricating a completely diamagnetic analogue, [LaY]. The latter complex indeed, only has lattice contributions to c_p, which should be virtually identical to these from any of the [Ln_2] family. The magnetic contribution to the specific heat, c_m, shows up below 10 K, and is well simulated as arising from a doublet ground state with strong anisotropy (using the parameters obtained from the magnetic study), as required to embody a qubit.

A similar study, including detailed EPR experiments, also provides evidence for a proper definition of qubit states at low temperatures for the Er(III) and Ce(III) ions in [LaEr] and [CeY], respectively [139].

7.4.4
Weak Coupling

The studies on [LaTb], [TbLu], [LaEr] and [CeY] established that the individual ions of the molecules [$Tb_2(HL)_2(H_2L)Cl(py)(H_2O)$] and [$CeEr(HL)_2(H_2L)(NO_3)(py)(H_2O)$] exhibit isolated, well-defined ground state doublets, thus leading to proper definitions of qubit states. The next step is to prove the existence of a weak coupling within each molecule conducive to the appropriate energy level spectrum for the realization of quantum gate operations.

The ac and dc magnetic susceptibility of [Tb_2] was determined from 300 K down to 13 mK (Figure 7.17). For the lowest frequency (0.0158 Hz) of the ac measurements, the molecules are in thermal equilibrium down to 100 mK, thus the in-phase component of the susceptibility, χ', gives the equilibrium paramagnetic response χ_M. At 300 K, the value of $\chi_M T$ (per mole of Tb_2) corresponds to that of two uncoupled Tb(III) ions (for $^7F_6$ and $g_J = 3/2$; 23.62 cm^3 K mol^{-1}). The drop of the $\chi_M T$ versus T curve observed between ~100 and 10 K is caused by the depopulation of the Stark levels within the $^7F_6$ manifolds. Above 10 K, the $\chi_M T$ versus T curve can be fit with Equation 7.2, giving, as for [LaTb], a second-order axial parameter $B_2^0 = 35.4\ cm^{-1}$. The $m_J = \pm 6$ ground state level doublet is therefore very well isolated, the first excited levels lying more than 180 K above it. Below 10 K, an additional drop of $\chi_M T$ is observed, resulting from the weak antiferromagnetic coupling between both magnetic moments within the molecule. This coupling represents the last requirement needed to be fulfilled to consider [Tb_2] as a potential 2qubit quantum gate. With the help of magnetic heat capacity $c_{m,p}$ data, this coupling was estimated to be $J_{ex} = -0.011\ cm^{-1}$ (considering the Hamiltonian $H = -2J_{ex}J_1J_2$ for this coupling) [25].

7.4.5
Asymmetry and Energy Diagrams

Interestingly, the value of χ' keeps increasing monotonically down to 100 mK, despite the presence of a negative J_{ex}. This corresponds to $\chi_M T$ reaching a plateau at temperatures between 300 and 100 mK (Figure 7.17, inset). The further decrease of $\chi' T$ below 100 mK is due to the blocking of the magnetic response caused by slow magnetic relaxation processes. The observation of this plateau can only be explained if the magnetic moments of both ions do not completely compensate each other in the (coupled) ground state of the dimer. This would be the case if the anisotropy axes are not parallel, which would cause both antiferromagnetic states ($|1,0\rangle$ and $|0,1\rangle$) to exhibit a net magnetization. The tilting angle $\delta = 66°$ between these axes was estimated by fitting magnetic susceptibility and heat capacity data. This amounts to having both ions coupling differently to an external magnetic field and, therefore, being described by two different effective gyromagnetic ratios, g_1 and g_2. As could be understood from the arguments given in Section 3.2, the magnetic non-equivalence, now confirmed, was expected from the fact that both anions occupy disparate coordination sites. Magnetization isotherms measured at $T = 0.26$ K (Figure 7.18) support independently both the cluster magnetic asymmetry and the antiferromagnetic coupling by showing a finite magnetic response starting from $H = 0$, and exhibiting an abrupt jump at slightly higher fields corresponding to the antiferromagnetic to ferromagnetic level crossing.

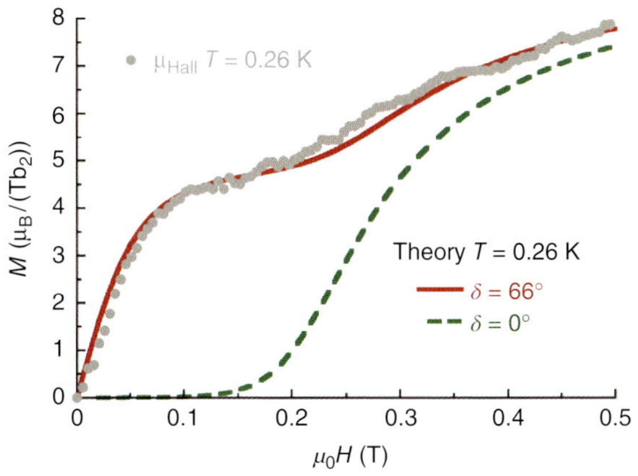

Figure 7.18 Magnetization isotherm of [Tb_2] at 0.26 K as measured through μHall magnetometry. Solid and dashed lines are the calculated behaviours for collinear and non-collinear (shown is the best agreement for $\delta = 66°$) orientations of the Tb(III) spins within the asymmetric dimer (see Figure 7.17).

Within the reduced subspace defined by the relevant 2qubit basis, the [Tb_2] system can be described by the Hamiltonian in Equation 7.3.

$$\hat{H} = -2J_{ex}J_{1,z}J_{2,z} - g_J\mu_B H_{1,z}J_{1,z} - g_J\mu_B H_{2,z}J_{2,z} + A_J(J_{1,z}I_{1,z} + J_{2,z}I_{2,z}) \tag{7.3}$$

In this Hamiltonian, eigenvalues of $J_{1,z}$ and $J_{2,z}$ adopt only +6 or −6 values, A_J is the hyperfine coupling constant and $H_{1,z}$ and $H_{2,z}$ are the (different) components of the magnetic field along the non collinear anisotropy axes, z_1 and z_2 of both Tb^{3+} ions. The energy spectrum ensuing from this Hamiltonian (Figure 7.19a) serves to discuss the performance of the molecule as 2qubit quantum gate. For clarity, only the levels with nuclear projections corresponding to $m_{I,1} = m_{I,2} = 3/2$ are considered. Any other projection would also be valid independently; note that the selection rule $\Delta m_I = 0$ must apply. This diagram shows how two qubits may be controlled simultaneously to perform quantum gate operations, by inducing specific transitions. For example, an external radiation source of 9.8 GHz as available from X-band EPR instruments, induces a transition between states $|0,1\rangle$ and $|0,0\rangle$, leaving the population at any of the other energy levels unaffected. This transition corresponds to a single-shot CNOT operation, where the first qubit exerts as the control qubit and the other as the target one. Tuning the magnetic field would allow one to perform the transition between states $|0,1\rangle$ and $|1,0\rangle$, which corresponds to the SWAP operation. The transitions involved in the gate operations just described were shown to be allowed (perhaps by the presence of non-diagonal terms of the magnetic anisotropy) by means of continuous-wave EPR spectroscopy measurements [25].

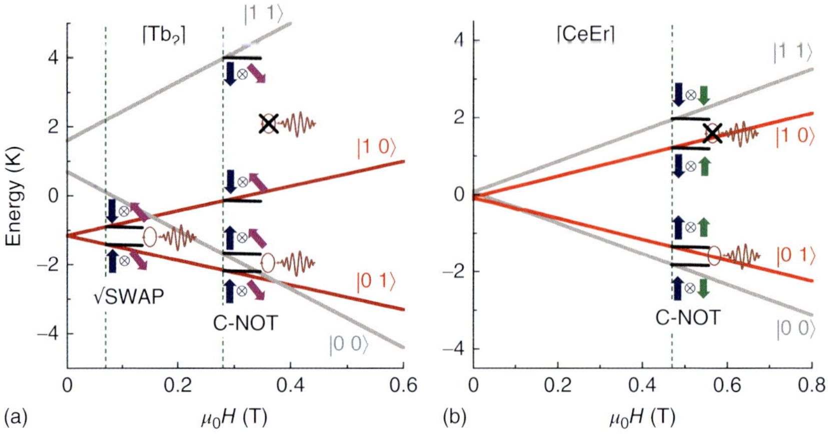

Figure 7.19 Field-dependent magnetic energy levels of $[Tb_2]$ (a) and [CeEr] (b) calculated as described in the text. Resonant transition with an X-band EPR radiofrequency excitation are shown, indicating that both asymmetric molecules provide the basis for 2qubit quantum gate operations.

The heterometallic complex [CeEr] provides a true chemical asymmetry in addition to the structural asymmetry present in the homometallic $[Tb_2]$. It is the other member of the family of dinuclear systems that has been studied in detail as quantum gate [139]. Magnetic susceptibility and heat capacity data of [CeEr] were obtained and compared with those from the analogues [CeY] and [LaEr], indicating the presence of a weak antiferromagnetic inter-qubit coupling. The corresponding splitting between the ferromagnetic and antiferromagnetic states, described by $J_{ex} = -0.005\,\text{cm}^{-1}$ was determined by fitting the zero-field susceptibility and specific heat data. Interestingly, the magnetic heat capacity of [CeEr] gradually approaches the sum of the [CeY] and [LaEr] heat capacities as the external magnetic field increases. This shows that the exchange interaction is gradually overcome by the magnetic field and confirms both its sign and magnitude. The EPR spectrum of [CeEr] also underscores the influence of the inter-qubit magnetic coupling, because it cannot be decomposed into a weighted sum of the spectra from [CeY] and [LaEr].

The energy level scheme of complex $[CeEr(HL)_2(H_2L)(NO_3)(py)(H_2O)]$, showing only its four lowest magnetic energy levels, is shown in Figure 7.19b. As with $[Tb_2]$, these four levels, which are non-degenerate thanks to the fact that both qubits are different, constitute a suitable basis for the realization of a molecular quantum processor. Given the range of energies, it appears that this processor can be operated at low magnetic fields and using conventional microwave technologies.

7.4.6
Decoherence of the Molecular Quantum Processor Prototypes

Following the demonstration of the basic requirements (see earlier text), a second, crucial step for ascertaining the suitability of $[Ln_2]$ complexes as quantum gates is to evaluate the coherence times associated with those transitions involved in the gate operation. As shown in Sections 3.1 and 3.2, studies of quantum coherence have already been performed successfully on lanthanide-based qubits [47, 106, 108, 111]. Pulsed EPR experiments have been performed on samples of complex $[CeEr(HL)_2(H_2L)(NO_3)(py)(H_2O)]$ dissolved in a mixture of MeOH and EtOH [139]. These experiments show that radio frequency pulses can generate coherent spin dynamics. The time decay of the echo signal was studied for $\mu_0 H = 0.47$ T. This resonance approximately corresponds to the $|0,1\rangle \leftrightarrow |0,0\rangle$ transition, associated to a CNOT gate operation, where the first qubit acts as control (see Figure 7.19b). From these experiments, a decoherence time T_2 of ~400 ns was estimated. Additional experiments will serve to identify the main sources of decoherence and to engineer adequate methods to reduce its effects.

7.5
Conclusions and Outlook

While the challenges lying ahead for the practical implementation of a universal quantum computer are still daunting, it is also clear that the last decade has witnessed spectacular advances towards this ambitious goal. Progress in this area has been possible undoubtedly thanks to the joint efforts of scientists from very diverse disciplines. Synthetic chemists have been among the latest communities to join this venture. Yet, as it has been argued in the previous sections, chemistry provides unique tools for the design, fabrication and fine tuning of the basic elements of a quantum computer. A major advantage of this approach, especially with the prospect of building complex and scalable computational architectures, is the possibility of producing as many *identical* qubits as necessary, something that is inaccessible to any other of the current existing proposals for QC hardware.

Lanthanide ions offer several salient properties that make them especially attractive as qubit candidates: (i) their magnetic states provide proper definitions of the qubit basis; (ii) they show reasonably long coherence times; (iii) important qubit parameters, such as the energy gap ΔE and the Rabi frequency Ω_R, can be chemically tuned by the design of the lanthanide co-ordination shell and (iv) the same molecular structure can be realized with many different lanthanide ions (e.g. with or without nuclear spin), thus providing further versatility for the design of spin qubits or hybrid spin registers.

A very promising prospect is also the possibility of working with molecules containing more than one lanthanide. The combination of weak magnetic interactions, typical of lanthanides and of dissimilar local coordination environments enables the design of multiple qubit spin Hamiltonians, which can mimic those required to perform specific logic operations. In this chapter, we have shown that dinuclear complexes of two weakly coupled, non-equivalent lanthanides fulfil the basic requirements to act as single-shot CNOT or $\sqrt{\text{SWAP}}$ quantum gates. The relevant transitions are well defined, that is, they can be separately addressed by either choosing the right frequency or tuning an external magnetic field, they are not forbidden and they can be coherently manipulated.

One of the most difficult future challenges is the development of suitable and robust techniques able to initialize, read out and control *individual* lanthanide qubits or quantum gates. A major hurdle will also be the transfer of quantum information between qubits and qugates in a scalable architecture. This will probably require the implementation of hybrid technologies, coupling such molecular 'processors' with either spintronic devices [112, 113] (point contacts, molecular nanojunctions, etc.) or superconducting quantum circuits [59] or to a combination of both. While this problem is probably not going to be addressed solely by chemical methods, the latter can certainly do much to alleviate it by optimizing the properties of the qubits themselves. In addition, the methods and techniques currently employed for depositing small molecular ensembles or individual molecules where required will need to be optimized. Chemistry can do a lot to help in this endeavour too. As an extension of these ideas, it is worth emphasizing that the rational design of molecular complexes offers alternative, and rather original, ways for scaling up computation *within* a molecule. For instance, molecular species capable of realizing more complex operations can be synthesized, using arrays of three, four or five appropriately disposed metals of different nature. This approach has the advantage of being highly reproducible and robust. However, the fact that inter-qubit couplings are always switched on might require the development of new computational protocols. Summarizing, the chemical engineering of artificial lanthanide molecular complexes provides an excellent platform for the development of scalable quantum architectures. The initial progress made in showing that molecules of lanthanides are suitable realizations of qubits and qugates should hopefully increase the interest, and synergic cooperation, of the diverse scientific communities working in this area.

References

1. Einstein, A., Podolsky, B. and Rosen, N. (1935) *Phys. Rev.*, **47**, 777–780.
2. Schrödinger, E. (1935) *Naturwissenschaften*, **23**, 807–812.
3. Bell, J.S. (1964) *Physics*, **1**, 195–200.
4. Nielsen, M.A. and Chuang, I.L. (2000) *Quantum Computation and Quantum Information*, Cambridge University Press.
5. Ladd, T.D., Jelezko, F., Laflamme, R., Nakamura, Y., Monroe, C. and O'Brien, J.L. (2010) *Nature*, **464**, 45–53.
6. Feynman, R.P. (1982) *Int. J. Theor. Phys.*, **21**, 467–488.
7. Feynman, R.P. (1985) *Opt. News*, **11**, 11–20.
8. Deutsch, D. (1985) *Proc. R. Soc. London, Ser. A*, **400**, 97–117.

9. Deutsch, D. and Jozsa, R. (1992) *Proc. R. Soc. London, Ser. A*, **439**, 553–558.
10. Shor, P.W. (1997) *SIAM J. Comput.*, **26**, 1484–1509.
11. Grover, L.K. (1997) *Phys. Rev. Lett.*, **79**, 325–328.
12. Brown, K.L., Munro, W.J. and Kendon, V.M. (2010) *Entropy*, **12**, 2268–2307.
13. DiVincenzo, D.P. (2000) *Fortschr. Phys.*, **48**, 771–783.
14. Cirac, J.I. and Zoller, P. (1995) *Phys. Rev. Lett.*, **74**, 4091–4094.
15. Monroe, C. and Kim, J. (2013) *Science*, **339**, 1164–1169.
16. Knill, E., Laflamme, R. and Milburn, G.J. (2001) *Nature*, **409**, 46–52.
17. Gershenfeld, N.A. and Chuang, I.L. (1997) *Science*, **275**, 350–356.
18. Clarke, J. and Wilhelm, F.K. (2008) *Nature*, **453**, 1031–1042.
19. Devoret, M.H. and Schoelkopf, R.J. (2013) *Science*, **339**, 1169–1174.
20. Kloeffel, C. and Loss, D. (2013) in *Annual Review of Condensed Matter Physics*, vol. 4 (ed. J.S. Langer), pp. 51–81.
21. Loss, D. and DiVincenzo, D.P. (1998) *Phys. Rev. A*, **57**, 120–126.
22. Awschalom, D.D., Bassett, L.C., Dzurak, A.S., Hu, E.L. and Petta, J.R. (2013) *Science*, **339**, 1174–1179.
23. Hanson, R. and Awschalom, D.D. (2008) *Nature*, **453**, 1043–1049.
24. Sato, K., Nakazawa, S., Rahimi, R., Ise, T., Nishida, S., Yoshino, T., Mori, N., Toyota, K., Shiomi, D., Yakiyama, Y., Morita, Y., Kitagawa, M., Nakasuji, K., Nakahara, M., Hara, H., Carl, P., Hofer, P. and Takui, T. (2009) *J. Mater. Chem.*, **19**, 3739–3754.
25. Luis, F., Repollés, A., Martínez-Pérez, M.J., Aguilà, D., Roubeau, O., Zueco, D., Alonso, P.J., Evangelisti, M., Camón, A., Sesé, J., Barrios, L.A. and Aromí, G. (2011) *Phys. Rev. Lett.*, **107**, 117203.
26. Lehmann, J., Gaita-Ariño, A., Coronado, E. and Loss, D. (2007) *Nat. Nanotechnol.*, **2**, 312–317.
27. Aromí, G., Aguilà, D., Gamez, P., Luis, F. and Roubeau, O. (2012) *Chem. Soc. Rev.*, **41**, 537–546.
28. Fischer, J. and Loss, D. (2009) *Science*, **324**, 1277–1278.
29. Barenco, A., Bennett, C.H., Cleve, R., Divincenzo, D.P., Margolus, N., Shor, P., Sleator, T., Smolin, J.A. and Weinfurter, H. (1995) *Phys. Rev. A*, **52**, 3457–3467.
30. Bell, J. (1987) *Speakable and Unspeakable in Quantum Mechanics*, Cambridge University Press, Cambridge.
31. DiVincenzo, D.P. (2009) *Phys. Scr.*, **T137**, 014020.
32. Knill, E. and Laflamme, R. (1997) *Phys. Rev. A*, **55**, 900–911.
33. Shor, P.W. (1995) *Phys. Rev. A*, **52**, R2493–R2496.
34. de Lange, G., Wang, Z.H., Riste, D., Dobrovitski, V.V. and Hanson, R. (2010) *Science*, **330**, 60–63.
35. Viola, L., Lloyd, S. and Knill, E. (1999) *Phys. Rev. Lett.*, **83**, 4888–4891.
36. Vandersypen, L.M.K., Steffen, M., Breyta, G., Yannoni, C.S., Sherwood, M.H. and Chuang, I.L. (2001) *Nature*, **414**, 883–887.
37. Monroe, C., Meekhof, D.M., King, B.E., Itano, W.M. and Wineland, D.J. (1995) *Phys. Rev. Lett.*, **75**, 4714–4717.
38. Walther, A., Ziesel, F., Ruster, T., Dawkins, S.T., Ott, K., Hettrich, M., Singer, K., Schmidt-Kaler, F. and Poschinger, U. (2012) *Phys. Rev. Lett.*, **109**, 080501.
39. Bowler, R., Gaebler, J., Lin, Y., Tan, T.R., Hanneke, D., Jost, J.D., Home, J.P., Leibfried, D. and Wineland, D.J. (2012) *Phys. Rev. Lett.*, **109**, 080502.
40. Chuang, I.L., Vandersypen, L.M.K., Zhou, X.L., Leung, D.W. and Lloyd, S. (1998) *Nature*, **393**, 143–146.
41. Negrevergne, C., Mahesh, T.S., Ryan, C.A., Ditty, M., Cyr-Racine, F., Power, W., Boulant, N., Havel, T., Cory, D.G. and Laflamme, R. (2006) *Phys. Rev. Lett.*, **96**, 170501.
42. Plantenberg, J.H., de Groot, P.C., Harmans, C. and Mooij, J.E. (2007) *Nature*, **447**, 836–839.
43. Leuenberger, M.N. and Loss, D. (2001) *Nature*, **410**, 789–793.
44. Meier, F., Levy, J. and Loss, D. (2003) *Phys. Rev. Lett.*, **90**, 047901.

45. Tejada, J., Chudnovsky, E.M., del Barco, E., Hernandez, J.M. and Spiller, T.P. (2001) *Nanotechnology*, **12**, 181–186.
46. Troiani, F., Ghirri, A., Affronte, M., Carretta, S., Santini, P., Amoretti, G., Piligkos, S., Timco, G. and Winpenny, R.E.P. (2005) *Phys. Rev. Lett.*, **94**, 207208.
47. Bertaina, S., Gambarelli, S., Tkachuk, A., Kurkin, I.N., Malkin, B., Stepanov, A. and Barbara, B. (2007) *Nat. Nanotechnol.*, **2**, 39–42.
48. Braunstein, S.L. and van Loock, P. (2005) *Rev. Mod. Phys.*, **77**, 513–577.
49. Knill, E., Laflamme, R., Martinez, R. and Negrevergne, C. (2001) *Phys. Rev. Lett.*, **86**, 5811–5814.
50. Politi, A., Matthews, J.C.F. and O'Brien, J.L. (2009) *Science*, **325**, 1221.
51. O'Brien, J.L. (2007) *Science*, **318**, 1567–1570.
52. O'Brien, J.L., Furusawa, A. and Vuckovic, J. (2009) *Nat. Photonics*, **3**, 687–695.
53. Wallraff, A., Schuster, D.I., Blais, A., Frunzio, L., Huang, R.S., Majer, J., Kumar, S., Girvin, S.M. and Schoelkopf, R.J. (2004) *Nature*, **431**, 162–167.
54. Schuster, D.I., Wallraff, A., Blais, A., Frunzio, L., Huang, R.S., Majer, J., Girvin, S.M. and Schoelkopf, R.J. (2005) *Phys. Rev. Lett.*, **94**, 123602.
55. Wallraff, A., Schuster, D.I., Blais, A., Frunzio, L., Majer, J., Devoret, M.H., Girvin, S.M. and Schoelkopf, R.J. (2005) *Phys. Rev. Lett.*, **95**, 060501.
56. Majer, J., Chow, J.M., Gambetta, J.M., Koch, J., Johnson, B.R., Schreier, J.A., Frunzio, L., Schuster, D.I., Houck, A.A., Wallraff, A., Blais, A., Devoret, M.H., Girvin, S.M. and Schoelkopf, R.J. (2007) *Nature*, **449**, 443–447.
57. Rabl, P., DeMille, D., Doyle, J.M., Lukin, M.D., Schoelkopf, R.J. and Zoller, P. (2006) *Phys. Rev. Lett.*, **97**, 033003.
58. Imamoglu, A. (2009) *Phys. Rev. Lett.*, **102**, 083602.
59. Jenkins, M., Hümmer, T., Martínez-Pérez, M.J., García-Ripoll, J.J., Zueco, D. and Luis, F. (2013) *New J. Phys.*, **15**, 095007.
60. Wesenberg, J.H., Ardavan, A., Briggs, G.A.D., Morton, J.J.L., Schoolkopf, R.J., Schuster, D.I. and Molmer, K. (2009) *Phys. Rev. Lett.*, **103**, 070502.
61. Hanson, R., Kouwenhoven, L.P., Petta, J.R., Tarucha, S. and Vandersypen, L.M.K. (2007) *Rev. Mod. Phys.*, **79**, 1217–1265.
62. Hennessy, K., Badolato, A., Winger, M., Gerace, D., Atature, M., Gulde, S., Falt, S., Hu, E.L. and Imamoglu, A. (2007) *Nature*, **445**, 896–899.
63. Morton, J.J.L., McCamey, D.R., Eriksson, M.A. and Lyon, S.A. (2011) *Nature*, **479**, 345–353.
64. Kane, B.E. (1998) *Nature*, **393**, 133–137.
65. Tyryshkin, A.M., Tojo, S., Morton, J.J.L., Riemann, H., Abrosimov, N.V., Becker, P., Pohl, H.J., Schenkel, T., Thewalt, M.L.W., Itoh, K.M. and Lyon, S.A. (2012) *Nat. Mater.*, **11**, 143–147.
66. Pla, J.J., Tan, K.Y., Dehollain, J.P., Lim, W.H., Morton, J.J.L., Jamieson, D.N., Dzurak, A.S. and Morello, A. (2012) *Nature*, **489**, 541–545.
67. van der Sar, T., Wang, Z.H., Blok, M.S., Bernien, H., Taminiau, T.H., Toyli, D.M., Lidar, D.A., Awschalom, D.D., Hanson, R. and Dobrovitski, V.V. (2012) *Nature*, **484**, 82–86.
68. Santini, P., Carretta, S., Troiani, F. and Amoretti, G. (2011) *Phys. Rev. Lett.*, **107**, 230502.
69. Lehmann, J., Gaita-Ariño, A., Coronado, E. and Loss, D. (2009) *J. Mater. Chem.*, **19**, 1672–1677.
70. Matsuda, K. and Irie, M. (2001) *Chem. Eur. J.*, **7**, 3466–3473.
71. Bogani, L. and Wernsdorfer, W. (2008) *Nat. Mater.*, **7**, 179–186.
72. Morton, J.J.L., Tyryshkin, A.M., Ardavan, A., Benjamin, S.C., Porfyrakis, K., Lyon, S.A. and Briggs, G.A.D. (2006) *Nat. Phys.*, **2**, 40–43.
73. Morita, Y., Yakiyama, Y., Nakazawa, S., Murata, T., Ise, T., Hashizume, D., Shiomi, D., Sato, K., Kitagawa, M., Nakasuji, K. and Takui, T. (2010) *J. Am. Chem. Soc.*, **132**, 6944–6946.

74. Bertaina, S., Gambarelli, S., Mitra, T., Tsukerblat, B., Muller, A. and Barbara, B. (2008) *Nature*, **453**, 203–207.
75. Timco, G.A., Faust, T.B., Tuna, F. and Winpenny, R.E.P. (2011) *Chem. Soc. Rev.*, **40**, 3067–3075.
76. Caciuffo, R., Guidi, T., Amoretti, G., Carretta, S., Liviotti, E., Santini, P., Mondelli, C., Timco, G., Muryn, C.A. and Winpenny, R.E.P. (2005) *Phys. Rev. B*, **71**, 174407.
77. Ardavan, A., Rival, O., Morton, J.J.L., Blundell, S.J., Tyryshkin, A.M., Timco, G.A. and Winpenny, R.E.P. (2007) *Phys. Rev. Lett.*, **98**, 057201.
78. Wedge, C.J., Timco, G.A., Spielberg, E.T., George, R.E., Tuna, F., Rigby, S., McInnes, E.J.L., Winpenny, R.E.P., Blundell, S.J. and Ardavan, A. (2012) *Phys. Rev. Lett.*, **108**, 107204.
79. Timco, G.A., Carretta, S., Troiani, F., Tuna, F., Pritchard, R.J., Muryn, C.A., McInnes, E.J.L., Ghirri, A., Candini, A., Santini, P., Amoretti, G., Affronte, M. and Winpenny, R.E.P. (2009) *Nat. Nanotechnol.*, **4**, 173–178.
80. Bellini, V., Lorusso, G., Candini, A., Wernsdorfer, W., Faust, T.B., Timco, G.A., Winpenny, R.E.P. and Affronte, M. (2011) *Phys. Rev. Lett.*, **106**, 227205.
81. Candini, A., Lorusso, G., Troiani, F., Ghirri, A., Carretta, S., Santini, P., Amoretti, G., Muryn, C., Tuna, F., Timco, G., McInnes, E.J.L., Winpenny, R.E.P., Wernsdorfer, W. and Affronte, M. (2010) *Phys. Rev. Lett.*, **104**, 037203.
82. Hill, S., Edwards, R.S., Aliaga-Alcalde, N. and Christou, G. (2003) *Science*, **302**, 1015–1018.
83. Costa, J.S., Barrios, L.A., Craig, G.A., Teat, S.J., Luis, F., Roubeau, O., Evangelisti, M., Camón, A. and Aromí, G. (2012) *Chem. Commun.*, **48**, 1413–1415.
84. Nguyen, T.N., Wernsdorfer, W., Abboud, K.A. and Christou, G. (2011) *J. Am. Chem. Soc.*, **133**, 20688–20691.
85. Novitchi, G., Costes, J.-P., Tuchagues, J.-P., Vendier, L. and Wernsdorfer, W. (2008) *New J. Chem.*, **32**, 197–200.
86. Aromí, G., Gamez, P. and Reedijk, J. (2008) *Coord. Chem. Rev.*, **252**, 964–989.
87. Barrios, L.A., Aguilà, D., Roubeau, O., Gamez, P., Ribas-Ariño, J., Teat, S.J. and Aromí, G. (2009) *Chem. Eur. J.*, **15**, 11235–11243.
88. Sañudo, E.C., Cauchy, T., Ruiz, E., Laye, R.H., Roubeau, O., Teat, S.J. and Aromí, G. (2007) *Inorg. Chem.*, **46**, 9045–9047.
89. Sañudo, E.C., Uber, J.S., Balagué, A.P., Roubeau, O. and Aromí, G. (2012) *Inorg. Chem.*, **51**, 8441–8446.
90. Aguilà, D., Barrios, L.A., Roubeau, O., Teat, S.J. and Aromí, G. (2011) *Chem. Commun.*, **47**, 707–709.
91. Stevens, K.W.H. (1952) *Proc. R. Soc. London, Ser. A*, **65**, 209–215.
92. Zhang, Y., Holzwarth, N.A.W. and Williams, R.T. (1998) *Phys. Rev. B*, **57**, 12738–12750.
93. Schlegel, C., van Slageren, J., Manoli, M., Brechin, E.K. and Dressel, M. (2008) *Phys. Rev. Lett.*, **101**, 147203.
94. Ishikawa, N., Sugita, M., Ishikawa, T., Koshihara, S.-Y. and Kaizu, Y. (2003) *J. Am. Chem. Soc.*, **125**, 8694–8695.
95. Jiang, S.D., Wang, B.W., Sun, H.L., Wang, Z.M. and Gao, S. (2011) *J. Am. Chem. Soc.*, **133**, 4730–4733.
96. Jiang, S.-D., Wang, B.-W., Su, G., Wang, Z.-M. and Gao, S. (2010) *Angew. Chem. Int. Ed.*, **49**, 7448–7451.
97. Cucinotta, G., Perfetti, M., Luzon, J., Etienne, M., Car, P.-E., Caneschi, A., Calvez, G., Bernot, K. and Sessoli, R. (2012) *Angew. Chem. Int. Ed.*, **51**, 1606–1610.
98. Wang, Y.-L., Ma, Y., Yang, X., Tang, J., Cheng, P., Wang, Q.-L., Li, L.-C. and Liao, D.-Z. (2013) *Inorg. Chem.*, **52**, 7380–7386.
99. Chilton, N.F., Langley, S.K., Moubaraki, B., Soncini, A., Batten, S.R. and Murray, K.S. (2013) *Chem. Sci.*, **4**, 1719–1730.
100. Menelaou, M., Ouharrou, F., Rodriguez, L., Roubeau, O., Teat, S.J. and Aliaga-Alcalde, N. (2012) *Chem. Eur. J.*, **18**, 11545–11549.
101. Liu, J.L., Yuan, K., Leng, J.D., Ungur, L., Wernsdorfer, W., Guo, F.S.,

Chibotaru, L.F. and Tong, M.L. (2012) *Inorg. Chem.*, **51**, 8538–8544.
102. Chen, G.J., Guo, Y.N., Tian, J.L., Tang, J.K., Gu, W., Liu, X., Yan, S.P., Cheng, P. and Liao, D.Z. (2012) *Chem. Eur. J.*, **18**, 2484–2487.
103. AlDamen, M.A., Cardona-Serra, S., Clemente-Juan, J.M., Coronado, E., Gaita-Ariño, A., Martí-Gastaldo, C., Luis, F. and Montero, O. (2009) *Inorg. Chem.*, **48**, 3467–3479.
104. AlDamen, M.A., Clemente-Juan, J.M., Coronado, E., Martí-Gastaldo, C. and Gaita-Ariño, A. (2008) *J. Am. Chem. Soc.*, **130**, 8874–8875.
105. Luis, F., Martínez-Pérez, M.J., Montero, O., Coronado, E., Cardona-Serra, S., Martí-Gastaldo, C., Clemente-Juan, J.M., Sese, J., Drung, D. and Schurig, T. (2010) *Phys. Rev. B*, **82**, 060403.
106. Martínez-Pérez, M.J., Cardona-Serra, S., Schlegel, C., Moro, F., Alonso, P.J., Prima-García, H., Clemente-Juan, J.M., Evangelisti, M., Gaita-Ariño, A., Sesé, J., van Slageren, J., Coronado, E. and Luis, F. (2012) *Phys. Rev. Lett.*, **108**, 247213.
107. Cardona-Serra, S., Clemente-Juan, J.M., Coronado, E., Gaita-Ariño, A., Camón, A., Evangelisti, M., Luis, F., Martínez-Pérez, M.J. and Sesé, J. (2012) *J. Am. Chem. Soc.*, **134**, 14982–14990.
108. Baldoví, J.J., Cardona-Serra, S., Clemente-Juan, J.M., Coronado, E., Gaita-Ariño, A. and Prima-Garcia, H. (2013) *Chem. Commun.*, **49**, 8922–8924.
109. Rinehart, J.D. and Long, J.R. (2011) *Chem. Sci.*, **2**, 2078–2085.
110. Baldoví, J.J., Cardona-Serra, S., Clemente-Juan, J.M., Coronado, E., Gaita-Ariño, A. and Palii, A. (2012) *Inorg. Chem.*, **51**, 12565–12574.
111. Ghosh, S., Datta, S., Friend, L., Cardona-Serra, S., Gaita-Ariño, A., Coronado, E. and Hill, S. (2012) *Dalton Trans.*, **41**, 13697–13704.
112. Urdampilleta, M., Klyatskaya, S., Cleuziou, J.P., Ruben, M. and Wernsdorfer, W. (2011) *Nat. Mater.*, **10**, 502–506.
113. Vincent, R., Klyatskaya, S., Ruben, M., Wernsdorfer, W. and Balestro, F. (2012) *Nature*, **488**, 357–360.
114. Khajetoorians, A.A., Wiebe, J., Chilian, B. and Wiesendanger, R. (2011) *Science*, **332**, 1062–1064.
115. Benelli, C. and Gatteschi, D. (2002) *Chem. Rev.*, **102**, 2369–2388.
116. Guo, Y.-N., Xu, G.-F., Wernsdorfer, W., Ungur, L., Guo, Y., Tang, J., Zhang, H.-J., Chibotaru, L.F. and Powell, A.K. (2011) *J. Am. Chem. Soc.*, **133**, 11948–11951.
117. Dei, A., Gatteschi, D., Massa, C.A., Pardi, L.A., Poussereau, S. and Sorace, L. (2000) *Chem. Eur. J.*, **6**, 4580–4586.
118. Kang, S.I., Ranganathan, R.S., Emswiler, J.E., Kumar, K., Gougoutas, J.Z., Malley, M.F. and Tweedle, M.F. (1993) *Inorg. Chem.*, **32**, 2912–2918.
119. Setyawati, I.A., Liu, S., Rettig, S.J. and Orvig, C. (2000) *Inorg. Chem.*, **39**, 496–507.
120. Piggot, P.M.T., Hall, L.A., White, A.J.P. and Williams, D.J. (2003) *Inorg. Chem.*, **42**, 8344–8352.
121. Bunzli, J.-C.G., Charbonniere, L.J. and Ziessel, R.F. (2000) *J. Chem. Soc., Dalton Trans.*, 1917–1923.
122. Deacon, G.B., Forsyth, C.M., Junk, P.C., Skelton, B.W. and White, A.H. (1999) *Chem. Eur. J.*, **5**, 1452–1459.
123. Evans, W.J., Greci, M.A. and Ziller, J.W. (1998) *Chem. Commun.*, 2367–2368.
124. Norton, K., Emge, T.J. and Brennan, J.G. (2007) *Inorg. Chem.*, **46**, 4060–4066.
125. Gröb, T., Seybert, G., Massa, W., Harms, K. and Dehnicke, K. (2000) *Z. Anorg. Allg. Chem.*, **626**, 1361–1366.
126. Aguilà, D., Barrios, L.A., Luis, F., Repollés, A., Roubeau, O., Teat, S.J. and Aromí, G. (2010) *Inorg. Chem.*, **49**, 6784–6786.
127. Aguilà, D., Barrios, L.A., Velasco, V., Arnedo, L., Aliaga-Alcalde, N., Menelaou, M., Teat, S.J., Roubeau, O., Luis, F. and Aromí, G. (2013) *Chem. Eur. J.*, **19**, 5881–5891.
128. D'Angelo, P., Zitolo, A., Migliorati, V., Chillemi, G., Duvail, M., Vitorge, P.,

Abadie, S. and Spezia, R. (2011) *Inorg. Chem.*, **50**, 4572–4579.

129. Seitz, M., Oliver, A.G. and Raymond, K.N. (2007) *J. Am. Chem. Soc.*, **129**, 11153–11160.
130. André, N., Jensen, T.B., Scopelliti, R., Imbert, D., Elhabiri, M., Hopfgartner, G., Piguet, C. and Bünzli, J.-C.G. (2003) *Inorg. Chem.*, **43**, 515–529.
131. Costes, J.P., Dahan, F., Dupuis, A., Lagrave, S. and Laurent, J.P. (1998) *Inorg. Chem.*, **37**, 153–155.
132. Jensen, T.B., Scopelliti, R. and Bünzli, J.-C.G. (2006) *Inorg. Chem.*, **45**, 7806–7814.
133. Andre, N., Scopelliti, R., Hopfgartner, G., Piguet, C. and Bunzli, J.-C.G. (2002) *Chem. Commun.*, 214–215.
134. Zhu, P., Pan, N., Li, R., Dou, J., Zhang, Y., Cheng, D.Y.Y., Wang, D., Ng, D.K.P. and Jiang, J. (2005) *Chem. Eur. J.*, **11**, 1425–1432.
135. Chabach, D., De Cian, A., Fischer, J., Weiss, R. and Bibout, M.E.M. (1996) *Angew. Chem. Int. Ed.*, **35**, 898–899.
136. Piguet, C. and Bunzli, J.-C.G. (1999) *Chem. Soc. Rev.*, **28**, 347–358.
137. Guerriero, P., Vigato, P.A., Bunzli, J.-C.G. and Moret, E. (1990) *J. Chem. Soc., Dalton Trans.*, 647–655.
138. Ishikawa, N., Iino, T. and Kaizu, Y. (2002) *J. Am. Chem. Soc.*, **124**, 11440–11447.
139. Aguilà, D., Barrios, L.A., Velasco, V., Roubeau, O., Repollés, A., Alonso, P.J., Sesé, J., Teat, S.J., Luis, F. and Aromí, G. (2014) *J. Am. Chem. Soc.*, **136**, 14215–14222.

8
Bis(phthalocyaninato) Lanthanide(III) Complexes – from Molecular Magnetism to Spintronic Devices

Yanhua Lan, Svetlana Klyatskaya, and Mario Ruben

8.1
Introduction

8.1.1
Molecular Magnetism

Magnetic phenomena and magnetic materials are everywhere. Since the invention of the compass, a great number of magnetic materials including frictionless bearings, medical devices, magnetic separators, loudspeakers, microphones, switches, sensors, data storage devices, motors and generators are used in daily life [1]. In contrast to these traditional magnetic materials, nano-scaled magnetic molecules are at the scale of a billionth of a metre and they provide a unique opportunity to observe the coexistence of classical and quantum properties. Integrating them into electronic circuits opens the possibility of processing or of storing information at the single-molecule level. Therefore, the search for molecular species with unusual magnetic properties which could lead to molecular devices is currently an important field of study in chemistry and physics called *molecular spintronics* [2–4].

In a specific class of magnetic molecules, often called *single-molecule magnets* (*SMMs*), an isolated molecule typically exhibits a footprint of around 1 – 2 nm^2 still behaving, under certain conditions, like a tiny magnet, in the sense that if magnetized by an applied field it retains the magnetization for days. In fact SMMs also give rise to magnetic hysteresis, which is one condition for storing information in a classical magnet. However, SMMs are still small enough to also show large quantum tunnelling effects (QTMs). On a fundamental level, this will allow the study of dynamic magnetic processes on perfectly monodisperse systems, which possess unprecedented cleanliness. QTM occurs when energy levels of two substrates coincide under an appropriate magnetic field and the states are brought to resonance (Figure 8.1a). The physical investigation on the magnetization dynamics of such systems could be featured as an impressive array of quantum effects, ranging from macroscopic quantum tunnelling of the magnetization, Berry phase

Lanthanides and Actinides in Molecular Magnetism, First Edition.
Edited by Richard A. Layfield and Muralee Murugesu.

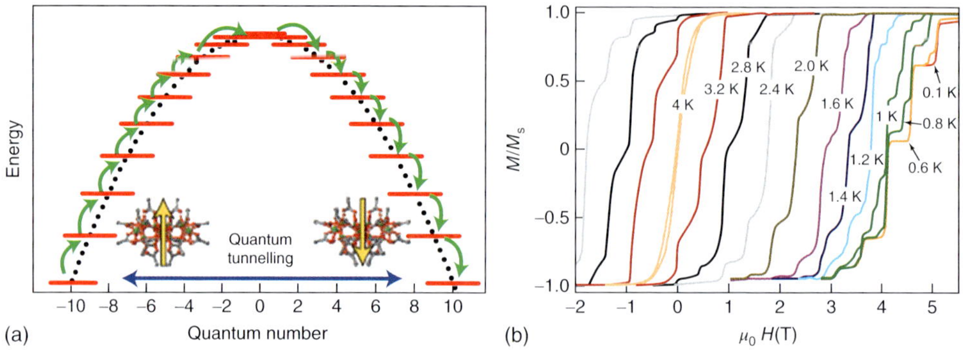

Figure 8.1 Dynamics of the magnetization in SMMs (here the prototype complex $[Mn_{12}O_{12}(O_2CCH_2{}^tBu)_{16}(CH_3OH)_4]$ [5]). (a) Schematic representation of the energy landscape of a SMM with a spin ground state $S = 10$. The magnetization reversal can occur via quantum tunnelling between energy levels (blue arrow) when the energy levels in the two wells are in resonance. Phonon absorption (green arrows) can also excite the spin up to the top of the potential energy barrier with the quantum number $M = 0$, and phonon emission descends the spin to the second well. (b) Hysteresis loops of single crystals of the prototype Mn_{12} SMM at different temperatures and a constant field sweep rate of $2\,mT\,s^{-1}$. (Data from Ref. [5].) The loops exhibit a series of steps, which are due to resonant quantum tunnelling between energy levels. As the temperature is lowered, there is a decrease in the transition rate due to reduced thermally assisted tunnelling. The hysteresis loops become temperature-independent below 0.6 K, demonstrating quantum tunnelling at the lowest energy levels. (Adapted from Ref. [6]. Copyright (2008) Nature Publishing Group.)

interference and quantum parity effects [6, 7]. Experimentally, the quantum nature of SMMs is manifested by a well-resolved staircase hysteresis loops (Figure 8.1b), temperature-independent relaxation and quantum phase interference. The discovery of these phenomena in molecular nanomagnets (MNMs) offers us promising candidates with potential applications in spintronics and quantum computing [8, 9]. Towards this goal, in the past 20 years MNMs have become the subject of intense investigation in the field of molecular magnetism, which makes a very important contribution to our understanding of the quantum world and leads to the discovery of new, intriguing properties of matter [3].

8.1.2
Multinuclear Versus Mononuclear: d- Versus f-Metal Ions

The discovery of the first SMM in the early 1990s stimulated a large number of research groups to pursue SMMs due to their potential use in high-density information storage. The bulk magnetization as induced by an external magnetic field is retained upon removal of the field, thanks to the presence of an energy barrier (ΔE) to the thermal equilibration of all possible quantized orientations of the molecular magnetic moment, giving rise to a hysteresis curve below a certain blocking temperature. The energy barrier ΔE was supposed to amount to $S^2|D|$ (integer spin) or $(S^2 - 1/4)|D|$ (half-integer spin), where S is the total spin of the molecule and D is the easy-axis anisotropy parameter [10]. Therefore, the motivation of this approach apparently is to achieve as much as high spin ground states (S) or the large easy-axis anisotropy parameter (D). Indeed, important advances have been made in these aspects; for example the ground state spin number has risen from $S_T = 10$ originally [11, 12], to a record value of $S_T = 83/2$ on a Mn_{19} cluster [13]. Later on, the family of SMMs has been extended to various other metals besides manganese, including iron, vanadium, cobalt and nickel ions. However, to date no new molecule of first row containing transition metal ions has been crystallized and structurally characterized as possessesing SMM properties at a significantly higher temperature than those of the original Mn_{12} systems. These results have demonstrated that the observation of SMM behaviour is no longer adequately described using simple D and S parameters; careful consideration of the overall symmetry of the complex as well as the orientation of the principal axis of anisotropy in individual ions should be made [14, 15]. Summing each tensor of individual single-ion anisotropy (D_i) up into the overall molecular anisotropy (D) based on the general criterion, Waldmann [16] theoretically pointed out that working on the local zero-field-splitting (ZFS) tensors to 'increase $|D|$' seems most promising and effective to enhance the energy barrier. The DFT (density functional theory) calculation on Mn_6 compounds by Ruiz and coworkers [17] also suggests that the magnitude of the anisotropy barrier is mainly determined by the strength of the spin–orbit coupling (SOC) and cannot be engineered by independently optimizing D and S, since the intrinsic relationship between these two parameters prevents this possibility.

In 2003, Ishikawa and coworkers observed slow magnetic relaxation for mononuclear Dy(III) and Tb(III) complexes with two phthalocyanine ligands. Particularly, the magnet-like behaviour during the slow relaxation of the magnetization is observable up to 50 K in the case of the terbium complex [18]. Since then, the use of lanthanides towards the preparation of SMMs has become of increasing interest [19, 20]. A huge number of lanthanide-containing SMMs in different nuclearities have been prepared, in which they almost exhibit magnetic hysteresis around liquid helium temperatures [4, 21–24]. The main reason for the rise to prominence of lanthanides is due to their large magnetic moments and, in most cases, huge magnetic anisotropy [25] Compared to 3d transition metals, the ligand field effect is almost negligible yielding large unquenched orbital angular momentum in most 4f systems; however, it can be regarded as a significant perturbation in the nature of their ground-state bistability which arises from the $[2J + 1]$ m_J microstates within the spin–orbit (SO)-coupled ground term. Recent developments in lanthanide chemistry show that even small ligand changes can drastically influence the overall physical properties of the molecule. Increasing nuclearity unavoidably results in increased complexity. Therefore, for polynuclearity of Ln systems, it would be very complicated to understand the nature and strength of interactions between lanthanide ions as well as possible alignment of spin vectors and anisotropy axes, especially when the molecule has a low symmetry, in which each Ln-ion is located in a different coordination environment [4, 21–24].

Bearing the general criteria of spin-reversal barrier in mind, researchers have also made efforts to increase *S* and the axial anisotropy *D* simultaneously by combining both 3d and 4f paramagnetic ions into one molecule. In this case, a large number of heterometallic SMMs are produced [26, 27], but unfortunately no substantial improvement has been made in terms of the blocking temperature compared with the mononuclear Ln-SMMs. In recent years, Long and coworkers [28] have incorporated high-anisotropy Dy(III) and Tb(III) ions into N_2^{3-} radical bridging ligands, leading to dinuclear molecules with a magnetic blocking temperature up to 8.3 and 13.9 K at a sweep rate of 0.08 Ts^{-1} for Dy(III) and Tb(III) binuclear complexes, respectively. These results demonstrate that the pronounced enhancement in spin-reversal barrier in this system has been achieved by controlling the magnetic coupling mediated by organic radicals in combination with f-element single-ion anisotropy, although the observed blocking temperatures remain low in view of potential applications and, at least so far, low compared to the mononuclear $TbPc_2$ SMM family.

The foregoing observations have led some researchers to focus on small, high-anisotropy molecules containing just one spin carrier. Indeed, mononuclear lanthanide complexes represent the simplest molecular units which allow the study of controlled entanglement of spins on neighbouring spin carriers, without causing decoherence of individual spins through dipolar exchange. Most importantly, mononuclear lanthanide SMMs are important model systems that can be used to answer fundamental questions regarding intrinsic single-ion relaxation. Moreover, such a simple system can be straightforwardly diluted in the yttrium(III) analogues to determine if the SMM behaviour observed was purely single-ion in origin or if it was due to a magnetically long-range coupled system. In a mononuclear system, the effects of direct interaction of the magnetic dipole may dominate. Measurements at sub-kelvin temperatures on oriented single-crystals which were magnetically diluted in the yttrium analogue, could easily confirm the SMM nature of these mononuclear complexes. The general method involved the doping of a diamagnetic yttrium matrix with varying percentages of paramagnetic ions. This dilution method has been extensively applied into the study of bis(phthalocyaninato) lanthanide(III) complexes by Ishikawa and Ruben *et al.* In recent years, Murugesu *et al.* [29] have investigated the use of doping to elucidate f–f magnetic interactions in order to gain insight into the slow magnetization relaxation of individual metal centres in Dy_2 SMMs. Roesky *et al.* [14] has demonstrated that the SMM behaviour of a mononuclear dysprosium compound can be switched off by incorporating two Mn(II) ions in close proximity to the Dy(III). With the dilution method, this quenching behaviour is ascribed to the presence of the weak ferromagnetic interactions between the Mn(II) and Dy(III) ions, which at $T > 2\,K$ act as a fluctuating field causing the reversal of magnetization on the dysprosium ion.

8.1.3
Molecular Versus Organic Spintronics

It was discovered in 1988 that, when two ferromagnetic layers are separated by a paramagnetic thin layer, electrical resistance of the multilayer changes depending on the orientations of the magnetization of the ferromagnetic thin layers. This effect was found to be much larger than other magnetoresistive effects that had ever been observed in metals and was, therefore, called *giant magnetoresistance* (*GMR*) (Figure 8.2a with an intercalated thin non-magnetic layer). When the directions of the magnetizations are the same, the possibility of electron scattering at the interface of paramagnetic/ferromagnetic layers becomes smaller, resulting in low electrical resistance. However, if the directions of the magnetizations of the two ferromagnetic layers are opposite, the electron with opposite spin orientation with respect to the magnetization of the electrode layer is scattered (spin-dependent electron scattering). The electrical resistance of the multilayer becomes higher than the case for the same directions of the magnetizations, and the devices thus obtained are called *spin valves.*

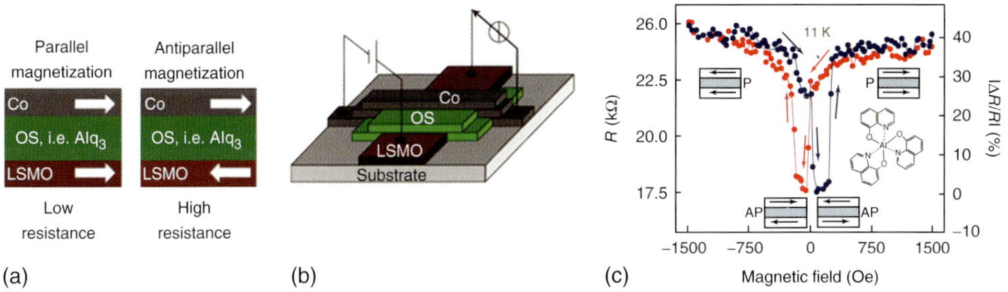

Figure 8.2 (a) A general schematic diagram of the change of electrical resistance in response to an applied magnetic field; OS standing for organic semiconductor, LMSO for $La_{0.7}Sr_{0.3}MnO_3$; and Alq_3 for tris(8-hydroxyquinolinato) aluminium (III). (b) Schematic of a typical vertical inorganic–organic spin valve. In this case, LSMO is the bottom electrode and Co is the top one. A typical magnetoresistive measurement is performed by applying a bias voltage to the metallic ferromagnetic electrodes and the current is measured as a function of the in-plane external magnetic field. (c) Magnetoresistance loop of a LSMO (100 nm)/Alq_3 (130 nm)/Co (3.5 nm) spin valve device measured at 11 K (Ref. [30]) As indicated in the figure, an antiparallel (AP) configuration of the magnetization of the electrodes corresponds to a low-resistance state. (Graphs (b) and (c) adapted from Ref. [31]. Copyright (2009) Nature Publishing Group.)

Based on this concept, the first report on experimental injection in organic spintronic devices (OSPDs) was published in 2002 [32]. The authors designed a lateral device that combined ferromagnetic electrodes and organic semiconductors, with the conducting channels being 100–500 nm in length. The selected ferromagnetic material was a highly polarized manganite ($La_{0.7}Sr_{0.3}MnO_3$ or LSMO). A strong magnetoresistive response was recorded up to room temperature in 100 and 200 nm channels, and was explained as a result of the conservation of the spin polarization of the injected carriers [31]. Following this encouraging report, an important step forward was the fabrication of a vertical spin valve device consisting of LSMO and cobalt electrodes, sandwiching a thick (100–200 nm) layer of tris(8-hydroxyquinoline)aluminium(III) (Alq_3) (Figure 8.2b). Alq_3 is by far the most successfully used organic semiconductor in the field of spintronics because its high-quality thin films can be grown on various ferromagnetic substrates by standard ultrahigh vacuum (UHV) evaporation. This spin valve showed magnetoresistance up to a temperature around 200 K, and for voltages below 1 V. Unlike standard spin valve devices, in which lower resistance corresponds to parallel electrode magnetization, the LSMO/Alq_3/Co devices showed lower resistance for an antiparallel magnetization configuration (Figure 8.2c). Several other groups have subsequently confirmed this so-called inverse spin-valve effect in LSMO/Alq_3/Co devices [33–35]. The result is one of the most widely reproduced effects in organic spintronics [31].

Generally speaking, spintronic systems exploit the fact that the electron current is composed of spin-up and spin-down carriers, which carry information encoded in their spin state and interact differently with magnetic materials. Information encoded in the spins persists when the device is switched off, and it can be manipulated without using magnetic fields, and it can be written using low energies, to cite just a few advantages of this approach [36]. New efforts are now directed at obtaining spintronic devices that preserve and exploit quantum coherence, and fundamental investigations are shifting from metals to semiconducting and organic materials. The latter is currently used in applications such as organic light emitting diodes (OLEDs), displays and organic transistors. Organic rare-earth metal complexes have been potential candidates as emitting materials due to extremely narrow emission line and high internal quantum efficiencies [37].

The trend towards developing ever-smaller electronic devices has already reached the nanoscale, and is now approaching the molecular scale limit. This development is critical due to the possibility that molecules could be bypassed by just single atoms, but at the same time offers the unique possibility of exploiting quantum effects if molecules are considered as quantum objects exhibiting strict monodispersity [6]. In this context, a new field of molecular spintronics is emerging that combines the ideas and the advantages of spintronics and molecular electronics. To realize molecular spintronics, we must be able to control the spin and electric conductance of the systems, and we must be able to study spin-transport at the single-molecule level. In contrast to organic spintronics, molecular spintronics use non-magnetic electrodes, and the transfer of the spin-response into the semiconducting layer between the electrodes is achieved by using magnetic molecules, among them SMMs, as active scatter units. The realization of this concept and its experimental consequences will be described in greater detail in the section on spintronic devices.

The rise of OLED devices/displays up to the commodity level has shown that these devices can be fabricated by two different techniques, namely: (i) vacuum deposition and (ii) solution techniques, including spin coating, inkjet technique, casting and so on [38]. These approaches can also provide fabrication criteria for spintronics applications. However, the key point is the creation of molecular devices using one or a few magnetic molecules. What can molecules bring to spintronics, or, in other words, what range of spintronics devices can be embraced? [39]. The concept underpinning such a device is that the interaction between the molecule and the transport channel (either in direct contact to the electrodes or, in an indirect way, through a contacted nanotube, nanowire or graphene sheet [40]) produces an electrical response which is sensitive to the magnetic state of the molecule (spin and spin orientation with respect to the channel). Also, in this case, one has to identify the strength of the interactions available. In general, both the spin of the molecules and its magneto-anisotropy should be as large as possible [39].

Within these requirements mononuclear terbium(III) bis(phthalocyaninato) complexes, a robust and evaporable SMM with very high anisotropy barriers, have attracted great interest as active units in single molecule electronics. Most importantly, the magnetic and electronic properties of this molecule remain relatively unchanged when grafting on different substrates. Therefore, this class of molecule offers great potential to be applied in nano-scale devices. Indeed, different members of this family have been successfully fabricated into different schemes of spintronic devices. In this chapter, we review the family of $LnPc_2$ compounds, with a focus on the terbium analogues, including synthetic and structural aspects; bulk magnetic properties; surface magnetic properties and applications in spintronic devices.

8.2 Synthesis and Structure of $LnPc_2$ Complexes

8.2.1 Synthesis of Bis(phthalocyaninato) Lanthanide(III) Complexes

Phthalocyanine (Pc) molecules consisting of four interconnected isoindole units are synthetic analogues of porphyrins. They represent planar aromatic macrocycles with an 18 π-electron aromatic system, which is delocalized over an arrangement of alternated carbon and nitrogen atoms. For years, phthalocyanines have

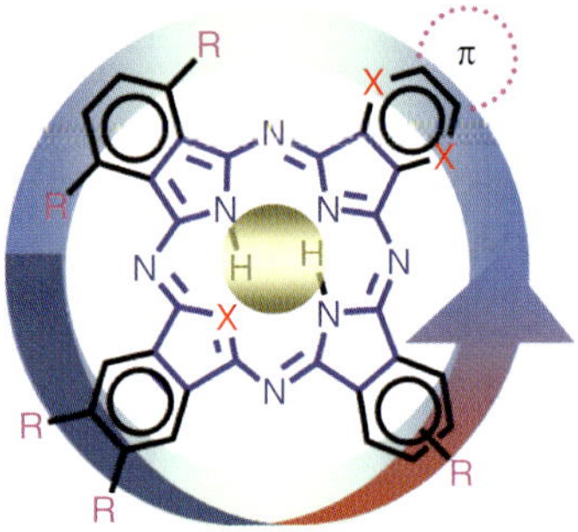

Figure 8.3 Structural formula of the phthalocyanine macrocycle and its chemically tuneable positions.

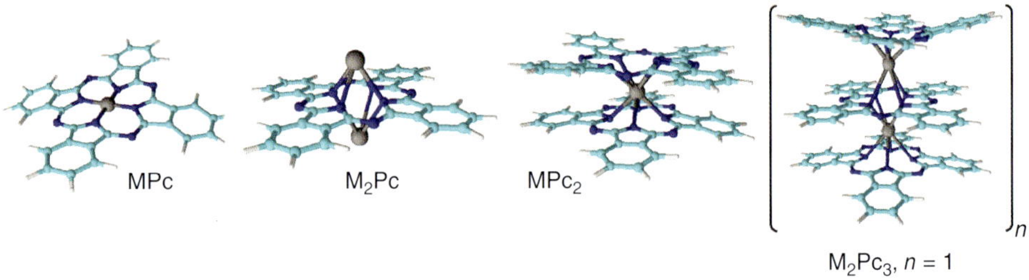

Figure 8.4 The structure of metal-phthalocyanine complexes, with increased complexity adapting to the requirement of different metal ionic species.

been important constituents of the textile industry as blue dyes pigment [41], and in medicine due to their antibacterial properties [42].

Over the last decade, interest in phthalocyanines has increased interest and they have undergone a renaissance in new branches of modern science. Along with the extreme thermal and chemical stability of phthalocyanine macrocycles, their chemical versatility has allowed them to play an exceptional role in nanotechnology in recent years. Organic chemistry thus offers versatile tools for tailoring the phthalocyanine ring periphery, furnishing a big library of Pc analogues (Figure 8.3). Phthalocyanine offers up to 16 sites for substitution. The most usual structural changes that provide modified Pc compounds are (i) substitution at the 3-, 4-, 5- and 6-positions of the benzene ring; (ii) extension of the π-system; (iii) variation in the number of isoindole units and (iv) replacement of some isoindole units by other heteroaromatic units.

The central cavity of Pc molecules enables the complexation of a range of metals, leading to a large variety of physical properties. Moreover, the phthalocyanine ligand adapts to the requirements of each metal species to form a complex of an appropriate structure (Figure 8.4). Among the metals able to form complexes with phthalocyanines, lanthanides are of particular interest. Possessing large ionic radii and high coordination numbers, lanthanides form compounds of both planar and sandwich structure with Pc ligands and their analogues. Lanthanide Pc complex give rise to unique characteristics, and they have opened up new avenues in molecular spintronics [6, 7], electronics and nonlinear optics [43]. The possibility of generating heteroleptic and mixed sandwich complexes based on phthalocyanines create a basis for the development of materials with specific properties [44, 45].

Several excellent reviews are available covering different scientific purposes and technological applications of phthalocyanines [46–51]. Here, we focus on synthetic aspects of one particular type of Pc-derivative, namely bis(phthalocyaninato) complexes of trivalent lanthanides, as well as analogous heteroleptic complexes containing porphyrin and porphyrin-like ligands.

It should be noted that the basic reactions used to prepare phthalocyanine derivatives today are fundamentally those developed by Linstead and coworkers in the 1930s [52–54]. Due to the large number of substituted phthalocyanines described in the literature, space limitations mean that a detailed review of synthetic aspects cannot be provided here. The following discussion is concerned with the synthesis of lanthanide phthalocyanines via (i) template tetramerization of phthalonitrile with lanthanide salts, (ii) direct metalation of the metal-free ligands by the salts or (iii) metal exchange of a labile metal ion or ions for a lanthanide.

Different literature sources use different terminology for naming metal-Pc complexes; in this review, we use the most common terms and notions. Homoleptic complexes comprise ligands of the same type. They are sub-classified as being symmetrical or unsymmetrical based on whether or not the substituents are the same or different, respectively. Systems with randomly distributed ligands along both phthalocyanine rings are called *mixed complexes*. Complexes with ligands of different types are called *heteroleptic*, for example, phthalocyanine–porphyrin or naphthalocyanine–porphyrin.

The key methods for the synthesis of bis(phthalocyanine) complexes can be divided into three groups, depending on the type of the initial compound. The homoleptic symmetrical complexes are prepared most often by the template methods (Scheme 8.1, A1, A2, A3) and by metallation of the free ligand or its dianion (Scheme 8.1, B). Thus, the simplest way of synthesizing the bis(phthalocyanine) complexes is via a template-assisted method, in which fusion of appropriate phthalonitriles with lanthanide salts is conducted at about 300 °C, followed by extraction of impurities (Scheme 8.1, A). The first representatives of metal bis (phthalocyanine) complexes were prepared and spectroscopically characterized by Kirin and Moskalev in 1965 [55–58]. $LnPc_2$ was synthesized by the thermal fusion (280–290 °C) in the solvent-free melt of phthalonitrile with lanthanide acetates in 1 : (4–8) molar ratios for 40–90 min, followed by sublimation, which resulted in bis(phthalocyanine) complexes in 10–15% yields. The yields of bis(phthalocyanine) complexes, $LnPc_2$, for the beginning of the lanthanide series were increased to 50–60% by performing the process in a sealed tube at 290 °C and using chromatography on alumina for purification of the resulting complexes instead of sublimation [59].

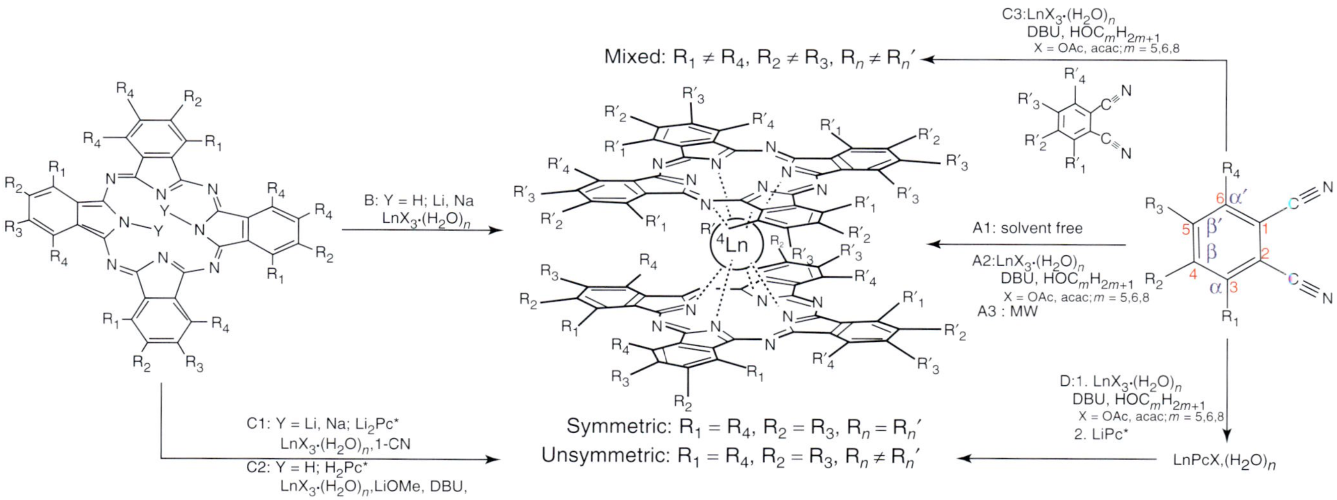

Scheme 8.1 Three key methods of the synthesis of lanthanide(III) bis(phthalocyanine) complexes. Homoleptic complexes: symmetrical (A, B) and unsymmetrical (C1 and C2, D1). Thereby, the thermal fusion group A can be subdivided into (A1) solvent free fusion; (A2) refluxing in the mixture of protic solvents in the presence of organic base – DBU and (A3) MW assisted fusion.

In recent years, the most often applied method of template synthesis of bis(phthalocyanine) complexes implies rather mild reaction conditions: refluxing of a mixture of the appropriate phthalonitrile and a lanthanide salt in the presence of DBU (1,8-diazabicyclo[5.4.0]undecen-7-ene) in a protic solvent (usually pentanol or hexanol) for 5–48 h. The target products are purified by filtering the reaction mixture to remove the insoluble impurities, followed by chromatography of the filtrate. The yields of unsubstituted and substituted bis(phthalocyanine) complexes are in the range of 6–79% the lowest values being typical of lutetium complexes and the highest corresponding to the middle of the lanthanide series (Eu, Gd). A vast library of complexes with different substituents has been synthesized from alkyl-, alkoxy-, alkylthio- and poly(oxoethylene) chains to exotic POSS (polyhedral oligomeric silsesquioxane) groups (see Table 8.2) [60–79]. It is noteworthy that the template synthetic approach can be applied to the phthalonitriles containing electron-withdrawing groups too, although its number is limited to a few examples. Thus, the reaction of 1,2,3,4-tetracyanobenzene with $Ln(OAc)_3$, (OAc, acetate ion) Ln = Nd, Dy, Y, in sulfolane with DBU at elevated temperatures afforded cyano-substituted complexes ($LnPc^{CN8}{}_2$) with the phthalocyanine in a close-shell configuration and doubly negatively charged, leading to an overall oxidation state of the complex of −1 [80, 81].

The template methods have also been used for the synthesis of a number of substituted Ln di(naphthalocyanine) complexes, $LnNc_2$ [82–88]. Apart from thermal fusion by conventional heating processes, complexation has been initiated by microwave radiation, although only a few publications are devoted to the template synthesis of lanthanide bis(phthalocyanine) complexes by this method [89, 90]. The use of microwave radiation (MW) reduces the reaction time from several hours to several minutes. Unsubstituted complexes $LnPc_2$ (Ln = Tb, Dy, Lu) were prepared [90] by irradiation (650–700 W) of a mixture of phthalonitrile with an appropriate lanthanide salt for 6–10 min (yields >70%).

To avoid side processes typical of template reactions, first of all, resinification of the starting phthalogens, the synthesis of lanthanide bis(phthalocyanine) complexes can be accomplished by using metal-free phthalocyanines (Scheme 8.1, B, Y = H). However, control of the transformation selectivity is often complicated in this case by steric factors arising in the complexation of the lanthanide mono phthalocyanine formed initially with the second free ligand. Therefore, bis(phthalocyanine) complexes often cannot be obtained in high yields, especially for the end of the lanthanide series, where the ionic radii of the metals are relatively small. Moreover, the insolubility of H_2Pc in common organic solvents requires the use of high-boiling aromatic solvents. Therefore, M_2Pc complexes, M = Li, Na, which are products of the reaction between phthalonitrile and lithium or sodium dissolved in high-boiling alcohols, are more useful intermediates due to their solubility in protic solvents. (Scheme 8.1, B, Y = Li). This method was subsequently optimized by Ng *et al.* [91]. And later on it was applied by a number of other researchers to prepare unsubstituted [68, 92, 93] and a series of bis(phthalocyanine) complexes bearing tetrasubstitutions at β/β′-positions or octasubstitutions at β and β′ positions of benzene rings in each of Pc macrocycles (4/8 Alk, OAlk, -SAlk) [68, 94–101].

Despite the fact that cyclic tetramerization of phthalonitriles catalysed with organic base such as DBU or, alternatively, reactions of [Li_2Pc] or H_2Pc with rare earth salts in a high-boiling solvent, were used successfully for the synthesis of homoleptic unsubstituted $LnPc_2$ (as well as for the homoleptic tetra- or octa-substituted at β-positions analogues, $LnPc^{(\beta\text{-R})_4}{}_2$ and $LnPc^{(\beta,\beta'\text{-R})_8}{}_2$ [98, 102] and even homoleptic tetra-α-substituted complexes $LnPc^{(\alpha\text{-R})_4}{}_2$ [98]), both methods failed to give homoleptic α,α′-octa-substituted complexes $LnPc^{(\alpha\text{-OR})_8}{}_2$. This is apparently due to the extremely non-planar saddle conformation adopted by the metal free α,α′-octa-substituted phthalocyanine molecule. Unexpectedly, in the presence of DBU and crown ether, the reaction between $H_2Pc^{(\alpha,\alpha'\text{-}OC_4H_9)_8}$ and $M(acac)_3{\cdot}nH_2O$ (M = Eu, Y) (acac, acetylacetonato) in refluxing *n*-octanol leads to the formation of $HLnPc^{(\alpha,\alpha'\text{-OR})_8}{}_2$, M = Eu, Y [103]. It was demonstrated by the same authors that the presence of α,α′-octa-substitutions stabilizes the protonated form usually existing as an intermediate, which is readily oxidized in air. Thus, a slipped pseudo-quadruple-decker as its bis-sodium salt was isolated and fully characterized [104].

A straightforward inefficient method for the preparation of unsymmetric complexes LnPcPc* involves the mixed condensation of two different phthalonitriles in the presence of a metal salt. As expected, the reactions give statistical mixtures of products containing a different number, type and position of substituents. For example, Liu *et al.* [105] fused an equimolar mixture of 4-propoxyphthalonitrile and 4-*tert*-butylphthalonitrile with lutetium acetate at 290 °C, resulting in a mixture of double-deckers from which $LuPc^*{}_2$, where Pc* = $Pc^{(OPr)_4(tBu)_4}$, can be isolated by repeated column and thin-layer chromatographic procedures in 4.6% yield. This purified product still exists as a mixture of many constitutional isomers. Formally, this reaction is expected to give nine randomers of bis(phthalocyanine) complexes. In addition, if 2 of 6 isoindole fragments differ from the other ones, up to 17 structural isomers may be formed. Thus, with the overall conditions being the same (approximately equal reactivities of the reacting dinitriles, the absence of steric factors, etc.), the total number of potentially formed complexes reaches 21.

An improved method to unsymmetrical complexes LnPcPc* has been reported by L'Her *et al.* [83] and references therein (Scheme 8.1, C). Reaction of $Lu(OAc)_3$ with Li_2Pc and Li_2Pc^* (4/8-Alk, -OAlk, -SAlk) [106] or Li_2Nc [83] or Li_2Nc^{tBu} [107] yields the unsymmetrical complex mixture LuPcPc*/LuPcNc*/$LuPcNc^{tBu}$ in 15–22% yield, along with the symmetrical complexes $LuPc_2$, $LuPc^*{}_2$, $LuNc_2$ and $Lu(Nc^{tBu})_2$, which can be separated by column chromatography. Reaction temperature appears to be the most critical parameter in these reactions. By this method, the unsymmetrical, mixed TbPcPc* complex, with Pc* of the A_3B-type ligand bearing (A) with six $-C_6H_{13}$ chains and (B) with one $-O(CH_2)_3$ Pyren substituent, was successfully synthesized in a total yield of 10% [108]. Tomilova *et al.* [109, 110] have reported the development of effective synthetic pathways to A_3B-type phthalocyanine-based homoleptic complexes bearing functional OH-groups. The reaction of $H_2Pc^{OBu,Bu}$ with $Eu(acac)_3{\cdot}nH_2O$ in *n*-hexadecanol, in the presence of lithium methoxide instead of DBU, selectively yielded the bis(phthalocyanine) complex. This allowed the reaction time to be decreased from 2–3 h to 20 min, generating a higher yield of the target compound. Moreover, this approach enabled the synthesis of nanosized structures based on planar binuclear phthalocyanines [111].

The unsymmetrical complexes LnPcPc* can also be prepared in a stepwise manner (Scheme 8.1, D, upper part) [87, 112–114]. Treatment of phthalonitrile with excess of $Lu(OAc)_3{\cdot}nH_2O$and DBU gives the half-sandwich compound $LuPc(OAc)(H_2O)_2$. The latter complex reacts with Na_2Nc in 1-chloronaphthalene leading to the formation of LuPcNc [87]. Similarly, the template reaction of unsubstituted and crown-substituted phthalonitriles with lutetium acetate was carried out in boiling *n*-hexanol in the presence of DBU [115–117]. Bi- and tri-nuclear [118] lutetium complexes could also be obtained following this approach [111, 119].

8.2.2 Synthesis of Heteroleptic Lanthanide(III) Complexes Containing Porphyrin-Based Ligands

Symmetrical bis(porphyrinato) lanthanide complexes $LnPor_2$ are generally prepared by treating $Ln(acac)_3{\cdot}nH_2O$ with metal-free porphyrins (Scheme 8.2, A, when $Por_1 = Por_2 = Por$) [120]. It is believed that the protonated species $LnHPor_2$ are initially formed, which then undergo deprotonation and oxidation to give $LnPor_2$. Ogawa *et al.* [121] synthesized a TPP-based (dianion of 5,10,15,20-tetraphenylporphine) terbium double-decker complexes by the reaction of $TPPH_2$, $Tb(acac)_3$ and DBU at 350 °C for 1.5 h as stable protonated form. By adding the organic proton acceptor, the deprotonation occurs to produce the anionic form. The analysis of the structure demonstrated for the first time in the tetrapyrrole based bis-complexes that the proton is located on the pyrrole ring nitrogen for charge balance reasons. Moreover, the authors investigated magnetization relaxation behaviour of both forms, demonstrating that the anionic form acts as an SMM.

$CeTPP_2$ was also synthesized in a novel, non-trivial route directly on a Ag (111) surface under UHV conditions by exposing a $TPPH_2$ precursor layer to an atomic beam of Ce, followed by a temperature-programmed reaction [122]. The nature of the formed mono- and binuclear compounds was studied by scanning tunnelling microscopy (STM) and scanning tunnelling spectroscopy (STS), including a comparison with $Ce(TPP)_2$ layers generated by molecular beam epitaxy.

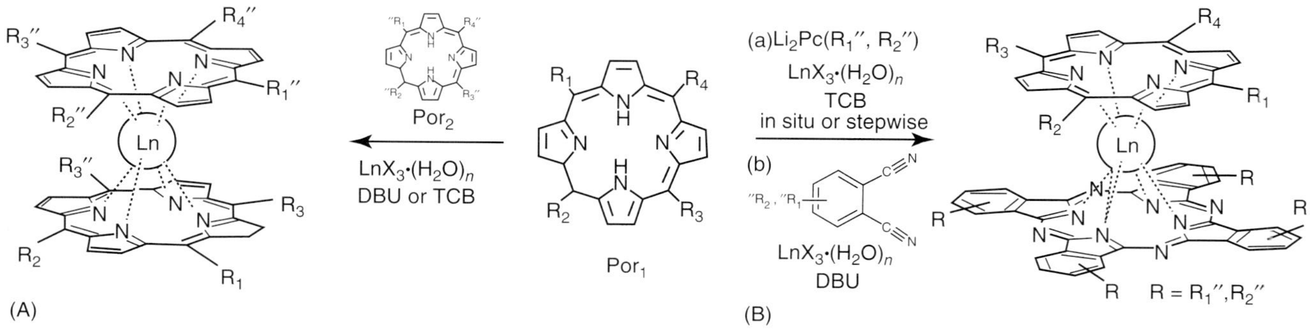

Scheme 8.2 (A,B) Synthesis of homoleptic $LnPor_2$ and heteroleptic (phthalocyaninato)(porphyrinato) lanthanides.

The solution-based methods can be extended to unsymmetrical $LnPor_1Por_2$. For example, treatment of $Ln(acac)_3 \cdot nH_2O$ with H_2OEP and H_2TPP in 1,2,4-trichlorobenzene (TCB) gives a mixture of products from which Ln(OEP)(TPP) can be separated by column chromatography [123–125]. But for Ln = La, the protonated LaH(OEP)(TPP) has been isolated instead, along with a small amount of $La_2(OEP)_2(TPP)$ [126]. The half-sandwich compounds Ln(TPP)(acac) (Ln = Sm [127], Eu [123–125], Dy, Yb [128], Lu [129]) also react with $Li_2(OEP)$, leading to the isolation of Eu(OEP)(TPP) or LnH(OEP)(TPP) (Ln = Sm, Dy, Yb, Lu). The latter species (for Ln = Sm, Lu) undergoes deprotonation in basic solvents such as DMF (*N*,*N*-dimethylformamide) and pyridine to form the corresponding anions $[Ln(OEP)(TPP)]^-$ [127, 129]. The number of homoleptic unsymmetrical $Ln(Por_1)(Por_2)$ synthesized is limited [130, 131], but among them are a few exotic complexes exhibiting unique rotational activities as a bevel-gear-shaped rotors [132–134].

There are two main synthetic pathways for preparing heteroleptic-(phthalocyaninato)(porphyrinato) rare-earth(III) complexes [114, 135–143]: (i) from the reaction between the corresponding half-complexes Ln^{III}(acac)[Porphyrin], generated *in situ* from $Ln(acac)_3 \cdot nH_2O$ and H_2Por, and phthalocyaninato dilithium (Li_2Pc) as starting materials in refluxing *n*-octanol and (ii) the reaction of Li_2Pc with $Ln(acac)_3 \cdot nH_2O$ followed by treatment with H_2Por (Scheme 8.2, B(a)). These reactions normally yield mixtures of the mixed double-deckers LnPcPor and the triple-deckers Ln_2PcPor_2, and Ln_2Pc_2Por the yields of which depend on the nature of the lanthanide metals and the macrocycles, and the ratio of these reactants.

Weiss *et al.* have employed the latter method to synthesize the gadolinium analogues [144]. Treatment of Li_2Pc with 2 equiv. of $Gd(acac)_3 \cdot nH_2O$ produces Gd(Pc)(acac), which further reacts with 1 equiv. of $H_2(TPP)$ giving GdPc(TPP) in 65% yield and a small amount of $Gd_2(Pc)(TPP)_2$. However, by using 4 equiv. of the metal salt and 2.5 of the metal-free porphyrin, the binuclear complex $Gd_2(Pc)(TPP)_2$ has been isolated in 78% yield.

It is interesting to note that direct synthesis gives bis(phthalocyanine) complexes of trivalent metals as anionic species, which are gradually transformed into neutral paramagnetic species under exposure to air. The stability of the anionic species decreases in the series of lanthanide complexes with a decrease in the ionic radii of the elements [135].

Although direct reaction of lanthanide mono-porphyrins with free phthalocyanine or its lithium derivatives is generally more efficient than the template synthesis, and gives rise to mixed-ligand complexes, the template strategy can also be applied for synthesis of phthalocyanine–porphyrin complexes, as in the case of unsymmetric bisphthalocyanine complexes (Scheme 8.2, B(b)) [106, 136, 145, 146]. Thus, metallation of free porphyrins with lanthanide salts in TCB or n-octanol leads to single-decker complexes, which then react with phthalonitriles under the action of DBU in alcoholic media to give the desired compounds.

The first double-decker complexes containing naphthalocyanine (Nc) and porphyrin ligands were prepared by Jiang *et al.* [145] by template tetramerization of unsubstituted and 6,7-bis(dodecylthio)-substituted naphthalonitriles in the presence of europium monoporphyrins, in 52 and 69% yields, respectively. Monoporphyrin complexes were also prepared by metallation of the ligand with europium acetylacetonate in boiling TCB, which was then evaporated [145], followed by subsequent reaction with naphthalonitriles. More recently, it was demonstrated, that direct reaction between free porphyrin, naphthalonitrile and a lanthanide salt gives mixed-ligand complexes in comparable yields [107, 139, 147–150].

In addition to the discussed cyclotetramerizations, direct metallation of the metal-free ligands or metal exchange of a labile metal ion or ions for one held more robustly, the desired complexes may be prepared by the direct substitution, exchange or modification of substituents on preformed phthalocyanine derivatives. However, a review of works carried out on these types of transformations lies out of the scope of this chapter.

As demonstrated above, lanthanide complexes containing phthalocyanines and various porphyrins as ligands have been studied in great detail. In contrast, the number of studies on analogous complexes containing phthalocyaninato and tetrapyrrole ligands like tetrabenzoporphyrinato-, tetraazaporphyrinato- or N-confused porphyrinato-, is rather limited [151–155], though such compounds also attract considerable interest.

Mixed-ligand heteroleptic lutetium, dysprosium, gadolinium, neodymium and lanthanum complexes containing tetrabenzoporphyrin and phthalocyanine fragments were synthesized in a stepwise manner. First, a mixture of tetrabenzoporphyrin with the corresponding lanthanide acetylacetonate was heated in boiling DMF over a period of 6 h. Upon disappearance from the electronic spectrum of the reaction mixture of the absorption bands typical for uncomplexed porphyrin ligands, and separation of intermediate mono-complex, the double-decker heteroleptic coordination compounds were obtained by heating with phthalocyanine dilithium salt in satisfactory yields of 50–70%. A considerable problem in the synthesis of tetrabenzoporphyrin metal complexes is the removal of the product from the pristine ligand [151–153]. The compound is very poorly soluble in organic solvents, thus it is difficult to remove from low-molecular weight organic impurities. As an extension of work in this area, Jiang *et al.* [154] described the first examples of bis(phthalocyaninato) rare earth complexes with regular/methylated N-confused porphyrinato ligands form of M(Pc)(HNTBPP) (M = La(III), Eu(III), Y(III), Lu(III)) and Y(Pc)[(CH_3)NTBPP] (NTBPP = N-confused 5,10,15,20-tetrakis[(4-*tert*-butyl)phenyl]porphyrinate). The synthesis was achieved: (i) by treatment of the mono-complex La(Pc)(acac), generated *in situ*, with the metal-free N-confused porphyrin ($NTBPPH_2$) in refluxing TCB for 4 h in 10% yield or (ii) stepwise by following the same procedure and using the isolated mono-complexes Ln(Pc)(acac) (M = Eu(III), Y(III), Lu(III)) as the starting materials, with yields of the Ln(Pc)(HNTBPP)

[M = Eu(III), Y(III), Lu(III)] 8–16%. These protonated compounds were isolated as the sole product, in contrast to the similar reactions with $TBPPH_2$, where the non-protonated analogues Ln^{III}(Pc)(TBPP) (TBPP, 5,10,15,20-tetrakis[(4-*tert*-butyl)phenyl]-porphyrin) were obtained instead [106]. These results clearly show that the inversion of a pyrrole ring in N-confused porphyrins can stabilize the protonated complexes.

A novel tetraazaporphyrinato (TAP) and phthalocyaninato mixed heteroleptic rare-earth compound, LnPc′ (TAP) with photochromic and electrochromic features has been readily synthesized by a one-pot reaction using $Eu(acac)_3{\cdot}nH_2O$, metal-free phthalocyanine H_2Pc' (Pc′ = 2,3,9,10,16,17,23,24-octakis(decyloxy)phthalo-
cyanine) and the photochromic precursor 1,2-dicyano-l,2-bis(2,3,5-trimethyl-3-thienyl)ethane as starting materials [154].

Recently, coordination compounds based on even more unusual derivatives – tetrabenzocorrins – were reported as well [156]. The synthesis of single and homoleptic lanthanide complexes with 1-methyltetrabenzooctadehydrocorrin (MeTBODC), in the presence of hydrazine hydrate in boiling butan-1-ol, was achieved with an excess or a deficiency of metal salt, correspondingly. To synthesize unsymmetrical corrin complexes the authors applied one of the most rational methods for the synthesis of heteroleptic sandwich-like complexes containing phthalocyanine fragments based on the reaction of the mono-lanthanide complex with excess phthalonitrile, which acts simultaneously as a reagent and a solvent at 290 °C over a period of 1.5 h. The desired heteroleptic complexes LnPc(MeTBODC) (Ln = La, Lu, Er) were isolated in good yields of 31–36%. No change of colour was found in the electronic absorption spectra of compounds by treatment with either oxidants (e.g. H_2O_2) or reducing agents (e.g. $NH_2NH_2{\cdot}H_2O$). Thus, homo- and heteroleptic tetrabenzocorrin-containing complexes do not exist in 'green', 'blue' or 'red' species as observed for $LnPc_2$ species.

8.2.3
Oxidation States of Bis(phthalocyaninato) Lanthanide(III) Complexes

Several groups [157–159] have reported the synthesis of blue and green compounds by the reaction of lutetium acetate with 1,2-dicyanobenzene, with their separation being achieved by chromatography. The nature of these compounds has been the subject of long debate [157], which ended only after X-ray studies of single crystals of all the forms. Firstly, Corker *et al.* [160] proposed that the green complex contains a one-electron ligand-oxidized, protonated form $[LuPcPc^{\bullet-}H]^+$, but subsequent mass spectrometric measurements made by Marchon [159] could not prove the presence of a proton. Furthermore, no anion is present in the structure by single crystal X-ray analysis (Figure 8.5b) first published by Weiss [161].

Later, the presence of an organic radical in the green form was revealed by means of a full set of physicochemical analytical tools, namely the UV/vis/NIR

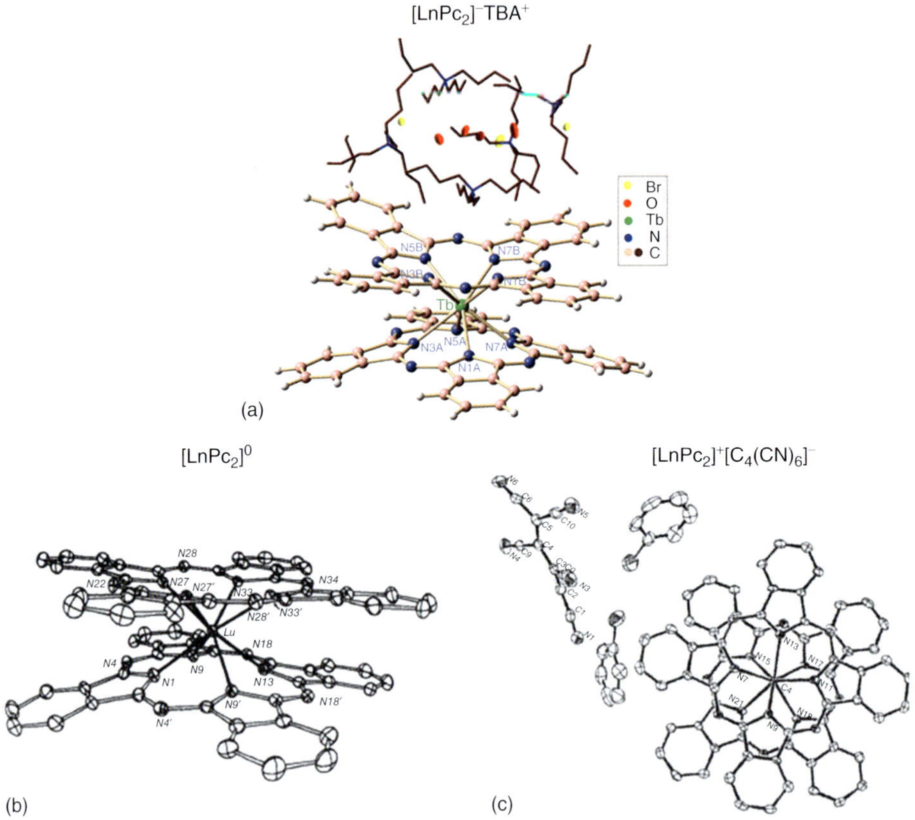

Figure 8.5 Structural analysis of $LnPc_2$ complexes: Single crystal X-ray analysis of (a) $[TbPc_2]^-TBA^+$ [90]; (b) $[LuPc_2]^0$ (Adapted from Ref. [161]. Copyright (1985) American Chemical Society) and (c) $(GdPc_2)^+[C_4(CN)_6]^-$. (Adapted from Ref. [162]. Copyright (2001) American Chemical Society.)

spectroscopy, magnetic susceptibility measurements and electron paramagnetic resonance (EPR) spectroscopy (Figure 8.6a,b). Thus, the green $[LuPc_2]^0$ has a one-electron ligand-oxidized, non-protonated structure $[Pc^{2-}LuPc^{\bullet-}]^0$ in resonance with $[Pc^{\bullet-}LuPc^{2-}]^0$ [165, 166], with an open shell $S = 1/2$ π-electronic system [167]. Upon reduction of the neutral compound to its monoanion, the complex can be described as $[Ln^{3+}(Pc^{2-})_2]^{1-}$, consisting of a trivalent lanthanide ion and two Pc ligands each with a formal charge of −2, leading to an overall closed shell π electronic system, which was first confirmed for the case of $[LuPc_2]^-$ by NMR (nuclear magnetic resonance) spectroscopy [168]. Moreover, NMR techniques are applicable even for complexes with magnetically active Ln as indicated by the paramagnetic contact shift of the signals in the spectra of $[TbPc_2]^-$ [90, 169] (Figure 8.6c).

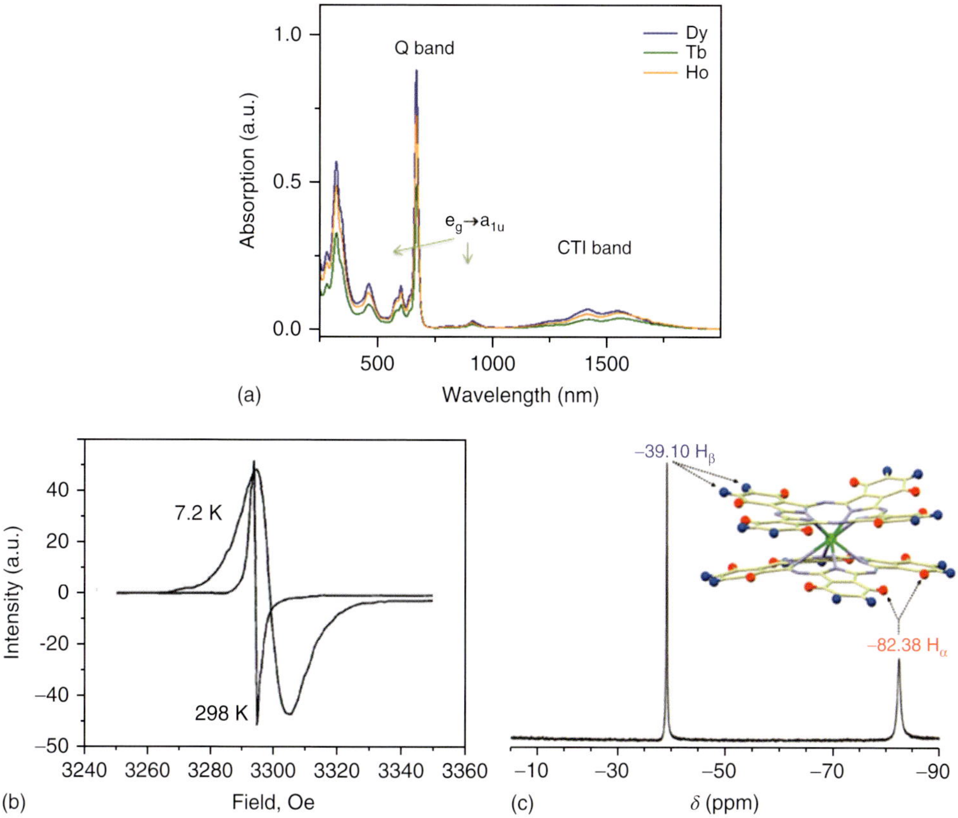

Figure 8.6 Physicochemical characteristics of the 'green form' of bis(phthalocyaninato) lanthanide complexes – $[LnPc_2]^0$. (a) UV–vis–NIR spectra in dichloromethane solution of $[LnPc_2]^0$, Ln = Tb, Dy, Ho showing the charge transfer intervalence (CTI) band in the NIR region due to the delocalization of the organic $S = 1/2$ radical over the two Pc rings [163, 108]; (b) EPR spectrum of lutetium bis(phthalocyanine) at 298 and 7.2 K (Adapted from Ref [164]. Copyright (1992) Published by Elsevier B.V.) showing the presence of an organic $S = 1/2$ radical and (c) ^1H NMR spectrum of terbium bis(phthalocyanine) upon adding hydrazine hydrate showing the chemical shifts for the H_α and H_β protons of the phthalocyanine rings. (Adapted from Ref. [90]. Copyright (2009) American Chemical Society.)

In contrast to the reduction process, oxidation to $[LnPc_2]^+$ is not evident and may lead to two possible electronic structures. Very few oxidized $LnPc_2$ species have been characterized; among them $[LuPc_2]^+[SbF_6]^-$ [159] and $[GdPc_2]^+[C_4(CN_6)]^-$ [162]. Based on the changes in the absorption spectrum as well on the single crystal X-ray analysis, the authors formulated the structure of oxidized species as $[GdPc^{\bullet-}{}_2]^+[(C_4(CN_6))]^-$with four paramagnetic centres.

In a very interesting semi-empirical quantum chemical study, Ishikawa *et al.* have shown that the molecular orbitals near the HOMO–LUMO (highest occupied molecular orbital–lowest unoccupied molecular orbital) energy region of bis(phthalocyaninato) lanthanides are essentially π orbitals on the Pc ligands [87, 170–172]: The π-electronic system of the bis(phthalocyaninato) lanthanide(III) complex is composed of the linear combination of the two Pc ligand π-orbitals. The HOMO and HOMO + 1 of the anionic form $[LnPc_2]^-$, which is the most stable oxidation state, are described as anti-bonding and bonding linear combinations of the HOMOs of the Pc ligands. When the complex is oxidized, the electrons in the antibonding HOMO are removed first. Two-electron oxidation of the anionic complex results in cationic form $[LnPc_2]^+$. More importantly, the calculations predicted that the two electron oxidation leads to a significantly shortened Pc–Pc distance, which strengthens the ligand-field potential under which the lanthanide ion is placed.

The electronic spectrum of of the bis(phthalocyaninato) lanthanide(III) complexes in different oxidation states demonstrates pronounced changes (Figure 8.7). Thus, $[GdPc_2]^+[C_4(CN)_6]^-$ in CH_2Cl_2 represents shifted absorptions with respect to the parent complex $[GdPc_2]^0$. Similar to $[GdPc_2]^0$, which exhibits an absorption band attributed to $Pc^{\bullet -}$, $[GdPc_2]^+$ exhibits a broad and shifted absorption at 489 and a shoulder at 518 nm due to the presence of $Pc^{\bullet -}$ in agreement with the proposal of four paramagnetic centres. Reduction of $[GdPc_2]^0$ to $[GdPc_2]^-$ is accompanied by the disappearance of the characteristic $Pc^{\bullet -}$ absorption band at 460 nm, consistent with the $[Pc^{2-}Gd^{3+}Pc^{2-}]^-$ representation.

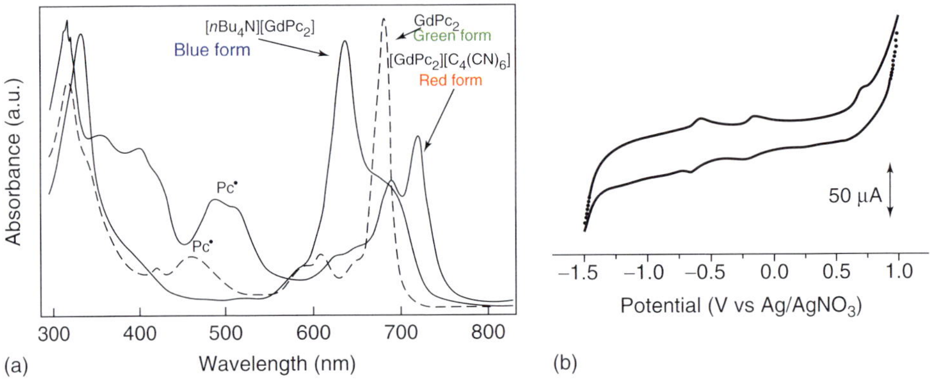

Figure 8.7 Redox properties of bis(phthalocyaninato) lanthanide complexes. (a) UV–vis–spectra in dichloromethane solution of $[GdPc_2]^{-/0/+}$ [162] showing pronounced changes in the electronic spectrum of the bis(phthalocyaninato) lanthanide(III) complexes in different oxidation states. (b) Cyclic voltammogram of 1.0 mM solutions of $[TbPc_2]^0$ in dichloromethane containing 0.1 M $TBA^+PF_6^-$. The experimental procedures were referred to those reported in Refs [150, 173].

The redox properties of $LnPc_2$ were intensively studied by cyclic voltammetry and hydrodynamic voltammetry in various solvents [139, 147, 148, 150, 173, 174]. It is believed that all redox exchanges occur on orbitals of the conjugated macrocyclic organic ligands, while the redox state of the trivalent lanthanide cation does not change [69], except for the case of the $CePc_2$, where change of the redox state of the Ce cation can take place [123]. Comparative electrochemical studies of unsubstituted and substituted bis(phthalocyaninato) lanthanide(III) complexes within the electrochemical window of CH_2Cl_2 revealed two quasi-reversible one-electron oxidations and one quasi-reversible one-electron reduction for the unsubstituted parent $[LnPc_2]^0$ complexes (Figure 8.7b), and up to five quasi-reversible one-electron reductions for complexes with substituted Pc-ligands [150]. The half-wave potentials of the first and second oxidations and first reduction for unsubstituted bis(phthalocyaninato) rare earth(III) complexes depend on the size of the metal centre. The difference between the redox potentials of the second and third reductions of $[LnPc_2]^0$, which represents the potential difference between the first oxidation and first reduction of $[LnPc_2]^-$, lies in the range 1.08–1.37 V and gradually diminishes along the lanthanide contraction, indicating enhanced $\pi-\pi$ interactions in the double-deckers connected by the smaller lanthanide ions.

As a result of their low redox potentials [173], bis(phthalocyaninato) lanthanide complexes are often inadvertently reduced or oxidized, and they are also very sensitive to acids and bases. In order to solve these problems, Veciana *et al.* achieved certain success on designing a series of novel compounds with characteristics that would give them improved redox stability. Electroactive ligands based on phthalocyaninato tetra dicarboximide [175] or perfluorinated phthalocyanine [176] were used to assemble the double-decker lanthanide complexes, with the effect of stabilizing the negative charge of the anionic state of the compounds, which resulted in a strong shift of ~0.7 V of their first oxidation potentials.

8.2.4
Rotation Angles and Skew Angles in $LnPc_2$ in Relation to the Lanthanide Contraction

Different structural analyses of the class of $LnPc_2$ complexes have been reported. From these structures, it is generally accepted that the central lanthanide(III) ions are sandwiched by the two Pc ligands and are eightfold coordinated by the isoindole nitrogen atoms (N_{iso}) of two Pc rings, which form a saucer shape, as shown in Figure 8.5. The plane composed of the four N_{iso} atoms of the Pc ligand is orthogonal to the fourfold axis. The bent angle is defined as the angle between the planes composed of the N_{iso} atoms and the vector connecting the N_{iso} atom and the centre of the central C–C bond of the isoindole moiety. The calculated bent angles [177] clearly indicate that the two Pc ligands have different conformations.

Several tetrabutylammonium salts of the $[LnPc_2]^{1-}$ complexes were prepared, and the crystal structures were analysed [166, 177–184]. The analysed structures

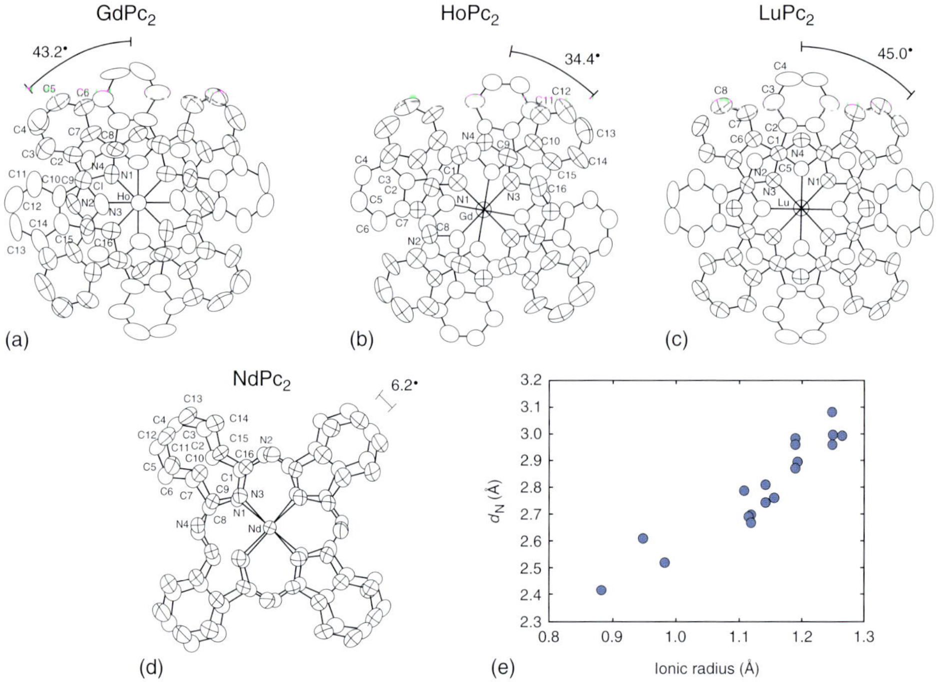

Figure 8.8 Top view of $[LnPc_2]^-$, Ln = Gd(III), Ho(III), Lu(III) and Nb(III) molecular structures (a–d) showing skew angles between the upper and lower Pc rings. (Adapted from Ref. [177]. Copyright (1996) American Chemical Society); (e) plot of the interplanar distances, d_N against the ionic radii of the central atoms [177].

showed that the conformations of $[LnPc_2]^-$ are not only staggered but also eclipsed if the Ln(III) ions are exchanged. The rotation angle of one Pc ring compared with another Pc ring, the skew angle, seems to be closely related to the ionic radius of the central Ln(III) ion (Figure 8.8a–d). The interplanar distance, d_N, between the two Pc ligands, defined as the distance between the upper and lower planes composed of four N_{iso} atoms, has been proposed as a parameter to relate the ionic radii of the central ion and skew angles [177, 185]. Table 8.1 shows the ionic radii and interplanar distances observed in the $[LnPc_2]^{-/0/+}$ complexes determined so far. The interplanar distance, d_N, was calculated from the distance between the upper and lower planes composed of the four N_{iso} atoms [166, 178–184]. Figure 8.8e shows the relationship between the ionic radii and the interplanar distances. The interplanar distances are linearly dependent on the ionic radii.

Table 8.1 also shows the skew angles Φ. Obviously there seems to be no relationship between the skew angle and the interplanar distance. This may be due to the fact that the calculated interplanar distances are too small, considering the

Table 8.1 Ionic radius, d_N[a], d_{int}[b] and skew angel[c] values in $[LnPc]^{-/0/+}$.

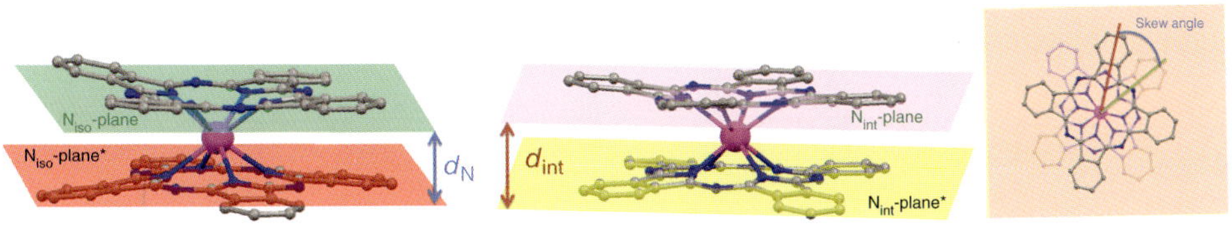

Ln	Ionic radius (Å)	d_N (Å)	d_{int} (Å)	Skew angle Φ (°)	References
$[YPc_2]^0$	1.119	2.760	3.07	45	[178]
$[YPc_2]^0$	1.116	2.746	3.62	41.3	[179]
$[YPc_2]^0$	1.118	2.791	3.23	41.2	[179]
$[YPc_2]^0$	1.116	—	3.27	45	[186]
$[PrPc_2]^0$	1.266	3.0	3.25	42	[180]
$[PrPc_2]^+$	1.266	3.08	3.21	41	[181]
$[NdPc_2]^0$	1.249	3.0	—	42.3	[184]
$[NdPc_2]^0$	1.249	2.96	3.29	38	[182]
$[NdPc_2]^-$	1.249	3.08	3.59	6.2	[177]
$[GdPc_2]^-$	1.193	2.89	3.48	34.5	[177]
$[TbPc_2]^-$	1.118	2.831	3.86	41.6	[90]
$[TbPc_2]^-$	1.118	2.825	3.76	46.7	[183]
$[DyPc_2]^0$	1.117	2.767	3.41	40.9	[179]
$[DyPc_2]^-$	1.117	2.785	3.81	46.6	[183]
$[HoPc_2]^-$	1.155	2.76	3.30	43.2	[177]
$[ErPc_2]^0$	1.144	2.74	3.06	41.4	[187]
$[LuPc_2]^-$	1.117	2.67	3.24	45	[177]
$[LuPc_2]^0$	1.117	2.68	3.08	41	[188]
$[LuPc_2]^-$	1.117	2.69	3.06	45	[161]
$[LuPc_2]^-$	1.117	2.676	3.05	41	[157]
$[LuHPc_2]^0$	1.117	2.701	3.21	43	[157]
$[Ce^{IV}Pc_2]^0$	1.11	2.79	3.27	38.8	[189]

a) d_N: Distance between the upper and lower planes composed of four N_{iso} atoms.
b) d_{int}: Distance between the upper and lower average planes of Pc ligands.
c) Φ: The skew angle is the rotation angle of one Pc ring compared with another one.

van der Waals radius of the aromatic carbon atom [190]; thus the ligand π–π interaction has to be taken into consideration. The π–π interaction distance was defined [177] as the distance between the average planes of the upper and lower Pc moieties. Each average plane is calculated from the atoms of four five-membered rings and the nitrogen atoms connecting the five-membered rings, since these atoms overlap with those of another Pc moiety. The π–π interaction distances calculated, d_{int}, are also given in Table 8.1. Most of the distances are >3.1 Å. This suggests that the d_{int} value is more appropriate for the interplanar distance than d_N if the π−π interaction is taken into account. As the d_{int} value increases, the skew angle gradually decreases. The linear relationship is changed when the d_{int} value is >3.5 Å, which is the sum of the van der Waals radii of the aromatic carbon atoms. This may suggest that, as the π−π interaction decreases, the skew angle decreases and approaches zero if the π−π interaction becomes sufficiently small, as can be seen in the case of $[NdPc_2]^-$ (Figure 8.8d).

Thus, the two Pc rings of the $LnPc_2$ complexes have a skew angle of about 45° only in the case of the fairly strong π−π contact between the two Pc rings. When the interplanar distance is greater than the usual van der Waals contacts, the skew angle is decreased dramatically.

8.3 Bulk Magnetism of $LnPc_2$ Complexes

8.3.1 Magnetism of Bis(phthalocyaninato) Lanthanide(III) Complexes

Solid state magnetism is a complicated phenomenon exhibiting many types of magnetic behaviours which arise from a variety of responses to temperature and applied magnetic field observed in adjacent magnetic spins of these materials. Therefore, the understanding of bulk magnetic properties has proven to be extremely difficult. In this section, we will briefly summarize the consequences of the coexistence of three spin systems such as an organic radical, electronic spin and nuclear spin in the same molecule before a thorough discussion on the bulk magnetism of bis(phthalocyaninato) lanthanide(III) complexes.

8.3.2 Three Spin Systems in $[TbPc_2]^0$ Single-Ion Molecular Magnets (SIMMs)

8.3.2.1 The Organic Radical (S)

The smallest spin system is the monoradical, which contains units of charged or uncharged organic magnetic materials. Organic radicals are one type of ligands which can promote magnetic interactions not only between the paramagnetic centres but also through the whole system. In such a system, the unpaired spin density is usually delocalized through the coordinating atoms [191]. One of the most attractive features of this system is the strength of magnetic interaction promoted by these radicals, which is much higher in magnitude than that of lanthanide–lanthanide. In such a case, it has been possible to overcome the drawback in using weakly interacting trivalent lanthanides. The coupled Ln-radical system can be treated as a giant spin (S) with the entire complex acting as a unit so as to elevate the energy barrier for the reversal of the magnetization [23, 28, 192].

Due to the unpaired electrons, free radicals do not have an electron octet. Therefore, they are usually unstable and highly reactive. Typically simple organic radicals are only stable at very low temperatures in solid matrices. The stability of radicals can be increased by aromatic substituents at the radical carbon atom. The central radical carbon atom of the triphenylmethyl radical, for instance, carries three phenyl groups. Therefore, the radical is highly resonance-stabilized. In other words, the stability of radicals can be enhanced by increasing the amounts of hyperconjugation. The more alkyl substituents a radical carbon atom possesses, the more stabilized it becomes from hyperconjugation. Among others, prototype organic radicals that are being very well studied are nitronyl-nitroxide (NITR), 7,7,8,8-tetracyanoquinodimethane (TCNQ) and phthalocyanine (Pc). NITR radicals have the capability to form both discrete and one-dimensional structure with metal ions and bridging NITR radicals in an alternating fashion [25]. The combination of trivalent lanthanide ions with TCNQ organic radicals yields extended structures showing magnetic ordering at a temperature up to 7.1 K [193]. However, trivalent lanthanide ions with coordinated Pc ligands can result in sandwich-type complexes in a face-to-face geometry, in which the π-radical delocalized over the two Pc ligands leads to a peripheric $S = 1/2$ (Figure 8.9a).

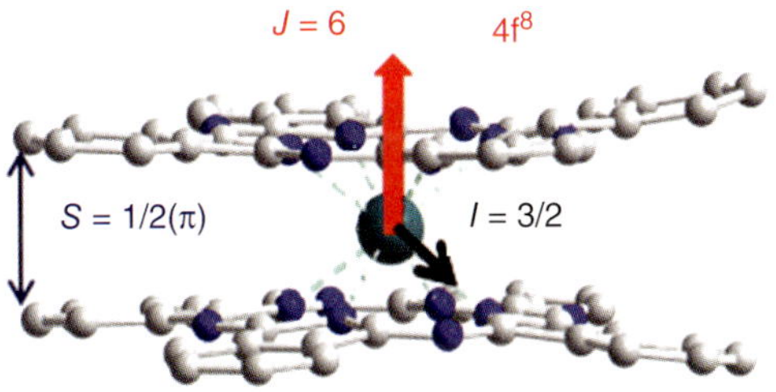

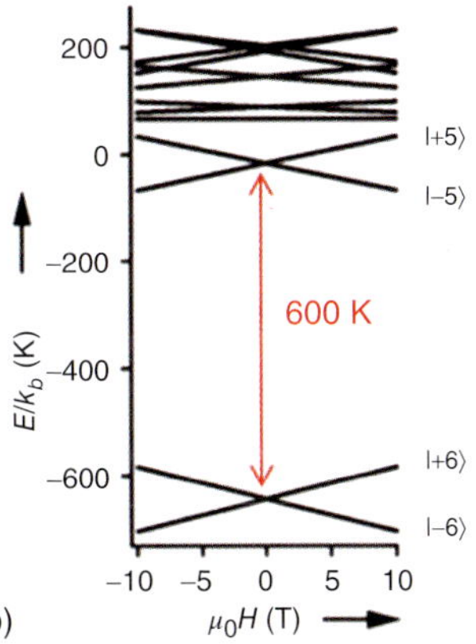

Figure 8.9 $TbPc_2$-based single ion magnets (SIMs) [90]: (a) Schematic representation molecular localization of the three spin-systems characteristics of the $[TbPc_2]^0$ complex: $J = 6$, the uniaxial magnetic moment of the $4f^8$ configuration; $I = 3/2$ nuclear spin of the only stable and naturally occurring ^{159}Tb isotope; $S = 1/2$ organic radical delocalized over the two Pc ligands. The radical is only present in $[TbPc_2]^0$, but quenched in the $[TbPc_2]^-TBA^+$, while both acting as SIMs. (b) Zeeman energy diagram as a function of longitudinal magnetic field for the $J = 6$ ground multiplet with the ligand-field parameters determined here for $TBA[TbPc_2]$ (in absence of the $S = 1/2$ radical) [90, 194, 195].

8.3.2.2 The Electronic Spin (J)

The spin–orbit interaction is also called *spin–orbit effect* or *spin–orbit coupling*, which is one cause of magnetocrystalline anisotropy. SOC, the intrinsic interaction between a particle spin and its motion, is responsible for various important phenomena, ranging from atomic fine structure to topological condensed matter physics. SOC plays a major role in many important condensed matter phenomena and applications, including spin and anomalous Hall effects, topological insulators, spintronics, spin quantum computation, and so on.

For 3d transition metals, SOC is subordinate to ligand field effect. In contrast, studies of the magnetic properties of lanthanide compounds are often based exclusively on the SOC scheme, with an implicit presumption of the ground electronic state having a strong unquenched orbital angular momentum. The trivalent ions are characterized by f^n configurations, which give rise to $^{2S+1}L_J$ multiplets, split by SOC to give J states, where J is defined by the angular momentum summation rules [196]. Lifting of the degeneracy by the ligand field is, however, not quite negligible in many lanthanide complexes, because the splitting can reach the order of several 100 K [197]. In such cases, information regarding the sublevel structure of the ground multiplet is vital in the study of their magnetic properties [198]. In the rare-earth ions with an even number of 4f electrons the $(2J + 1)$-fold degeneracy, for instance Tb(III) ion ($4f^8$), the total angular momentum J of 6 results (Figure 8.9a). The ligand field potential around the Tb(III) ion with a $J = 6$ splits the ground multiplet so that the lowest sublevel has the largest J_z value ($|J_z| = 6$, (corresponding to up/down spin states) and large energy gaps to the remaining sublevels (about 600 K) (Figure 8.9b).

Tunnelling may occur not only between the lowest-lying states $M_S = \pm S$, but also between pairs of degenerate excited states. This phenomenon is called the *phonon-assisted* (*or thermally activated*) *tunnelling mechanism*, as phonons need to be adsorbed to populate the higher M_S states involved in the tunnelling process. This mechanism is very important at intermediate temperature because the tunnelling frequency is expected to increase on decreasing $|M_S|$. It offers therefore a shortcut for the relaxation of the molecules. In principle a molecule may not need to go over the maximum of the barrier even at relatively high temperatures, but may find a shortcut and tunnel [199]. Based on the study of both Mn_{12}-Ac and Mn_{12}-^{t}BuAc SMMs (Figure 8.1), it has been found that the mechanism exhibits QTM between different substrates $|S_z|$ due to dipolar coupling, easy-axis tilting and distributions of molecular environments, in particular with asymmetric exchange due to SO interactions [5]. However, there is a fundamental difference in the mechanism of QTM between the lanthanide SMMs and the previously known transition-metal-cluster SMMs such as Mn_{12}. The quantum process in these single-ion lanthanide SMMs is a result of the resonant quantum tunnelling between entangled states of electron and nuclear spin systems [9, 200].

8.3.2.3 The Nuclear Spin (I)

The magnetic moments of nuclear spins are of the order of 10^{-3} of the electron magnetic moment, and their mutual interaction is weak. As a result, the entropy of nuclear spins stays finite well below the temperatures where the electron and lattice degrees of freedom are quenched [197]. The interaction between a nucleus and its surrounding environment is known as a *hyperfine interaction*, which can be visualized as the motion of the electron in the magnetic dipole field of the nucleus. These interactions are very small compared to the energy levels of the nucleus itself but the extreme energy resolution of the Mössbauer effect enables these interactions to be observed. The hyperfine interactions may shift energy levels or lift their degeneracy. Both of these variations will affect the shape of a Mössbauer spectrum. Through hyperfine interaction the nuclear spins feel the enhanced external field induced by the Van Vleck susceptibility. This process brings about the enhanced indirect coupling between the nuclear spins, leading to nuclear magnetic ordering around 1 mK in various substances.

Isotopes are variants of a particular chemical element such that, while all isotopes of a given element share the same number of protons and electrons, each isotope differs from the others in its number of neutrons. The number of protons within the nucleus of an atom uniquely identifies an element, but a given element may, in principle, have any number of neutrons [201]. For example, the ground state of the free Tb(III) ion is $4f^8$, $^7F_6$. Terbium-159 has a nucleus with $I = 3/2$ spin in a natural abundance of 100% together with a single electronic transition splitting into four hyperfine lines (Figure 8.9a). It is therefore necessary to take into account the interaction between the $4f^8$ system and the nucleus. The interaction of each ground state with the four nuclear-spin states of the Tb(III) ion nuclear spin ($I = 3/2$) leads to several energy-level crossings and therefore QTM [9]. The situation for the dysprosium-162 complex is more complicated than that for the Tb complex. There are seven naturally occurring isotopes of Dy, namely, ^{156}Dy, ^{158}Dy, ^{160}Dy, ^{161}Dy, ^{162}Dy, ^{163}Dy and ^{164}Dy, with a natural abundance of 0.06%, 0.01%, 2.34%, 18.91%, 25.51%, 24.90% and 28.18%, respectively. ^{161}Dy and ^{163}Dy have a nuclear spin of $I = 5/2$ while other nuclear species have $I = 0$ [9]. Holmium-165 has a nucleus with $I = 7/2$ spin in a natural abundance of 100%. Each sublevel of the $J = 8$ ground multiplet is split into an octet by the hyperfine interaction between the $4f^{10}$ system and the nucleus [202].

8.3.3 Further SIMs of $LnPc_2$ with Ln = Tb, Dy and Ho

The bis(phthalocyaninato) lanthanide(III) complexes have generated a great deal of interest because they are found to show electrochromic and semiconducting properties [87, 203]. They have been characterized by a range of spectroscopic methods, but much research interest has been directed towards understanding the f-electronic structures and their magnetic properties. In order to understand the electronic states and the nature of the magnetic interaction between the f-system and the π-hole, Ishikawa *et al.* [170, 194, 204, 205] so far have determined sublevel structures of ground multiplets of bis(phthalocyaninato) lanthanide(III) complexes of lanthanide ions with $4f^8-4f^{13}$ configurations, from terbium to ytterbium. In these complexes, the multiplet splittings are due to the presence of the non-zero orbital angular momentum in the ground states, which can interact directly with the ligand field potential. Due to the multiplet splittings, these complexes show characteristic paramagnetic shifts in NMR signals, and temperature dependences of magnetic susceptibilities. In addition, a computational approach has been employed to determine the ligand field parameters that reproduce simultaneously the experimental data of the isostructural complexes with different 4f electronic configurations from $4f^8$ to $4f^{13}$. With those detailed investigations both experimentally and theoretically, a clear picture on the magnetic properties of this family has been developed.

Generally speaking, for the negatively $[LnPc_2]^-$ TBA^+ species (i.e. in the absence of the $S = 1/2$ radical), the magnetic susceptibilities of powder samples in the range 1.8–300 K show characteristic temperature dependences, which result from splittings of the ground-state multiplets [195]. However, the magnetic behaviour of the neutral $[LnPc_2]^0$ compounds, in which one of the phthalocyanine ligands is considered to be a singly oxidized radical $S = 1/2$ with a charge of -1, reveals that there is a strong magnetic interaction between the Ln f-electrons and the phthalocyanine radical electron. Depending on the symmetry and occupancy of the magnetic orbitals on the Ln ion, both antiferro- and ferromagnetic ground states are observed in this series of compounds [164]. When the f electron shell is more than half-filled, antiferromagnetic exchange is observed in these systems. The temperature dependence of the effective magnetic moment, μ_{eff}, at room temperature is significantly smaller than the Curie constants for those species. In the case of praseodymium, where the f electron shell is less than half-filled, ferromagnetic coupling with the Pc radical electron results. However, thulium and ytterbium compounds exhibit the presence of both the antiferro- and ferromagnetically coupled states of the lanthanide and phthalocyanine radical electrons. In these cases, the energy separation between states was evaluated from the magnetic data [164]. The presence of strong antiferromagnetic coupling of the ligand radical electron with the lanthanide f-electrons is also confirmed by EPR measurements, in which there is no transition observed at or near $g = 2$ (indicative of a free radical). Instead, the very complicated EPR spectra are indicative of very significant deviations from g [164, 206].

Six $[Ln(Pc)_2]^-$ complexes with heavy lanthanide ions (Ln = Tb, Dy, Ho, Er, Tm or Yb) were investigated by the measurements of alternating current (AC) magnetic susceptibility [18]. Out of the six compounds, $[TbPc_2]^-$ and $[DyPc_2]^-$ were found to show temperature and frequency dependence on AC magnetic susceptibility similar to that observed for the transition-metal SMMs, while the rest did not. Their SMM behaviour have been observed either in bulk, in dilute solid solutions

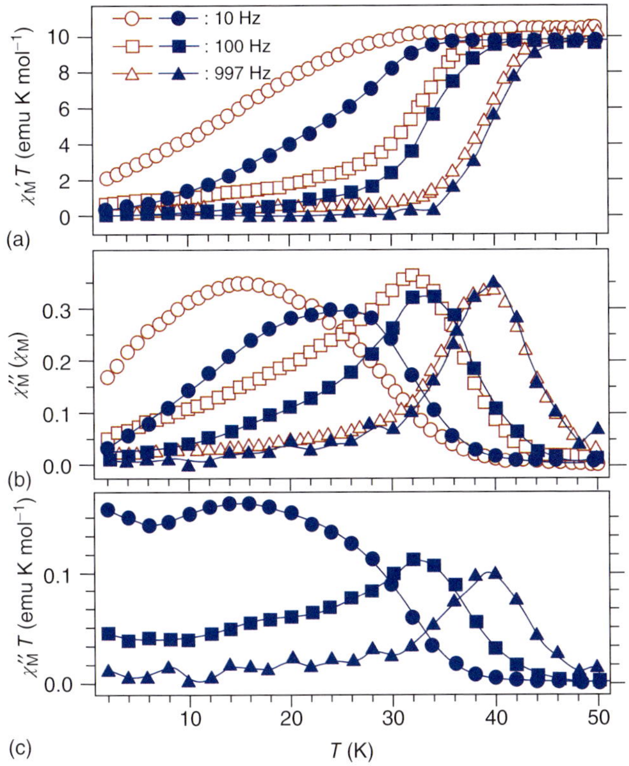

Figure 8.10 Plots of (a) $\chi'_M T$, (b) χ''_M/χ_M and (c) χ''_M against temperature T, where χ'_M, χ''_M, and χ_M are in-phase-AC, out-of-phase-AC and DC molar magnetic susceptibilities, respectively, for a powder sample (open marks) and that 80% diluted in $[YPc_2]^-$ TBA^+ (filled marks) measured in a 3.5 G AC magnetic field oscillating at indicated frequencies. (Adapted from Ref. [194]. Copyright (2003) American Chemical Society.)

and even in a frozen glass solution. The $[TbPc_2]^-$ and $[DyPc_2]^-$ complexes exhibit out-of-phase AC susceptibility χ'' peaks at 40 and 10 K with a 10^3 Hz AC field, respectively (Figure 8.10). Apart from the slow magnetization relaxation originating from single-molecular properties, they also exhibit hysteresis around 2 K (Tb, Dy) and 0.5 K (Ho), and resonant quantum tunnelling of magnetization between hyperfine levels. Hence, these compounds are the first lanthanide complexes functioning as a magnet at the single-molecular level, namely single-ion magnets (SIMs) [90]. One of the most remarkable observations about these SIMs is that the slow magnetization relaxation is observed in temperature ranges that are significantly higher than those of previously known transition-metal SMMs [194].

To ensure that the slow magnetization relaxation is an intrinsic molecular property, the measurement could be carried out for a sample diluted in a diamagnetic matrix. Indeed $[TbPc_2]^-$ and $[DyPc_2]^-$ compounds were investigated along with a diluted form using various percentages of the diamagnetic component. In both Tb and Dy diluted cases, the Arrhenius analysis suggests that the main relaxation path in the higher temperature range is the Orbach process, in which a transition between the two lowest sublevels occurs through absorption of a phonon to reach an excited state, followed by emission of the second phonon. In other words, the SMM behaviour of these lanthanide complexes deeply depends on the energy gap between the lowest and second lowest sublevels in the ground multiplet, which approximately corresponds to the energy barrier for spin reversal [194, 207]. The values of U_{eff} for the Tb(III) and Dy(III) complexes diluted in a 80% matrix of the isostructural $[YPc_2]^-TBA^+$ complex were determined to be 260 and 31 cm^{-1}, respectively, which are in fairly good agreement with the energy separations between the first and second lowest sublevels (about 600 and 60 K, respectively) (Figure 8.9b). It is worth mentioning that Ruben and coworkers [90] have synthesized a series of magnetically active $[TbPc_2]^-$ anions diluted through the use of the diamagnetic $[Bu_4N]^+Br^-$ matrix materials. The spin dynamics of the diamagnetic diluted molecules are probed by solid-state ^1H NMR techniques, with a much larger energy barrier (584–641 cm^{-1}) with respect to those previously estimated for the $[Pc_2Tb]^-[Bu_4N]^+$ system (~230 cm^{-1}) and even one order of magnitude larger than in any other SMM system. Interestingly, it has been shown that the barrier height and tunnelling rates depend on the magnetic dilution employed [90] and on the experimental methods employed [208] (see Table 8.2).

Moreover, the spin dynamics in $TbPc_2$ and $DyPc_2$ SMMs have been investigated by means of muon and nuclear spin-lattice relaxation rate measurements. Salman *et al.* [214] reported zero-field muon spin relaxation measurements of the magnetic properties of thin film $TbPc_2$ SIMs evaporated onto a gold substrate, and compare them to the bulk susceptibility measurements. They found that the SIM properties of $TbPc_2$ are clearly maintained in thin film, and at low temperature the fluctuations in films are much faster than in bulk and depend strongly on the distance between the molecules and the Au substrate.

In addition to the parent unsubstituted bis(phthalocyaninato) complexes, complexes with a variety of substituents on the periphery of the phthalocyanine macrocycles have also been widely studied. The most important concern is that, by introducing different ligand substituents in a rational manner, the intrinsic properties of the molecule can be altered, which may eventually tailor the bulk characteristics of the material. Another point is that substitution may also increase the solubility of complexes, which facilitates the fabrication process. In general, the substitution can influence the energy barrier for spin reversal magnetization slightly, but not significantly. The characteristic parameters of the slow magnetization relaxation for all reported bis(phthalocyaninato) Ln(III) complexes with different charges and different substituents are summarized in Table 8.2. In the following paragraph, we will select a couple of examples to be discussed in more detail.

Table 8.2 Characteristic parameters of the slow magnetization relaxation observed in $[LnPc_2]^{-/0/+}$ complexes.

Compound[a)]	U_{eff} (cm^{-1}) H_{dc} (Oe)[a)]	Tunnelling parameters ($1/\tau_0$) (s^{-1})	Hysteresis (K)	References
$[Tb_{0.2}Y_{0.8}Pc_2][Bu_4N]$	230	1.6×10^7	—	[194]
$[Dy_{0.2}Y_{0.8}Pc_2][Bu_4N]$	28	1.6×10^5	—	[194]
$[Tb_{0.02}Y_{0.98}Pc_2][Bu_4N]$	260	5.0×10^7	1.7	[9, 18]
$[Dy_{0.02}Y_{0.98}Pc_2][Bu_4N]$	31	3.0×10^5	1.7	[9, 18]
$[TbPc_2][Bu_4N]$	584[b)]	—	—	[90]
$[TbPc_2][Bu_4N]\times143$ $[Bu_4N]Br$	641[b)]	—	—	[90]
$[Dy\{Pc(CN)_8\}_2][H]$	—	—	—	[81]
$[Dy_{0.1}Y_{0.9}\{Pc(CN)_8\}_2][H]$	5.1, 40.3 (3000)	—	1.8, 1.6[c)]	[81]
$[Tb\{Pc(phth^1)\}_2][Bu_4N]$	445	1.6×10^{10}	2	[175]
$[Tb\{Pc(phth^2)\}_2][Bu_4N]$	428	7.5×10^9	2	[175]
$[Tb\{Pc(phth^3)\}_2][Bu_4N]$	463	4.5×10^{10}	2	[175]
$[TbPc_2]$	410	6.8×10^8	—	[209]
$[Tb\{Pc(ODd)_8\}_2]$	—	—	—	[209]
$[Tb\{Pc(S\text{-}DOP)_8\}_2]$, crystalline	480	1.44×10^{11}	—	[210]
$[Tb\{Pc(S\text{-}DOP)_8\}_2]$, disordered	422	5.85×10^9	—	[210]
$[Tb\{Pc(OEt)_8\}_2][SbCl_6]$	550	2.4×10^{10}	—	[211]
$[Tb\{Pc(OEt)_8\}_2][Bu_4N]$	509	—	—	[211]
$[Dy\{Pc(OEt)_8\}_2][SbCl_6]$	55	—	1.8	[167]
$[Dy\{Pc(OEt)_8\}_2][Bu_4N]$	27	—	1.8	[167]
$[Tb\{Pc(IPD)_4\}_2]^-$	—	—	1.5[c)]	[212]
$[Tb\{Pc(IPD)_4\}_2]$	—	—	1.5[c)]	[212]
$[Tb\{Pc(IPD)_4\}_2]^+$	—	—	1.5[c)]	[212]
[Dy(Pc)(TClPP)]	16	4.0×10^6	—	[213]
$[Dy\{Pc(\alpha\text{-}OC_5H_{11})_4\}(TClPP)]$	30 (2000)	1.4×10^7	—	[213]
$[DyH\{Pc(\alpha\text{-}OC_5H_{11})_4\}(TClPP)]$	40 (2000)	2.6×10^7	—	[213]
$[Tb\{Pc(F_{32})\}_2]$	365	7.5×10^9	2, 1.8[c)]	[176]
$[Ho_{0.02}Y_{0.98}Pc_2][Bu_4N]$	—	—	0.04–0.50[d)]	[202]

a) U_{eff} values are extracted from measurements in zero applied dc field, unless followed by a number in parentheses to indicate the strength of the applied field. Lattice solvent molecules are not listed.
b) Solid-state ^{1}H NMR spectroscopy.
c) Magnetic circular dichroism spectroscopy.
d) Single-crystal micro-SQUID.
S-DOP, (*S*)-1-bromo-2-dodecyloxypropane; IPD, 4,5-isopropylidenedioxyphthalonitrile; phth^1, *N,N,N,N*-tetrapentadecyl-29H,31H-2,3,9,10,16,17,23,24-phthalocyaninatotetradicarboximide; phth^2, *N,N,N,N*-tetrapropyl-29H,31H-2,3,9,10,16,17,23,24-phthalocyaninatotetradicarboximide; phth^3, *N,N,N,N*-tetra((*S*)-methyl(phenyl)methyl)-29H,31H-2,3,9,10,16,17,23,24-phthalocyaninatotetradicarboximide and ODd, 2,3,9,10,16,17,23,24-octakis(dodecoxy)-phthalocyanine.

Ishikawa and coworkers used peripherically substituted $Pc(OEt)_8$, 2,3,9,10,-16,17,23,24-octaethoxyphthalocyanine, instead of unsubstituted Pc for the ligand to overcome the extremely low solubility of the cationic complex $[LnPc_2]^+$ in organic solvents. $[Pc(OEt)_8]$ ligand can exhibit two-electron oxidation, leading to a longitudinal contraction of the coordination space of the $LnPc_2$ SMM. The magnetization-reversal barrier energy of the cation $[TbPc_2]^+$ was determined to be 550 cm^{-1}, which is much higher than that for the unsubstituted $[TbPc_2]$ using the same experimental technique. This example suggests that the U_{eff} can be increased by oxidizing the ligand from the anion $[TbPc_2]^-$ to the neutral form $[TbPc_2]^0$ and even further on to the cation $[TbPc_2]^+$. This is also the first evidence that the dynamic magnetism of 4f-SMMs can be controlled by a redox reaction on the ligand side without introducing any additional magnetic site or spin system [211]. However, to avoid their redox instability in the presence of mild oxidizing and reducing agents, Veciana and coworkers designed, prepared and characterized phthalimide substituted bis(phthalocyaninato) Tb(III) SIMs bearing electron-withdrawing groups, which resulted in increased redox stability. The magnetic behaviour of these complexes is virtually identical to that of the parent unsubstituted compounds. The observed effective barrier heights for these compounds were found to be about 440 cm^{-1}, which is again in good agreement with those estimated from the f-electronic structure of the unsubstituted anionic $TbPc_2$ complex (about 400 cm^{-1}) [175]. Furthermore, a super-reduced $TbPc_2$ species has been characterized recently. The phthalocyanine ligands in this new $LnPc_2$ SIM are completely fluorinated and do not contain any hydrogen atom. The magnetic relaxation of this compound is similar to that of other $TbPc_2$ compounds and follows a thermally activated Orbach process at high temperatures with a pre-exponential factor of $\tau_0 = 1.33 \times 10^{-10}$ s and an activation barrier $U_{eff} = 365\,cm^{-1}$ [176] (see Table 8.2).

$LnPc_2$ molecules showing SMM behaviour have significantly large axial magnetic anisotropy, which occurs by a different mechanism than those for known transition-metal-cluster SMMs [215]. In the mononuclear lanthanide magnets, the origin of the magnetism is both orbital and spin angular momenta of a single ion, whilst in transition-metal SMMs the spin angular momenta of multiple magnetic ions are responsible. The long magnetization relaxation times of the Tb(III) and Dy(III) ions are caused by the ligand field potential that controls the anisotropy and brings the substrates with a large $|J_z|$ value to the lowest energy level, whereas those of the transition-metal SMMs are caused by the magnetic interactions between component metal ions, leading to a negative ZFS constant. Here we take the $[TbPc_2]^{0/-}$ compound as an example to elaborate the mechanism in a little more detail. The ligand field potential around the Tb(III) ion ($4f^8$) with a total angular momentum (J) of six splits the ground multiplet so that the lowest sublevel has the largest J_z value ($|J_z| = 6$, corresponding to up/down spin states) and large energy gaps to the remaining sublevels (about 600 K) (Figure 8.9b). Thus, there is a small probability for a transition between $J_z = +6$ (up-spin state) and

−6 (down-spin state) substates and hence a slow magnetization response to an applied magnetic field [215]. The $TbPc_2$ is a rare-earth SMM in which the magnetic moment is carried by a single Tb(III) ion sandwiched between two organic Pc ligand planes. The $S = 1/2$ π-radical in $[TbPc_2]^-$ is quenched, whereas it is delocalized over the Pc ligand planes in the neutral species. Therefore, the $[TbPc_2]^0$ contains two spin systems: a central $J = 6$ Ising-type high-spin system arising from the Tb 4f-electrons, and a peripheral $S = 1/2$ π-radical delocalized over the two Pc ligands (Figure 8.9a) [90, 195, 209]. As a result of π−π interaction, this radical can easily hybridize with the p-electrons of any form of sp^2 carbon without affecting the magnetic properties of the Tb(III) ion. The highly anisotropic 4f shell of the Tb(III) ion and its intrinsically strong SOC result in a magnetic ground state of $J = 6$ and a pronounced uniaxial magnetic anisotropy. The ground state doublet $J_z = \pm 6$ is separated from the excited states by several hundreds of kelvin, which makes the $TbPc_2$ an Ising-like spin system at low temperatures ($T < 5$ K) and small magnetic fields ($B < 10$ T). A hyperfine interaction with the nuclear spin $I = 3/2$ of the Tb(III) ion splits the ground-state doublet $J_z = \pm 6$ into four states each. Finally, the ligand field generates a small transverse magnetic anisotropy resulting in avoided level crossings (Figure 8.11) [9].

The terbium and dysprosium complexes exhibit large axial magnetic anisotropies and behave as SIMs. The change from easy axis to easy plane magnetic anisotropy occurs on moving from the $HoPc_2$ to the $ErPc_2$ complex in the series of the investigated six bis(phthalocyaninato) lanthanide(III) complexes [202]. It has, therefore, been of great interest to show whether the holmium(III) complex with an axial anisotropy can show SMM behaviour at low temperatures. The Ho complex $[TBA][HoPc_2]^-$ doped in diamagnetic $[TBA][YPc_2]$ not only exhibited hysteresis loops in the sub-kelvin temperature range, but also resonant QTM was found. Hysteresis loops were clearly detected in the temperature range of 0.04–0.5 K at the field scan rate of $0.28\ T\ s^{-1}$ using micro-SQUID (superconducting quantum interference device) techniques, in which the field was aligned parallel to the easy axis of magnetization (see Table 8.2). The SMM behaviour of the $[HoPc_2]^-$ complex anion was explained by QTM between entangled states of the electronic and nuclear spin systems [202]. It has to emphasized that this mechanism is essentially the same as that of the terbium and dysprosium complexes but different from that of the transition metal cluster SMMs [195, 204].

8.3.4 Internal Kondo in $LnPc_2$ Complexes with Ln = Ce, Yb

Lanthanides with fractional valences have II, III and IV valences, as well as mixed II/III and III/IV valences. Depending on temperature and pressure, the degree of oxidation can change. This effect may result in a change in the different properties of nanoparticles, such as the stability, heat capacity, conductivity and magnetic susceptibility [218]. Valence fluctuation phenomena have been reported to occur

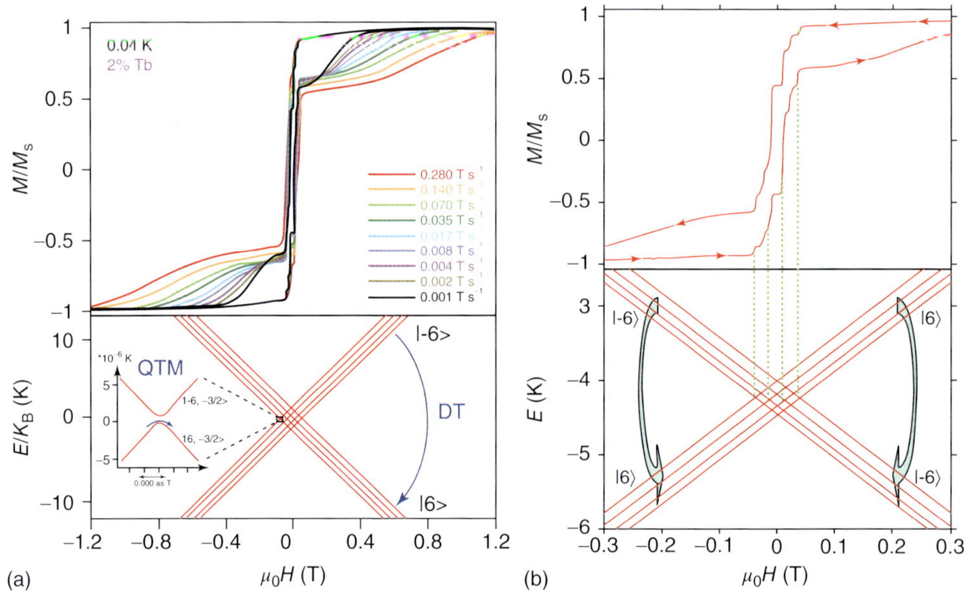

Figure 8.11 (a) Hysteresis loops at 0.04 K for a single crystal of TBA[$(Pc)_2Tb_{0.02}Y_{0.98}$] measured at several field scan rates (the applied magnetic field was aligned along the easy axis of magnetization) (top) and enlargement of the area around zero field of Zeeman energy diagrams illustrating that the $J_z = -6$ substrates are brought to resonance at $\mu_0 H = 0$ T, thus giving rise to an 'avoided level crossing' which allows QTM to occur (down) [9, 216]. (b) A close view of the hysteresis loop with steps corresponding to the four avoided level crossings [216, 217].

in rare-earth compounds containing the early (Ce), mid (Sm and Eu) and late (Tm and Yb) lanthanides, in which the proximity of the 4f level to the Fermi energy leads to instabilities of the charge configuration (valence) and/or of the magnetic moment [219, 220].

Valence fluctuation takes place between one electron and the empty 4f-shell states. In the case of cerium, the ambivalent tendency arises from the fact that the 4f orbital is more spatially extended than for the other lanthanides. Indeed Ng and coworkers [139] have employed a range of spectroscopic, electrochemical and structural methods to elucidate the valence of the cerium centre in a series of bis(phthalocyaninato) and mixed (porphyrinato) (phthalocyaninato) lanthanide complexes. Depending on the electronic nature of the ligands, the cerium centre adopts an intermediate valence, ranging from predominantly trivalent, corresponding to the f^1 configuration, to tetravalent, corresponding to the f^1 electronic configuration, by simultaneous quenching of the $S = 1/2$ to $S = 0$ on the ligand systems. The results have been supported by XANES (X-ray absorption near edge structure) studies. Instead of the sextet of $^2F_{5/2}$ for Ce(III), the ground-state multiplet of Yb(III) with $4f^{13}$ is the octet of $^2F_{7/2}$, for which the mixing matrix elements between the 4f states and the plane-wave states are easily obtained. Therefore, the arguments of fluctuating-valence states hold for both cerium and ytterbium compounds in which the valence fluctuates between one hole and the fulfilled 4f-shell states, while the peripheral radical systems on the ligands act as buffers for electrons, causing the valence fluctuation of ± 1. For example, in the 4f shell, Yb(III) gives 13 electrons and Yb(II) gives fully occupied 14 electrons. This was then reflected in the typical behaviour of a compound with the valence fluctuation between Yb(III) and Yb(II). Due to this mutual exchange interaction between a localized $S = 1/2$ spin and the metal ion based f^1 electron, the sandwich-like lanthanide complexes of cerium and ytterbium were defined as *internal Kondo systems*.

The Kondo resonance is a many-electron phenomenon resulting from the exchange interaction between a localized spin and the conduction electrons in metallic electrodes, and it appears only when the electronic state of a quantum dot has non-zero spin or degeneracy. In the late 1960s, measurements on several superconductors containing cerium impurities under pressure revealed that the impurities, which were magnetic and produced a Kondo effect at atmospheric pressure, appeared to undergo a continuous demagnetization under pressure [221]. Likewise, substituting ytterbium into a cerium-based Kondo lattice leads to a linear decrease of the unit cell volume, indicating that ytterbium ions are divalent, forming Kondo holes and leads to a strong suppression of the superconducting properties and Kondo coherence. These results, combined with measurements of the Hall effect, indicate that ytterbium ions act as nonmagnetic impurity scatters in the coherent Kondo lattice without serious suppression of the antiferromagnetic fluctuations. These results are in stark contrast to studies performed using bulk single crystals, which claim the importance of valence fluctuations of ytterbium ions.

8.3.5
Stable Organic Radicals $S = 1/2$ in $LnPc_2$ with Ln = Y, Lu

Among various rare-earth bis(phthalocyaninato) complexes, the derivatives containing diamagnetic ions such as yttrium and lutetium are usually taken as examples to compare with other paramagnetic cases so that they are extensively studied by different physical techniques. In particular, the lutetium case is primarily studied because a high-purity lutetium compound can be obtained by a relatively simple method. Neutral $LuPc_2$ has been reported as the first stable free radical phthalocyanine and the first molecular semiconductor [59]. In the case of the lutetium bis(phthalocyanine) system $[LuPc_2]^0$ with an $S = 1/2$ radical, the effects of the orbital moment are negligible. Since lutetium(III) has a $^1S_0$ ground state, only the singly oxidized phthalocyaninato-radical would be expected to contribute to the magnetic susceptibility in this system. Indeed the magnetic susceptibility measurement shows a typical Curie–Weiss behaviour with a room temperature effective magnetic moment of about 1.67 μB, corresponding to one unpaired electron [164, 206]. Furthermore, the room-temperature EPR measurements confirm the nature of a free radical, in which they show a single strong resonance as small deviations (from g_e) at or near $g = 2$ (Figure 8.6b) [206].

Owing to the flat shape of phthalocyanine molecules, the crystal structure of Pc-based materials is typically characterized by chains along which Pc molecules tend to stack. The neutral bis(phthalocyaninato) yttrium $[YPc_2]^0$ compound can be considered to be the parent compound of the $[LnPc_2]^0$ molecular magnets. In fact, it is characterized by the absence of localized f electrons and the microscopic properties are mainly associated with the presence of an unpaired electron delocalized in the $a_2\pi$ orbital, due to the one-electron oxidation of the $[YPc_2]^-$ unit [222]. Thus, $[YPc_2]^0$ allows investigations of the spin dynamics associated only with this unpaired electron spin, independent of the one due to f-electrons. One of the most suitable tools to address this aspect is the NMR technique. Branzoli and Ruben [223] presented an experimental study of the magnetic properties of a $[YPc_2]^0$ compound by means of magnetization and NMR measurements. The temperature dependence of the macroscopic susceptibility, of the NMR shift, and of the spin-lattice relaxation rate $1/T_1$ clearly show that the $[YPc_2]^0$ system is a prototype of a $S = 1/2$ antiferromagnetic chain, characterized by a diffusive low-frequency spin dynamics and by the presence of a low-energy spin gap.

8.3.6
A Special Case: Half-Filling of the f-Orbitals in $GdPc_2$ and its Consequences

The gadolinium(III) ion, with a half-filled $4f^7$ configuration, is quite different from the other cases. It is known that ligand-field potential does not directly lift the degeneracy of the multiplet because the orbital angular momentum of the $^8S_{7/2}$ ground state is zero [203]. Therefore, compounds of Gd(III) in general exhibit an extremely small temperature dependence of the effective magnetic moment μ_{eff}. In other words, the curve of μ_{eff} versus T almost always obeys the Curie law and provides the best example of the free ion model. However, it is also worth pointing out that a Gd(III) ion in an actual crystalline environment shows a ZFS which could result in mixing of the $^8S_{7/2}$ state and excited states with a non-zero orbital angular momentum [203].

From the experimental data of $[GdPc_2]^-$ taken from 2 to 300 K, the μ_{eff} versus T curve almost remains constant up to 300 K, and the μ_{eff} value corresponds to that for a free Gd(III) ion ($\mu_{eff} = 7.94$ BM) over the entire temperature range [164]. For the neutral $[GdPc_2]^0$, it can be seen that μ_{eff} levels off at $T < 10$ K with a value of 6.9 BM, which remains constant up to 100 K. The μ_{eff} value is less than that for a free ion due to the presence of strong antiferromagnetic interactions between the Gd(III) ion and the phthalocyanine radical electron. All attempts to fit the data to the Curie–Weiss law with $J = 7/2$ and $g = 2.00$ were fruitless, indicating that the ground state of the Gd(III) ion is not $^8S_{7/2}$ anymore. Two good fits are obtained: (i) when all parameters were allowed to vary, the parameters obtained were $g = 2.01$ and $\theta = -0.51$ K with an R value of 0.0118 and (ii) with $J = 3$, and g fixed at 2.00, a value of $\theta = -0.47$ K was obtained with an R value for the fit of 0.0122. Both fits show excellent agreement to the data and appear to be equivalent within the limits of error. Therefore, it can be concluded that the ground state for $GdPc_2$ is a $^7S_3$ state, which has non-zero orbital angular momentum. The ferromagnetically coupled $^9S_4$ excited state does not contribute to the magnetism of this compound in the temperature range studied.

8.4
Surface Magnetism of $LnPc_2$ Complexes

Before entering the world of molecular spintronic devices, investigating the magnetic properties of the molecules in the device environment, for example, on surfaces or close to conducting electrodes, is an indispensable step. Ideally, in order to develop molecular devices, the magnetic hysteresis on individual molecules organized on surfaces should be retained when the grafting is performed. Under such a requirement for technological inputs, material scientists are trying to fabricate functionalized magnetic molecules onto different substrates, and to investigate to which degree the surface changes the magnetism relative to that observed for the original molecules in the bulk phase. Indeed the impact of these new ideas towards the fabrication of spintronic devices is immediate and tangible. The results of surface magnetism obtained from the study of classic SMMs such as Mn_{12} and Fe_4 suggest that the quantum properties are strongly affected by the orientation and interaction of the adsorbed molecules on/with the substrate. In addition, most probably owing to the chemical instability of clusters on surfaces, magnetic hysteresis has been hardly retained for monolayers of d-metal based SMMs on various non-magnetic substrates. Therefore, from the viewpoint of surface science, it is difficult to control the individual magnetism of the surface structures and the challenges still remain to be solved, and some concerns have to be considered [224].

There are two main approaches for grafting molecules onto surfaces, using either direct deposition of properly functionalized molecules on native surfaces or pre-functionalization of the surface and subsequent grafting of magnetic molecules. In general, how the deposition procedure, as well as the kind of interaction with the conductive metallic surface, influence the properties could be revealed by characterization tools such as morphological techniques, in particular STM, STS and non-magnetic spectroscopic measurements such as X-ray absorption spectroscopy (XAS) and X-ray photoelectron spectroscopy (XPS) [225, 226]. The chemical composition of the adsorbates can be elucidated by XPS, although an unambiguous determination of oxidation states of metal ions is difficult using this technique alone. Although being chemically insensitive at the local level, scanning probe methods have become 'the new eyes' of the scientists and mandatory tools to characterize substrates and deposition processes. Of them, STM is commonly utilized because this local technique reveals not only the electronic structure of the surface-deposited magnetic molecules with unprecedented resolution, but it also allows one to probe indirectly their magnetic properties by STS via resonance phenomena as, for example, Kondo-effects and inelastic scattering events [227]. STM topography, together with STS, can provide information on the geometry of the deposition and scattering information of the magnetic centres with partially submolecular resolution. In particular, flat molecules such as $TbPc_2$ molecules seem to adapt to such investigations due to their intrinsic fit to the geometry of scanning tunnelling experiments, implying very flat but conducting surfaces. However, the latter point, the requirement of conducting surface, represents a certain experimental challenge, since the molecule-based electronic (and spin) information has to be carefully disentangled from the electronic states of the surface due to very often observed hybridization effects.

However, the key property to be measured for the molecules fabricated on the surface is the magnetization, which, in the case of (sub)monolayers of SMMs, is not directly accessible to standard magnetometry and therefore requires other techniques. Due to their high sensitivity, XAS and X-ray magnetic circular dichroism (XMCD) are most commonly utilized methods to characterize the magnetic properties of surface-supported molecules. XAS is a well-established method that was first developed on X-ray tubes but became an easy-to-perform technique with the development of synchrotron-light sources. XMCD stems from XAS, so it possesses most of its characteristics but with much higher sensitivity at low temperature, but uses in addition circular polarized radiation which gives access to magnetic properties. XMCD is a fundamental tool to evidence if molecules deposited as a monolayer or a submonolayer keep their electronic and magnetic properties after the deposition process [226]. Low temperature XMCD experiments permit the study of problems related to surface deposition that cannot be identified by techniques like STM. Due to its high sensitivity, XMCD permits the study of very small amounts of magnetic material down to a few atoms, and its chemical selectivity allows investigations of the interactions between different elements as well as the influence of the substrate [228]. In the case of anisotropic magnetic molecules that adopt a preferential orientation on the surface precious information can be obtained by measuring the field dependence of the XMCD signal at different orientations of the magnetic field. For example, for isolated $TbPc_2$ molecules evaporated on copper or gold, the XMCD saturates at a lower field when the magnetic field is applied perpendicular to the surface, where the easy axis of the molecule orients. More details are discussed in the following sections. Therefore, XAS/XMCD can be considered as the ultimate and compulsory tests needed to confirm the preservation of magnetic properties of SMMs on the surface before pursuing the investigation and development of high-tech nanodevices for spintronics purposes [228, 229].

In recent years, attempts have been made to try transition metal SMMs such as Fe_4 [224] and Mn_{12} [226] for surface science with implications for the use of new experimental techniques and with the development of new synthetic approaches. Apart from these magnetic materials, the heterometallic molecular rings as Cr_7Ni were also successfully grafted on different substrates, including gold and graphite, which were shown to be robust enough to suffer only minor changes in the pattern of its low-lying levels when single units are anchored on the surface [230–232]. Double- and triple-decker Pc molecules with lanthanides have high blocking temperatures, and their thermal stability allows deposition by sublimation. Another important advantage is owing to the possibility of fine-tuning the chemical properties of the organic moiety that surrounds the inorganic core without significantly altering the magnetic properties. Therefore, grafting molecules of the $LnPc_2$ family onto different substrates has been explored thoroughly [233]. Of the many classes of compounds, $TbPc_2$ and its derivatives remain to date the family of SMMs that has been studied most intensely. In the following paragraphs, we discuss some experimental results in which attempts were made to graft the $TbPc_2$ molecules on different characteristics of substrates, including non-magnetic substrates such as highly oriented pyrolitic graphite (HOPG), Au(111), Cu(111), Cu(100) and magnetic substrates such as nickel films, ultrathin cobalt and nickel films, a perovskite manganite of LSMO as well as with antiferromagnetic manganese and cobalt oxide layers.

8.4.1 Deposition of $[TbPc_2]^0$ SIMMs on Nonmagnetic Substrates

8.4.1.1 Highly Oriented Pyrolitic Graphite

HOPG, like carbon nanotubes (CNTs), gives noncovalent $\pi-\pi$ stacking interactions with polycyclic aromatic groups (e.g. pyrene), which can be easily incorporated in molecular structures. In order to deposit $TbPc_2$ molecules onto the HOPG surface by supramolecular interactions, the negatively charged $TbPc_2$ SMM is decorated with alkoxy chains. Indeed well-ordered 2D self-assembled domains were formed when the functionalized derivatives of $TbPc_2$ SMMs were deposited on HOPG via drop-casting technique. A drop of a toluene solution of $TbPc_2$ decorated with alkoxy chains, heated to 55 °C in advance, was cast onto a freshly cleaved HOPG surface and was left to dry at room temperature. Immediately afterwards, large-scale STM images under ambient conditions were obtained over three independent surface areas. Additionally, the STM images over the same region of the HOPG surface were also taken at different intervals of time. All results obtained reveal that this SMM is an excellent candidate for deposition on inert HOPG. First, the presence of alkoxyl chains increases the solubility and favours 2D self-assembly on the surface without the need for supplementary lithographic techniques. Second, clear observation of individual molecules is achieved, a fact that may facilitate their manipulation or analysis. And finally, the presence of alkoxyl chains on the $TbPc_2$ and its planar character favour 2D self-assembly on HOPG with the molecular plane parallel to the graphite surface, enabling control and/or prediction of the orientation of the preferential magnetization axis on the surface [234, 235].

The 2D self-assembly of a $TbPc_2$ on HOPG was also investigated by atomic force microscopy (AFM), and it was shown that it forms highly regular, rectangular two-dimensional nanocrystals on the surface, aligned with the graphite symmetry axes, in which the molecules are organized in a rectangular lattice as shown by STM. Molecular dynamics simulations were run in order to model the behaviour of a collection of the $TbPc_2$ complexes on HOPG. The results were in excellent agreement with the experiment, showing that – after diffusion on the graphite surface – the molecules self-assemble into nanoscopic islands, which align preferentially along the three main graphite axes. These low dimensional assemblies of independent magnetic centres are mainly monolayers (as shown by AFM) and are therefore very interesting nanoscopic magnetic objects, in which all of the molecules interact with the graphite substrate and might therefore be affected by it. The magnetic properties of these self-assembled bar-shaped islands on HOPG were studied by XAS/XMCD (Figure 8.12), confirming that the compounds maintain their properties as SMMs when they are in close interaction with the graphite surface [234].

8.4.1.2 **Au(111)**

$[TbPc_2]^0$ complexes were deposited onto a clean Au(111) surface at room temperature by using a sublimation method under UHV. An STM image taken on an isolated $TbPc_2$ molecule that was adsorbed onto the Au(111) surface showed eight lobes from the two Pc ligands. XPS experiments showed that $TbPc_2$ did not interact strongly with the Au(111) surface, thus leading to a smaller degree of charge transfer. In addition, conduction measurements have proven thin-film field-effect transistor (FET) properties [179]. The appearance of FET properties was related to the electronic structures of the molecule, which represents one of exciting bulk-phase electron-transport properties for use in applications such as FET devices.

Furthermore, it was shown the unpaired spin $S = 1/2$, which is delocalized over the two Pc rings, still remained in the π-orbitals after absorption on Au(111). Consequently, STS measurements also provided direct observation of the $S = 1/2$ radical on the $TbPc_2$ molecules on Au(111); whereby the indicative Kondo-peak could be switched off by tunnelling current pulses [215]. Indeed the tunnelling conductance (dI/dV) was analysed from STS experiments of $TbPc_2$ on Au(111) near the Fermi level showed a zero-bias peak (ZBP) in the spectra, which could be assigned as a Kondo resonance. Clear Kondo features for the molecules with $\theta = 45°$ were observed when the tip was positioned over one of the lobes of $TbPc_2$.

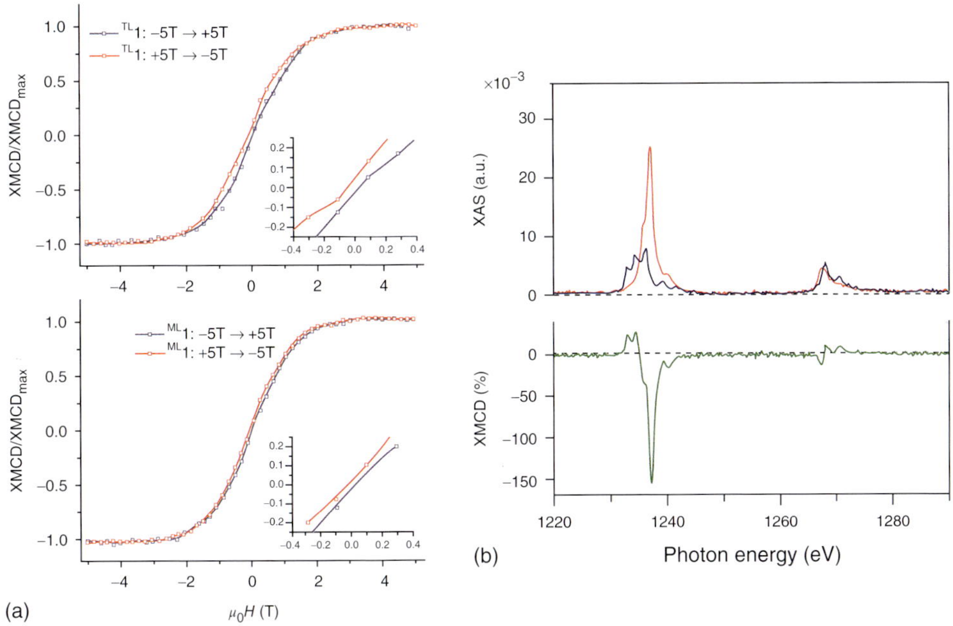

Figure 8.12 (a) XMCD detected hysteresis of magnetization for the thick layer sample $^{TL}1$ (above) and the monolayer sample $^{ML}1$ (below) on HOPG measured at 7 K. The inset shows the enlargement of the central part of both hysteresis. (b) XAS spectra at the $M_{4,5}$ edge of Tb for $^{ML}1$ with helicities parallel ($I^{\uparrow\uparrow}$, in red) and antiparallel ($I^{\uparrow\downarrow}$, in blue) to an applied magnetic field of 5 T at 7 K (above), and the XMCD spectrum given by their difference normalized to the maximum intensity of the unpolarized spectrum $1/2\ (I^{\uparrow\downarrow} + I^{\uparrow\uparrow})$ (below in green). (Adapted from Ref. [234]. Copyright (2011) American Chemical Society.)

However, when the tip was placed over the centre of the molecule, only a small feature was observed. Therefore, the Kondo feature of $TbPc_2$ on the Au(111) surface could be switched on and off by changing the azimuthal rotational angle (θ) – the twist angle between the two Pc rings [233]. The upper ligand could be rotated in a controlled manner, thereby switching the ligand spin on and off. Thus, information could be encoded at the single-molecule level by controlling the molecular spin through an electric current. Such an ability to control the molecular spin of $TbPc_2$-based SMMs holds perspectives for the new design spintronics devices.

8.4.1.3 **Cu(111)**

Vitali *et al.* [236] reported for the first time the electronic structure of $[TbPc_2]^0$ SIMs that were deposited onto a Cu(111) surface in UHV conditions by using a dry-imprint technique. A fibreglass bundle coated with fine-grained powder of the $TbPc_2$ crystals was brought in gentle contact with the copper surface previously cleaned by cycles of Ar-sputtering and annealing, leading to direct

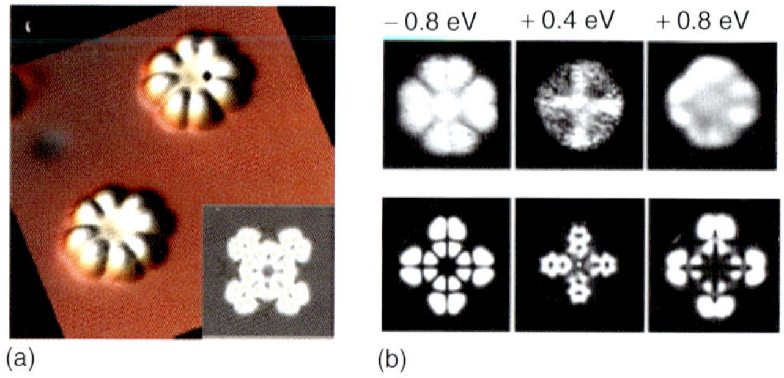

Figure 8.13 (a) Constant current topograph of two isolated $TbPc_2$ molecules supported by the Cu(111) surface (image size, 7 nm × 7 nm; tunnelling conditions, −0.8 eV, 0.7 nA); the right bottom corner is a simulated STM image of an isolated $TbPc_2$ molecule. (b) STS measured (top) and DFT calculated (bottom) energy-resolved conductance maps of the $TbPc_2$ molecule on the Cu(111) surface at energies corresponding to the electron states as shown in label. (Adapted from Ref. [236]. Copyright (2008) American Chemical Society.)

deposition of single molecules on the Cu(111) surface. Topographic images confirmed that the molecular structure was unchanged by the interaction with the surface Cu(111) (Figure 8.13a). The ligand spin was suppressed and quenched due to strong metal-molecule hybridization, but the 4f electron states were not perturbed by the adsorption and the SIM character of the surface supported $TbPc_2$ molecules was preserved. Compared to a spin one-half ($S = 1/2$) Kondo system for $TbPc_2$ deposited on Au(111) [233], this is clearly different behaviour, and illustrates the importance of the choice of the device environment.

In order to support this conclusion, DFT calculations were carried out for the unsupported molecule. The STS experimental images of the surface supported $TbPc_2$ molecule are directly compared with the corresponding charge density calculated for the unsupported molecule (Figure 8.13b). Both the experimental and theoretical maps reflect a pronounced spatial variation of electron densities across the molecule with distinctive features at characteristic energies. The dI/dV maps display the same trend as a function of increasing energy. They show four pairs of lobes delocalized over the π-molecular orbitals of the Pc ligands at −0.8 eV (A), a weak cross-like feature centred on the Tb atom and developing along the Tb–N bonds at 0.4 eV (B), and again four pairs of lobes delocalized over the Pc ligand at 0.8 eV (C). This substantiates the presence of electron states highly localized on the f-metal centre, which appear as peaks with minor intensity at intermediate energies obtained from experimental dI/dV spectrum for the surface-supported molecule measured at the dot position in Figure 8.13a. In short, the DFT calculations did not show a spin-splitting for the ligands for free and negatively charged $TbPc_2$, and for molecules in contact with Cu(111) surfaces. These observations indicate that the inherent f-orbital spin magnetic moment of the molecule is only weakly affected by the interaction with the surface.

8.4.1.4 Cu(100)

The samples were prepared *in situ* under UHV conditions by thermal evaporation of unsubstituted $[TbPc_2]^0$ molecules onto the clean Cu(100) surface. An STM topograph of mono-dispersed $TbPc_2$ molecules on the Cu(100) surface showed that single $TbPc_2$ units adsorbed intact onto the substrate with the Pc plane parallel to the surface. The molecules do not aggregate due to the strong interaction of the lower Pc ligand with the substrate. The Tb(III) magnetic moment retains Ising-like magnetization. The Tb(III) ion presents the full orbital and spin magnetic moment expected for a $J = 6$ ground state along the easy axis, and vanishing magnetization in the hard Pc plane (Figure 8.14a) [237].

XMCD spectroscopy allows separation of the orbital and spin magnetic moments of the Tb(III) ion as a probe of their extraordinary anisotropic behaviour. The magnetic properties of isolated $TbPc_2$ molecules supported on a Cu(100) surface were investigated by XMCD at 8 K in magnetic fields up to 5 T [237]. The XMCD spectrum, shown at incidence angles $\theta = 0°$, 45° and 70° after normalization to the integrated XAS intensity (Figure 8.14b), is strongly anisotropic. While the XAS line shape varies with θ, the XMCD line shape does not change apart from a scaling factor, contrary to what is observed in planar molecular coordination networks. This finding suggests that the magnetic moment of the Tb(III) ion follows an Ising-like behaviour; that is, it aligns either parallel or antiparallel to the surface normal, depending on the sign of the applied field, and the angle-dependent XMCD scales as $\cos\theta$, that is, as the projection of a perpendicular moment onto the X-ray direction. Such a scaling is also found for the high field values of the magnetization reported in Figure 8.14c. Despite the very strong anisotropy, neither sample shows magnetic hysteresis down to 8 K, far below the blocking temperature in bulk of 25–40 K reported in Ref. [194] on the basis of AC susceptibility measurements. This is not surprising given that the time (~20 min) required to measure a magnetization

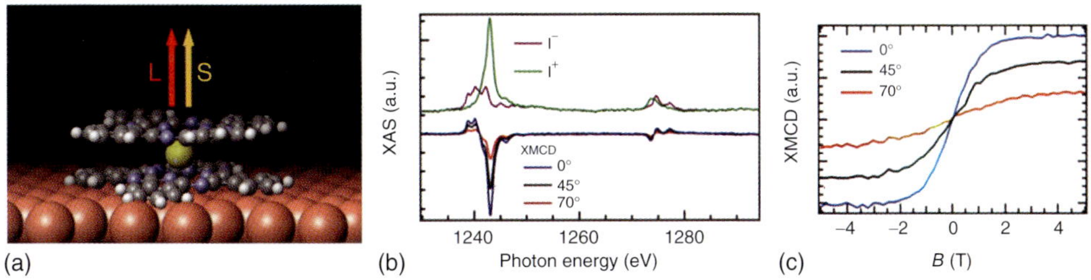

Figure 8.14 (a) Schematic view of the $TbPc_2$ magnetization on Cu(100) surface with the Pc plane parallel to the surface. (b) Circular polarized XAS of $TbPc_2$ on Cu(100) at 0° (normal) incidence (I^+, I^-) and XMCD at 0°, 45° and 70° incidence ($T = 8$ K, $B = 5$ T). (c) Magnetization curves obtained by the Tb M_5 XMCD intensity for $TbPc_2$ on Cu(100), but without hysteresis observed at 8 K. (Adapted from Ref. [237]. Copyright (2010) American Chemical Society.)

curve by XMCD is four to seven orders of magnitude larger compared to AC measurements, and that the susceptibility is still appreciable below 10 K. The experimental results revealed that, although the interaction strength of the lower Pc ligand with the substrate was sufficient to impede diffusion of the molecules, their ligand field, and hence magnetic properties, remained unaltered (Figure 8.14c).

It is worthwhile pointing out that the deposition of single bis(phthalocyaninato)-neodymium(III) molecules onto Cu(100) substrates was also recently attempted. Compared to Tb(III) ion ([Xe]$4f^8 6s^0$), the Nd(III) ion ([Xe]$4f^3 6s^0$) has a larger ionic radius (112 pm versus 104 pm) with more extended 4f-electrons and, thus, stronger hybridization with the phthalocyanine ligands. Thereby, as shown previously in Ref. [236], the 4f orbitals seem to be accessible here also to transport experiments, while largely maintaining the relative energetic positions of spin-up/-down states and consequentially the magnetic moment. The $NdPc_2$ molecule appears in STM topography images as an eight-lobe shape with a fourfold symmetry as previously observed for that of $TbPc_2$, which indicates that for $NdPc_2$/Cu(100), the chemisorption does not significantly alter the molecular electronic features of the upper Pc ligand. The Nd 4f-states can not only be directly accessed by STM, but can also be unambiguously identified based on their spectral position and spatial features by comparing STS experiments and *ab initio* calculations. They are highly spin-polarized and, thus, carry the spin magnetic moment of the electrode-adsorbed molecule. Accessing these states by means of transport is expected to be a general feature of early Ln-based molecular magnets, where the 4f-states are closer to the Fermi level and can be adjusted via coordination and surface interactions. These results showcase how tailoring molecular orbitals can yield all-electrically controlled spintronic device concepts [238].

Based on the aforementioned results, we could conclude that, in general, the electrical features of the nanoconstrictions are not altered by the deposition of molecules. These studies reveal that the interaction of the $[TbPc_2]^0$ molecules with the metal surface preserves both the molecular structure and the large spin magnetic moment of the metal centre. The 4f electron states are not perturbed by the adsorption, while a strong molecular/metal interaction can induce the suppression of the minor spin contribution delocalized over the molecular ligands. The DFT calculations for the cases of the copper substrate [236, 237] show that the inherent spin magnetic moment of the molecule is only weakly affected by the interaction with the surface and suggest that the SIM character might be preserved. In addition, the Ising-like uniaxial anisotropy of the $TbPc_2$ SMM is found to be orthogonal to the Pc rings, whereas a hard axis is found along with the surfaces of substrates [237]. In such a case, the easy axis of magnetization of the $TbPc_2$ molecule can be easily defined when the molecule is deposited on substrates. In contrast, the peripheral $S = 1/2$ organic radical could be quenched depending on the work function of the chosen substrate as observed in the case of Cu(111) and Cu(100) or preserved as seen for Au(111).

8.4.2 Deposition of $[TbPc_2]^0$ SIMs on Magnetic Substrates

As different substrates have different chemical reactivity, a careful design of molecular structure and a proper choice of deposition techniques are essential to achieve the required chemical complementarity between molecule and substrate. For example, the grafting of thiols on diamagnetic gold substrate induces the production of molecular hydrogen in a redox reaction during the formation of S–Au bond [239]. However, organizing magnetic molecules on a solid paramagnetic substrate should introduce new ingredients in the self-assembly process because a magnetic coupling will surely be mediated between the molecules and the substrate. This type of surface coupling (i.e. indirect, superexchange interaction) will affect or even change the magnetic properties of these molecules so that their magnetic stability and their anisotropic nature will not be independent. Therefore, the organization of SIMs onto paramagnetic substrates opens up an avenue for spin-dependent molecular electronics.

In view of the operation of $TbPc_2$ SMMs via STM [228], conductive ultra flat non-magnetic substrates such as HOPG, Au(111), Cu(111) and Cu(100) have been most commonly used, which have already been discussed in earlier sections. In the following section, we discuss some experimental results on the investigation of $TbPc_2$ SIMs sublimated on some commonly used paramagnetic substrates such as ferromagnetic nickel substrate, ultrathin cobalt and nickel films and a perovskite manganite of LSMO as well as with antiferromagnetic manganese and cobalt oxide layers. In general, the experimental results have revealed that the magnetic properties of the original SIMs are entirely altered by these substrates, and the magnetization of these molecules along with different anisotropic axes are significantly different depending on the steric or electronic influence of the substrate.

8.4.2.1 Nickel Thin Films

The samples were prepared in situ by molecular beam evaporation of $[TbPc_2]^0$ on thin Ni films on Cu(100) in UHV conditions. Gambardella and Ruben [240] investigated the interaction of $TbPc_2$ SMMs with out-of-plane (OP) ferromagnetic Ni substrates, in which case the molecule and substrate easy axes are collinear (Figure 8.15a). The interaction between the SIM and the substrate magnetic anisotropy was clearly observed, and the magnetic moment of $TbPc_2$ could be effectively stabilized against thermal spin fluctuations, while preserving typical SIM features. XAS/XMCD experiments exhibited that both Ni and Tb had a strong remanent XMCD intensity, which remained stable over the time scale of the measurements. The sign of the Tb XMCD, however, was opposite to Ni, indicating that M_{Tb} and M_{Ni} were antiferromagnetically coupled. Element-resolved XMCD measurements showed that $TbPc_2$ coupled antiferromagnetically to Ni films through ligand-mediated superexchange. This coupling was strongly anisotropic and could be tuned by oxidizing or reducing the ferromagnetic substrate.

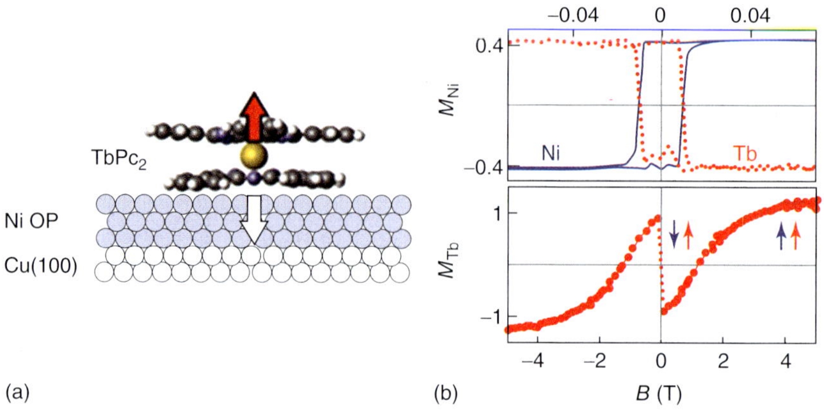

Figure 8.15 (a) Schematic view of the $TbPc_2$ SIMM (single-ion molecular magnet) and Ni magnetization on out-of-plane Ni/Cu(100) films. (b) Element-resolved hysteresis loops of Ni (top) and Tb (bottom) for $TbPc_2$/Ni/Cu(100) measured at out-of-plane ($\theta = 0°$) incidence at $T = 8$ K. The units of M_{Ni} and M_{Tb} correspond to the XMCD/XAS ratio at the L_3 and M_5 absorption edges, respectively. The dashed lines show the Tb magnetization at low field normalized to M_{Ni}. The solid lines superposed to M_{Tb} are fits according to the equation discussed in the text [240].

Element-resolved hysteresis curves revealed that the SIM magnetization depended critically on the alignment of the molecule and substrate easy axes as well as on the balance between interface-dependent superexchange and applied magnetic field. Figure 8.15b shows that the OP ($\theta = 0$) magnetization loops of the OP$TbPc_2$/Ni/Cu(100) sample were measured by recording the XMCD intensity at the Ni L_3 and Tb M_5 edge as a function of the applied field, in which a clear hysteresis loop was observed in the timescale ($t \sim 10^3$ s) of XMCD magnetization measurements. However, $[TbPc_2]^0$ complexes deposited on perpendicularly magnetized Ni films exhibited enhanced magnetic remanence compared to SIMs in the bulk. Contrary to paramagnetic molecules strongly pinned to a ferromagnetic support layer, weak coupling enabled $TbPc_2$ to be magnetized parallel or antiparallel to the resulting magnetic moment of the substrate, opening the possibility of exploiting SIMs in spin valve devices. These results showed that $TbPc_2$ SIMs behaved as coupled but separate magnetic units from an underlying ferromagnetic surface. The enhanced thermal stability of the $TbPc_2$ magnetic moment and the possibility of orienting it parallel or antiparallel to a macroscopic ferromagnetic layer make $TbPc_2$ very interesting for applications in hybrid devices. An independent experiment confirmed the antiferromagnetic coupling between the molecular and substrate magnetic vector, however, with slightly different internal parameters, for example, even weaker coupling. This observation emphasizes reproducibility but also the dependence of the coupling parameters on the local epitaxy at the substrate–molecule interface [241].

8.4.2.2 **Cobalt Thin Films**

Another type of ferromagnetic substrate studied is the ultrathin Co film thatwas epitaxially grown onto a Cu(100) substrate. A submonolayer of $TbPc_2$ molecules was deposited *in situ* onto the substrate and the magnetic and electronic properties of these $TbPc_2$ SMMs were studied by the element specificity of XAS and XMCD methods [241]. On such substrates the $[TbPc_2]^0$ SIMs were antiferromagnetically coupled to the ferromagnetic films, which was possibly due to a superexchange interaction via the phthalocyanine ligand that was in contact with the magnetic surface. It was not possible to detect remnant magnetization at a temperature of 8 K in the XMCD measurements, as a result of an antiferromagnetic coupling to the Co substrate. In the case of that on the Co substrate, the magnetic easy axis was in the plane and thus perpendicular to the one of the terbium(III) total angular momentum of $J = 6$. Thus, due to this misalignment, the antiferromagnetic contribution was much smaller than that on the Ni film, partially due to the higher measurement temperature, but primarily because of the large magnetic anisotropy of the molecule. Therefore the magnetization curve of the $[TbPc_2]^0$ SIMs on the Co substrate was dominated by the paramagnetic signal at this temperature. These experiments also demonstrated that the magnetic coupling of the $[TbPc_2]^0$ SIMs on the Co films is much more complex compared to the previously described situation on Ni-films (as also shown for the noncollinear case in Ref. [240]); under slightly different experimental situations, this situation could lead to the eventual absence of any magnetic coupling [242].

8.4.2.3 **LSMO**

Very recently, Sessoli and coworkers [242] investigated structural features and magnetic behaviour of $[TbPc_2]^0$ thin films sublimated on $La_{0.7}Sr_{0.3}MnO_3$ (LSMO). As another commonly used ferromagnetic substrate, LSMO is one of the few ferromagnetic electrodes that can be prepared *ex situ* with an excellent magnetic polarization at room temperature. Using synchrotron-based X-ray natural linear dichroism (XNLD) and XMCD techniques, they detected two different orientations of the molecules for the substrate depending on the starting surface. It is worthwhile pointing out that, in contrast to what was found in the case of ferromagnetic substrates mentioned earlier [240, 241], no sizable antiferromagnetic coupling between $[TbPc_2]^0$ SIMs and the LSMO was detected; moreover $TbPc_2$ hysteresis loops showed no correlation with those of the substrate.

8.4.2.4 **Manganese and Cobalt Oxide Layers**

With regard to antiferromagnetic substrates, Gambardella [243] reported an exchange bias in a system of $[TbPc_2]^0$ SIMs and antiferromagnetic Mn and CoO layers. Both the presence and absence of unidirectional exchange anisotropy for $[TbPc_2]^0$ SIMs deposited on antiferromagnetic Mn and CoO layers were investigated by element-resolved XMCD measurements. $[TbPc_2]^0$ deposited on Mn thin films presented magnetic hysteresis and a negative horizontal shift of the Tb magnetization loop after field cooling, consistent with the observation of pinned spins in the Mn layer coupled parallel to the Tb magnetic moment. Conversely, molecules deposited on CoO substrates presented paramagnetic magnetization loops with no indication of an exchange bias. That is, the $TbPc_2$ molecules were able to polarize the pinned spins of an AFM during field cooling, which induced exchange bias of the molecules to the substrate. The bias field was found to reach a maximum when the cooling field was set parallel to the SIM easy axis. From the shape of the $[TbPc_2]^0$ SIMs magnetization curve, they inferred that the exchange bias occurred at the level of single molecules. Control over the origin of the pinned spins and positioning of the molecules may result in new applications that exploit the interaction between SIM and AFM, such as spin valves and spin filters, where the molecular magnetic moment is simultaneously stabilized and biased by unidirectional exchange coupling.

8.4.2.5 Spin Polarized Scanning Tunnelling Microscopy (SP-STM) on Co/Ir(111)

We now briefly mention another type of microscopic technique, namely spin-polarized scanning tunnelling microscopy (SP-STM). SP-STM represents one of the most direct ways to measure the spin-polarization of the electron current emerging from a magnetic surface through a single molecule. SP-STM can provide direct proof of the manipulation of a spin-current by chemical bond tailoring, that is, it is a preferential tool of spin interface science. Within the last decade, SP-STM has become a mature tool for the microscopic investigation of surface magnetic domain structures. The imaging of these sample systems is particularly difficult with other magnetically sensitive microscopic techniques since the net magnetization vanishes on a length scale larger than the atomic scale [244]. For $[TbPc_2]^0$ SIMs, using SP-STM technique, Schwöbel *et al.* [245] directly observed the spin-splitting of molecular orbitals of $TbPc_2$ molecules deposited on a ferromagnetic cobalt support grown on an iridium(111) substrate (Figure 8.16). The spin polarization of individual molecules was mapped on a sub-molecular scale by low-temperature SP-STM, and the spin-splitting of the LUMO was determined quantitatively. The $[TbPc_2]^0$ SIMs were deposited on a ferromagnetic cobalt nanostructure exhibiting a magnetization along the surface normal. As mentioned before, in neutral $TbPc_2$, the HOMO, which was delocalized over both Pc rings, was therefore just singly occupied. With SP-STM, this ligand spin should be observable as follows: as the singly occupied molecular orbital of $TbPc_2$ was also the LUMO, it should be visible both below as well as above the Fermi level. To unambiguously identify the spin state of an adsorbed $TbPc_2$ molecule, a direct spin-resolved study of specific ligand orbitals was required. Spin-resolved investigations of simple Pc molecules on ferromagnetic substrates by SP-STM showed that the spin structure of molecule–substrate hybrid systems could indeed be mapped with high spatial and energy resolution.

The understanding and control of the magnetism at the electrode interfaces is crucial for the design of SMM-based devices [108, 246]. Although STM is a

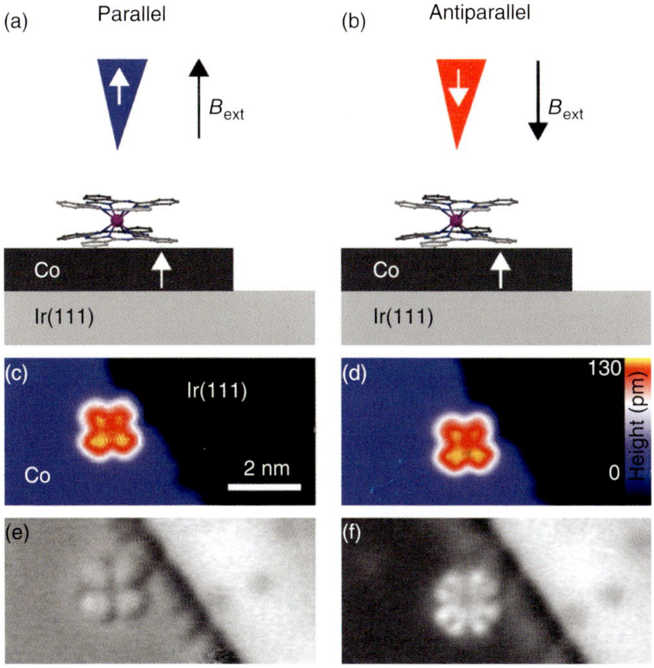

Figure 8.16 Schematic diagram of SP-STM experiments as performed [245]. To obtain spin sensitivity, the magnetization direction of the tip is aligned either parallel (blue/dark) or antiparallel (red/light grey) by applying an external magnetic field of $B = \pm 1\,T$. The Co nanostructure is a hard magnet with a magnetization that does not rotate with the field applied here. Because of the tunnel magnetoresistance effect, the tunnelling current depends on the relative alignment of tip and sample magnetization directions, thereby giving spin sensitivity. To isolate the spin-polarized component, signals from measurements with parallel (a) and antiparallel (b) alignments of tip and sample magnetization are subtracted from each other. (c–f) Example of a spin-resolved measurement: SP-STM topographs (c, parallel and d, antiparallel) and spin-resolved maps of differential conductance (e, parallel and f, antiparallel) taken at $U = -0.5\,V$. The white scale bar corresponds to 2 nm. The topographs (c, d) show little difference, the maps of spin-resolved differential conductance (e, f) do: the Co island appears brighter for parallel alignment of tip and sample magnetization. Not only the Co island, but also the $TbPc_2$ molecule shows magnetic contrast, that is, an eight-lobe structure for antiparallel and a cross-shape for parallel alignment.

powerful tool for understanding the basic mechanism of the magnetic interaction between a molecule and a substrate, it is not a device fabrication platform. In contrast, two- and three-terminal junctions incorporating magnetic molecules are closer to real devices since both the spin and charging state of the molecule can be altered. These are usually fabricated by combining breaking junction type technologies with wet chemistry and, as such, one does not need any longer UHV growth conditions [247]. However, when moving away from planar molecules and UHV, other problems emerge. A rather fundamental one is the fact that, although many classes of magnetic molecules can be chemically synthesized, most of them are extremely fragile away from a solution and often react on a metallic surface. Furthermore, some molecules can lose their magnetic moment entirely by coupling with the electrodes, even if they remain intact. This creates a substantial uncertainty because it is difficult to establish whether the molecule entering a device is the same one that was intentionally designed for that device. Therefore, the primary goal is to construct robust magnetic molecules that can survive on surfaces and preserve both their spin-state and most of the magneto-crystalline anisotropy [36]. Last but not least, it is also important to externally control the magnetization of a molecular magnet by tailoring magnetic interactions between the molecules and the supporting substrates.

8.5 Molecular Spintronic Devices on the Base of $[TbPc_2]^0$ SIMs

The integration of magnetic materials into nanoelectronic devices enables the use of the electron spin, as well as its charge, for carrying information. This new paradigm in information processing devices has been called *spintronics* in analogy to electronics [248]. The principal trend for the development of spintronic devices is to miniaturize components and to seek improved and novel performance; therefore, using SMMs as electronic components is a powerful new direction in the science and technology of nano-scale systems. SMMs, and in particular smaller SIMs, belong to a size domain (that of a few nanometres) where the quantum and the classical regimes meet, and their properties are potentially suitable for applications in electron transport, data storage, spintronics and quantum-computing technologies [225].

Various designs for spintronic devices were proposed over the past decade [6]. Different possible connectors for bridging between the molecule-based nano-scale and the electronic circuit-based micro-scale (nanowires, CNTs, quantum dots, molecules) have been considered. Among these, the CNT stands out due to its unique structural (in particular their high aspect ratio combining few nanometre-diameters with micrometer-lengths), mechanical (high Young modulus) and electronic properties (variable choice between semiconducting and metallic tubes). Additionally, the $TbPc_2$ molecule grafted onto CNTs should be orientated with the Pc rings parallel to the nanotube, which is in accordance with the supramolecular design maximizing the $\pi-\pi$ interactions between the Pc ring and the nanotube [217]. Several designs of supramolecular quantum spintronic devices have been recently proposed based on three main experimental geometries (Figure 8.17), where a CNT is used in an indirect coupling scheme to probe individual molecular spins via different coupling mechanisms (magnetic

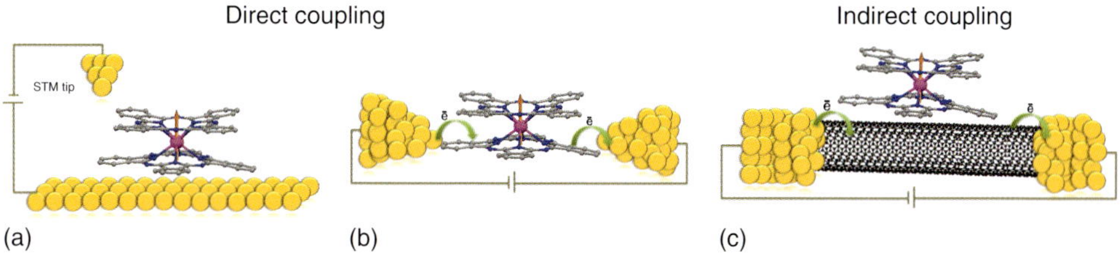

Figure 8.17 The main experimental geometries [6] in molecular spintronics via *Direct coupling*: (a) asymmetric, using a conducting tip and a planar substrate and (b) symmetric, using molecular junction device and *Indirect coupling*: (c) double magnetic quantum dot (QD) scheme based on sp^2-carbon materials, such as SW-CNT, graphene.

flux, electronic or mechanical) [249]. A CNT-SQUID magnetometer, the first spintronic device, for instance is able to probe the magnetic flux emanating from the set-up of a SQUID with single-walled carbon nanotube (SW-CNT) Josephson junctions at cryogenic temperatures [250]. The CNT-SQUIDs should, in principle, be sensitive enough to act as a local magnetometer, which is very promising for the study of magnetization reversal of an individual magnetic particle or molecule placed on one of the two CNT Josephson junctions. The nanotube SQUID allows one to explore two important problems in physics: one can use the SQUID as a magnetometer to measure the total spin of a magnetic molecule and understand its dependence on the field, and can also observe the π-phase shift in the superconducting current produced by the spin of a single electron sitting in the nanotube. In two quite different ways, one familiar, the other more subtle, the nanoSQUID enables spins to be counted [251]. However, no magnetism of SMMs has been determined with nanoSQUID so far.

In this context, in 2008, Bogani and Wernsdorfer [6] already proposed three basic molecular schemes of spintronic devices, which are based on the results obtained from Mn_{12} [252], $Co(TerPy)_2$ [253] and divanadium (V_2) molecules [(N,N',N''-trimethyl-1,4,7-triazacyclononane)$_2$-$V_2(CN)_4(\mu\text{-}C_4N_4)$] [254]. However, the realization of spintronic devices was not feasible due to some drawbacks with these molecules, such as the low operational temperatures, the decomposition of the molecule upon heating or in vacuum, and so on. Towards this goal, the $TbPc_2$ molecule class has undoubtedly become an excellent candidate for the assembly of molecular spintronic devices. There are several reasons for this particular SIM to have been utilized. Firstly, the $TbPc_2$ SIMs have not only the magnetic moment carried by a single Tb(III) ion but also an $S = 1/2$ radical delocalized over the peripheral Pc ligand planes. Secondly, the flat shape of the molecules with a pronounced uniaxial magnetic anisotropy promotes a preferential lying-down geometry of absorption of the molecules on substrates. Thirdly, the unsubstituted pristine molecules are sublimable and can be thermally evaporable under UHV conditions [255]. Pyrene-substituted $TbPc^*_2$ SIMs

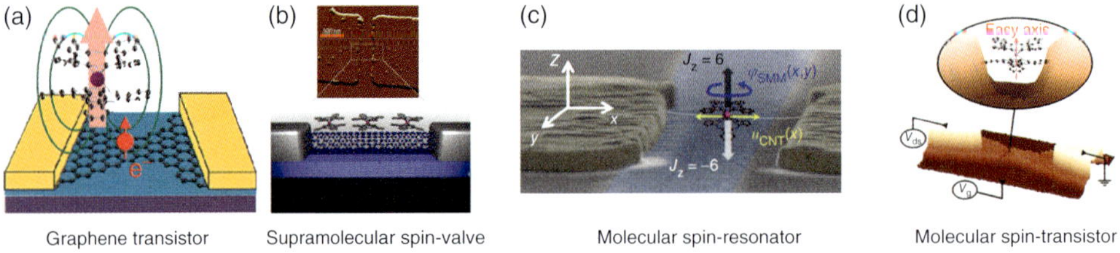

Figure 8.18 $TbPc_2$-based molecular spintronic devices. (a) Graphene nanotransistor with sensitivity at the single-molecule level [39, 256]. (b) Scheme of the supramolecular spin valve architecture [217]. (c) Scheme of the carbon nanotube NEMS. The magnetization reversal of a $TbPc^*_2$ SIMM from a spin state $J_z = +6$ to $J_z = -6$ results in a rotation of the SIM in the *xy* plane. The rotation by an angle φSMM (*x*,*y*) generates a longitudinal stretching mode in the SW-CNT with an amplitude *u*(*x*) [249]. (d) Three-dimensional extrapolation of a scanning-electron microscope image showing the most favourable structure of the single molecule-based spin transistor [257].

remain structurally magnetically robust when attached out of solution through supramolecular $\pi-\pi$ interactions to sp^2-carbon materials, such as CNTs [108] and graphene [39]. That is to say, its peripheral radical can easily hybridize with the π-electrons of any form of sp^2-carbon without affecting the magnetic properties of the Tb(III) ion [233, 255], but coupling the magnetically active Tb(III) ion (see following text) close to the conduction channel of the respective substrate of electrodes, which is a very important prerequisite for its application in spintronic nanodevices.

In recent years these molecular schemes have been experimentally achieved by grafting of different kinds of $TbPc_2$ molecules onto CNT-based materials. The fascinating experimental results obtained from these set-ups open up a bright new world of nanospintronics [6, 7]. In the following we provide a detailed overview of four different schemes of molecular spintronic devices including spin valve, spin transistor and spin resonator (Figure 8.18). In addition, we have also summarized electronic transport characteristics of these three types of devices in Figures 8.19–8.21.

8.5.1
Graphene Transistor

Graphene is one type of sp^2 carbon substrates that has been probed in the fabrication of spintronic nanodevices. Candini *et al.* [256] has presented the design and the realization of a novel hybrid device made by a graphene nanoconstriction decorated with pyrene-substituted $TbPc^*_2$ molecules, to electrically detect the magnetization reversal of the molecules in proximity with graphene (Figure 8.18a). An identical magnetic hysteresis was observed in the magnetoconductance of graphene devices. The emergence of hysteresis in the

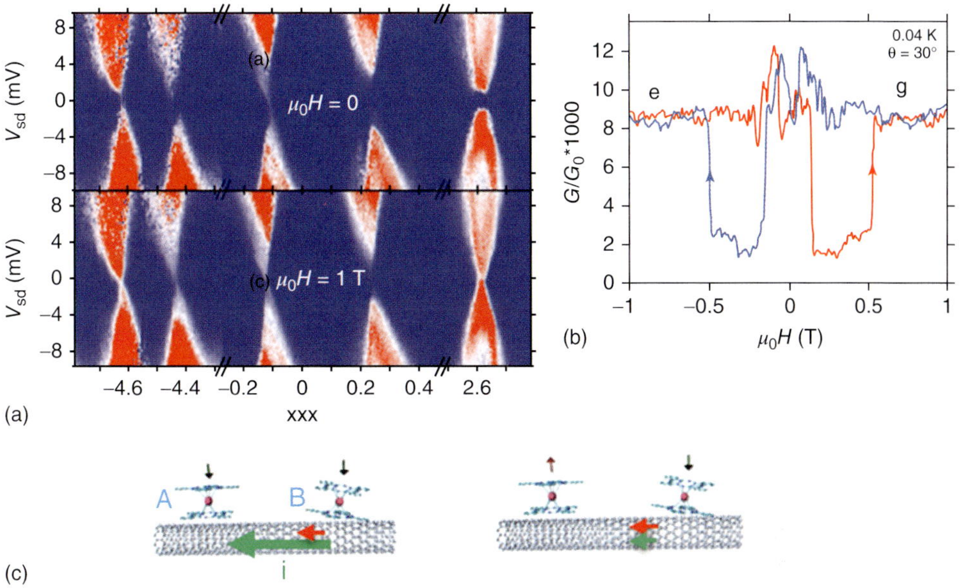

Figure 8.19 Supramolecular spin valve [217]. (a) Coulomb maps of the differential conductions as a function of source–rain voltage V_{sd} and backgate voltage V_g without magnetic field and under a magnetic field of 1 T. A gap Δ is induced at the degeneracy point when all the magnetic moments are randomly oriented (c, right). Magnetic molecules behave as magnetic impurity centres. The gap is closed under applied field, and the standard degeneracy points are recovered. It corresponds to the situation where all molecules are polarized in the same direction (c, left). (b) Zero bias conductance measured as a function of the magnetic field under increasing field (−1 to +1 T, red curve) and under decreasing field (blue curve). The conductance jumps around zero-field are attributed to molecules experiencing quantum tunnelling due to tunnel splitting. The last jumps around ±500 mT are attributed to a direct relaxation process of a single molecule. When all the magnetic momenta are parallel, only one spin carrier can be backscattered, whereas when the magnetic moments are antiparallel, both spin carriers can be backscattered. (c) Scheme of the scattering mechanism involving two $TbPc_2$*-SIMM-molecules (A and B) grafted on a SW-CNT. The molecules play the role of spin-polarizer and spin-analyser. For parallel spin alignment (left), one of the spin orientations is back scattered. With increasing magnetic field, molecule A switches first (right), rendering an anti-parallel configuration of the spin valve with lowest conductance, that is, the electrons in both spin channels are back scattered. When molecule B switches, a parallel configuration is recovered with high conductance (hexyl and pyrenyl groups are omitted for clarity).

magnetoconductivity indicates the presence of 'external' magnetic moments coupled with the graphene device, and it is a signature that the transport can be effectively tuned by the deposition of the $TbPc^*_2$ SIMs. These results depict the behaviour of multiple-field-effect nanotransistors with sensitivity at the single-molecule level.

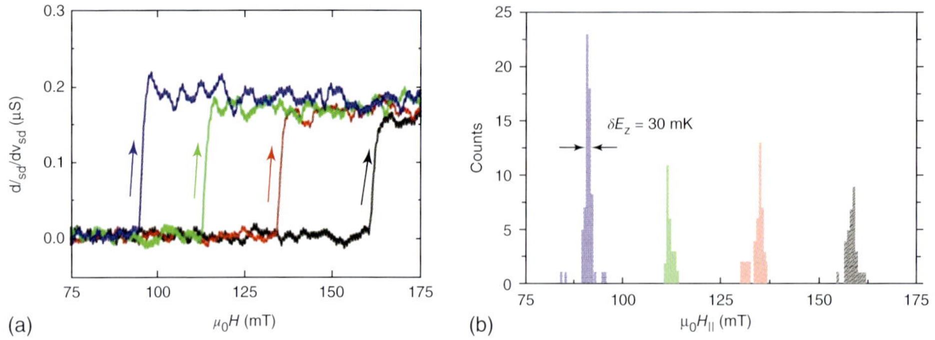

Figure 8.20 TbPc*$_2$ SIM and NEMS based resonator [258]. (a) The magnetization reversal of the Tb^{3+} ion induced by sweeping the magnetic field parallel to the easy axis of the SMM, resulted in an abrupt increase in the differential conductance through the SW-CNT of NEMS. The switching field depends on the nuclear spin state occupied before magnetization reversal. (b) Histogram of the switching fields obtained for 200 field sweeps with a zero transverse magnetic field component for one sweep direction and a sweep rate of 50 mT s^{-1}. Four dominant switching events were observed at 88, 112, 137 and 160 mT corresponding to the switching fields for the +3/2, +1/2, −1/2 and −3/2 nuclear spin states, respectively.

8.5.2 Supramolecular Spin Valve

A molecular spin valve contains nonmagnetic electrodes connected by a SW-CNT and implies at least two magnetically active molecules exhibiting an intrinsic Ising-type anisotropy. Supramolecular spin valves change their electrical resistance for different mutual alignments of the magnetizations of the different molecules, analogous to a polarizer–analyser set-up. Conventional spin valves consist of a diamagnetic molecule in between two magnetic leads, which can be metallic or semiconducting [6]. The first experiments sandwiched a C_{60} fullerene between Ni electrodes, and exhibited a very large negative magnetoresistance effect [259]. A standard GMR spin valve is an electronic device in which two conducting magnetic layers are separated by a nonmagnetic layer. In order to integrate and conduct the two magnetic layers, sp^2 carbon substrates such as CNTs are of particular interest because of their unique conductivity properties and the expected long spin coherence lifetimes and lengths. The active layers are usually classical magnets, but in recent years an alternative experimental set-up proposes to use quantum nanomagnets. In view of this point, the $TbPc_2$ SIM represents the most promising system for this purpose.

A supramolecular spin valve is a multi-terminal device, which consists of quantum magnets $TbPc_2$ and SW-CNT components (Figure 8.18b) [36, 217]. The grafting was achieved by anchoring modified $TbPc_2^*$ SIMs on the sidewall of a chemical vapour deposition (CVD) grown SW-CNT [216]. To increase the non-invasive

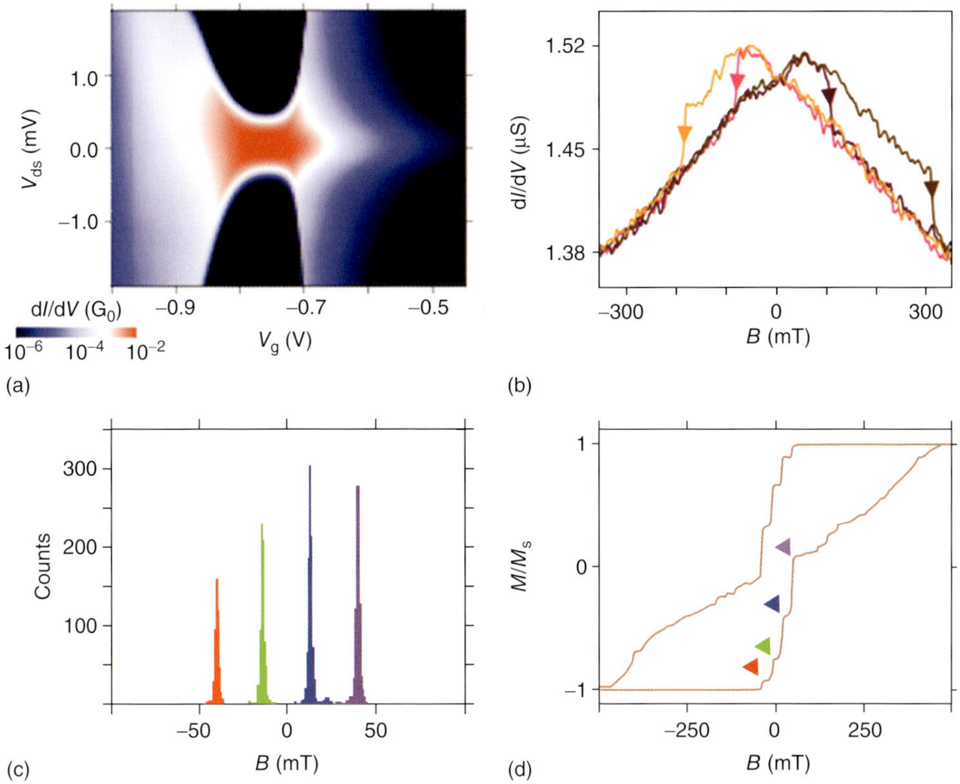

Figure 8.21 Molecular spin transistor [257]. (a) Stability diagram exhibiting the differential conductance d*I*/d*V* in units of the quantum of conductance G_0 as a function of the gate voltage V_g and of the bias voltage V_{ds}. (b) d*I*/d*V* measurements for a given working point ($V_g = -0.9$ V; $V_{ds} = 0$ V) as a function of the magnetic field *B*, the arrows indicating the sweeping direction. Abrupt steps in the differential conductance attributed to the reversal of the magnetization are clearly visible for both traces and retraces of *B* showing a clear hysteresis in the d*I*/d*V* characteristic. (c) Histogram of the step positions in magnetic field for the trace showing four preferential positions for the reversal to occur corresponding to QTM events. This measurement can be performed for both trace and retrace on a larger magnetic field range and integrated in order to reconstruct the hysteresis of a single $TbPc_2$-SMM. (d) Reconstructed hysteresis loop representing the normalized magnetization of the single $TbPc_2$-SIMM as function of the magnetic field. The four arrows on the trace curve spot the characteristic steps associated to QTM.

attachment to SW-CNTs at very low concentrations, Ruben and coworkers [108] have modified the Pc ligand by introducing one pyrene group and six hexyl groups into one of the two Pc rings. The pyrene arm and the six hexyl groups added to one of the Pc planes are known for an attractive van der Waals interaction with sp^2-carbon and further enhance the grafting efficiency. Moreover, the hexyl groups induce a steric hindrance in the molecule which prevents the reaggregation of the $TbPc_2$ in a solution and the formation of $TbPc_2$ clusters on the sidewall of the CNT.

Ruben and Wernsdorfer [217] demonstrated that two SIMs, coupled to the CNT via $\pi-\pi$ interactions, act as spin polarizer and analyser for the conduction electrons in the CNT channel. Mediated by exchange interactions, the magnetic moment of each molecule induces a localized spin-polarized dot in the CNT quantum dot, which can be controlled by a magnetic field (Figure 8.19). At large negative transverse fields, both molecular spins are oriented in parallel to each other and the SW-CNT quantum dot is in a high conductance state. Upon increasing the magnetic field, one molecular spin is reversed by quantum tunnelling of magnetization close to zero field, resulting in an antiparallel spin orientation and a current blockade in the quantum dot. By further increasing the field, the uniaxial spin of a second (or further) $TbPc_2^*$ SIM is reversed by a direct transition, restoring a parallel spin orientation and the high conductance regime in the quantum dot. After reversing the sweep direction, one obtains the characteristic butterfly hysteresis loop of a spin valve device with a magnetoresistance ratio up to 300% at temperatures <1 K. It was also shown that the magnetic anisotropy of the above-described spin valve effect is in very good agreement with the uniaxial anisotropy of the Tb(III) spin and therefore provides a magnetic fingerprint for the $TbPc_2^*$ SIM. A detailed description of the mechanism, as is known so far, can be found in Ref. [249]. Furthermore, the Landau–Zener (LZ) quantum tunnelling probability was studied as a function of field sweep rate, establishing a good agreement with the LZ equation and yielding the tunnel splitting Δ. It was found that Δ increased linearly as a function of the transverse field. These studies are an essential prerequisite for the coherent manipulation of a single nuclear spin in $TbPc_2$ (see above) [260].

This example demonstrates the functionality of a supramolecular spin valve without magnetic leads, and opens a pathway towards the design of operable molecular spintronic devices. Most importantly, Ruben and Wernsdorfer used this spin valve effect to confirm the strong uniaxial anisotropy and the superparamagnetic blocking temperature ($T_B \sim 1$ K) of isolated $TbPc_2^*$ SIMs. For the first time, the strength of exchange interaction between the different SMMs of the molecular spin valve geometry could be determined. The results introduce a new design for operable molecular spintronic devices using the quantum effects of individual SMMs [216].

8.5.3 Molecular Spin Resonator

Recent progress in nanofabrication has shown that SW-CNTs have become an essential building block for nanoelectromechanical systems (NEMSs) and

SW-CNT NEMS can be used for ultrasensitive mass sensing or as magnetic torque detectors for single spin systems [261]. As described above, chemical engineering allows the integration of SMMs and SIMs into nanoscale devices such as electromigrated junctions [257] or CNT transistors [216]. Recently, Wernsdorfer and Ruben [249] have realized a spin-resonator system with a pyrene-substituted $TbPc_2^*$ SIM rigidly grafted to a suspended SW-CNT (Figure 8.18c). They studied the magnetization reversal of this system and provided the first experimental evidence for a strong spin–phonon coupling between a single molecule spin and a quantized, high-Q mechanical motion of a SW-CNT resonator. It has been predicted that a coupling of this magnitude would induce strong nonlinearities in the mechanical motion of a CNT, which can, for instance, be used to enhance the sensitivity of SW-CNT-based magnetometers [261]. A sensitivity of a few Bohr magneton (μB) is to be expected. Such a strong coupling between a quantum CNT NEMS and a single electron spin would enable coherent spin manipulation and the implementation of quantum entangled states. Moreover, the strong spin–phonon coupling would enable the electric manipulation of a single molecular spin by microwave irradiation [262]. Likewise, a strong spin–phonon interaction would allow the spin-based detection of the mechanical motion of the nanotube or the ground-state cooling of the resonator by manipulating the molecular spin via electron spin resonance. Ultimately, these results open the path to the quantum entanglement of a single spin and a quantized mechanical motion [258].

In 2013, Wernsdorfer and Ruben [258] further studied the grafting process of the spin-resonator system. They concluded that grafting pyrene-substituted $TbPc_2^*$ SIMs from low concentration $TbPc_2$ solutions onto CNT NEMS conserves both the magnetic properties of the $TbPc_2$ part (uniaxial magnetic anisotropy, QTM, etc.) and the mechanical properties of the resonator (high quality factors, nonlinearities, single-electron tunnelling (SET) modulation, etc.) and is therefore a key component for the fabrication of CNT-based supramolecular spintronic device. The noninvasive nature of the supramolecular grafting process subsequently allows for a study of the SIM magnetization reversal via the influence on the nanotube's mechanical oscillation at cryogenic temperatures. They found evidence that the mechanical bistability of a CNT NEMS in a nonlinear regime can be used to probe the magnetization reversal of the $TbPc_2$. The detection principle and the sensitivity of such a torque magnetometer have also been described. Furthermore, they have provided an overview on the fingerprint-like characteristics of a pristine CNT NEMS. In particular, they have addressed the modulation of the resonance frequency and the nonlinearity induced by an electric and magnetic field (Figure 8.20). Those results provide the first experimental evidence for the detection of a single spin by a mechanical degree of freedom on a molecular level and the first implementation of a molecular magnetometer based on a CNT NEMS [263].

8.5.4 Molecular Spin Transistor

A spin transistor, a magnetically sensitive transistor, represents another type of spintronic device. It was originally proposed in 1990 and is still being developed [264]. A molecular spin-transistor is a magnetic molecule attached between two nonmagnetic electrodes [6]. So far, only few atoms on surfaces have been probed in this way, mainly by STM, revealing interesting Kondo effects [265, 266] and single-atom magnetic anisotropies [267].

Transport measurements through a single $TbPc_2$ SIM have been taken previously by STS experiments, for which electronic transport occurred through the Pc ligands and exhibited Coulomb blockade and Kondo effects depending on its charge state (spin states $S = 0$ or $1/2$). However, no signature of the magnetic moment carried by the Tb(III) ion was observed in these experiments [268]. To detect the reversal of the magnetic moment, and to produce the electronic read-out of the Tb(III) nuclear spin states, Ruben and Balestro inserted the $TbPc_2$ SIM directly into a gold junction obtained by the electromigration technique. In this way, a three-terminal molecular spin transistor is built up, in which the individual $TbPc_2$ SIM bridges the gap between two nonmagnetic leads (Figure 8.18d). In such a transistor-like configuration, the electric current flows directly through the molecule, leading to strong coupling between the electrons and the magnetic core. This direct coupling thus enables a read-out of the magnetic properties of the molecule via a coupling to a spin cascade consisting of the three spin systems of $TbPc_2$ SIMs (see following text) [257].

Electronic transport studies under a variable magnetic field have revealed a hysteretic behaviour that is reminiscent of the stepped magnetic hysteresis due to resonant quantum tunnelling between hyperfine split energy levels. Indeed the experimental read-out procedure relies on the highly efficient detection of the QTM of the electronic magnetic moment at particular values of the magnetic field corresponding to the four avoided energy-level crossings, which is in perfect quantitative agreement with theoretical predictions (Figure 8.21). Remarkably, by recording the conductance as a function of time, the dynamics of the four different nuclear spin states were determined and energy-level lifetimes in the range of tens of seconds were found. This is an important result for future quantum information processing experiments, which opens the way to coherent manipulations of a single nuclear spin [257, 269]. Very recently, Balestro [270] further presented the electrical read-out of time-trajectories obtained from the isolated nuclear spin $I = 3/2$. The device detects the four different nuclear spin states of the Tb(III) ion with high fidelities of 95%, allowing the measurement of individual relaxation times (T_1) of several tens of seconds. The experiment is found to be in good agreement with quantum Monte Carlo simulations, suggesting that the relaxation times are limited by the current traversing the transistor, which creates the possibility of tuning T_1 electrically by means of the bias voltage. Thus, the possibility of addressing and detecting single nuclear spin states using the QTM of SMMs, in combination with the observation of long energy-level lifetimes, opens up a bright new world of nanospintronics with integrated memory, logics and possibly quantum logics.

Furthermore, because of the different energy-level spacing that originates from the nuclear quadrupole interaction term, this particular device is presumed to be an excellent candidate to perform the Grover algorithm. Indeed it has been proposed that the coherent manipulation of a nuclear spin $I = 3/2$ can be performed through a multiphoton process using this algorithm [271]. Therefore, the fabricated $TbPc_2$-SIM spin transistor realizes this algorithm and opens the way to the study of quantum computing at the single-molecule level [269].

Currently there has been enormous progress in the understanding of the electronic and magnetic properties of isolated molecular systems and in the fabrication of different molecular schemes, revealing intriguing new physics. In particular, the fundamental physics behind these devices remains largely unexplored. Although these works demonstrate convincingly that electronic transport measurements could be performed at the molecular level with reasonably high efficiency, there is a general lack of control over the magnetic response of the devices [248, 272]. As we know, the physical laws governing the functioning of any device with sizes below 10 nm change dramatically because energy levels are quantized and quantum effects tend to become dominant. At the same time, the classical – and currently used – Boolean logic can be no more valid at nm scale or, alternatively, other logics are possible and even more appropriate with quantum devices [248]. In addition, low yield in the device production, severe temperature and bias-dependence of the magneto-transport properties, and large resistances [272, 273] are serious limiting factors for future applications. Consequently, the utilization of those fabricated molecular-scaled devices will be expected to face challenges. However, the field of spintronics is still likely to be the focus of considerable attention in the near future.

8.6 Conclusion and Outlook

In this chapter, we have reviewed the family of $LnPc_2$ compounds with an emphasis on its terbium analogue, which covers their synthetic approaches, bulk magnetism, magnetism on different substrates and the fabrication towards different molecular schemes of spintronic devices.

The observation of the SIM behaviour of $TbPc_2$ and of $TbPc^*_2$ derivatives obtained by substituting the periphery of the phthalocyanine macrocycles combined with a considerable blocking temperature attracted much attention from both chemists and physicists. It was soon realized that these systems proved to be ideal testing grounds for theories of the coexistence of quantum and classical effects in magnets; in particular they provided evidence of quantum tunnelling of the magnetization and of oscillations of the tunnel splitting. Even more exciting, these $TbPc_2$ SIMs present a number of appealing features to be exploited at the single molecule level or in nanodevices, such as its evaporability and flat shape, and the conjugated structure of the phthalocyanine ligands particularly suitable for electron transport. In addition, because these $TbPc_2$ systems have very strong couplings between electron and nuclear spins, they exhibit enough particular characteristics that can be used as fingerprints for the presence of the molecule in the junctions or on the nanotubes. At this point, the $TbPc_2$ SIM offers a fundamental link between two novel disciplines of spintronics and molecular electronics. In fact, different molecular schemes of spintronic devices such as spin valve, spin transistor and spin resonator have been fabricated in recent years, and the experimental results have demonstrated the feasibility of the design of operable molecular spintronic devices. The success of $TbPc_2$ SIM utilized in the fabrication of nanodevices directs a new research field, namely molecular spintronics.

There are quite a few advantages for SIMs and SMMs, to become the next generation of technological devices. Most importantly, carefully designed SMMs offer the possibility to magnetically process information at a molecular level, making the development of SMM-based quantum computation devices possible. Secondly, the weak SO and hyperfine interactions of organic molecules could help preserve electron-spin polarization over a much greater distance than in conventional semiconductor systems. Thirdly, because a spin-polarized current, passing through an SMM, should allow the reading and reversing of its magnetization, such devices can be a breakthrough for quantum information processing. Fourthly, as molecular spintronic devices require only small amounts of power to operate, sustainable devices based on SMMs will be easily achievable. Finally, specific functions (for example, switchability with light and electric field) could be directly integrated into the molecule [6]. However, in the perspective of molecular spintronics, not every SMM molecule can be applied into devices. One of the grand challenges in this field is still to design and to synthesize efficient SMMs that function at temperatures likely to be of practical use, or which show physics that goes beyond what can be achieved with classical magnets. The observation of magnetic remanence of surface-coupled $TbPc_2$ up to $T = 100\,K$ consists of a step in this discretion [240]. Furthermore, the stepwise extension of the SIM concept to first binuclear and then later again polynuclear SMMs, but now under strict control of the anisotropy vectors as well as under simultaneous engineering of the conduction channels through the molecule remains an enormous challenge for the chemist of the future [274]. Moreover, depositing and addressing individual molecules of SMMs onto surfaces have only been explored with very few examples so far. The experimental observations usually indicate that most of the molecules are extremely fragile away from solution and often react on a metallic surface. Some molecules can lose their magnetic moment entirely by coupling with the electrodes, even if they remain intact. This creates a substantial uncertainty because it is difficult to establish whether the molecule entering a device is the same one that was intentionally designed for that device. More detailed current challenges have already been outlined in the section on surface magnetism and spintronic devices.

In summary, SMM-based molecular spintronics is an emerging multidisciplinary field full of fascination for theoreticians, experimental physicists and chemists alike. Chemists are trying to design new classes of SMMs with enhanced properties, physicists are fascinated by the wealth of different new magnets on which they can measure new properties and test theories and theoreticians are trying to understand the spin and electronic response of the devices. Apart from its fundamental importance, these devices based on $TbPc_2$ SMM are exciting experimental playgrounds for testing quantum effects at the single-molecule level and could eventually lead to important consequences for applications. The field of molecular spintronics has currently been strongly addressed as a new promising science towards information processing technology, which might produce a breakthrough and diffuse new technological products in the market.

Abbreviations

15C5	crown ether
acac	acetylacetonato
DBU	1,8-diazabicyclo[5.4.0]undecen-7-ene
DMF	*N,N*-dimethylformamide
MeTBODC	1-methyltetrabenzooctadehydrocorrin
Nc	2,3-naphthalocyanine
NTBPP	N-confused 5,10,15,20-tetrakis[(4-*tert*-butyl) phenyl]porphyrinate
(CH_3)NTBPP	methylated N-confused 5,10,15,20-tetrakis[(4-Bu^t) phenyl] porphyrinate
OAc	acetate ion
OEP	dianion of 2,3,7,8,12,13,17,18-octaethyl porphyrin
Por	dianion of porphyrin
POSS	polyhedral oligomeric silsesquioxane
Pc	dianion of phthalocyanine
Pc*	dianion of phthalocyanine, substituted
TBA	Tetrabutylammonium anion
TBPP	dianion of 5,10,15,20-tetrakis[(4-*tert*-butyl)phenyl]-porphyrin
TCB	1,2,4-trichlorobenzene
TPP	dianion of 5,10,15,20-tetraphenylporphine
UHV	ultra-high vacuum
$phth^1$	*N,N,N,N*-tetrapentadecyl-29H,31H-2,3,9,10,16,17,23,24-phthalocyaninatotetradicarboximide

$phth^2$	*N*,*N*,*N*,*N*-tetrapropyl-29H,31H-2,3,9,10,16,17,23,24-phthalocyaninatotetradicarboximide
$phth^3$	*N*,*N*,*N*,*N*-tetra((S)-methyl(phenyl)methyl)-29H,31H-2,3,9,10,16,17,23,24-phthalocyaninatotetradicarboximide
ODd	2,3,9,10,16,17,23,24-octakis(dodecoxy)-phthalocyanine
S-DOP	(*S*)-1-bromo-2-dodecyloxypropane
IPD	4,5-isopropylidenedioxyphthalonitrile
TClPP	meso-tetrakis-(4-chlorophenyl)porphyrinate
$F_{64}PcH_2$	1,4,8,11,15,18,22,25-octakis-fluoro-2,3,9,10,16,17,23,24-octakis-perfluoro (isopropyl)phthalocyanine
LSMO	$La_{0.7}Sr_{0.3}MnO_3$, a highly polarized manganite

References

1. Kahn, O. (1993) *Molecular Magnetism*, VCH Publishers, New York.
2. Miller, J.S. and Epstein, A.J. (1994) *Angew. Chem., Int. Ed. Engl.*, **33**, 385–415.
3. Dei, A. and Gatteschi, D. (2011) *Angew. Chem. Int. Ed.*, **50**, 11852–11858.
4. Sorace, L., Benelli, C. and Gatteschi, D. (2011) *Chem. Soc. Rev.*, **40**, 3092–3104.
5. Wernsdorfer, W., Murugesu, M. and Christou, G. (2006) *Phys. Rev. Lett.*, **96**, 057208(4).
6. Bogani, L. and Wernsdorfer, W. (2008) *Nat. Mater.*, **7**, 179–186.
7. Jiang, S., Goß, K., Cervetti, C. and Bogani, L. (2012) *Sci. China Chem.*, **55**, 867–882.
8. Winpenny, R.E.P. (2008) *Angew. Chem. Int. Ed.*, **47**, 7992–7994.
9. Ishikawa, N., Sugita, M. and Wernsdorfer, W. (2005) *Angew. Chem. Int. Ed.*, **44**, 2931–2935.
10. Aromi, G. and Brechin, E.K. (2006) *Struct. Bond.* doi: 10.1007/430_022
11. Caneschi, A., Gatteschi, D. and Sessoli, R. (1991) *J. Am. Chem. Soc.*, **113**, 5873–5874.
12. Sessoli, R., Gatteschi, D., Caneschi, A. and Novak, M.A. (1993) *Nature*, **365**, 141–143.
13. Ako, A.M., Hewitt, I.J., Mereacre, V., Clérac, R., Wernsdorfer, W., Anson, C.E. and Powell, A.K. (2006) *Angew. Chem.*, **118**, 5048–5051.
14. Bhunia, A., Gamer, M.T., Ungur, L., Chibotaru, L.F., Powell, A.K., Lan, Y., Roesky, P.W., Menges, F., Riehn, C. and Niedner-Schatteburg, G. (2012) *Inorg. Chem.*, **51**, 9589–9597.
15. Oshio, H. and Nakano, M. (2005) *Chem. Eur. J.*, **11**, 5178–5185.
16. Waldmann, O. (2007) *Inorg. Chem.*, **46**, 10035–10037.
17. Ruiz, E., Cirera, J., Cano, J., Alvarez, S., Loose, C. and Kortus, J. (2008) *Chem. Commun.*, **2**, 52–54.
18. Ishikawa, N., Sugita, M., Ishikawa, T. and Koshihara, S. (2004) *J. Phys. Chem. B*, **108**, 11265–11271.
19. Leuenberger, M.N. and Loss, D. (2001) *Nature*, **410**, 789–793.
20. Romeike, C., Wegewijs, M., Hofstetter, W. and Schoeller, H. (2006) *Phys. Rev. Lett.*, **96**, 196601(4).
21. Rinehart, J.D. and Long, J.R. (2011) *Chem. Sci.*, **2**, 2078–2085.
22. Guo, Y.-N., Xu, G.-F., Guo, Y. and Tang, J. (2011) *Dalton Trans.*, **40**, 9953–9963.
23. Habib, F. and Murugesu, M. (2013) *Chem. Soc. Rev.*, **42**, 3278–3288.
24. Woodru, D.N., Winpenny, R.E.P. and Layfield, R.A. (2013) *Chem. Rev.*, **113**, 5110–5148.
25. Sessoli, R. and Powell, A.K. (2009) *Coord. Chem. Rev.*, **253**, 2328–2341.

26. Abbas, G., Lan, Y., Mereacre, V., Wernsdorfer, W., Clérac, R., Buth, G., Sougrati, M.T., Grandjean, F., Long, G.J., Anson, C.E. *et al.* (2009) *Inorg. Chem.*, **48**, 9345–9355.
27. Mereacre, V., Ako, A.M., Clérac, R., Wernsdorfer, W., Hewitt, I.J., Anson, C.E. and Powell, A.K. (2008) *Chem. Eur. J.*, **14**, 3577–3584.
28. Rinehart, J.D., Fang, M., Evans, W.J. and Long, J.R. (2011) *Nat. Chem.*, **3**, 538–542.
29. Habib, F., Lin, P., Long, J., Korobkov, I., Wernsdorfer, W. and Murugesu, M. (2011) *J. Am. Chem. Soc.*, **133**, 8830–8833.
30. Xiong, Z.H., Wu, D., Vardeny, Z.V. and Shi, J. (2004) *Nature*, **427**, 821–824.
31. Dediu, V.A., Hueso, L.E., Bergenti, I. and Taliani, C. (2009) *Nat. Mater.*, **8**, 707–716.
32. Dediu, V., Murgia, M., Matacotta, F.C., Taliani, C. and Barbanera, S. (2002) *Solid State Commun.*, **122**, 181–184.
33. Wang, F., Yang, C., Vardeny, Z. and Li, X. (2007) *Phys. Rev. B*, **75**, 245324.
34. Dediu, V., Hueso, L., Bergenti, I., Riminucci, A., Borgatti, F., Graziosi, P., Newby, C., Casoli, F., De Jong, M., Taliani, C. *et al.* (2008) *Phys. Rev. B*, **78**, 115203.
35. Xu, W., Szulczewski, G.J., LeClair, P., Navarrete, I., Schad, R., Miao, G., Guo, H. and Gupta, A. (2007) *Appl. Phys. Lett.*, **90**, 072506.
36. Sanvito, S. (2011) *Chem. Soc. Rev.*, **40**, 3336–3355.
37. Thejo Kalyani, N. and Dhoble, S.J. (2012) *Renew. Sustain. Energy Rev.*, **16**, 2696–2723.
38. Sasabe, H. and Kido, J. (2013) *J. Mater. Chem. C*, **1**, 1699–1707.
39. Lopes, M., Candini, A., Urdampilleta, M., Reserbat-Plantey, A., Bellini, V., Klyatskaya, S., Marty, L., Ruben, M., Affronte, M., Wernsdorfer, W. *et al.* (2010) *ACS Nano*, **4**, 7531–7537.
40. Fert, A. (2008) *Angew. Chem. Int. Ed.*, **47**, 5956–5967.
41. Shirai, H., Kimura, M. (2003) The Porphyrin Handbook, (ed. K.M. Kadish), Elsevier Science, San Diego, pp. 151–178.
42. Shirai, H. and Hojo, N. (1987) *Functional Monomers and Polymers*, (ed. K. Takemoto, Y. Inaki and R. Ottenbrite) Marcel Dekkers Inc., New York, 49–148.
43. Nalwa, H.S. and Shirk, J.S. (1993) *In Phthalocyanines, Properties and Applications*, VCH Publishers, New York.
44. Huang, W., Xiang, H., Gong, Q., Huang, Y., Huang, C. and Jiang, J. (2003) *Chem. Phys. Lett.*, **374**, 639–644.
45. Casilli, S., De Luca, M., Apetrei, C., Parra, V., Arrieta, Á.A., Valli, L., Jiang, J., Rodríguez-Méndez, M.L. and De Saja, J.A. (2005) *Appl. Surf. Sci.*, **246**, 304–312.
46. McKeown, N.B. (2004) *Science of Synthesis: Product Class 9: Phthalocyanines and Related Compounds 613-640*, Georg Thieme Verlag.
47. Ishikawa, N. and Kaizu, Y. (2002) *Coord. Chem. Rev.*, **226**, 93–101.
48. Jiang, J., Kiu, W. and Arnold, D.P. (2003) *J. Porphyrins Phthalocyanines*, **7**, 459–473.
49. Pushkarev, V.E., Tomilova, L.G. and Tomilov, Y.V. (2008) *Russ. Chem. Rev.*, **77**, 875–907.
50. de la Torre, G., Claessens, C.G. and Torres, T. (2007) *Chem. Commun.*, 2000–2015.
51. Nemykin, V.N. and Lukyanets, E.A. (2010), *ARKIVOC*, Arkat USA, Inc., pp. 136–208.
52. Linstead, R.P.J. (1934) *Chem. Soc.*, 1022.
53. Dent, C.E., Linstead, R.P. and Lowe, A.R.J. (1934) *Chem. Soc.*, 1033.
54. Robertson, J.M., Linstead, R.P. and Dent, C.E. (1935) *Nature*, **135**, 506.
55. Kirin, I.S., Moskalev, P.N. and Makashev, Y.A. (1965) *Zh. Neorg. Khim.*, **10**, 1951–1953.
56. Kirin, I.S., Moskalev, P.N. and Makashev, Y.A. (1967) *Zh. Neorg. Khim.*, **12**, 707–709.
57. Moskalev, P.N., Shapkin, G.N. and Darovskih, A.N. (1979) *Zh. Neorg. Khim.*, **24**, 340–346.

58. Moskalev, P.N. and Kirin, I.S. (1971) *Zh. Neorg. Khim.*, **16**, 110–115.
59. Markovitsi, D., Tran-Thi, T.-H., Even, R. and Simon, J. (1987) *Chem. Phys. Lett.*, **137**, 107–112.
60. Piechocki, C., Simon, J., Andr, J., Guillon, D., Petit, P., Skoulios, A. and Weber, P. (1985) *Chem. Phys. Lett.*, **122**, 124–128.
61. Komatsu, T., Ohta, K., Watanabe, T., Ikemoto, H., Fujimotob, T. and Yamamotob, L. (1994) *J. Mater. Chem.*, **4**, 537–540.
62. Binnemans, K., Sleven, J., De Feyter, S., De Schryver, F.C., Donnio, B. and Guillon, D. (2003) *Chem. Mater.*, **15**, 3930–3938.
63. Belarbi, Z., Sirlin, C., Simon, J. and Andre, J.-J. (1989) *J. Phys. Chem.*, **93**, 8105–8110.
64. Warman, J.M., Hasebe, H., Naito, R., Ohta, K. and van de Craats, A.M. (1997) *J. Phys. Chem. B*, **101**, 9224–9232.
65. Sleven, J., Gorller-Walrand, C. and Binnemans, K. (2001) *Mater. Sci. Eng. C*, **18**, 229–238.
66. Naito, R., Ohta, K. and Shirai, H. (2001) *J. Porphyrins Phthalocyanines*, **5**, 44–51.
67. Maeda, F., Hatsusaka, K., Ohta, K. and Kimura, M. (2003) *J. Mater. Chem.*, **13**, 243–251.
68. Liu, W., Jiang, J., Pan, N. and Arnold, D.P. (2000) *Inorg. Chim. Acta*, **310**, 140–146.
69. Jiang, J., Liu, R.C.W., Mak, T.C.W., Chan, T.W.D. and Ng, D.K.P. (1997) *Polyhedron*, **16**, 515–520.
70. Ivanov, A.V.V., Svinareva, P.A.A., Zhukov, I.V.V., Tomilova, L.G.G. and Zefirov, N.S.S. (2006) *Russ. Chem. Bull. Int. Ed.*, **55**, 281–286.
71. Pushkarev, V.E.E., Ivanov, A.V.V., Zhukov, I.V.V., Shulishov, E.V.V. and Tomilov, Y.V.V. (2004) *Russ. Chem. Bull. Int. Ed.*, **53**, 554–560.
72. Ban, K., Nishizawa, K., Ohta, K., van de Craats, A.M., Warman, J.M., Yamamoto, I. and Shirai, H. (2001) *J. Mater. Chem.*, **11**, 321–331.
73. Yoshino, K., Lee, S.B., Sonoda, T., Kawagishi, H., Hidayat, R., Nakayama, K., Ozaki, M., Ban, K., Nishizawa, K., Ohta, K. *et al.* (2000) *J. Appl. Phys.*, **88**, 7137–7143.
74. Toupance, T., Bassoul, P., Mineau, L. and Simon, J. (1996) *J. Phys. Chem.*, **100**, 11704–11710.
75. Ceyhan, T., Özdağ, M.A., Salih, B., Erbil, M.K., Elmali, A., Özkaya, A.R. and Bekaroğlu, Ö. (2008) *Eur. J. Inorg. Chem.*, **2008**, 4943–4950.
76. Yabaş, E., Sülü, M., Dumludağ, F., Özkaya, A.R., Salih, B. and Bekaroğlu, Ö. (2012) *Polyhedron*, **42**, 196–206.
77. Kandaz, M., Bilgiçli, A.T. and Altindal, A. (2010) *Synth. Met.*, **160**, 52–60.
78. Yaraşir, M., Kandaz, M., Koca, A. and Salih, B. (2006) *J. Porphyrins Phthalocyanines*, **10**, 1022–1034.
79. Lv, W., Zhu, P., Bian, Y., Ma, C., Zhang, X. and Jiang, J. (2010) *Inorg. Chem.*, **49**, 6628–6635.
80. Yonekura, T., Ohsaka, T., Kitamura, F. and Tokuda, K. (2005) *J. Porphyrins Phthalocyanines*, **9**, 54–58.
81. Waters, M., Moro, F., Krivokapic, I., McMaster, J. and van Slageren, J. (2012) *Dalton Trans.*, **41**, 1128–1130.
82. Passard, M., Blanc, J.P. and Maleysson, C. (1995) *Thin Solid Films*, **271**, 8–14.
83. Guyonj, F., Pondavenj, A., Guenot, P. and L'Her, M. (1994) *Inorg. Chem.*, **33**, 4787–4793.
84. Bouvet, M. and Simon, J. (1990) *Chem. Phys. Lett.*, **172**, 299–302.
85. Bouvet, M., Bassoul, P. and Simon, J. (1994) *Mol. Cryst. Liq. Cryst. Sci. Technol., Sect. A. Mol. Cryst. Liq. Cryst.*, **252**, 31–38.
86. Clerrnont, P. (1996) *Synth. Met.*, **80**, 25–28.
87. Ishikawa, N., Ohno, O. and Kaizu, Y. (1991) *Chem. Phys. Lett.*, **180**, 51–56.
88. Ishikawa, N., Ohno, O., Kaizu, Y. and Kobayashi, H. (1992) *J. Phys. Chem.*, **96**, 8832–8839.
89. Kogan, E.G., Ivanov, A.V., Tomilova, G. and Zefirov, N.S. (2002) *Mendeleev Commun.*, **12**, 54–55.
90. Branzoli, F., Carretta, P., Filibian, M., Zoppellaro, G., Graf, M.J.,

Galan-Mascaros, J.R., Fuhr, O., Brink, S. and Ruben, M. (2009) *J. Am. Chem. Soc.*, **131**, 4387–4396.

91. Jiang, J., Xie, J., Choi, M.T.M., Yan, Y., Sun, S. and Ng, D.K.P. (1999) *J. Porphyrins Phthalocyanines*, **3**, 322–328.
92. HERR, V.W. (1953) *Angew. Chem.*, **65**, 303–304.
93. L'Her, M., Cozien, Y. and Courtot-Coupez, J. (1983) *J. Electroanal. Chem.*, **157**, 183–187.
94. Lu, F., Cui, J. and Yan, X. (2006) *Spectrochim. Acta A Mol. Biomol. Spectrosc.*, **63**, 550–555.
95. Bao, M., Wang, R., Rintoul, L., Arnold, D.P. and Jiang, J. (2006) *Vib. Spectrosc.*, **40**, 47–54.
96. Lapkina, L.A., Gorbunova, Y.G., Nefedov, S.E. and Tsivadze, A.Y. (2003) *Russ. Chem. Bull. Int. Ed.*, **52**, 1633–1636.
97. Tomilova, L.G., Chernykh, E.V., Loffe, N.T. and Luk'yanets, E.A. (1983) *Zh. Obshch. Khim.*, **53**, 2594–2601.
98. Wang, R., Li, R., Bian, Y., Choi, C.-F., Ng, D.K.P., Dou, J., Wang, D., Zhu, P., Ma, C., Hartnell, R.D. *et al.* (2005) *Chem. Eur. J.*, **11**, 7351–7357.
99. Gürol, İ., Durmuş, M. and Ahsen, V. (2012) *J. Porphyrins Phthalocyanines*, **16**, 907–916.
100. Nekelson, F., Monobe, H. and Shimizu, Y. (2007) *Mol. Cryst. Liq. Cryst.*, **479**, 205/[1243]–211/[1249].
101. Nekelson, F., Monobe, H. and Shimizu, Y. (2006) *Chem. Commun.*, 3874–3876.
102. Gümrükçü, G., Özgür, M.Ü., Altindal, A., Özkaya, A.R., Salih, B. and Bekaroğlu, Ö. (2011) *Synth. Met.*, **161**, 112–123.
103. Gao, Y., Li, R., Dong, S., Bian, Y. and Jiang, J. (2010) *Dalton Trans.*, **39**, 1321–1327.
104. Zhang, H., Wang, R., Zhu, P., Lai, Z., Han, J., Choi, C.-F., Ng, D.K.P., Cui, X., Ma, C. and Jiang, J. (2004) *Inorg. Chem.*, **43**, 4740–4742.
105. Liu, Y., Shigehara, K., Hara, M. and Yamada, A. (1991) *J. Am. Chem. Soc.*, **113**, 440–443.
106. Jiang, J., Liu, W., Law, W.-F., Lin, J. and Ng, D.K.P. (1998) *Inorg. Chim. Acta*, **268**, 141–144.
107. Wang, R., Li, Y., Li, R., Cheng, D.Y.Y., Zhu, P., Ng, D.K.P., Bao, M., Cui, X., Kobayashi, N. and Jiang, J. (2005) *Inorg. Chem.*, **44**, 2114–2120.
108. Kyatskaya, S., Mascarós, J.R.G., Bogani, L., Hennrich, F., Kappes, M., Wernsdorfer, W. and Ruben, M. (2009) *J. Am. Chem. Soc.*, **131**, 15143–15151.
109. Pushkarev, V.E., Tolbin, A.Y., Borisova, N.E., Trashin, S.A. and Tomilova, L.G. (2010) *Eur. J. Inorg. Chem.*, **2010**, 5254–5262.
110. Pushkarev, V.E., Tolbin, A.Y., Zhurkin, F.E., Borisova, N.E., Trashin, S.A., Tomilova, L.G. and Zefirov, N.S. (2012) *Chem. Eur. J.*, **18**, 9046–9055.
111. Pushkarev, V.E., Tolbin, A.Y., Ryabova, A.V. and Tomilova, L.G. (2009) *Mendeleev Commun.*, **19**, 24–26.
112. Zhang, H., Lai, Z., Cui, X., Jiang, J. and Machida, K.-I. (2006) *J. Alloys Compd.*, **408-412**, 1041–1045.
113. Bian, Y., Wang, R., Jiang, J., Lee, C.-H., Wang, J. and Ng, D.K.P. (2003) *Chem. Commun.*, 1194–1195.
114. Lu, G., Bai, M., Li, R., Zhang, X., Ma, C., Lo, P.-C., Ng, D.K.P. and Jiang, J. (2006) *Eur. J. Inorg. Chem.*, **2006**, 3703–3709.
115. Ishikawa, N. and Kaizu, Y. (1998) *Chem. Lett.*, 183–184.
116. Ishikawa, N. and Kaizu, Y. (2000) *J. Phys. Chem. A*, **104**, 10009–10016.
117. Sheng, N., Li, R., Choi, C.-F., Su, W., Ng, D.K.P., Cui, X., Yoshida, K., Kobayashi, N. and Jiang, J. (2006) *Inorg. Chem.*, **45**, 3794–3802.
118. Altindal, A., Abdurrahmanoğlu, Ş., Bulut, M. and Bekaroğlu, Ö. (2005) *Synth. Met.*, **150**, 181–187.
119. Şen, P., Dumludağ, F., Salih, B., Özkaya, A.R. and Bekaroğlu, Ö. (2011) *Synth. Met.*, **161**, 1245–1254.
120. Buchler, J.W., Kapellmann, H.-G., Knoff, M., Lay, K.-L. and Pfeifer, S. (1983) *Z. Naturforsch, Part. B*, **38**, 1339–1345.
121. Tanaka, D., Inose, T., Tanaka, H., Lee, S., Ishikawa, N. and Ogawa, T. (2012) *Chem. Commun.*, **48**, 7796–7798.

122. Écija, D., Auwärter, W., Vijayaraghavan, S., Seufert, K., Bischoff, F., Tashiro, K. and Barth, J.V. (2011) *Angew. Chem. Int. Ed.*, **50**, 3872–3877.
123. Buchler, J.W., Cian, A.D., Fisherb, J., Hammerschmitt, P., Loffler, J., Scharbert, B. and Weiss, R. (1989) *Chem. Ber.*, **122**, 2219–2228.
124. Bilsel, O., Rodriguez, J. and Holten, D. (1990) *J. Phys. Chem.*, **94**, 3508–3512.
125. Buchler, J.W. and Löffler, J. (1990) *Z. Naturforsch, Part. B*, **45**, 531–542.
126. Buchler, J.W., Kihn-Botulinski, M., Löffler, J. and Scharbert, B. (1992) *New J. Chem.*, **16**, 545–553.
127. Spyroulias, G.A.A., Raptopoulou, C.P.P., Coutsolelos, A.G. and Terzis, A. (1995) *Inorg. Chem.*, **34**, 2476–2479.
128. Bertini, I., Coutsolelos, A., Dikiy, A., Luchinat, C., Spyroulias, G.A., Troganis, A. and Capponi, G. (1996) *Inorg. Chem.*, **14**, 9851–9858.
129. Spyroulias, G.A. and Coutsolelos, A.G. (1996) *Inorg. Chem.*, **35**, 1382–1385.
130. Miyake, K., Fukuta, M., Asakawa, M., Hori, Y., Ikeda, T. and Shimizu, T. (2009) *J. Am. Chem. Soc.*, **131**, 17808–17813.
131. Otsuki, J., Kawaguchi, S., Yamakawa, T., Asakawa, M. and Miyake, K. (2006) *Langmuir ACS J. Surf. Colloids*, **22**, 5708–5715.
132. Ogi, S., Ikeda, T., Wakabayashi, R., Shinkai, S. and Takeuchi, M. (2010) *Chem. Eur. J.*, **16**, 8285–8290.
133. Ogi, S., Ikeda, T. and Takeuchi, M. (2012) *J. Inorg. Organomet. Polym. Mater.*, **23**, 193–199.
134. Otsuki, J., Komatsu, Y., Kobayashi, D., Asakawa, M. and Miyake, K. (2010) *J. Am. Chem. Soc.*, **132**, 6870–6871.
135. Chabach, D., Tahiri, M., De Cian, A., Fischer, J., Weiss, R. and Bibouts, M.E.M. (1995) *J. Am. Chem. Soc.*, **117**, 8548–8556.
136. Lu, F., Sun, X., Li, R., Liang, D., Zhu, P., Choi, C., Ng, D.K.P., Fukuda, T., Kobayashi, N., Bai, M. *et al.* (2004) *New J. Chem.*, **28**, 1116–1122.
137. Zhang, X., Muranaka, A., Lv, W., Zhang, Y., Bian, Y., Jiang, J. and Kobayashi, N. (2008) *Chem. Eur. J.*, **14**, 4667–4674.
138. Wu, X., Lv, W., Wang, Q., Wang, H., Zhang, X. and Jiang, J. (2011) *Dalton Trans.*, **40**, 107–113.
139. Bian, Y., Jiang, J., Tao, Y., Choi, M.T.M., Li, R., Ng, A.C.H., Zhu, P., Pan, N., Sun, X., Arnold, D.P. *et al.* (2003) *J. Am. Chem. Soc.*, **125**, 12257–12267.
140. Sheng, N., Zhu, P., Ma, C. and Jiang, J. (2009) *Dyes Pigm.*, **81**, 91–96.
141. Jiang, J., Mak, T.C.W. and Ng, D.K.P. (1996) *Chem. Ber.*, **129**, 933–936.
142. Dong, S., Zhang, X., Zhou, Y., Jiang, J. and Bian, Y. (2011) *Dyes Pigm.*, **91**, 99–104.
143. Zhou, Y., Zhang, Y., Wang, H., Jiang, J., Bian, Y., Muranaka, A. and Kobayashi, N. (2009) *Inorg. Chem.*, **48**, 8925–8933.
144. Tran-Thi, T.-H., Mattioli, T.A., Chabach, D., De Cian, A. and Weiss, R. (1994) *J. Phys. Chem.*, **98**, 8279–8288.
145. Jiang, J., Du, D., Choi, M.T.M., Xie, J. and Ng, D.K.P. (1999) *Chem. Lett.*, 261–262.
146. Bai, M., Bao, M., Ma, C., Arnold, D.P., Choi, M.T.M., Ng, D.K.P. and Jiang, J. (2003) *J. Mater. Chem.*, **13**, 1333–1339.
147. Jiang, J., Liu, W., Cheng, K., Poon, K. and Ng, D.K.P. (2001) *Eur. J. Inorg. Chem.*, **2001**, 413–417.
148. Furuya, F., Kobayashi, N., Bian, Y. and Jiang, J. (2001) *Chem. Lett.*, 944–945.
149. Jiang, J., Bian, Y., Furuya, F., Liu, W., Choi, M.T.M., Kobayashi, N., Li, H.-W., Yang, Q., Mak, T.C.W. and Ng, D.K.P. (2001) *Chem. Eur. J.*, **7**, 5059–5069.
150. Zhu, P., Lu, F., Pan, N., Arnold, D.P., Zhang, S. and Jiang, J. (2004) *Eur. J. Inorg. Chem.*, **2004**, 510–517.
151. Kudrik, E.V., Islyaikin, M.K. and Shaposhnikov, G.P. (2000) *Russ. J. Gen. Chem.*, **70**, 767.
152. Galanin, N.E., Yakubov, L.A. and Shaposhnikov, G.P. (2008) *Russ. J. Org. Chem.*, **44**, 921–926.
153. Galanin, N.E. and Shaposhnikov, G.P. (2012) *Russ. J. Org. Chem.*, **48**, 851–857.
154. Zhang, Y., Qi, D. and Jiang, J. (2009) *J. Porphyrins Phthalocyanines*, **13**, 1197–1206.

155. Cao, W., Wang, H., Wang, X., Lee, H.K., Ng, D.K.P. and Jiang, J. (2012) *Inorg. Chem.*, **51**, 9265–9272.
156. Galanin, N.E., Yakubov, L.A. and Shaposhnikov, G.P. (2011) *Russ. J. Org. Chem.*, **47**, 941–947.
157. Kasuga, K. and Tsutsui, M. (1980) *Coord. Chem. Rev.*, **32**, 67–95.
158. Collins, G.C.S. and Schiffrin, D.J. (1982) *J. Electroanal. Chem. Interfacial Electrochem.*, **139**, 335–369.
159. Chang, A.T. and Marchon, J.-C. (1981) *Inorg. Chim. Acta*, **53**, L241–L243.
160. Corker, G.A., Grant, B. and Clecak, N.J. (1979) *J. Electrochem. Soc.*, **126**, 1339–1343.
161. De Cian, A., Moussavi, M., Fischer, J. and Weiss, R. (1985) *Inorg. Chem.*, **24**, 3162–3167.
162. Rittenberg, D.K., Sugiura, K., Sakata, Y., Arif, A.M. and Miller, J.S. (2001) *Inorg. Chem.*, **40**, 3654–3655.
163. Katoh, K., Komeda, T. and Yamashita, M. (2010) *Dalton Trans.*, **39**, 4708–4723.
164. Trojan, K.L., Kendall, J.L., Kepler, K.D. and Hatfield, W.E. (1992) *Inorg. Chim. Acta*, **198-200**, 795–803.
165. La Mar, G.N., de Ropp, J.S., Smith, K.M. and Langry, K.C. (1980) *J. Am. Chem. Soc.*, **102**, 4835–4836.
166. Moussavi, M., De Cian, A., Fischer, J. and Weiss, R. (1988) *Inorg. Chem.*, **27**, 1287–1291.
167. Ishikawa, N., Mizuno, Y., Takamatsu, S., Ishikawa, T. and Koshihara, S. (2008) *Inorg. Chem.*, **47**, 10217–10219.
168. Iwase, A., Harnoode, C. and Kameda, Y. (1993) *J. Alloys Compd.*, **192**, 280–283.
169. Konami, H., Hatano, M. and Tajiri, A. (1989) *Chem. Phys. Lett.*, 163–167.
170. Ishikawa, N., Ohno, O. and Kaizu, Y. (1993) *J. Phys. Chem.*, **97**, 1004–1010.
171. Ishikawa, N. and Kaizu, Y. (1999) *J. Porphyrins Phthalocyanines*, **3**, 514–521.
172. Ishikawa, N. (2001) *J. Porphyrins Phthalocyanines*, **5**, 87–101.
173. Kadish, K.M., Nakanishi, T., Gürek, A., Ahsen, V. and Yilmaz, I. (2001) *J. Phys. Chem. B*, **105**, 9817–9821.
174. Ostendorp, G. and Homborg, H. (1996) *Z. Anorg. Allg. Chem.*, **622**, 1222–1230.
175. Gonidec, M., Amabilino, D.B. and Veciana, J. (2012) *Dalton Trans.*, **41**, 13632–13639.
176. Gorun, S.M., Veciana, J., Gonidec, M., Krivokapic, I., Vidal-Gancedo, J., Davies, E.S. and McMaster, J. (2013) *Inorg. Chem.*, **52**, 4464–4471.
177. Koike, N., Uekusa, H., Ohashi, Y., Harnoode, C., Kitamura, F., Ohsaka, T. and Tokuda, K. (1996) *Inorg. Chem.*, **35**, 5798–5804.
178. Paillaud, J.L., Drillon, M., Decian, A., Fischer, J., Weiss, R., Poinsot, R. and Herr, A. (1991) *Phys. B*, **175**, 337–348.
179. Katoh, K., Yoshida, Y., Yamashita, M., Miyasaka, H., Breedlove, B.K., Kajiwara, T., Takaishi, S., Ishikawa, N., Isshiki, H., Zhang, Y.F. *et al.* (2009) *J. Am. Chem. Soc.*, **131**, 9967–9976.
180. Darovsky, A., Wu, L.Y., Lee, P. and Sheu, H.S. (1991) *Acta Crystallogr., Sect. C: Cryst. Struct. Commun.*, **47**, 1836–1838.
181. Haghighi, M.S., Rath, M., Rotter, H.W. and Humborg, H. (1993) *Z. Anorg. Allg. Chem.*, **619**, 1887–1896.
182. Darovskikh, A.N., Tsishchenko, A.K., Frank-Kamenetskaya, O.V. and Fundamenskii, V.S. (1984) *Kristallografiya*, **29**, 455.
183. Loosli, C., Liu, S.-X., Neels, A., Labat, G. and Decurtins, S. (2006) *Z. Kristallogr. - New Cryst. Struct.*, **221**, 135–141.
184. Mossoyan-Deneux, M., Pierrot, M., Sorbier, J.-P., Fournel, A. and Benlian, D.C. (1986) *R. Acad. Sci., Ser. II*, **303**, 669.
185. Gieren, A. and Hoppe, W. (1971) *Chem. Commun.*, 413–415.
186. Paillaud, J.L., Drillon, M., De Cian, A., Fischer, J., Weiss, R. and Villeneuve, G. (1991) *Phys. Rev. Lett.*, **67**, 244–247.
187. Ostendorp, G., Werner, J.P. and Homborg, H. (1995) *Acta Crystallogr., Sect. C: Cryst. Struct. Commun.*, **51**, 1125–1128.
188. Darovskikh, A.N., Frank-Kamenetskaya, O.V., Fundamenskii, V.S. and Golubev,

A.M. (1986) *Sov. Phys. Crystallogr.*, **31**, 165.
189. Haghighi, M.S., Teske, C.L. and Homborg, H. (1992) *Z. Anorg. Allg. Chem.*, **608**, 73–80.
190. Bondi, A. (1964) *J. Phys. Chem.*, **68**, 441–451.
191. Powell, A.K. (2010) *Nat. Chem.*, **2**, 351–352.
192. Rinehart, J.D., Fang, M., Evans, W.J. and Long, J.R. (2011) *J. Am. Chem. Soc.*, **133**, 14236–14239.
193. Ballesteros-Rivas, M., Zhao, H., Prosvirin, A., Reinheimer, E.W., Toscano, R.A., Valdés-Martínez, J. and Dunbar, K.R. (2012) *Angew. Chem. Int. Ed.*, **51**, 5124–5128.
194. Ishikawa, N., Sugita, M., Ishikawa, T., Koshihara, S. and Kaizu, Y. (2003) *J. Am. Chem. Soc.*, **125**, 8694–8695.
195. Ishikawa, N., Sugita, M., Okubo, T., Tanaka, N., Iino, T. and Kaizu, Y. (2003) *Inorg. Chem.*, **42**, 2440–2446.
196. Benelli, C. and Gatteschi, D. (2002) *Chem. Rev.*, **102**, 2369–2388.
197. Ishii, H. (2004) *J. Low Temp. Phys.*, **135**, 579–620.
198. Ishikawa, N. (2003) *J. Phys. Chem. A*, **107**, 5831–5835.
199. Gatteschi, D. and Sessoli, R. (2003) *Angew. Chem. Int. Ed.*, **42**, 268–297.
200. Giraud, R., Wernsdorfer, W., Tkachuk, A.M., Mailly, D. and Barbara, B. (2001) *Phys. Rev. Lett.*, **87**, 057203(4).
201. Isotope *http://en.wikipedia.org/wiki/Isotope* can be found under *http://en.wikipedia.org/wiki/Isotope* (accessed 13 August 2014).
202. Ishikawa, N., Sugita, M., Wernsdorfer, W., Ne, L.L., Martyrs, A.V. and Cedex, G. (2005) *J. Am. Chem. Soc.*, **127**, 3650–3651.
203. Ishikawa, N., Tanaka, N. and Kaizu, Y. (2004) *Inorg. Chim. Acta*, **357**, 2181–2184.
204. Ishikawa, N., Iino, T. and Kaizu, Y. (2002) *J. Am. Chem. Soc.*, **124**, 11440–11447.
205. Ishikawa, N., Otsuka, S. and Kaizu, Y. (2005) *Angew. Chem. Int. Ed.*, **44**, 731–733.
206. Trojan, K.L., Hatfield, W.E., Kepler, K.D. and Kirk, M.L. (1991) *J. Appl. Phys.*, **69**, 6007–6009.
207. Abragam, A. and Bleaney, B. (1970) *Electron Paramagnetic Resonance*, Clarendon Press, Oxford.
208. Branzoli, F., Filibian, M., Carretta, P., Klyatskaya, S. and Ruben, M. (2009) *Phys. Rev. B*, **79**, 220404(4).
209. Ishikawa, N., Sugita, M., Tanaka, N., Ishikawa, T., Koshihara, S. and Kaizu, Y. (2004) *Inorg. Chem.*, **43**, 5498–5500.
210. Gonidec, M., Luis, F., Vílchez, A., Esquena, J., Amabilino, D.B. and Veciana, J. (2010) *Angew. Chem. Int. Ed.*, **49**, 1623–1626.
211. Takamatsu, S., Ishikawa, T., Koshihara, S. and Ishikawa, N. (2007) *Inorg. Chem.*, **46**, 7250–7252.
212. Gonidec, M., Davies, E.S., McMaster, J., Amabilino, D.B. and Veciana, J. (2010) *J. Am. Chem. Soc.*, **132**, 1756–1757.
213. Wang, H., Wang, K., Tao, J. and Jiang, J. (2012) *Chem. Commun.*, **48**, 2973–2975.
214. Hofmann, A., Salman, Z., Mannini, M., Amato, A., Malavolti, L., Morenzoni, E., Prokscha, T., Sessoli, R. and Suter, A. (2012) *ACS Nano*, **6**, 8390–8396.
215. Katoh, K., Isshiki, H., Komeda, T. and Yamashita, M. (2011) *Coord. Chem. Rev.*, **255**, 2124–2148.
216. Urdampilleta, M., Nguyen, N.-V., Cleuziou, J.-P., Klyatskaya, S., Ruben, M. and Wernsdorfer, W. (2011) *Int. J. Mol. Sci.*, **12**, 6656–6667.
217. Urdampilleta, M., Klyatskaya, S., Cleuziou, J.-P., Ruben, M. and Wernsdorfer, W. (2011) *Nat. Mater.*, **10**, 502–506.
218. Tsivadze, A.Y., Ionova, G.V. and Mikhalko, V.K. (2010) *Prot. Met. Phys. Chem. Surf.*, **46**, 149–169.
219. Lawrence, J.M., Riseborough, P.S. and Parks, R.D. (1981) *Rep. Prog. Phys.*, **44**, 1–85.
220. Hanzawa, K. (2002) *J. Phys. Soc. Jpn.*, **71**, 1481–1494.
221. Shimozawa, M., Watashige, T., Yasumoto, S., Mizukami, Y., Nakamura, M., Shishido, H., Goh, S.K., Terashima,

T., Shibauchi, T. and Matsuda, Y. (2012) *Phys. Rev. B*, **86**, 144526(5).
222. Rousseau, R., Aroca, R. and Rodriguez-Mendez, M.L. (1995) *J. Mol. Struct.*, **356**, 49–62.
223. Branzoli, F., Carretta, P., Filibian, M., Klyatskaya, S. and Ruben, M. (2011) *Phys. Rev. B*, **83**, 174419(5).
224. Mannini, M., Pineider, F., Sainctavit, P., Danieli, C., Otero, E., Sciancalepore, C., Talarico, A.M., Arrio, M.-A., Cornia, A., Gatteschi, D. *et al.* (2009) *Nat. Mater.*, **8**, 194–197.
225. Pineider, F., Mannini, M., Sessoli, R., Caneschi, A., Barreca, D., Armelao, L., Cornia, A., Tondello, E. and Gatteschi, D. (2007) *Langmuir*, **23**, 11836–11843.
226. Mannini, M., Sainctavit, P., Sessoli, R., Cartier dit Moulin, C., Pineider, F., Arrio, M.-A., Cornia, A. and Gatteschi, D. (2008) *Chem. Eur. J.*, **14**, 7530–7535.
227. Eigler, D.M. and Schweizer, E.K. (1990) *Nature*, **344**, 524–526.
228. Cornia, A., Mannini, M., Sainctavit, P. and Sessoli, R. (2011) *Chem. Soc. Rev.*, **40**, 3076–3091.
229. Mannini, M., Pineider, F., Sainctavit, P., Cartier dit Moulin, C., Arrio, M.-A., Cornia, A., Gatteschi, D. and Sessoli, R. (2009) *Eur. Phys. J. Spec. Top.*, **169**, 167–173.
230. Corradini, V., Biagi, R., del Pennino, U., De Renzi, V., Gambardella, A., Affronte, M., Muryn, C.A., Timco, G.A. and Winpenny, R.E.P. (2007) *Inorg. Chem.*, **46**, 4937–4943.
231. Corradini, V., Moro, F., Biagi, R., De Renzi, V., del Pennino, U., Bellini, V., Carretta, S., Santini, P., Milway, V., Timco, G. *et al.* (2009) *Phys. Rev. B*, **79**, 144419(8).
232. Ghirri, A., Corradini, V., Cervetti, C., Candini, A., del Pennino, U., Timco, G., Pritchard, R.J., Muryn, C.A., Winpenny, R.E.P. and Affronte, M. (2010) *Adv. Funct. Mater.*, **20**, 1552–1560.
233. Katoh, K., Isshiki, H., Komeda, T. and Yamashita, M. (2012) *Chem. Asian J.*, **7**, 1154–1169.
234. Gonidec, M., Biagi, R., Corradini, V., Moro, F., De Renzi, V., Pennino, U., Summa, D., Muccioli, L., Zannoni, C., Amabilino, D.B. *et al.* (2011) *J. Am. Chem. Soc.*, **133**, 6603–6612.
235. Gómez-Segura, J., Díez-Pérez, I., Ishikawa, N., Nakano, M., Veciana, J. and Ruiz-Molina, D. (2006) *Chem. Commun.*, 2866–2868.
236. Vitali, L., Fabris, S., Conte, A.M., Brink, S., Ruben, M., Baroni, S. and Kern, K. (2008) *Nano Lett.*, **8**, 3364–3368.
237. Stepanow, S., Honolka, J., Gambardella, P., Vitali, L., Abdurakhmanova, N., Tseng, T., Rauschenbach, S. and Tait, S.L. (2010) *J. Am. Chem. Soc.*, **132**, 11900–11901.
238. Fahrendorf, S., Atodiresei, N., Besson, C., Caciuc, V., Matthes, F., Blügel, S., Kögerler, P., Bürgler, D.E. and Schneider, C.M. (2013) *Nat. Commun.*, **4**, 2425.
239. Ulman, A. (1996) *Chem. Rev.*, **96**, 1533–1554.
240. Lodi Rizzini, A., Krull, C., Balashov, T., Kavich, J.J., Mugarza, A., Miedema, P.S., Thakur, P.K., Sessi, V., Klyatskaya, S., Ruben, M., Stepanow, S. and Gambardella, P. (2011) *Phys. Rev. Lett.*, **107**, 177205(5).
241. Klar, D., Klyatskaya, S., Candini, A., Krumme, B., Kummer, K., Ohresser, P., Corradini, V., de Renzi, V., Biagi, R., Joly, L. *et al.* (2013) *Beilstein J. Nanotechnol.*, **4**, 320–324.
242. Malavolti, L., Poggini, L., Margheriti, L., Chiappe, D., Graziosi, P., Cortigiani, B., Lanzilotto, V., de Mongeot, F.B., Ohresser, P., Otero, E., Choueikani, F., Sainctavit, P., Bergenti, I., Dediu, V.A., Mannini, M. and Sessoli, R. (2013) *Chem. Commun.*, **49**, 11506–11508.
243. Lodi Rizzini, A., Krull, C., Balashov, T., Mugarza, A., Nistor, C., Yakhou, F., Sessi, V., Klyatskaya, S., Ruben, M., Stepanow, S. and Gambardella, P. (2012) *Nano Lett.*, **12**, 5703–5707.
244. Wiesendanger, R. and Bode, M. (2001) *Solid State Commun.*, **119**, 341–355.
245. Schwöbel, J., Fu, Y., Brede, J., Dilullo, A., Hoffmann, G., Klyatskaya, S., Ruben, M. and Wiesendanger, R. (2012) *Nat. Commun.*, **3**, 1–5.
246. Gambardella, P., Stepanow, S., Dmitriev, A., Honolka, J., de

Groot, F.M.F., Lingenfelder, M., Sen Gupta, S., Sarma, D.D., Bencok, P., Stanescu, S. *et al.* (2009) *Nat. Mater.*, **8**, 189–193.
247. Moroni, R., Buzio, R., Chincarini, A., Valbusa, U., de Mongeot, F.B., Bogani, L., Caneschi, A., Sessoli, R., Cavigli, L. and Gurioli, M. (2008) *J. Mater. Chem.*, **18**, 109–115.
248. Troiani, F. and Affronte, M. (2011) *Chem. Soc. Rev.*, **40**, 3119–3129.
249. Ganzhorn, M., Klyatskaya, S., Ruben, M. and Wernsdorfer, W. (2013) *ACS Nano*, **7**, 6225–6236.
250. Cleuziou, J.-P., Wernsdorfer, W., Bouchiat, V., Ondarçuhu, T. and Monthioux, M. (2006) *Nat. Nanotechnol.*, **1**, 53–59.
251. Aprili, M. (2006) *Nat. Nanotechnol.*, **1**, 15–16.
252. Jo, M.-H., Grose, J.E., Baheti, K., Deshmukh, M.M., Sokol, J.J., Rumberger, E.M., Hendrickson, D.N., Long, J.R., Park, H. and Ralph, D.C. (2006) *Nano Lett.*, **6**, 2014–2020.
253. Park, J., Pasupathy, A.N.N., Goldsmith, J.I.I., Chang, C., Yaish, Y., Petta, J.R., Rinkoski, M., Sethna, J.P., Abruna, H.D., McEuen, P.L. *et al.* (2002) *Nature*, **417**, 722–725.
254. Liang, W., Shores, M.P., Bockrath, M., Long, J.R. and Park, H. (2002) *Nature*, **417**, 725–729.
255. Malavolti, L., Mannini, M., Car, P.-E., Campo, G., Pineider, F. and Sessoli, R. (2013) *J. Mater. Chem. C*, **1**, 2935–2942.
256. Candini, A., Klyatskaya, S., Ruben, M., Wernsdorfer, W. and Affronte, M. (2011) *Nano Lett.*, **11**, 2634–2639.
257. Vincent, R., Klyatskaya, S., Ruben, M., Wernsdorfer, W. and Balestro, F. (2012) *Nature*, **488**, 357–360.
258. Ganzhorn, M., Klyatskaya, S., Ruben, M. and Wernsdorfer, W. (2013) *Nat. Nanotechnol.*, **8**, 165–169.
259. Pasupathy, A.N., Bialczak, R.C., Martinek, J., Grose, J.E., Donev, L.A.K., McEuen, P.L. and Ralph, D.C. (2004) *Science*, **306**, 86–89.
260. Urdampilleta, M., Klyatskaya, S., Ruben, M. and Wernsdorfer, W. (2013) *Phys. Rev. B*, **87**, 195412(6).
261. Lassagne, B., Ugnati, D. and Respaud, M. (2011) *Phys. Rev. Lett.*, **107**, 130801(5).
262. Pályi, A., Struck, P.R., Rudner, M., Flensberg, K. and Burkard, G. (2012) *Phys. Rev. Lett.*, **108**, 206811(5).
263. Winpenny, R.E.P. (2013) *Nat. Nanotechnol.*, **8**, 159–160.
264. Spain Transistor *http://en.wikipedia.org/wiki/Spin_transistor*, can be found under *http://en.wikipedia.org/wiki/Spin_transistor* (accessed 13 August 2014).
265. Bogani, L., Cavigli, L., Gurioli, M., Novak, R.L., Mannini, M., Caneschi, A., Pineider, F., Sessoli, R., Clemente-León, M., Coronado, E. *et al.* (2007) *Adv. Mater.*, **19**, 3906–3911.
266. Wahl, P., Simon, P., Diekhöner, L., Stepanyuk, V., Bruno, P., Schneider, M. and Kern, K. (2007) *Phys. Rev. Lett.*, **98**, 056601(4).
267. Hirjibehedin, C.F., Lin, C.-Y., Otte, A.F., Ternes, M., Lutz, C.P., Jones, B.A. and Heinrich, A.J. (2007) *Science*, **317**, 1199–1203.
268. Komeda, T., Isshiki, H., Liu, J., Zhang, Y.-F., Lorente, N., Katoh, K., Breedlove, B.K. and Yamashita, M. (2011) *Nat. Commun.*, **2**, 1–7.
269. van der Zant, H.S.J. (2012) *Nat. Nanotechnol.*, **7**, 555–556.
270. Thiele, S., Vincent, R., Holzmann, M., Klyatskaya, S., Ruben, M., Balestro, F. and Wernsdorfer, W. (2013) *Phys. Rev. Lett.*, **111**, 037203(5).
271. Leuenberger, M.N. (2003) *Phys. Rev. B*, **68**, 165317(11).
272. Rocha, A.R., García-Suárez, V.M., Bailey, S.W., Lambert, C.J., Ferrer, J. and Sanvito, S. (2005) *Nat. Mater.*, **4**, 335–339.
273. Petta, J., Slater, S. and Ralph, D. (2004) *Phys. Rev. Lett.*, **93**, 136601(4).
274. Wagner, S., Kisslinger, F., Ballmann, S., Schramm, F., Chandrasekar, R., Bodenstein, T., Fuhr, O., Secker, D., Fink, K., Ruben, M. and Weber, H.B. (2013) *Nat. Nanotechnol.*, **8**, 575–579.

9
Lanthanides and the Magnetocaloric Effect

Joseph W. Sharples and David Collison

9.1
Applications of Magnets

The most studied applications of molecular magnets relate to their potential employment as single-molecule magnets [1] and in spintronic components [2]. While these are relatively new fields with real-world usage perhaps several years away, the field of magnetocaloric materials has been established since the 1930s [3] and, via adiabatic demagnetization, is employed in many ways around the world as an alternative refrigeration technique.

This chapter mainly examines how cages and polymers containing lanthanide(III) ions, specifically gadolinium(III), can be employed as magnetocaloric materials which show strong performance in the low temperature millikelvin regime, rivalling ^{3}He/^{4}He methods, such as dilution refrigeration.

There has been a great number of mainly molecular magnetocaloric materials investigated and reported over the last few decades, although the best performing, as seen more recently, are polymeric structures and MOFs (metal-organic frameworks). Between these we see further divisions into homometallic 3d or 4f compounds and those which combine the two into heterometallic 3d–4f compounds. In order to disentangle the myriad strategies, we will attempt to describe the ideal magnetocaloric material for low temperature applications, working through the importance of several factors by means of published examples and showing why these contribute to a larger magnetocaloric effect. Then we present in more detail what we consider to be the research highlights in this field, although this is by no means an exhaustive review. Several alternative reviews may help to fill in some of the gaps [4–6]. After reading this, we hope all of the related chemical literature will be fairly straightforward to understand, too.

Lanthanides and Actinides in Molecular Magnetism, First Edition.
Edited by Richard A. Layfield and Muralee Murugesu.

9.2 Cold Reasoning

First though, we will consider why this is at all a profitable strategy. What is wrong with using helium to cool, and why are lanthanide-based materials of interest to an increasing number of researchers?

The simple answer is that helium on Earth, as many researchers in the field of molecular magnetism will know, is increasing in cost, and supply is becoming increasingly difficult. This is despite the fact that in the Universe as a whole, helium is plentiful and makes up around a quarter of the matter we can detect. On our Moon [7], too, it is abundant, including in its most useful form to us, ^{3}He. Sadly, on Earth, the relatively small gravitational attraction is unable to hold it in significant quantities. Disregarding the purely intellectual motivation, alternative cooling methods should be developed.

Of all the metals in the Periodic Table, lanthanide-based coolants are amongst the most suitable as replacements for helium systems as they can operate best in the low-temperature regime. We now describe some theoretical ideas showing how any paramagnet is able to act as a refrigerant, what makes lanthanide(III) ions special for this application and also a brief summary of current technologies.

9.3 Current Technologies

Simply transferring helium into a cryostat is a common enough technique in many physics or chemistry laboratories, to give temperatures of 5 K and below for routine magnetic measurements, for example, with a SQUID (superconducting quantum interference device) magnetometer or for EPR (electron paramagnetic resonance) and NMR (nuclear magnetic resonance) experiments.

The established methods for cooling below 1.8 K tend to be more involved, with dilution refrigeration [8] being the main method. This uses both ^{3}He and ^{4}He to potentially reach 2 mK. Spontaneous separation of these isotopes creates a 'two-phase' system, where the crossing of the boundary by ^{3}He atoms into the ^{4}He layer (which actually contains ^{3}He as an 'impurity') absorbs energy from the surroundings. This technique is used everywhere from laboratories on Earth to far beyond, such as the Planck Observatory [9] (Satellite). For a while it was thought that such technology was unsuitable for space, because of the lack of gravity and the time limitations of an open system, with the mixed ^{3}He$-^{4}$He being unable to be separated practically, although these issues have been addressed. Nevertheless, magnetic cooling is a viable alternative.

9.4
How Paramagnets Act as Refrigerants

The term *magnetocaloric* refers to a change in the temperature (*calor* = heat) of a paramagnet when a magnetic field (*magneto*) is applied. If the field is applied adiabatically, that is, under constant entropy, then the material increases in temperature. Removing this field allows the paramagnet to return to its starting temperature. This process defines the *most useful* parameter for characterizing a paramagnet in terms of its magnetocaloric effect, ΔT_{AD} (K), or adiabatic temperature change, a direct measure of a material's cooling power. Different paramagnets will have very different responses, though, in both the magnitude and temperature range at which they can operate best. For example, gadolinium metal [10] is best suited to cooling at room temperature, whereas, say, the dimetallic gadolinium acetate molecule $[Gd^{III}_2(H_3CCO_2)_6(H_2O)_4]\cdot 4H_2O$ [11] works better at 2 K. Physicists and chemists can understand why this occurs and therefore design and synthesize new materials for certain applications.

The origin of this heating is the changing distribution of entropy within the paramagnet with the applied field [12–14]. We can write the sum total of entropies according to Equation 9.1, where S_{total} is the total entropy, S_M is the magnetic entropy, $S_{lattice}$ is the lattice entropy and $S_{electronic}$ is the electronic entropy of the system.

$$S_{total} = S_M + S_{lattice} + S_{electronic} \tag{9.1}$$

When the spin angular momenta, *s*, of each paramagnet is aligned with the field, the magnetic entropy S_M decreases. This is illustrated in Figure 9.1, whereby spins (shown as arrows) align with a magnetic field, where *H* is the magnetic field strength. Chemists may imagine this as follows: under adiabatic conditions the total entropy is constant and $S_{lattice}$ increases to compensate, whereby the atoms are less ordered or more energetic, and hence the material is warmer. Physicists may prefer to consider this in terms of phonon exchange; as the spin system gives up energy, the lattice is raised to a higher vibrational mode. When the field is removed the entropy changes are reversible and any

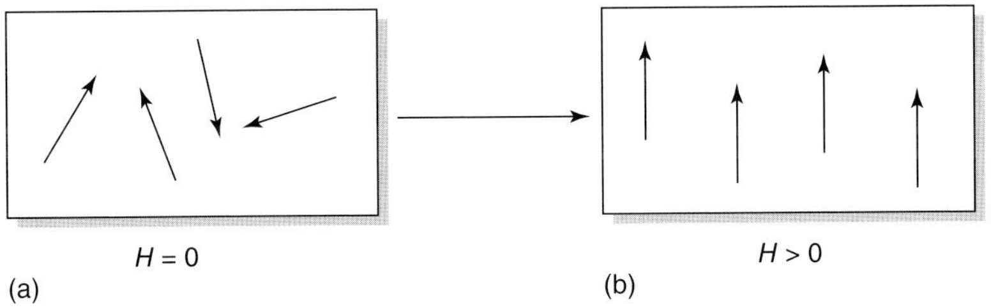

Figure 9.1 Spin angular momenta, *s*, depicted as arrows, are randomly aligned in zero field, $H = 0$ (a), and align with the applied field, $H > 0$ (b), thereby decreasing the magnetic entropy of the system.

heat dissipates to the surroundings. The change in S_M between the two states, magnetized ($H > 0$) and in zero field ($H = 0$), $-\Delta S_M$, is the *most commonly used* parameter to characterize the magnetocaloric effect amongst molecular magnetism practitioners and can be obtained indirectly via magnetic measurements or heat capacity, with the former method being more common for practical reasons.

This is done by measuring the magnetization of a sample, using a SQUID magnetometer, for example, at various fields and temperatures, with intervals between 0–7 T and 1.8–300 K as a rough standard in the literature. These data are then used with the following Maxwell Equation (Equation 9.2) [13], where T is the temperature, and H_f and H_i are the final and initial magnetic fields ($H_f > H_i$):

$$\Delta S_M (T, \Delta H) = \int_{H_i}^{H_f} \left(\frac{\partial M(T,H)}{\partial T} \right)_H \mathrm{d}H \tag{9.2}$$

Heat capacity measurements can also be used to extract ΔT_{AD} in addition to $S_M(T)$ (Equations 9.3a and 9.3b), where C_M is the heat capacity of magnetization for a more complete analysis:

$$\Delta T_{AD}(T, \Delta H) = \int_{H_i}^{H_f} \left(\frac{T}{C(T,H)} \right)_H \left(\frac{\partial M(T,H)}{\partial T} \right)_H \mathrm{d}H \tag{9.3a}$$

$$S_M (T) = \int_0^T \frac{C_M(T)}{T} \mathrm{d}T \tag{9.3b}$$

A further illustration and summary of the entropy and temperature changes involved upon moving between two magnetic fields is shown in Figure 9.2 [13].

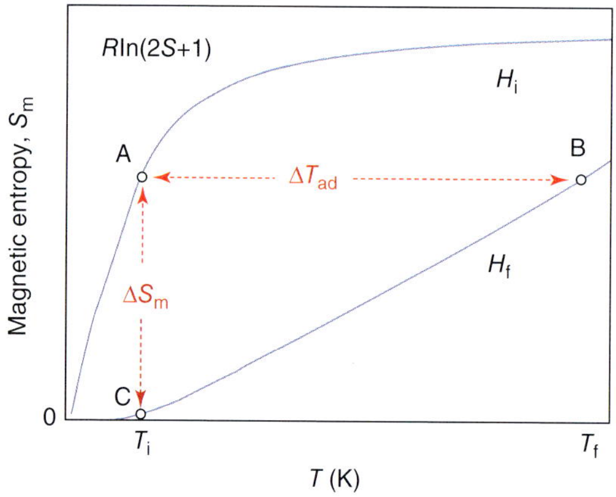

Figure 9.2 Magnetic entropy versus temperature for a hypothetical paramagnet at two distinct magnetic fields. H_f represents the final field and H_i the initial, although this nomenclature is arbitrary and there is reversibility between the two ($H_f > H_i$). ΔT_{AD} is the adiabatic temperature change and $-\Delta S_M$ the magnetic entropy change. (Taken from Ref. [13], with permission from Royal Society of Chemistry.)

Moving from A to B in Figure 9.2, an increased field is applied, the total entropy of the system is constant under adiabatic conditions and the temperature of the paramagnet rises, giving ΔT_{AD}. The process from A to C is something different; under isothermal conditions the temperature of the paramagnet stays the same and ΔS_{total} decreases, due to an uncompensated fall in the magnetic entropy, ΔS_M. One can make predictions as to which materials will have a large ΔS_M, by using Equation 9.4, where R is the gas constant, S is the spin of each paramagnetic component and n is the number of spins of the given s (note that we use s for individual spins and S for the total spin, to distinguish from the entropy S):

$$S_M = nR\ln(2s + 1) \tag{9.4}$$

This becomes a sum when we are considering a heterometallic system. In reality, there are several factors why this theoretical maximum may not be attainable, such as anisotropy or the magnetic field being too small.

9.5
More Parameters

There are two main caveats to Equation 9.4; firstly, that this formula gives S_M in units of joule per kilogram per kelvin, meaning no account is taken of the volume of the material under test. For this, the density is required and these two parameters combine ($-\rho^* \Delta S_M$) to give a better characterization, with units of millijoule per cubic centimetre per kelvin, acknowledging the composition of the material in bulk. Densities can readily be calculated from X-ray crystallographic data so that this presents no extra hardship. Secondly, one should be aware that the highest possible spin is not always attained, and, particularly in 3d systems, it may be impossible to saturate the spin system in moderate fields [15].

One final parameter, occasionally encountered in the literature, is the refrigerant capacity [16], RC, given in Equation 9.5, where T_1 and T_2 are the positions of the full-width-half-maxima in the $-\Delta S_M$ versus Field plots, where $T_1 < T_2$, and S_M is the maximum value found.

$$\text{RC} = S_M(T_1 - T_2) \tag{9.5}$$

Although this is calculated in principal from magnetization data, it can be difficult to obtain in practice because it requires a maximum in $-\Delta S_M$ versus T to be observed. This is uncommon in gadolinium(III) compounds under certain laboratory conditions because of their weak interactions. RC characterizes the performance of a material more fully across a greater range of temperatures than simply $-\Delta S_M$, although the two are obviously related.

Practically, the paramagnet can be employed as a magnetic refrigerant in the following way, the main feature being termed *adiabatic demagnetization*. A magnetic field is applied to the material, warming it relative to the surroundings. Still in field, a fluid is passed over, removing the excess heat, such that the paramagnet has an ordered spin system but is at ambient temperature. The adiabatic demagnetization process now takes place. Removal of the paramagnet from the field (or switching off the field) cools it relative to its starting temperature and surroundings by ΔT_{AD} with $S_{lattice}$ converted back to S_M. This represents a lower vibrational state, as phonons are exchanged from the lattice to the spin system. The paramagnet can now act as a refrigerant, absorbing heat from its surroundings.

9.6 Aims

We now need a target at which chemists and physicists can aim. Commercial applications for low-temperature 'adiabatic coolers', as they are known, utilize gadolinium(III)–gallium(III)–garnet abbreviated as GGG [17], with the composition [$Gd(III)_5Ga(III)_5O_{12}$]. Although there is a degree of tuning and flexibility possible with this structure, for example, substituting Ga(III) atoms for Fe(III) or Al(III) [18], molecular systems are more well suited to modification, Olivier Kahn's vision for molecular magnetism [19] being one of tailored compounds and extensive chemical engineering. With this in mind, we now turn to the factors which make up a useful magnetocaloric material.

9.7 Important Concepts for a Large Magnetocaloric Effect

Here, we examine six concepts of importance for a large magnetocaloric effect and illustrate each with some examples, to show the advantages and disadvantages of some reported approaches. We show that a combination of these factors can produce the best performing refrigerant materials, such as $[Gd^{III}(HCO_2)_3]_n$ [20] and $[Mn^{II}(H_2O)_6][Gd^{III}Mn^{II}(oda)_3]_2{\cdot}6H_2O$ [21], where oda is oxydiacetate.

9.7.1 Spin

We have seen how $-\Delta S_M$ is related proportionally to s, the spin of the metal cage, so the most direct strategy is simply to construct a cage with lots of high-spin ions, for example, gadolinium(III). These spins should easily saturate in magnetic fields achievable in the laboratory, maximizing the difference between the magnetic entropy under zero and applied fields. Lanthanide(III) ions [22] can have a larger spin than transition metals on account of their 4f orbitals, which can accommodate up to 14 electrons, thus giving a maximum spin for Gd(III) of $s = 7/2$ when these are half-filled.

9.7.1.1 Examples

High-nuclearity cages were made by Blagg *et al.* [23] and Sharples *et al.* [24] around the same time in 2013. These are made up of metal ions connected by organic ligands. The first is the pentametallic square-based pyramid $[Gd(III)_5O(O^iPr)_{13}]$. Each ion is six-coordinate with a distorted octahedral geometry and bonded to an unusual central μ_5-oxide. There are three varieties of isopropoxide: five are terminally bonding, four bond μ_3 across the triangular faces and four bridge μ_2 between the ions on the square base. The second compound is the heptametallic, disc-like $[Gd(III)_7(OH)_6(tpa)_6(thmeH_2)_5(thmeH)(MeCN)_2](NO_3)_2$ where tpa is triphenylacetate, $thmeH_3$ is tris(hydroxymethyl)ethane and MeCN is acetonitrile. This is made up of a ring of six metal ions, in one plane, bridged by a tripodal alcohol ligand centred on each, which are capped by a terminal tpa group with both oxygen atoms. The μ_3-OH groups connect two of these outer metal ions with the inner, out-of-plane one, supporting a triangular-based arrangement. Their metal cores are highlighted in Figure 9.3a,b respectively.

The 'headline' $-\Delta S_M$ figures of 34 and 23 J kg^{-1} K^{-1} (each for $\Delta H = 0 - 7$ T and 3 K) were impressive at the time and result from almost fully saturated, for $\{Gd(III)_5\}$ or fully saturated, for $\{Gd(III)_7\}$, spin systems under applied fields of 7 T and 2 K, which are achievable in the laboratory with standard set-ups.

We can also show that relying on only a large spin can give compounds that do not reach their potential: comparison of the experimental values of $-\Delta S_M$ with the theoretical values reveals that, while the numbers exceed those expected for simple paramagnets of the appropriate spin, that is, 35/2 for five Gd(III) ions giving about 20 J kg^{-1} K^{-1} and 49/2 for seven Gd(III) ions giving about 8.5 J kg^{-1} K^{-1}, compared to fully decoupled spin systems the actual numbers are feeble, the maximum values being 55 and 31.5 J kg^{-1} K^{-1} for $\{Gd(III)_5\}$ and $\{Gd(III)_7\}$, respectively.

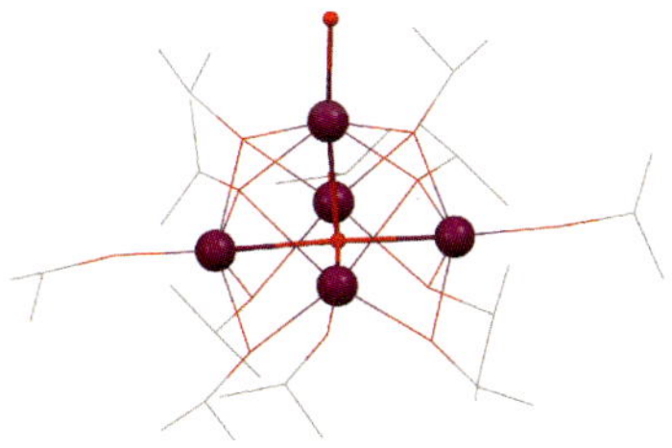
(a)

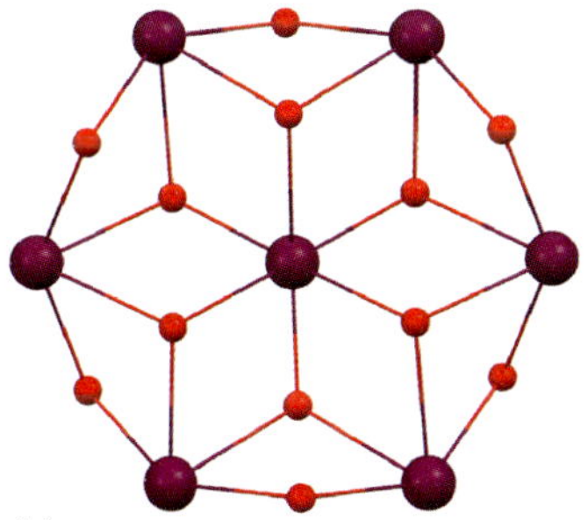
(b)

Figure 9.3 (a) X-ray structure of $[Gd(III)_5O(O^iPr)_{13}]$ [35] with permission from the University of Manchester. (b) $\{Gd(III)_7O_{12}\}$ [4, 21] with permission from Elsevier core of $[Gd(III)_7(OH)_6(tpa)_6(thmeH_2)_5(thmeH)(MeCN)_2]$, where H atoms and counter ions are removed for clarity. Key: Gd(III) purple; O red; C grey.

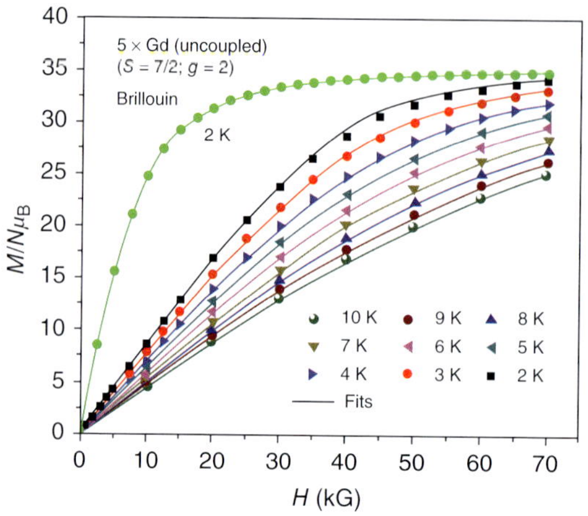

Figure 9.4 Magnetization versus field plots for $\{Gd(III)_5\}$ for a range of temperatures. The green line is the Brillouin function for five uncoupled spins, each of $s = 35/2$ and $g = 2.0$. (Taken from Ref. [23] with permission from Royal Society of Chemistry.)

Why should this be so? To understand this, we must examine the magnetization data. For $\{Gd(III)_5\}$ the comparison of $M(H)$ with the Brillouin function at 2 K (Figure 9.4) for fully uncoupled spins is striking [23], the theoretical value being far above the equivalent 2 K experimental data, until around 7 T. These data, in conjunction with a model using the spin Hamiltonian below (Equation 9.6), reveals an exchange interaction of $J = -0.085\,\text{cm}^{-1}$ between all spins, being antiferromagnetic in this scheme.

$$H = -2J_1 s_1(s_2 + s_3 + s_4 + s_5) - 2J_2(s_2s_3 + s_3s_4 + s_4s_5 + s_2s_5) - 2J_3(s_2s_4 + s_3s_5) \tag{9.6}$$

This tells us that the reduced magnetic entropy change will be lower in an antiferromagnetic system than in an uncoupled one, as can be seen below, as the spins are not fully saturated below the conditions of lowest temperature and strongest field. The antiferromagnetic coupling in $\{Gd(III)_7\}$ is also a hindrance, with behaviour similar to $\{Gd(III)_5\}$, when compared to the relevant Brillouin functions.

It should be noted that these compounds were published alongside their isostructural Dy(III) SMM (single-molecule magnet) counterparts. While this tuning of the magnetic properties of a single structure by simple switching of the lanthanide(III) ion is appealing from a research point of view, it is unproductive in the search for the best magnetocaloric materials, which we will see in polymeric structures.

A large nuclearity cluster, with lots of high-spin ions, is not necessarily the route to the largest magnetocaloric effect. There are deeper factors, such as exchange interactions, which we now explore.

9.7.2 Nature of Exchange Coupling

To verify these thoughts on the nature of the exchange coupling being an important factor in magnetocaloric materials, we can use a model system where everything but this is fixed.

Using the program PHI, developed by Chilton *et al.* [25], the $-\Delta S_M$ values of a model system at 5 and 7 T were calculated for a dimetallic of two $s = 7/2$ spins, with $g = 2.0$, and assigned a molecular weight of 1300 Da. The results are given in Figure 9.5.

If we first consider the three scenarios at the same field, we can immediately see how the paramagnetic material shows the largest magnetocaloric effect, followed by a ferromagnetically coupled system, and then the antiferromagnetic case.

9.7.2.1 Paramagnetism

Qualitatively, we should prepare systems with uncoupled spins for the largest magnetocaloric effect (MCE) at the lowest temperatures. The best examples [26] of paramagnetic, or almost paramagnetic, systems, also demonstrating the wonderful potential of chemical design in this area, are $Na_9[Gd(III)(W_5O_{18})_2]\cdot35H_2O$ and $K_{12}[Gd(III)(P_5W_{30}O_{110})]\cdot54H_2O$. In an effort to reduce the interaction between metals the central gadolinium(III) ion is entombed in a diamagnetic coffin made

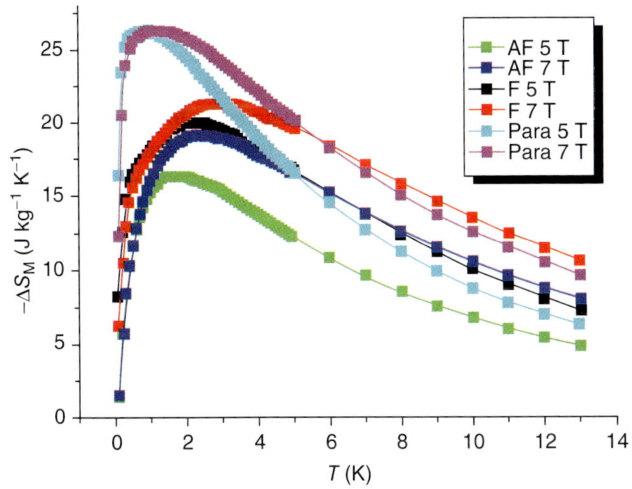

Figure 9.5 PHI [25] simulation of the $-\Delta S_M$ for an arbitrary system of two $s = 7/2$ spins with a M_W of 1300 Da. Key as shown, where Para is a paramagnetic system, F is a ferromagnetic interaction between spins and AF an antiferromagnetic interaction of $\pm0.15\,cm^{-1}$, at 5 and 7 T fields. (Taken from Ref. [35] with permission from the University of Manchester.)

up of a polyoxowolframate capping group, above and below the ion. The smallest Gd(III)–Gd(III) distances from the crystal structure are 1.12 and 1.56 nm, as shown in Figure 9.6.

From susceptibility experiments, each of these was found to show Curie (paramagnetic) behaviour down to 0.1 K, with magnetic susceptibility being constant. Only below this temperature were interactions in the former seen, with a maximum in the susceptibility at 36 mK assigned to a dipole–dipole interaction. $K_{12}[Gd(III)P_5W_{30}O_{110}]\cdot 54H_2O$ was found to be paramagnetic at 10 mK (note that these temperatures were achieved using ^{3}He–^{4}He dilution) and theory suggested this would persist down to 4 mK. From magnetization and heat capacity data it was found that $-\Delta S_M$ were only 4.7 and 1.9 J kg^{-1} K^{-1}, for {Gd(III)W_{10}} and {Gd(III)W_{30}} (1.8 and 1.3 K, $\Delta H = 7$ T) which, while low, are more useful than they first appear.

It is known that below the ordering temperature, the $-\Delta S_M$ of a paramagnet will quickly reduce to zero, thus defining the temperature range in which the material can be usefully employed. Note that gadolinium metal works best at room temperature, which constitutes a separate and productive area of research with

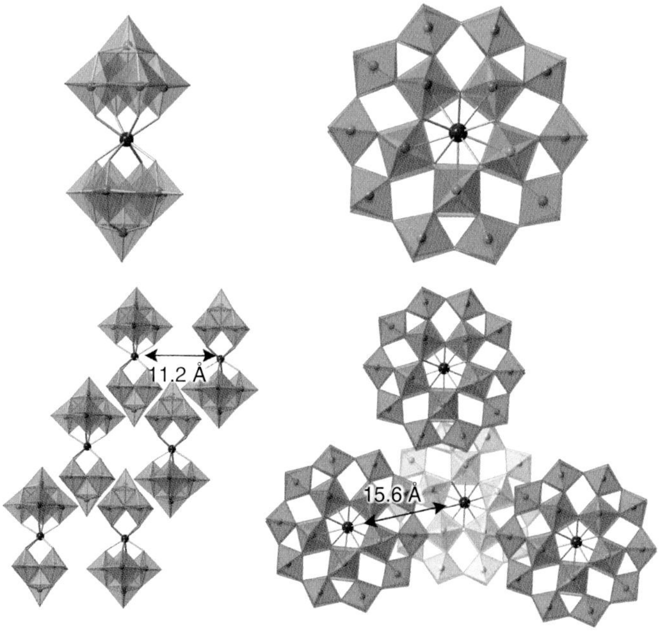

Figure 9.6 (a) $Na_9[Gd(III)(W_5O_{18})_2]\cdot 35H_2O$ with the nearest distance between adjacent Gd(III) ions shown. (b) The same for $K_{12}[Gd(III)(P_5W_{30}O_{110})]\cdot 54H_2O$. Key: Gd(III) purple; W grey spheres with polyhedra shown in grey. Solvent is omitted. (Taken from Ref. [26] with permission from Wiley-VCH Verlag GmbH & Co. KGaA.)

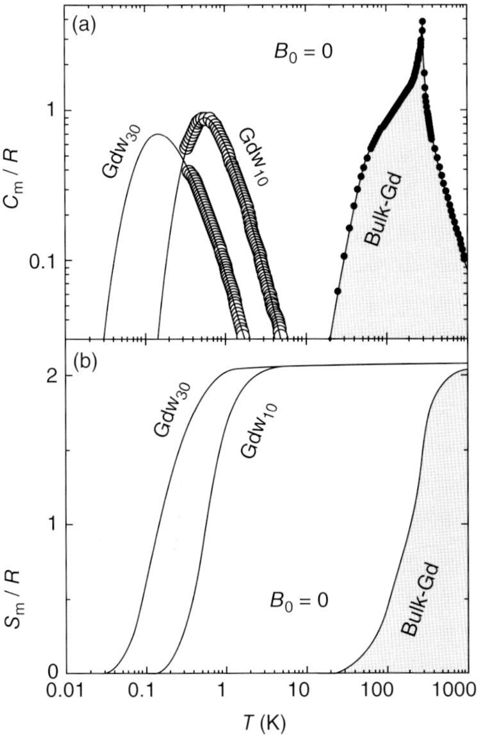

Figure 9.7 Upper graph (a) showing zero field magnetic contribution to the heat capacity for gadolinium metal, {Gd(III)W_{10}} and {Gd(III)W_{30}}. Lower graph (b) showing zero field magnetic entropy for the same. (Taken from Ref. [26] with permission from Wiley-VCH Verlag GmbH & Co. KGaA.)

working prototypes [27]. Using {Gd(III)W_{10}} and {Gd(III)W_{30}} therefore extends the working range deep into the millikelvin regime, this utility overcoming the rather low $-\Delta S_M$ in these situations. As shown in Figure 9.7, the magnetic contribution to the entropy is still substantial even at low temperatures, in contrast to that for Gd(0).

9.7.2.2 Ferromagnetism

Having seen the usefulness of paramagnetic systems we now demonstrate how, despite resulting in a lower MCE, at higher temperatures, ferromagnetic systems can be useful. A useful example is to compare two very similar compounds [28], the gadolinium(III) chains $[Gd(III)(O_2CCH_3)_3(MeOH)]_n$ and $[Gd(III)(O_2CCH_3)_3(H_2O)_{0.5}]_n$. These are one-dimensional compounds, where adjacent metals are bridged by an acetate ligand, and which display different exchange interactions, and this may be attributable to the reduced Gd–O–Gd bond angles in the latter. For the following Hamiltonian (Equation 9.7), for nearest ion interactions, it was found that $J = 0.03\,\mathrm{cm}^{-1}$ ($g = 1.99$) for the former (ferromagnetic) and $-0.01\,\mathrm{cm}^{-1}$ ($g = 2.00$) for the latter (antiferromagnetic), as fitted to magnetization data.

$$H = -J\Sigma s_n s_{n+1} \tag{9.7}$$

From the magnetization data $[Gd(III)(O_2CCH_3)_3(H_2O)_{0.5}]_n$ actually has a larger MCE, which we address shortly, but this is because of the differing relative molecular masses of the solvent, compared to $[Gd(III)(O_2CCH_3)_3(MeOH)]_n$, at 47.7 to $45.0\,\mathrm{J\,kg^{-1}\,K^{-1}}$ ($1.8\,\mathrm{K}$, $\Delta H = 0-7\,\mathrm{T}$). However, despite having the same spin of a single Gd(III) ion, in units of R, $-\Delta S_M$ is indeed larger for the ferromagnetic compound. The usefulness of a ferromagnetic system is due to the way in which the spins are more easily aligned with the applied field, so the magnetization saturates in lower fields or at higher temperatures. The consequence for the MCE is a material that is more usable over both a wider range of temperatures and fields, and with a potentially larger magnetocaloric effect under less forcing conditions than materials with higher peak $-\Delta S_M$ values, as we have here. For example, $[Gd(III)(O_2CCH_3)_3(MeOH)]_n$ shows larger $-\Delta S_M$ values at the more accessible conditions of 5 T and 5 K (approximately double!) than $[Gd(III)(O_2CCH_3)_3(H_2O)_{0.5}]_n$.

9.7.2.3 Antiferromagnetism

Finally, in this section we address antiferromagnetic interactions using a fascinating 3d–4f system [29], $[Gd(III)_4M(II)_8(OH)_8(L)_8(O_2CR)_8]ClO_4$, where M(II) can be Ni(II), Cu(II) or Zn(II); R is CH_2Me for Ni(II) and $CHMe_2$ for the latter two; and LH is 2-(hydroxymethyl)pyridine, found by Hooper *et al.* The metal ions are arranged as a gadolinium(III) square with each side making up a cubane as two transition metal ions lie above and below the lanthanide plane.

If we focus on the Cu(II) version of this we find a mixture of competing interactions, with fits of the magnetization data using the following Hamiltonian (Equation 9.8) solved by ITO-MAGFIT, so this is a ferrimagnetic compound; more precisely, antiferromagnetic coupling is found between the Cu(II) spins ($J_{CuCu} = 11.84\,\mathrm{cm}^{-1}$, in this scheme) and Gd(III) spins ($J_{GdGd} = 0.2\,\mathrm{cm}^{-1}$) but ferromagnetic coupling exists between heterospins ($J_{CuGd} = -1.38\,\mathrm{cm}^{-1}$). This scheme takes account of the nearest metal ion–metal ion interactions where s_{1-4} are the Gd(III) ions in the square, which interact with two transition metal ions in each adjacent cubane.

$$\begin{aligned}H = {} & g\mu_B B_0 \Sigma s_i + \Sigma\{J_{CuCu}(s_5 s_6 + s_7 s_8 + s_9 s_{10} + s_{11} s_{12})\} \\ & + \Sigma\{J_{CuGd}[s_1(s_5 + s_6 + s_{11} + s_{12}) + s_2(s_5 + s_6 + s_7 + s_8) \\ & + s_3(s_7 + s_8 + s_9 + s_{10}) + s_4(s_9 + s_{10} + s_{11} + s_{12})\} \\ & + \Sigma\{J_{GdGd}(s_1 s_2 + s_2 s_3 + s_3 s_4 + s_4 s_1)\}\end{aligned} \tag{9.8}$$

This gives a moderate MCE of $-\Delta S_M = 14.6\,\mathrm{J\,kg^{-1}\,K^{-1}}$ ($5.6\,\mathrm{K}$, $\Delta H = 0-7\,\mathrm{T}$). A higher value of $-\Delta S_M = 18\,\mathrm{J\,kg^{-1}\,K^{-1}}$ is found when M(II) = Zn(II), that is, for a diamagnetic metal. It would seem that antiferromagnetically coupled transition metal ions have a negative effect on the MCE here, as {Gd(III)$_5$} and {Gd(III)$_7$} suggest. Indeed, from the low field $-\Delta S_M$ plots, shown in Figure 9.8 by the red data points in the central panel, at 1 T field and at 0.9 K, $-\Delta S_M = -0.7\,J\,kg^{-1}\,K^{-1}$, this compound is a 'cryogenic heater'. Only with larger fields is the MCE positive in this region and so the compound behaves as a refrigerant. As shown, only the Cu(II) variant displays this behaviour.

We can explain this quite simply; if the exchange between spins is antiferromagnetic then the applied field is required first to 'flip' some of the spin to align with this field in order to achieve saturation. The antiferromagnetic coupling seen in {Gd(III)$_4$(Zn(II)$_8$)} is between Gd(III) ions only, and consequently much weaker; hence even the smallest applied field of 1 T is sufficient to induce a positive MCE by aligning the spins (also known as *decoupling*). We also see how the maximum MCE of this is at a lower temperature than for the Cu(II) system because the exchange interactions are much weaker.

A more general point is that the maximum spin is often not attained due to antiferromagnetic coupling. In Gd(III) systems, M_{sat} can usually be attained in moderate fields, whereas when transition metals are involved the spin state can be less than the maximum possible. Here, {Gd(III)$_4$Cu(II)$_8$} has a spin of $S = 18$, the maximum possible (albeit field induced).

Although quite subtle, the difference between paramagnetic, antiferromagnetic and ferromagnetic exchange can be significant, even in homometallic gadolinium(III) compounds, where this can affect usability at less harsh conditions of field and temperature than those more readily attainable in the laboratory. For transition metals, where the exchange is stronger, the effect is more marked, with even negative magnetocaloric effects observed as a consequence. For the largest $-\Delta S_M$ values at the lowest temperatures, a paramagnetic system is preferred.

9.7.3
Active Metal Percentage

It was noted above that despite having seven gadolinium(III) ions, [Gd(III)$_7$(OH)$_6$(tpa)$_6$(thmeH$_2$)$_5$(thmeH)(MeCN)$_2$](NO$_3$)$_2$ had only a moderate $-\Delta S_M$ [24]. A disadvantage of this compound, and others, is the relatively low percentage of metal. Here, about 71% is made up by diamagnetic ligands which cannot contribute to the magnetic entropy change; maximizing $-\Delta S_M$ means maximizing the active metal percentage. This has led to a new strategy based on much smaller ligands such as carbonate, formate and acetate, resulting in structures with more than double the active content of {Gd(III)$_7$} and much larger $-\Delta S_M$ values as a consequence. Of course, this is only appreciated when the units of joule per kilogram per kelvin or similar are used instead of units of R, but the former are more relevant to practical work.

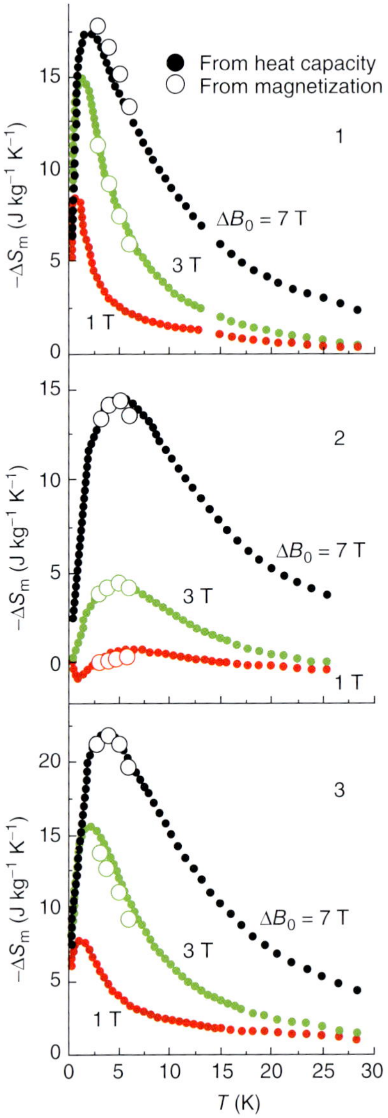

Figure 9.8 $-\Delta S_M$ versus temperature data for {Gd(III)$_4$Ni(II)$_8$} (3), {Gd(III)$_4$Cu(II)$_8$} (2) and {Gd(III)$_4$Zn(II)$_8$} (1), at various fields. Key: Results from magnetization data shown as open circles and from heat capacity data as solid circles. (Taken from Ref. [29] with permission from Wiley-VCH Verlag GmbH & Co. KGaA.)

Table 9.1 List of selected compounds showing the percentage of metal in each and their $-\Delta S_M$ values.

Compound	Dimensionality	ΔH (T)	T (K)	$-\Delta S_M$ ($J\,kg^{-1}\,K^{-1}$)	Gd(III) (%)
$[Gd(HCOO)_3]_n$ [20]	Three	7	3	55	54
$[Gd(OAc)_3(H_2O)_{0.5}]_n$ [28]	One	7	1.8	47.7	46
$[Gd(cit)(H_2O)]_n$ [30]	Two	7	2	43.6	43
$[Gd_2(OAc)_6(H_2O)_4]\cdot 4H_2O$ [11]	Zero	7	1.8	41.6	39
$[Gd_5(O)(O^iPr)_{13}]$ [23]	Zero	7	3	34	39
$[Gd_2(N\text{-}BDC)_3(dmf)_4]$ [31]	Three	7	1.8	29	27
$[Gd_7(OH)_6(thmeH_n)_6(tpa)_6(MeCN)_2](NO_3)_2$ [24]	Zero	7	3	23	29
$K_{12}[GdP_5W_{30}O_{110}]\cdot 54H_2O$ [26]	Zero	7	1.3	1.9	2

OAc, acetate; cit, citrate; N-BDC, 2-amino-1,4-benzenedicarboxylate and dmf, dimethylformamide.

The effect described above can be seen in the examples [26] $Na_9[Gd(III)(W_5O_{18})_2]\cdot 35H_2O$ and $K_{12}[Gd(III)(P_5W_{30}O_{110})]\cdot 54H_2O$, amongst others, rather clearly, considering that they are paramagnetic down to extremely low temperatures; the former has a much higher MCE resulting from the larger metal percentage, confirming theoretical expectations.

More widely, from Table 9.1, which shows the best performing gadolinium(III)-only magnetic coolers, along with the metal percentage, it appears that this factor is the most significant in determining whether a material has a large $-\Delta S_M$; there is an almost direct relation between the two factors, and the small discrepancies can be attributed to factors of exchange (see above). The metal percentage becomes less important as 3d metals are incorporated, up to the case of homometallic 3d compounds, where exchange is more important.

9.7.4 Density

A commonly forgotten property of 'good' magnetocaloric materials is their density, particularly as this cannot be easily controlled. This is important as it dictates the amount of a given material that can fit into a particular volume. This practical aspect may explain why it has been ignored by scientists for the most part, although a brief recap of the 'industry standard' makes it obvious. GGG [17] or $Gd(III)_3Ga(III)_5O_{12}$, has a $-\Delta S_M$ of $25\,J\,kg^{-1}\,K^{-1}$, hitherto the most commonly used units by which chemists/physicists quantify their discoveries. Unfortunately, the reason GGG is the industry standard stems not from this but from its extremely high density of $7.1\,g\,cm^{-3}$. Examination of Table 9.2 shows how the majority of lanthanide cages have densities only around $2\,g\,cm^{-3}$. Hence, when

Table 9.2 List of selected compounds with $-\Delta S_M$, their density and adjusted magnetocaloric effect.

Compound	Dimensionality	ρ (g cm^{-3})	$-\rho\Delta S_M$ (mJ cm^{-3} K^{-1})
$[Gd(HCOO)_3]_n$ [20]	Three	3.9	212
$[Gd(OAc)_3(H_2O)_{0.5}]_n$ [28]	One	2.2	106
$[Gd(HCOO)(C_8H_4O_4)]_n$ [32]	Three	2.7	125
$[Gd(cit)(H_2O)]_n$ [30]	Two	2.6	113
$[Gd_{24}(DMC)_{36}(CO_3)_{18}(H_2O)_2]\cdot 6H_2O$ [33]	Zero	2	90
$[Gd(HCOO)(OAc)_2(H_2O)_2]_n$ [31]	Three	2.4	110
$[Gd(OAc)_3(MeOH)]_n$ [11]	One	2.2	97
$[Gd_2(OAc)_6(H_2O)_4]\cdot 4H_2O$ [28]	Zero	2	85
$[Gd_4(OAc)_4(acac)_8(H_2O)_4]$ [28]	Zero	1.9	70
$[Gd_5(O)(O^iPr)_{13}]$ [23]	Zero	1.9	65
$\{Gd_{38}\}$ [34]	**Zero**	**2.7**	**102**
$\{Gd_{48}\}$ [34]	**Zero**	**2.8**	**121**
$[Gd_2(N\text{-}BDC)_3(dmf)_4]$ [31]	Three	1.4	4.1
$[Gd_2(OAc)_2(Ph_2acac)_4(MeOH)_2]$ [28]	Zero	1.5	36
$[Gd_7(OH)_6(thmeH_n)_6(tpa)_6(MeCN)_2](NO_3)_2$ [24]	Zero	1.8	41
$[Gd_2(fum)_3(H_2O)_4]\cdot 3H_2O$ [16]	Three	2.5	52
$[Gd_2(teaH_2)_2(tpa)_2(NO_3)_2]$ [35]	Zero	1.7	35
$[Gd_3Ga_5O_{12}]_n$ [17]	Three	7.1	191

DMC, *N,N′*-dimethylcarbamate; Ph_2acac, diphenylacetonacetonate; fum, fumarate; $teaH_3$, triethanolamine.

converted to the practical unit of millijoule per cubic centimeter per kelvin, the true utility is exposed.

Several cages have escaped this limitation, via 'anion templating'. Highlighted in Table 9.2 are two recent large cages [34] (bold) that have abnormally high densities compared to those of other zero-dimensional structures, which makes them competitive with polymeric materials. $\{Gd(III)_{38}\}$ and $\{Gd(III)_{48}\}$ rely on the templating of large numbers of metal ions around small anionic ligands such as ClO_4^- in the former and Cl^- and NO_3^- in the latter.

While it is hard to deliberately engineer increased densities, metal-organic frameworks represent an improvement over molecular cages, although methods such as anion templating can help. Reducing the size of ligands, too, can assist. A large $-\Delta S_M$, though, without a high density, is not enough to challenge commercial compounds.

9.7.5
Anisotropy or Spin: What Kind?

Of the lanthanide(III) ions, gadolinium(III) is exceptional, being the only isotropic paramagnetic ion of this series. Anisotropy, *D*, in 3d systems has been shown to hinder a large magnetocaloric effect at low temperatures as demonstrated by

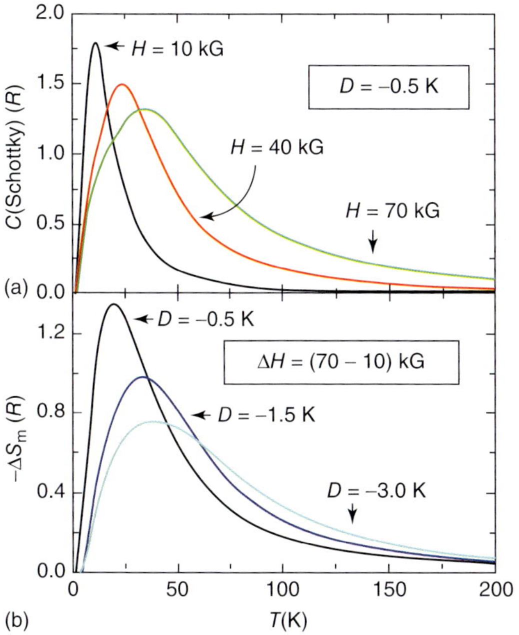

Figure 9.9 (a) Calculated Schottky heat capacity shown for three different fields of 1, 4 and 7 T for a hypothetical S = 10 and $D = -0.5$ K. (b) $-\Delta S_M$ values for a hypothetical S = 10 spin, with $D = -0.5$, -1.5 and -3.0 K for $\Delta H = 0 - 7$ T. (Taken from Ref. [13] with permission Royal Society of Chemistry.)

Evangelisti and Brechin [13]. The results of their work can help determine which of the d-transition metals are most suitable as cometals, if appropriate to the synthetic routes pursued.

This study [13] calculated the Schottky heat capacity for a hypothetical spin of $s = 10$, with an axial anisotropy, $-D$, of -0.5 K. Across three different fields, Figure 9.9 shows that the larger the field, the smaller the contribution of this heat capacity. More relevant here is that the derived $-\Delta S_M$ values showed the maximum $-\Delta S_M$ shifts and decreases with increasing anisotropy from $D = -0.5$ to -1.5 K and -3.0 K, as shown in Figure 9.9, for $\Delta H = 0 - 7$ T up to 200 K.

A larger anisotropy may actually be desirable for certain applications, though providing a possible route to the much sought after 'tunability' promised for molecular magnets in general. By tailoring the point at which the maximum $-\Delta S_M$ occurs by modification of the anisotropy of the metal, either by changing it or by changing its ligand environment, such control could be achieved [36]. We can relate this to Gd(III) in particular, as, while the anisotropy in lanthanides is described by more than a single term, it is dominated by B_2^0.

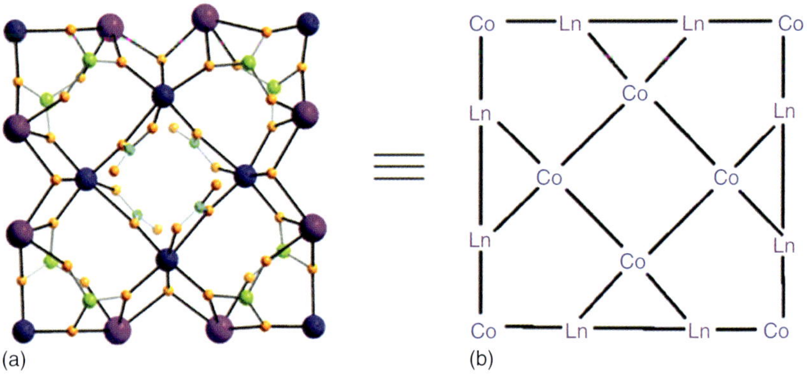

Figure 9.10 (a) Core of $[Ln(III)_8Co(II)_8(OH)_4(NO_3)_4(O_3P^tBu)_8(O_2C^tBu)_{16}]$ with connecting O-based ligands. Key: Ln(III) purple; Co(II) blue; N turquoise; P green; O yellow. (b) Cartoon of the same. (Taken from Ref. [32] with permission American Chemical Society.)

Such modification of the anisotropy would be hard to achieve, though, with Gd(III) ions, as their anisotropy would be minimal even with engineered ligand changes; so this approach would be better suited to 3d chemistry, where the metal ion orbitals can interact more strongly with those of the ligands. Figure 9.9 tells us that isotropic 3d-metals are better for lower temperature work than anisotropic ones.

The unsuitability of non-gadolinium(III) lanthanide(III) ions has been demonstrated empirically by Zheng *et al.* [32]. The isostructural pair of compounds $[Ln(III)_8Co(II)_8(OH)_4(NO_3)_4(O_3P^tBu)_8(O_2C^tBu)_{16}]$, where Ln(III) is either Gd(III) or Dy(III), are 3d–4f phosphonates, with the core structure shown in Figure 9.10. See the comprehensive review in [32] from whence Figure 9.10 is taken for a huge study of lanthanide–cobalt–phosphonate compounds.

For the Gd(III) compound $-\Delta S_M$ is $21.4\,J\,kg^{-1}\,K^{-1}$ ($\Delta H = 0 - 7\,T, 3\,K$), whereas for Dy(III) this is only $11.6\,J\,kg^{-1}\,K^{-1}$. This is accounted for by the lower spin of Dy(III), but what is interesting is the fact that $-\Delta S_M$ has already passed through a maximum at 4 K, compared to the Gd(III) plot, which is still rising at 3 K, which is qualitatively in line with the work of Evangelisti and Brechin [13] for 3d metals. From this, we can say that the best lanthanide(III) for low-temperature MCE is gadolinium(III); indeed, there are only a handful of compounds using Dy(III) reported for this application.

9.7.6
Dimensionality

The majority of MCE materials have been zero dimensional. More recently, MOFs have emerged as superior alternatives, relative to lanthanide cages, as magnetocaloric materials on account of their improved stability, increased density and metal percentage.

Tables 9.1 and 9.2 show how non-zero-dimensional compounds can have the largest MCEs. Even amongst this non-exhaustive selection, these compounds have larger Gd(III) percentages and densities. It is possible, of course, to synthesize polymeric compounds with *inferior* performance. However, these represent a new tool in the arsenal of magnetocaloric research, which, as we will see below, has been extremely successful, including in 3d–4f materials.

There is nothing intrinsically superior about non-zero-dimensional materials compared to lanthanide cages as magnetocaloric materials, but, thus far, synthetic chemists have been unable to realize as many of the required properties in cages simultaneously, as they have done with some chains and lattices. The key advantage is cramming in as many metals as possible into a structure with as few ligands as possible.

9.8 High-Performance MCE Materials

Here, we examine some of the most significant compounds reported to date. One of the best and most well-investigated homometallic lanthanide compounds is $[Gd^{III}{}_2(H_3CCO_2)_6(H_2O)_4]\cdot 4H_2O$ [11]. This comprises two metal ions linked by two acetate groups ($\eta^2:\eta^1:\mu_2$ or 2 : 2 : 1 in Harris notation [37]) with further capping acetate groups and water molecules. Thus, it exemplifies a molecular lanthanide system with a large metal content of 39%, with a small ferromagnetic interaction of $J/k_B = 0.068(2)$ K and $g = 2.01(1)$ fitted to susceptibility data with the Hamiltonian given in Equation 9.9.

$$H = -J(s_{Gd1} \cdot s_{Gd2}) \tag{9.9}$$

Interestingly, this compound was known for some years [38, 39] before MCE research came back into vogue. Here, the maximum $-\Delta S_M$ for a decoupled system is $42\,J\,kg^{-1}\,K^{-1}$ and is almost met for $\Delta H = 0 - 7$ T and 1.8 K. So, we have a reasonably high metal content, with a small, though ferromagnetic, interaction, with the appropriate high spin metals. Heat capacity data allow the adiabatic temperature change to be calculated; here, this was found to be 12.7 K below 2 K, one of the best by this measure until recently.

This has been improved upon by various compounds, including $[Mn^{II}(H_2O)_6][Gd^{III}Mn^{II}(oda)_3]_2\cdot 6H_2O$ [21], a 3D network where each metal is linked via ligands to six heterometals. Despite the seeming advantages of Gd(III) outlined above in spin, isotropy and weak exchange, 'chemical engineering' or even serendipity should allow sufficiently useful exchange to be found in 3d materials. It is worth remembering that Mn(II) and not Gd(III) has the highest $-S_M$ per ion (271.1 versus $109.9\,J\,kg^{-1}\,K^{-1}$). The success of this alternation of metal ions approach in

{Mn(II)Gd(III)} is evident, as $-\Delta S_M$ approaches the maximum value achievable for uncoupled ions: 50.1 $J\,kg^{-1}\,K^{-1}$ (at 1.8 K and $\Delta H = 0-7$ T) to compared to the maximum possible of 53.3 $J\,kg^{-1}\,K^{-1}$. Furthermore, no maxima in $-\Delta S_M$ were seen at any field (1.8 K), validating this strategy, but this does require a degree of fortune in the minimization of interactions and in the way the metals are arranged.

The best performing homometallic magnetocaloric material is currently $[Gd(III)(HCO_2)_3]_n$ [20]: it has a three-dimensional network with only a few lightweight ligands (Gd(III)% = about 54%) with only very weak antiferromagnetic interactions because of the choice of formate as a ligand. This has also been seen in a significant number of MOFs, which show a large magnetocaloric effect [42]. Resulting from this ligand choice, here [20], is a very large density of 3.9 $g\,cm^{-3}$, higher than all reported MCE materials, excepting GGG [17]. This has 55 $J\,kg^{-1}\,K^{-1}$ (3 K, $\Delta H = 0-7$ T) and a huge value of 212 $mJ\,cm^{-3}\,K^{-1}$, surpassing even GGG (191 $mJ\,cm^{-3}\,K^{-1}$). Impressively, $\Delta T_{AD} = 25$ K at around 2 K, surpassing $[Gd^{III}{}_2(H_3CCO_2)_6(H_2O)_4]\cdot 4H_2O$ [11] significantly.

We now return to the theme of Mn(II), though, in a brief departure from our main focus of Ln(III) materials, but which may point the way to future advances.

If one takes only the 'headline' figure of $-\Delta S_M$ as a guide one may suspect that lanthanide magnetocaloric materials are outperformed by a recent 3d molecule $[Mn(II)(HOCH_2CO_2)_2(H_2O)_2]$ ($-\Delta S_M = 60\,J\,kg^{-1}\,K^{-1}$ at 1.8 K and for $\Delta H = 0-7$ T). Indeed, this work by Chen *et al.* [40] may reopen avenues into 3d transition-metal magnetocaloric materials. The strategy involves the separation of Mn(II) ions in a 3D lattice, via addition of water molecules, into discrete units, which are more isolated magnetically. This is an extension of the same idea as that used in the commercial ferric ammonium alum (FAA) [41], which may be used in conjunction with GGG.

One problem with this approach, though, is that it reduces the density of the material such that, in the adjusted units of millijoule per cubic centimetre per kelvin, the magnitude of $-\Delta S_M$ is only half as effective as that of gadolinium formate.

9.9
Outlook

By now, chemists and physicists know what they are competing against, namely, the commercial materials GGG (and FAA), which are used to reach millikelvin temperatures routinely. For their own research to be productive, significant improvements need to be made to molecular-based systems, though Gd(III) MOF materials have leap-frogged these considerably. It is possible that the 'ideal' material has already been made; numerous gadolinium extended materials lying in the crystallographic databases have not been investigated for their magnetocaloric properties, for instance, gadolinium fluoride, $[Gd(III)F_3]$.

Hopefully, this chapter has provided a concise guide to the main points to consider when making and measuring such materials, and has illustrated the best successes in this field recently.

References

1. Sessoli, R., Gatteschi, D., Caneschi, A. and Novak, M.A. (1993) *Nature*, **365**, 141–143.
2. Wernsdorfer, W. (2010) *Int. J. Nanotechnol.*, 7, 497–522.
3. Giauque, W.F. and MacDougall, D.P. (1933) *Phys. Rev.*, **43**, 0768–0768.
4. Sharples, J.W. and Collison, D. (2013) *Polyhedron*, **54**, 91–103.
5. Zheng, Y.Z., Zhou, G.J., Zheng, Z.P. and Winpenny, R.E.P. (2014) *Chem. Soc. Rev.*, **43**, 1462–1475.
6. Bartolomé, J., Luis, F. and Fernández, J.F. (eds) (2014) *Molecular Magnets, NanoScience and Technology*, Springer-Verlag, Berlin, Heidelberg, pp. 365–387.
7. Fa, W.Z. and Jin, Y.Q. (2010) *Chin. Sci. Bull.*, **55**, 4005–4009.
8. Brisson, J.G. (1998) *J. Low Temp. Phys.*, **111**, 181–199.
9. Martin, F., Vermeulen, G., Camus, P. and Benoit, A. (2010) *Cryogenics*, **50**, 623–627.
10. Yu, B.F., Gao, Q., Zhang, B., Meng, X.Z. and Chen, Z. (2003) *Int. J. Refrig.-Revue Int. Du Froid*, **26**, 622–636.
11. Evangelisti, M., Roubeau, O., Palacios, E., Camon, A., Hooper, T.N., Brechin, E.K. and Alonso, J.J. (2011) *Angew. Chem. Int. Ed.*, **50**, 6606–6609.
12. Gschneidner, K.A., Pecharsky, V.K., Pecharsky, A.O. and Zimm, C.B. (1999) *Rare Earths*, **98**, 315–313, 69–76.
13. Evangelisti, M. and Brechin, E.K. (2010) *Dalton Trans.*, **39**, 4672–4676.
14. Gschneidner, K.A. and Pecharsky, V.K. (2000) *Annu. Rev. Mater. Sci.*, **30**, 387–429.
15. Langley, S., Helliwell, M., Sessoli, R., Teat, S.J. and Winpenny, R.E.P. (2008) *Inorg. Chem.*, **47**, 497–507.
16. Sedlakova, L., Hanko, J., Orendacova, A., Orendac, M., Zhou, C.L., Zhu, W.H., Wang, B.W., Wang, Z.M. and Gao, S. (2009) *J. Alloys Compd.*, **487**, 425–429.
17. McMichael, R.D., Ritter, J.J. and Shull, R.D. (1993) *J. Appl. Phys.*, **73**, 6946–6948.
18. Hamilton, A.C.S., Lampronti, G.I., Rowley, S.E. and Dutton, S.E. (2014) *J. Phys. Condens. Matter*, 26.
19. Kahn, O., Cador, O., Larionova, J., Mathoniere, C., Sutter, J.P. and Ouahab, L. (1997) *Mol. Cryst. Liq. Cryst. Sci. Technol., Sect. A: Mol. Cryst. Liq. Cryst.*, **305**, 1–16.
20. Lorusso, G., Sharples, J.W., Palacios, E., Roubeau, O., Brechin, E.K., Sessoli, R., Rossin, A., Tuna, F., McInnes, E.J.L., Collison, D. and Evangelisti, M. (2013) *Adv. Mater.*, **25**, 4653–4656.
21. Guo, F.-S., Chen, Y.-C., Liu, J.-L., Leng, J.-D., Meng, Z.-S., Vrabel, P., Orendac, M. and Tong, M.-L. (2012) *Chem. Commun. (Camb.)*, **48**, 12219–12221.
22. Cotton, S. (2006) *Lanthanide and Actinide Chemistry*, John Wiley and Sons, Ltd, Chichester. doi: 10.1002/0470010088.
23. Blagg, R.J., Tuna, F., McInnes, E.J.L. and Winpenny, R.E.P. (2011) *Chem. Commun.*, **47**, 10587–10589.
24. Sharples, J.W., Zheng, Y.Z., Tuna, F., McInnes, E.J.L. and Collison, D. (2011) *Chem. Commun.*, **47**, 7650–7652.
25. Chilton, N.F., Anderson, R.P., Turner, L.D., Soncini, A. and Murray, K.S. (2013) *J. Comput. Chem.*, **34**, 1164–1175.
26. Martinez-Perez, M.-J., Montero, O., Evangelisti, M., Luis, F., Sese, J., Cardona-Serra, S. and Coronado, E. (2012) *Adv. Mater.*, **24**, 4301–4305.
27. Pecharsky, V.K. and Gschneider, K.A. Jr., (1999) *J. Magn. Magn. Mater.*, **200**, 44–56.

28. Guo, F.S., Leng, J.D., Liu, J.L., Meng, Z.S. and Tong, M.L. (2012) *Inorg. Chem.*, **51**, 405–413.
29. Hooper, T.N., Schnack, J., Piligkos, S., Evangelisti, M. and Brechin, E.K. (2012) *Angew. Chem. Int. Ed.*, **51**, 4633–4636.
30. Liu, S.J., Xie, C.C., Jia, J.M., Zhao, J.P., Han, S.D., Cui, Y., Li, Y. and Bu, X.H. (2014) *Chem. Asian J.*, **9**, 1116–1122.
31. Lorusso, G., Palacios, M.A., Nichol, G.S., Brechin, E.K., Roubeau, O. and Evangelisti, M. (2012) *Chem. Commun.*, **48**, 7592–7594.
32. Zheng, Y.-Z., Evangelisti, M., Tuna, F. and Winpenny, R.E.P. (2012) *J. Am. Chem. Soc.*, **134**, 1057–1065.
33. Chang, L.X., Xiong, G., Wang, L., Cheng, P. and Zhao, B. (2013) *Chem. Commun.*, **49**, 1055–1057.
34. Guo, F.-S., Chen, Y.-C., Mao, L.-L., Lin, W.-Q., Leng, J.-D., Tarasenko, R., Orendac, M., Prokleska, J., Sechovsky, V. and Tong, M.-L. (2013) *Chem. Eur. J.*, **19**, 14876–14885.
35. Sharples, J.W. (2014) Cooling rapidly and relaxing slowly. PhD thesis. The University Of Manchester.
36. Luzon, J. and Sessoli, R. (2012) *Dalton Trans.*, **41**, 13556–13567.
37. Coxall, R.A., Harris, S.G., Henderson, D.K., Parsons, S., Tasker, P.A. and Winpenny, R.E.P. (2000) *J. Chem. Soc., Dalton Trans.*, 2349–2356.
38. Hatscher, S.T. and Urland, W. (2003) *Angew. Chem. Int. Ed.*, **42**, 2862–2864.
39. Favas, M.C., Kepert, D.L., Skelton, B.W. and White, A.H. (1980) *J. Chem. Soc., Dalton Trans.*, 454–458.
40. Chen, Y.C., Guo, F.S., Liu, J.L., Leng, J.D., Vrabel, P., Orendac, M., Prokleska, J., Sechovsky, V. and Tong, M.L. (2014) *Chem. Eur. J.*, **20**, 3029–3035.
41. Shirron, P.J. (2007) *J. Low Temp. Phys.*, **148**, 915–920.
42. Sibille, R., Mazet, T., Malaman, B. and Francois, M. (2012) *Chem. Eur. J.*, 18.

10
Actinide Single-Molecule Magnets

Stephen T. Liddle and Joris van Slageren

10.1
Introduction

This book documents recent progress in the field of molecular nanomagnetism, focusing on the f-elements, that is, the lanthanides and actinides. Research in molecular nanomagnets has radically shifted its focus in recent years and the f-elements have become a major interest. In order to understand the reasons behind this change of direction, as well as the nomenclature and approaches used, it is helpful to outline the history of the field [1]. It all started with the discovery that certain polynuclear molecular transition metal coordination clusters display slow magnetization dynamics at low temperatures, which survives in dilute conditions, for example, in frozen solution or polymer films [2]. This means that the observed slow dynamics of the magnetization must be a molecular property, since all other mechanisms of slow dynamics (long-range magnetic ordering, spin-glass behaviour) rely on interparticle interactions. It turns out that the magnetization dynamics can often be well described considering the ground spin multiplet of the exchange coupled cluster only (the giant spin approximation). The ground spin multiplet possesses magnetic anisotropy, which in certain cases causes the magnetic moment to be preferentially aligned along a certain axis, called the *easy axis of magnetization*. Whether this type of magnetic anisotropy occurs depends on a multitude of factors, such as the coordination geometry, symmetry of the single ions and the symmetry of the cluster itself. Furthermore, thermal relaxation must occur in a series of steps via a number of intermediate (quantized) orientations of the magnetic moment, because of the selection rule for spin–phonon interaction. This effectively creates an energy barrier for inversion of the magnetic moment, and hence the temperature dependence of the relaxation rate follows the Arrhenius law, $\tau = \tau_0 \exp(\Delta E/k_B T)$. The rate of tunnelling of the magnetic moment under the energy barrier is often rather limited, especially for larger values of the ground state spin. For the determination of relatively rapid relaxation rates, alternating

Lanthanides and Actinides in Molecular Magnetism, First Edition.
Edited by Richard A. Layfield and Muralee Murugesu.

current (ac) magnetic susceptibility is the method of choice, exploiting the fact that the product of the angular frequency of the applied magnetic field and the relaxation time is 1 ($\omega\tau = 1$) at the maximum in the frequency-dependence of the out-of-phase component of the ac magnetic susceptibility χ'' (see, for example, Figure 10.1). In 2003, slow relaxation of the magnetic moments in the complexes $(NBu_4)[Ln(Pc)_2](H_2Pc$ = phthalocyanine, Ln = Dy^{3+}, Tb^{3+}) was reported [3]. The spin dynamics of lanthanides in extended lattices had been investigated in detail in the 1960s and even before [4], but this was the first time that molecular compounds had been considered in this context. Interestingly, assuming an Arrhenius-type temperature dependence of the relaxation time, much higher energy barriers were found for these and other lanthanide complexes than for transition-metal-based systems [5].

However, the Arrhenius plot of $\ln(\tau)$ versus $1/T$ is seldom a straight line in f-element-based single molecule magnets (see, for example, Figure 10.2), indicating the contribution of different types of mechanisms to the relaxation of the magnetic moment. Indeed, this points to a fundamental difference between the spin dynamics of exchange coupled clusters and (effective) single-ion systems. Spin relaxation processes, in which energy is exchanged with the surroundings (lattice) are collectively known as *spin–lattice relaxation*, with a characteristic spin lattice relaxation time T_1 [4a]. There are at least four mechanisms, each with its own frequency- and temperature dependence (Figure 10.3) [7]. The simplest of these is the direct process, in which the spin system absorbs or emits a phonon from or to the lattice, which has exactly the same energy as the spin transition. This is the only mechanism which is operative at the lowest temperatures. The selection rule requires that $\Delta m_J = \pm 1, \pm 2$ [1], which makes the exact composition of the wavefunctions of the magnetic states very important. On the other hand, the relaxation rate at low frequencies is rather small because of the small phonon density of states. In Raman processes, a phonon is scattered inelastically, with the energy difference being taken up or released by the spin system, either directly (first order Raman) or via an intermediate virtual spin state (second order Raman). Both of these processes can be viewed as two-phonon processes. Hence, phonons need to be present in the system, which requires finite temperatures. The temperature dependence of the Raman relaxation rate follows a power law $k_{Raman} \alpha T^n$ ($n = 5 - 9$, depending on the details of the system). Finally, the spin lattice relaxation can proceed via real intermediate states, both of the spin and phonon systems, which is called the *Orbach process*. This process can be viewed as a one-step two-phonon process or as two consecutive direct processes, where Orbach preferred the former. Only this last process gives rise to an exponential temperature dependence, with a functional form similar to the Arrhenius equation. In addition to spin lattice relaxation, quantum tunnelling can occur, where the magnetic moment tunnels under the barrier, if the quantum mechanical eigenstates are superpositions of up and down ($\pm m_J$) states. Interactions that give rise to such superposition

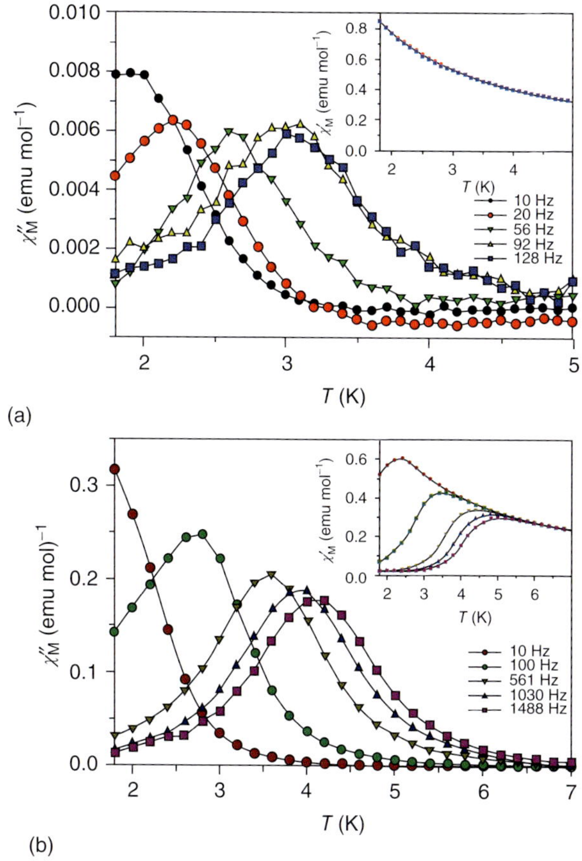

Figure 10.1 Temperature dependence of imaginary parts of the ac susceptibility (real parts in insets) recorded on powder samples of **1** at different frequencies as indicated, at external dc fields of 0 Oe (a) and 1000 Oe (b). (Reprinted from Ref. [6]. Copyright (2009) American Chemical Society.)

states include the crystal field, transverse Zeeman and hyperfine interactions. In systems with an odd number of unpaired electrons (Kramers systems), the crystal field cannot give rise to such superposition states, which partly explains the success of the use of dysprosium in lanthanide-based single molecule magnets. Perhaps a word on terminology here: It is clear from recent literature that for the large majority of f-element-based molecular systems that show slow relaxation of the magnetic moment, the slow relaxation is a single-ion or slightly modified single-ion property, which has led to the term *single-ion magnet*. Here, we will refer to any system with a relaxation rate of the magnetic moment that is small enough to be measured by ac susceptibility at a frequency of the applied ac field of 10^3 Hz – a single-molecule magnet, irrespective of the presence or absence of an

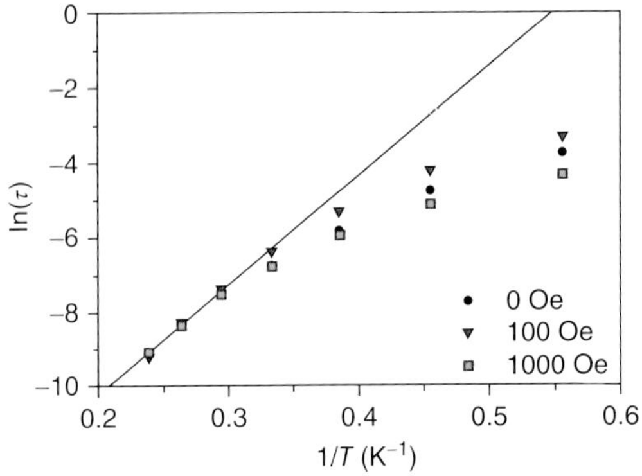

Figure 10.2 Arrhenius plot of the natural logarithm of the relaxation time extracted from the ac susceptibility data as a function of the inverse temperature for **1** at different external fields as indicated. (Reprinted from Ref. [6]. Copyright (2009) American Chemical Society.)

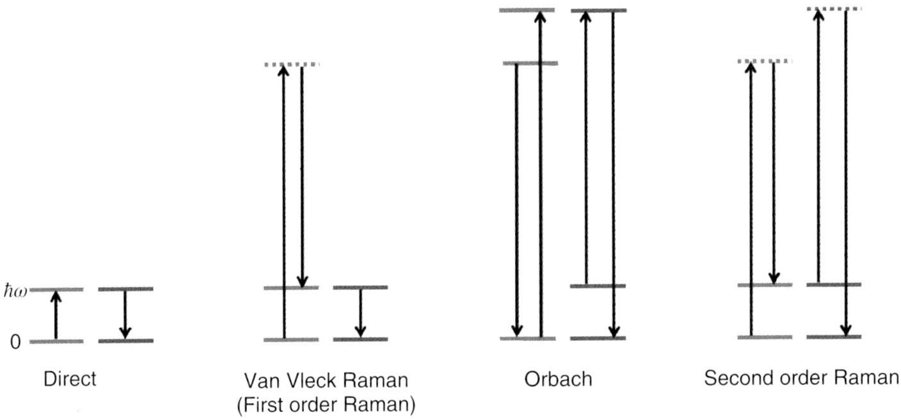

Figure 10.3 Schematic representation of the different mechanisms of spin–lattice relaxation, where it is assumed that initially the spin system is in an excited state. The change(s) in the phonon system is/are depicted in blue on the left, and the change(s) in the spin system on the right. Solid lines are real states, and dotted lines are virtual states. Quantum tunnelling processes are not shown.

external field, and independent of the number of metal ions present. A final time constant in the dynamics of the magnetic moment is the thermalization of the phonons inside the sample with the surroundings of the sample. This relaxation time can be surprisingly long, because of the small phonon density of states at low energies, compared to the spin density in a nondiluted sample. The resulting effect is called the *phonon bottleneck*, which may very well play a significant role in the dynamics of the magnetic moment, given that it leads to butterfly-shaped

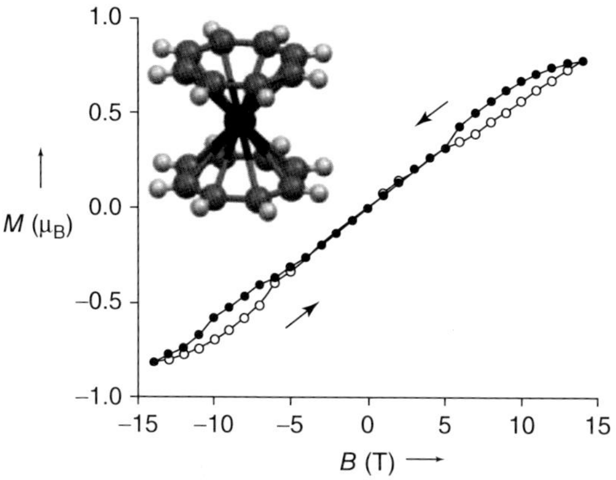

Figure 10.4 Magnetic hysteresis curve recorded on a powder sample of 10 at 1.8 K. (Reprinted from Ref. [9]. Copyright (2011) Wiley-VCH.)

hysteresis curves (for example, see Figure 10.4) [8], often also observed in f-element single-molecule magnets (SMMs). Only detailed studies of the field-, frequency- and temperature dependence can allow determination of the principal operative mechanisms of relaxation of the magnetic moment.

In transition metal complexes, the electron–electron and crystal field interactions are typically of similar magnitudes, while spin–orbit coupling is much weaker. In contrast, for lanthanide complexes, the weakest interaction (neglecting hyperfine interactions and applied fields) is the crystal field due to the limited radial extension of the f-orbitals [10]. The crystal field will therefore determine the energy spectrum at lowest energies ($10^2\,cm^{-1}$). Larger than the crystal field interaction is spin–orbit coupling ($\sim 10^3\,cm^{-1}$), and larger still (but potentially of similar size) is the electron–electron interaction ($\sim 10^4\,cm^{-1}$). Actinides are different from lanthanides, however. For the development of novel molecular nanomagnets, these differences may or may not be advantageous. For *trivalent* actinides, the Coulomb (electron–electron) interaction is about 60% of that of the lanthanide ion with the same electronic configuration, the spin–orbit coupling constant (ζ) is about twice as large, while the crystal field splitting is more than double that in the corresponding lanthanide [11]. Table 10.1 reports typical values for the electron–electron correlation parameters F^k, the spin–orbit coupling constant ζ, as well as the crystal field splitting parameters B_k^q. Of course, the last are strongly dependent on the host lattice, both on account of the coordination symmetry and the nature of the ligands. The values are meant as crude indications. What can be learned from the values in this table is that the electron–electron interaction (absent for the f^1 configuration) increases slightly with oxidation state and atomic number, and the spin orbit coupling constant increases with atomic number, but only very slightly, if at all, with oxidation state. In contrast, crystal field splitting is very strongly dependent on oxidation state (as indeed for the

Table 10.1 Magnitudes of electron–electron interaction, spin–orbit coupling and crystal field splitting interactions.

Configuration, ion	Ionic radius (nm)[a)]	max(F^k)	ζ	max($\|B_k^q\|$)
f^3: U(III)[b)]	0.1025	40×10^3	1.6×10^3	1.3×10^3
f^2: U(IV)[c)]	0.089	43×10^3	1.8×10^3	2.8×10^3
f^1: U(V)[d)]	0.076	0	1.6×10^3	13×10^3
f^3: Np(IV)[e)]	0.087	45×10^3	2.1×10^3	3.5×10^3
f^2: Np(V)[f)]	0.075	49×10^3	2.2×10^3	15×10^3
f^1: Np(VI)[g)]	0.072	0	2.4×10^3	45×10^3
f^3: Nd(III)[h)]	0.0995	72×10^3	0.88×10^3	0.7×10^3
f^2: Pr(III)[h)]	0.101	68×10^3	0.75×10^3	0.6×10^3
f^1: Ce(III)[h)]	0.103	0	0.64×10^3	1.0×10^3

All values in per centimetre.
a) Coordination number 6, from Ref. [12].
b) In $LaCl_3$ [11, 13].
c) In $ThCl_4$ [14].
d) In UCl_5 [11].
e) In $ThSiO_4$ [14].
f) As $CsNpF_6$ [11].
g) As NpF_6 [11].
h) Mean free ion parameters [13].

transition metals) and also increases with atomic number. Because, in actinides, the relative strengths of spin–orbit coupling and electron–electron interaction are much closer, neither the Russell–Saunders coupling scheme, which assumes weak spin–orbit coupling, nor *jj* coupling (weak electron–electron interaction) is valid, and both interactions must be considered at the same time. In addition, configuration interaction is important. This leads to 20 free ion parameters and between 2 and 27 crystal field splitting parameters, depending on the molecular symmetry and total angular momentum J. Hence, full analysis of the electronic structure of actinides is a formidable task. The electronic structure is, however, closely related to the magnetic properties that are of interest in this chapter. A final difference between lanthanides and actinides is that the radial extension of the f-orbitals is larger for the latter than for the former. This leads to larger crystal field splittings, and potentially stronger exchange interactions between the f-ion and other magnetic moments. Both of these can 'improve' SMM properties. However, at the same time, spin–phonon coupling becomes stronger, which leads to faster spin-lattice relaxation, hence faster relaxation of the magnetic moment.

The general magnetic properties of ionic and molecular actinide compounds have been reviewed recently [15], and only the briefest of overviews is given here. The $5f^1$ ions U^{5+} and Np^{6+} have $^2F_{5/2}$ ground multiplets in the Russell–Saunders coupling scheme. For these ions, electron–electron interactions are absent, which greatly simplifies the analysis. Under the influence of the crystal field the $J = 5/2$ ground multiplet splits into a fourfold degenerate state (Γ_8 in Bethe notation, $G_{3/2}$ in Mulliken notation, O′ octahedral double group symmetry) [16] and a twofold degenerate state ($\Gamma_7 = E_{5/2}$). This ground doublet is not a pure $J = 5/2$ state but contains a contribution from the $J = 7/2$ excited state. In addition, covalency has also been found to be important in determining the magnetic properties. In practice, magnetic moments of the order of $1\,\mu_B$ are found. The ground doublet is electron paramagnetic resonance (EPR)-active if it has a contribution from the $m_J = \pm 1/2$ states (crystal field quantum number $\mu = 1/2$), and many EPR studies have been performed on $5f^1$ compounds [4a, 17]. For the ground doublet, g values between 0.7 and 2.1 are found [17b, 18]. For weak-field ligands, the crystal-field-induced J-mixing is limited and quite strong g value anisotropies are observed [19].

The magnetic properties of $5f^2$, that is, U^{4+} and Np^{5+} systems are more complicated. The ground multiplet is $^3H_4$. In relatively high symmetries (tetragonal or higher), the ground state is often (in O_h always, but not in cubic coordination geometry) a singlet with zero first-order Zeeman coefficient [15, 20], and, hence, temperature-independent magnetism only. In symmetries lower than tetragonal, the $J = 4$ state will split into nine singlets, and because they are generally thermally accessible, the compounds display temperature-dependent paramagnetic susceptibilities. EPR has been less used in $5f^2$ systems. Finally, the f^3 ions U^{3+} and Np^{4+} have $^5I_{9/2}$ Russell–Saunders ground states. Typically, magnetic moments of $\mu_{eff} \sim 2.5 - 3.5\,\mu_B$ are found, which decrease to $0.4 - 0.6\,\mu_B$ at liquid helium temperatures due to depopulation of the crystal field levels. EPR spectra reveal anisotropic g values in the range 0.7–3.6. In the context of single-molecule magnets, it is interesting to note here the report of measured spin–lattice relaxation times of up to a second in U^{3+} : $LnCl_3$ [21].

EPR investigations are limited to the ground (quasi)doublet of an ion. Spectroscopically, the energy gap to the first excited crystal field states of the ground doublet can be investigated by optical spectroscopy. For lower valency ions, the corresponding transitions are in the far infrared (FIR), while for the higher valency ions, such as U^{5+}, the lowest transitions are in the infrared (IR) or near-infrared (NIR) [17b]. For the former energy range, either direct absorption of electromagnetic radiation (FIR spectroscopy) or inelastic neutron scattering would be amenable. However, while IR/NIR/vis/UV studies of the ionic actinide compounds are well established [11], FIR and inelastic neutron scattering (INS) appear to have been used less frequently [22]. The crystal field splitting can also be investigated by studying the splitting of formal intermultiplet transitions by electronic absorption or magnetic circular dichroism spectroscopies, which is a well-developed area [11, 23]. Finally, luminescence spectroscopy has also been used to investigate the electronic structure of actinide complexes [24].

10.2 Literature Survey of Published Actinide Single-Molecule Magnets

To summarize, compared to lanthanides, actinides offer larger spin–orbit couplings, leading to larger anisotropies, and concurrently potentially larger energy barriers. In addition, the larger radial extension of the 5f compared to 4f orbitals potentially leads to stronger exchange interactions in polynuclear systems, leading to diminished quantum tunnelling rates. Both of these features are positive when it comes to improving single-molecule magnets. In spite of these attractive features, only 15 molecular actinide compounds that display slow dynamics of the magnetic moment (single-molecule magnets) have been reported to date (end of 2013, Table 10.2). Most examples are based on f^3 actinides (U^{3+} and Np^{4+}), although f^1 and mixed-valent systems have also been reported.

10.2.1 Single-Molecule Magnets of f^3 Actinides (U^{3+}, Np^{4+})

The best studied f^3 actinide single-molecule magnets are pyrazolylborate complexes of U^{3+}, entries **1–5** in Table 10.2, whose magnetic properties have recently been modelled [25]. The first published such complex was $[U(Ph_2BPz_2)_3]$ (**1**), where Ph_2BPz^- is diphenylbis(pyrazolyl)borate [6]. The stated rationale behind this investigation was that the ligand might stabilize an oblate f-electron density distribution, because the ligand electron density can be expected to be concentrated above and below the pseudo symmetry axis, rather than in the plane perpendicular to that axis. The pseudo symmetry axis is assumed to be the crystal field quantization axis. Because for an f^3 configuration, crystal-field doublets of high m_J values possess such an oblate electron density distribution [26], effectively, states with large magnetic moments would be stabilized. Inspection of the crystal structure [27] reveals that the distance of the coordinating nitrogen atoms to the pseudo C_3 axis (crystal space group is $P_{\bar{1}}$) is of the order of 2 Å, while that between the nitrogen atoms above and below the plane is about 3 Å, which fulfils the requirement of stabilizing an oblate electron density. Indeed, ac susceptibility measurements at frequencies of up to around 1500 Hz demonstrated that measurably slow relaxation of the magnetic moment does occur (Figure 10.1). In zero applied static magnetic field, the fraction of the sample involved is of the order of only 1%. Applying a small static field in the range of 100–1000 Oe causes the whole sample to display slow relaxation dynamics with a very small distribution in relaxation times. Interestingly, while the relaxation times at low temperature ($T < 3$ K) change by a factor of 2 depending on static field, at higher temperatures they are almost temperature independent. The fit to the Arrhenius equation $\ln \tau = \ln \tau_0 + U_{eff}/kT$ yielded an effective energy barrier $U_{eff} = 20\,cm^{-1}$ and a pre-exponential factor of $\tau_0 = 1 \times 10^{-9}$s (Figure 10.2). These values would turn out to be quite typical

Table 10.2 Overview of energy barriers and pre-exponential factors derived from Arrhenius fits of the ac susceptibility data for single-molecule magnets 1–15.

#	Year	Compound	H_{dc}(Oe)	U_{eff}/k_B(K)	τ_0(s)	References
f^3: Uranium(III), Neptunium(IV)						
1	2009	$[U(Ph_2BPz_2)_3]$	0	29	1×10^{-9}	[6]
2	2010	$[U(H_2BPz_2)_3]$	100	12	1.2×10^{-6}	[29]
		Diluted in $[Y(H_2BPz_2)_3]$	100	23	6×10^{-8}	
3	2011	$[U(TpMe_2)_2(bipy)]I$	500	26.2	1.4×10^{-7}	[31]
4	2012	$[U(TpMe_2)_2I]$	500	30.3	1.8×10^{-7}	[32]
5	2012	$[UTp_3]$	100	5.5	7.0×10^{-5}	[33]
6	2011	$[U(BIPM^{TMS})(I)_2(\mu_2\text{-}\eta^6{:}\eta^6\text{-}C_6H_5CH_3)]$	—	—	—	[34]
7	2013	$[U(I)_3(THF)_4]$	2 000	18.6 ± 0.7	$(6.4 \pm 1.8) \times 10^{-7}$	[35]
		Frozen solution	2 000	11.5 ± 1.3	$(6 \pm 6) \times 10^{-6}$	
8	2013	$[U(N(SiMe_3)_2)_3]$	2 000	31 ± 3	10^{-11}	[35]
		Frozen solution	2 000	13.9 ± 0.6	$(1.5 \pm 0.4) \times 10^{-6}$	
9	2013	$[U(BIPM^{TMS})(I)_2(THF)]$	2 000	23.4 ± 0.8	$(2.9 \pm 0.8) \times 10^{-7}$	[35]
		Frozen solution	2 000	12.2 ± 0.9	$(2.8 \pm 1.0) \times 10^{-6}$	
10	2011	$[Np(COT)_2]$	30 000	41	1.1×10^{-5}	[9]
f^1: Uranium(V), Neptunium(VI)						
12	2012	$[\{[UO_2(salen)]_2Mn(Py)_3\}_6]$	0	142	3×10^{-12}	[36]
13	2013	$[UO(TREN^{TIPS})]$	1 000	21.5 ± 0.7	2.6×10^{-7}	[37]
14	2014	$[\{[UO_2(salen)(py)][Mn(py)_4](NO_3)\}]_n$	0	134	3.1×10^{-11}	[38]
Miscellaneous						
15	2010	$\{Np^{VI}O_2Cl_2\}\{Np^{V}O_2Cl(THF)_3\}_2$	0	140	—	[39]

for U^{3+} (Table 10.2). The second noteworthy observation is that the Arrhenius plot of $\ln(\tau)$ versus $1/T$ is not linear. This is a common observation in f-element single-molecule magnets. Indeed, there is no a priori reason why this plot should be linear. Only the Orbach mechanism of spin-lattice relaxation would be expected to give rise to a linear Arrhenius plot, and clearly other relaxation mechanisms are operative here. In fact, the question of whether Orbach relaxation plays a role at all has not really been addressed in recent f-element single-molecule magnet literature. The presence of low-lying excited crystal field levels certainly makes it possible for the Orbach mechanism to occur.

Other relaxation mechanisms can be expected to be operative, especially at low temperatures where the phonon density is low, decreasing the efficiency of the Orbach process. These processes will have different, if any, temperature dependencies from the Orbach process and, hence, lead to curvature of the Arrhenius plot. Such processes are direct relaxation and quantum tunnelling, as the Long group [28] themselves together with Atanasov have recently quantitatively addressed in low-coordinate transition metal compounds. For **1**, the deviation from Arrhenius-type behaviour was attributed to ground state tunnelling, which is consistent with the relaxation time increasing upon application of a static field. The static field removes the degeneracy between the two states with opposite projections of the magnetic moment ($\pm m_J$ in the simplest case), decreasing the effectiveness of interactions mixing these states, and hence decreasing quantum tunnelling efficiency. The question then is what interaction can cause quantum tunnelling. For half-integer angular momentum ions, such as U^{3+}, the crystal field cannot lift the degeneracy of the $\pm m_J$ states (Kramers' theorem). For lanthanides that often possess nuclear spins that are strongly coupled to the electron spin, the hyperfine interactions can cause efficient tunnelling. However, the depleted uranium used for these investigations does not possess a nonzero nuclear spin. On the other hand, an effective magnetic field such as the magnetic dipolar interaction between molecules in a solid could cause an appreciable tunnelling efficiency. Surprisingly, ac susceptibility measurements on frozen solutions of **1**, where intermolecular dipolar interactions are expected to be strongly diminished, did not show a large change in the temperature of the maximum in χ'' at a given ac frequency. This suggests that at the 1000 Oe applied dc field, tunnelling due to intermolecular interactions does not play a significant role.

Less than a year later, a second derivative $[U(H_2BPz_2)_3]$ (**2**) was published [29a], which has a very similar molecular symmetry, but a slight elongation (by almost 0.2 Å, in spite of the lower temperature used for the crystal structure determination) along the pseudo C_3 axis. Such an elongation might be expected to decrease the crystal field splitting of the ground multiplet and hence the effective energy barrier. However, the distances between the central U^{3+} ion and the coordinating ligands do not change appreciably. In any case, no slow relaxation of the magnetic moment was observed in zero applied dc field, but in a relatively small dc field of 50 Oe, slow relaxation is observed. An Arrhenius fit yields $U_{\text{eff}} = 8\,\text{cm}^{-1}$ and $\tau_0 = 1.2 \times 10^{-6}\,\text{s}$. While the former value is appreciably smaller, the latter is several orders of magnitude larger. Interestingly, at applied fields above 500 Oe, two clearly distinct relaxation processes are observed. The faster process ('A') is also observed at lower dc fields, but appears to lose its temperature dependence. The latter observation suggests that a temperature-independent relaxation process such as direct relaxation or tunnelling becomes dominant, and may also contribute at lower fields. This is supported by the fact that the Cole–Cole plots are asymmetric and have relatively large distribution parameters α compared to **1**. The second relaxation process is three orders of magnitude slower than the first, and of unclear origin. In f-element single-molecule magnets, the magnetization dynamics have often been found to change drastically going from the pure, magnetically dense solid to dilute conditions in dilute solids or frozen solutions [30].

Long *et al.* [29b] studied the influence of dilution on the relaxation of the magnetization in **2**. They found that even a moderate 1 : 1 dilution in the isostructural Y^{3+} complex results in a drastic increase of the effective energy barrier in a small applied dc field of 100 Oe as determined by the fit of the high-temperature part of the Arrhenius plot ($U_{\text{eff}} = 16\,\text{cm}^{-1}$ versus $U_{\text{eff}} = 8\,\text{cm}^{-1}$). Further dilution does not appear to affect the energy barrier. Intriguingly, the pre-exponential factor τ_0 decreases by a factor of 20 on dilution from 1.2×10^{-6} to 6×10^{-8} s, meaning that at some temperature the relaxation times will cross. However, in the studied temperature range, the relaxation time in the dilute system is clearly longer. The slow process that was found previously in the undiluted system at relatively high dc fields can no longer be observed in the dc measurements, even at 1 : 1 dilution. In view of the observed magnetic hysteresis (without coercivity, Figure 10.5), the authors conclude that the process has become even slower. Intriguingly, the hysteresis disappears for higher dilutions than 1 : 13, and the hysteresis is attributed to field-induced ferromagnetic short range order [29b].

The electronic structures of both **1** and **2** were studied by crystal field theory methods [25]. The authors first assess the deviation from idealized D_{3h} symmetry, which they find to be very small. The employed crystal field Hamiltonian, formulated in terms of Stevens operator equivalents, contains average radial factors $\langle r^k \rangle$, as well as structural factors A_k^q formulated in terms of spherical harmonics Y_k^q. The fit parameters were the effective point charge Z_i of the coordinating ligand atom, and a radial displacement vector, here taken as a scalar D_r, which is an effective metal–ligand distance. The nitrogen lone pairs can be expected to point towards the metal, so no displacement of the ligand electron density perpendicular to the metal–ligand axis was considered. Two further parameters were used, namely, an offset in the susceptibility (correcting for temperature-independent magnetism) and an overall scaling factor (presumably

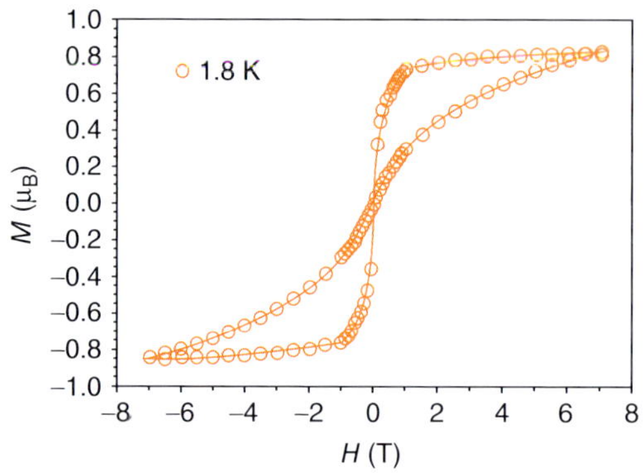

Figure 10.5 Magnetic hysteresis curve recorded on a 1 : 1 diluted powder sample in the isostructural Y^{3+} compound at 1.8 K. (Reprinted from Ref. [29b]. Copyright (2011) American Chemical Society.)

of the susceptibility temperature product χT). Good fits of the experimental χT data are obtained for $D_r = 1.30$Å and $Z_i = 0.0247$ for **1**, and $D_r = 1.40$Å and $Z_i = 0.0203$ for **2**. In addition, values of -1.5×10^{-3} emu mol^{-1} and 0.95 were found for temperature-independent paramagnetism (TIP) and the overall scaling factor. No error bars are given. Clearly, D_r and Z_i are strongly correlated [25], so the assessment of the uniqueness of this solution is not trivial. Intriguingly, the first excited crystal field doublet is found at about 200 cm^{-1}, which is nowhere near the experimental energy barriers of 20 and 8 cm^{-1} for **1** and **2**, respectively. This poses the question: to what extent is the Orbach mechanism of spin-lattice relaxation, which proceeds via a real intermediate state, operative at all? The second interesting outcome of the crystal field study is the nature of ground doublet in **1** and **2**, which in both cases is dominated by contributions from the $m_J = \pm 5/2$ levels, which are intermediate levels of the $J = 9/2$ Russell–Saunders ground multiplet. It will be interesting to see how the higher m_J levels can be stabilized to give a ground doublet with a larger magnetic moment.

Another question that remains is how the symmetry influences the relaxation of the magnetic moment. A first answer was given in 2011, when Almeida *et al.* [31] published a magnetic study of a further pyrazolylborate U^{3+} complex, $[U(TpMe_2)_2(bipy)]I$ (**3**). The uranium ion possesses approximately C_2 symmetry [31], while the coordination sphere can be described as either a distorted triangular dodecahedron or a strongly distorted square antiprism. Ac magnetic susceptibility measurements reveal that **3** shows slow relaxation of the magnetic moment both in zero applied dc field and at a small dc field of $H = 500$ Oe. The 500 Oe data were analysed and the fit to the Arrhenius law, giving an energy barrier of $U_{eff} = 18.2$ cm^{-1} and $\tau_0 = 1.4 \times 10^{-7}$ s. The Cole–Cole plot of $\chi''(\chi')$ revealed a small distribution in relaxation times with $\alpha = 0.173$ (Figure 10.6). Interestingly, this energy barrier is very close to that of **1**, suggesting that

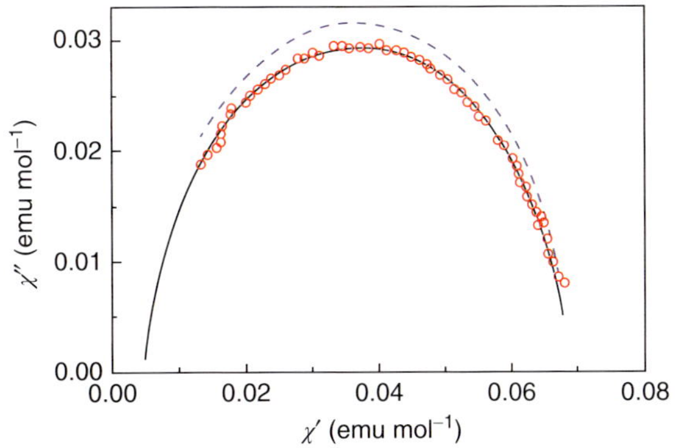

Figure 10.6 Cole–Cole (or Argand) plot of χ'' versus χ' recorded on a powder sample of 3 at 4.5 K at an external dc field of 500 Oe. The dashed line indicates the expected behaviour for a single relaxation time, while the solid line was calculated for a distribution parameter $\alpha = 0.173$. (Reprinted from Ref. [31]. Copyright (2011) American Chemical Society.)

symmetry does not play a decisive role in determining the relaxation properties of pyrazolylborate U^{3+} complexes. In this study, the $\chi''(T)$ data were divided by the equilibrium (dc)-susceptibility for deriving the relaxation times. This is formally more correct than using the $\chi''(T)$ data, because the Bloch relaxation equation predicts that the inverse of the ac frequency corresponds to the relaxation time at the maximum of $\chi''(T)/\chi(T)$, and not that of $\chi''(T)$ [30]. Clearly, the error made depends on the width of the peak in $\chi''(T)$. This procedure has the added advantage of dividing some of the χ'' signal due to further relaxation processes at lower temperature, thus making the peak in $\chi''(T)$ more evident. The starting compound for the synthesis of **3**, $[U(TpMe_2)_2I]$ (**4**), was also reported, a year later, to display slow relaxation of the magnetic moment, with similar parameters from the Arrhenius fit of the ac data (Table 10.2) [32]. This last publication also features CASPT2 calculations of the electronic structure of both **3** and **4**, where the energy gap between the ground and first excited doublet was found to be much larger than the experimental energy barriers, at 146 and 138 cm^{-1} for **3** and **4**, respectively. Interestingly and intriguingly, these values are very close to those obtained from the point dipole crystal field calculations of Gaita-Ariño *et al.* [25] (110 and 136 cm^{-1} for **3** and **4**, respectively).

The experimental energy barrier is much lower than the experimentally determined crystal field splitting also in $[UTp_3]$ (**5**), where the Arrhenius fit gives a tiny energy barrier Ueff = 5.5 K, while the pre-exponential factor is rather large at $\tau_0 = 7.0 \times 10^{-5}$ s. In contrast, from optical measurements [40], the expected lowest gap (between $m_J = \pm 5/2$ and $m_J = \pm 3/2$) is >250 cm^{-1} (corresponding to >350 K). In fact, the authors suggest that the Orbach mechanism does not play a major role at all in these complexes. At this point, the symmetry of the crystal field clearly becomes a factor of minor importance, as long as the ground doublet

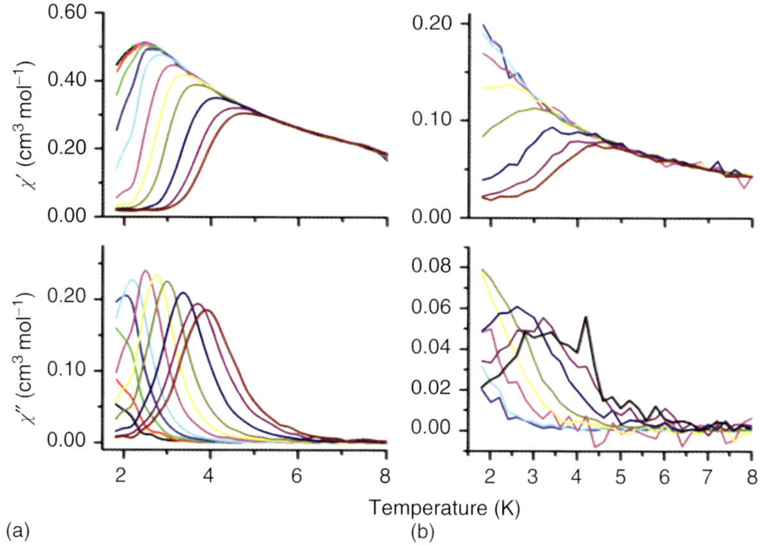

Figure 10.7 In-phase (top) and out-of phase (bottom) components of the ac susceptibility measured in an applied dc field of 2 kOe and an ac field of 3.5 Oe for powder samples of **9** in the solid state (a) and in frozen THF solution (b) at different ac frequencies between 1 and 1500 Hz. (Adapted from Ref. 35. Copyright (2013) Wiley-VCH.)

does not contain the $m_J = \pm 1/2$ levels, which would increase the efficiency of the direct process of spin-lattice relaxation (Figure 10.3).

This conclusion was also drawn from the study of three completely unrelated U^{3+} SMMs, $[U(I)_3(THF)_4]$ (**7**), $[U(N(SiMe_3)_2)_3]$ (**8**), $[U(BIPM^{TMS})(I)_2(THF)]$ [**9**, $BIPM^{TMS} = CH(PPh_2NSiMe_3)_2$] (THF, tetrahydrofuran) (Figure 10.7) [35]. They have no particular symmetry and even their idealized symmetries are very different (C_s, C_{3v}, C_1 for **7**–**9**, respectively). In addition, these complexes are also the first mononuclear U^{3+} SMMs without pyrazolyl borate ligands. The Arrhenius plots are reasonably linear and energy barriers in the 20–31 K range are obtained (Table 10.2). For compound **8**, the pre-exponential factor is rather small, $\tau_0 = 7.0 \times 10^{-5}$ s. Therefore, the complexes were also studied in frozen solution (Figure 10.7), where it was found that SMM behaviour persists under these conditions, thus ruling out other causes for slow relaxation of the magnetic moment, such as spin-glass behaviour and phonon bottleneck effects. The effective energy barriers are lowered, compared to the solid state, presumably due to either changes in the crystal field splitting, in the phonon spectrum or in spin–phonon coupling. Interestingly, even in dilute solution, no coercive field is observed in the magnetic hysteresis loops. This may be due to the ground doublet having significant $m_J = \pm 1/2$ character, increasing the spin-lattice relaxation (Figure 10.3) [25].

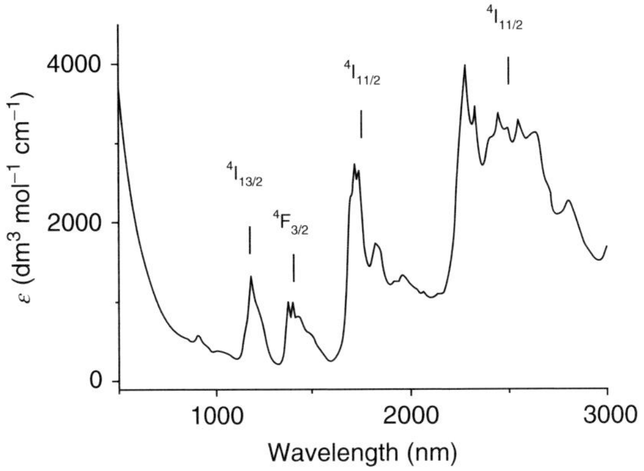

Figure 10.8 Room temperature electronic absorption spectrum recorded on a THF solution of **6**. Bands are marked by the nominal Russell–Saunders multiplet to which the excitation occurs. (Adapted from Ref. [34]. Copyright (2011) Nature Publishing Group.)

The first polynuclear U^{3+} SMM was dinuclear $[U(BIPM^{TMS})(I)_2(\mu_2\text{-}\eta^6{:}\eta^6\text{-}C_6H_5CH_3)]$ (**6**), published in 2011 [34]. Although it breaks no energy barrier records, the ac susceptibility measurements show a clear out-of-phase signal, evidencing slow relaxation of the magnetization in this compound. The origin of this effect must be SMM-like behaviour, because the signal persists in frozen solution (S.T. Liddle and J. van Slageren, unpublished results). The main question of interest is to what extent the electronic structure and magnetic properties are influenced by the polynuclear nature of the molecule. The magnetization relaxation itself can be explained by single ion properties alone. However, the NIR electronic absorption spectrum shows several absorption bands attributed to f→f-transitions, but with extraordinary extinction coefficients ε of several thousand per mole per centimetre (Figure 10.8). This suggests that the effective inversion centre of the uranium ion has been lifted due to strong overlap between f- and ligand p-orbitals, and the electronic f → f (u → u) transitions have become allowed as a consequence. Secondly, density functional theory calculations yield a set of eight singly occupied orbitals as highest energy occupied frontier orbitals. Of these, four have strongly mixed character, involving both uranium centres and the bridging arene ligand. Interestingly, the uranium orbitals that are involved are the f-orbitals, and not the s- or d-orbitals. The other four have predominant f-character, again involving both ions. Delocalization of f-electrons was also recently observed in $Ln^{3+}-U^{4+}-Ln^{3+}$ trimers [41].

The actinide SMMs discussed so far were based on uranium. In 2011, the first mononuclear transuranic SMM in the form of the Np^{4+} complex $[Np(COT)_2]$ (**10**, COT = cyclooctatetraene, $C_8H_8^{\ 2-}$), was published [9]. Although first published in 1970, the dynamics of its magnetic moment had never been investigated. The relaxation is slow enough to be measured when applying an external dc magnetic field of $H \geq 1000$ Oe. At low applied fields ($H < 20$ kOe), and temperatures above 10 K, the plot of ln τ versus $1/T$ is essentially linear, and an energy barrier of Ueff = 41 K and $\tau_0 = 1.1 \times 10^{-5\text{s}}$ was extracted. The value of the energy barrier is intriguing, because it is much lower than that expected for the excited crystal field states (>1400 cm^{-1}). Hence the occurrence of the Orbach mechanism of spin-lattice relaxation is not expected to be operative in this regime. The authors suggest that below 20 kOe, the relaxation is mediated by interaction with the nuclear spin, whereas above this field no field induced crossings between electron–nuclear spin states are expected. At low temperatures, the plot levels off, suggesting the occurrence of quantum tunnelling of the magnetic moment (or direct spin-lattice relaxation, given the presence of an external field). For very large fields (>50 kOe), the relaxation slows down and becomes independent of frequency, supporting the interpretation of nuclear spin-induced quantum tunnelling as the main relaxation pathway at low temperatures. It is only at these very high fields that magnetic hysteresis is observed (Figure 10.4). Intriguingly, the corresponding f^3 U^{3+} complex K[U(COT)$_2$] does not display slow relaxation of the magnetic moment [6], possibly due to its lower symmetry or to its smaller crystal field splitting.

10.2.2
Single-Molecule Magnets of f^1 Actinides (U^{5+})

The first mononuclear $5f^1$ single-molecule magnet was reported in 2013. The complex [U(O){N(CH$_2$CH$_2$NSi^{i}Pr$_3$)$_3$}] was found to display slow relaxation of the magnetization in an applied dc field of 1 kOe. The Arrhenius fit gave $U_{\text{eff}} = 21.5$ K and $\tau_0 = 2.6 \times 10^{-7}$ s, which is quite similar to values found for U^{3+}. The distribution factor α lies in the range 0.07–0.11, depending on temperature, which signifies a rather small distribution of relaxation times. The magnetic moment is quite high ($\mu_{\text{eff}} = 1.79\,\mu_B$ at room temperature, decreasing to $1.2\,\mu_B$ at 1.8 K), suggesting that the ground doublet is $m_J = \pm 3/2$. This is supported by the absence of an EPR signal, given that in the (idealized) threefold symmetry of the complex, the $m_J = \pm 3/2$ doublet does not mix with the $m_J = \pm 1/2$ doublet, and the direct $-3/2$ to $+3/2$ transition is forbidden by the $\Delta m_J = \pm 1$ selection rule of EPR.

Next, we discuss an aesthetically pleasing molecule (Figure 10.9), which is also the first heteronuclear actinide SMM (**12**) [36]. It consists of a ring of six UO_2^+ (U^{5+} ions) bridged by six Mn^{2+} ions, where the latter are also coordinated by a terminal U^{5+} each. The two uranyl ions are bridged by the salen ligand. The dc magnetic susceptibility shows a decrease in magnetic moment on cooling from room temperature leading to a minimum at around 50 K. To lower temperatures, the magnetic moment increases, giving rise to a cusp at 20 K followed by a final decrease. These observations appear reminiscent of ferrimagnetic coupling, where the antiferromagnetic exchange coupling leads to a ground state with a large magnetic moment. The low temperature magnetization measurements

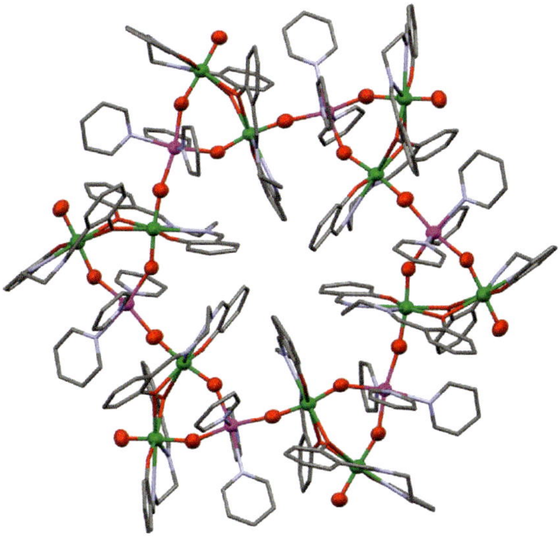

Figure 10.9 Crystal structure of **12**, with the following colour scheme: C (grey), O (red), N (blue), Mn (magenta), U (green). (Reprinted from Ref. [36]. Copyright (2012) Nature Publishing Group)

display a strong rise of the magnetization, followed by a weaker, linear increase, with a hint of an increase in the slope, which could possibly indicate a transition to a magnetic state with a larger moment. This is supported by the authors' suggestion that the lowest magnetic state has antiparallel uranyl and manganese moments, while in the excited state they are parallel. The dynamic susceptibility data are quite remarkable. In zero field, a clear out-of-phase signal is observed, whose maximum is frequency dependent. The Arrhenius plot of $\ln \tau$ versus $1/T$ shows a clear straight line, where the fit yields an energy barrier of 142 K, which is five times that typically observed for uranium SMMs. A small pre-exponential factor of 3×10^{-12} s is obtained, but the shift in peak maximum temperature per decade in frequency lies in the range observed for SMMs, ruling out spin glass behaviour. The distribution in relaxation times increases towards lower temperatures. Interestingly, magnetic hysteresis can be observed below 4 K (Figure 10.10), a clear indication of the multispin nature of the system, which suppresses tunnelling.

Finally, the first actinide single-chain magnet was recently published, with an energy barrier of 134 K and a pre-exponential factor of 3×10^{-11} s [38]. No estimate of the relative contributions of the exchange interaction and the local anisotropy to the energy barrier was made. The fact that these $U^{4+}-Mn^{2+}$ systems appear to possess very favourable magnetic properties raises the question of exchange coupling in 5f systems and 5f–3d systems. This question is addressed in Section 10.3.

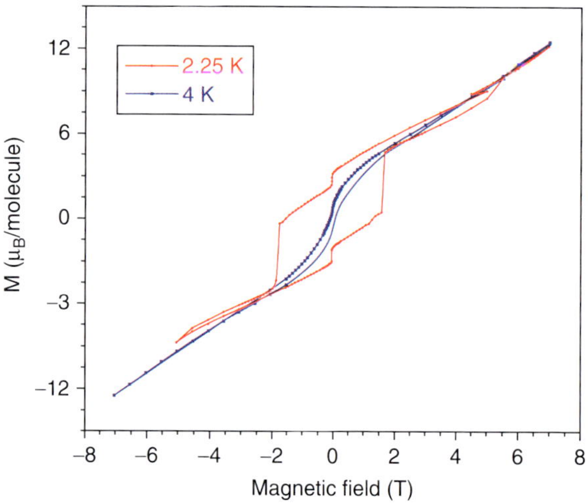

Figure 10.10 Magnetic hysteresis recorded on a powder sample of **12** at different temperatures, showing clear coercivity. (Reprinted from Ref. [36]. Copyright (2012) Nature Publishing Group.)

10.2.3
Miscellaneous: $\{Np^{VI}O_2Cl_2\}\{Np^{V}O_2Cl(THF)_3\}_2$ (15)

The title compound of this section can be considered an isosceles triangle of one f^1 Np^{6+} and two f^2 Np^{5+} ions [39]. From the susceptibility temperature product χ^T, and the strongly axial nature of the neptunyl ions, it was deduced that the Np^{5+} ions possess $m_J = \pm 3$ ground quasi doublets, while the Np^{6+} ions have an $m_J = \pm 1/2$ ground doublet. A rather large antiferromagnetic coupling strength of $J = 10.8\,K$ between Np^{5+} and Np^{6+} ions was found, while the $Np^{5+}-Np^{5+}$ interaction were much weaker at $J' = 0.56\,K$. In spite of the cluster ground state predicted to have a small magnetic moment, ac susceptibility measurements show clear out-of-phase signals. This is reminiscent of $[Dy_3(\mu_3\text{-}OH)_2L_3Cl(H_2O)_5]Cl_3{\cdot}4H_2O{\cdot}2MeOH{\cdot}0.7MeCN$ (HL = *o*-vanillin) [42]. The Arrhenius plot is linear in the limited temperature range accessible, and the fit yields quite a high energy barrier of 140 K, with no pre-exponential factor given.

10.3
Magnetic Coupling in Actinides

One of the prime reasons to consider (early) actinides for the development of SMMs with stable magnetizations at high temperatures is that they can potentially display much stronger interactions to neighbouring spin carriers due to the larger spatial extension of the 5f orbitals compared to the 4f orbitals of the lanthanides [43]. Constructing larger ensembles of coupled paramagnetic ions is important to prevent the efficient quantum tunnelling of the magnetization seen in all mononuclear f-element SMMs. Indeed, the strongly coupled compounds $[\{[(Me_3Si)_2N]_2Ln(THF)\}_2(N_2^{3-})]^-$ (Ln = Tb, Dy) show pronounced hysteresis, for example [44]. In the following section, we give a brief overview of the investigations into magnetic coupling in discrete molecular actinide complexes. In these systems, the exchange coupling is weak compared to the crystal field splitting for the f-ion(s). If the crystal field splitting is not much larger than the thermal energy at room temperature, this means that both effects must be considered at the same level. The beauty of chemistry is that it is often possible to substitute a paramagnetic ion for a diamagnetic one, in, for example, a dinuclear complex. Assuming the substitution leaves all else unchanged (geometric and electronic structures), these magnetically monomeric complexes allow the determination of the single ion properties. When subtracting the magnetic data for the diamagnetically substituted complexes from those for the original dinuclear complex, the effects of the magnetic coupling may be extracted. The diamagnetic substitution procedure for f-elements was recently discussed [45].

10.3.1 5f–5f Couplings

Already several decades ago, the first report on magnetic exchange coupling between actinide ions appeared [46]. The bis-amidobridged U^{5+} dinuclear complexes $[(MeC_5H_4)_3U]_2(L)$ (L = μ-1,4-$N_2C_6H_4$, μ-1,3-$N_2C_6H_4$) show clearly different magnetic susceptibilities. The magnetic susceptibility of the 1,3-derivative increases monotonously with decreasing temperature. At low temperatures, a small Weiss temperature ϑ is obtained from the fit to the Curie–Weiss law $\chi = C/(T - \vartheta)$. At higher temperatures, the Weiss temperature is much higher, possibly due to the presence of TIP. In contrast, the 1,4-derivative displays a clear maximum in the magnetic susceptibility at ~20 K, which is ascribed to magnetic ordering, but more likely due to intramolecular exchange coupling. The absence of an EPR signal led the authors to conclude that the ground state of the uranium ions is $m_J = \pm 3/2$ with the crystal quantum number $\mu = 3/2$. Assuming an Ising-type exchange between $S_{\text{eff}} = 1/2$ spins, that is, considering the z-component of the exchange interaction only, a large antiferromagnetic exchange coupling of $J = -19\ \text{cm}^{-1}$ was found. It was speculated that the strong coupling in the 1,4-derivative originates in the possibility to formulate resonance structures which feature both unpaired electrons on one uranium ion. Alternatively, the spin-polarization mechanism of exchange interaction could be operative, which for a 1,4-coupling would indeed predict antiferromagnetic coupling [47]. A similar system is the doubly bridged $[\{U(NtBu)_2(I)(tBu_2bpy)\}_2]$ [48].

Another interesting example of coupling between U^{5+} ions is the triangular triuranyl(V) complex $[UO_2L]_3$ (L = 2-(4-tolyl)-1,3-bis(quinolyl)malondiiminate) [49]. Dc susceptibility measurements revealed a maximum in the susceptibility, χ, at 11 K, with low temperature χ-values tending to zero. Intriguingly, the magnetization versus field measurements showed a sigmoid-like shape indicative of a field-induced transition to an excited state with larger magnetic moment. The observation of an EPR signal proves that the uranyl ions possess $\mu = 3/2$ ground doublets, in accordance with the strong axial ligand field stabilizing the $m_J = \pm 5/2$ levels of the $f^1 J = 5/2$ ground multiplet [26]. The O–U–O axes, and therefore, presumably, the magnetic moments in the ground state of the three ions lie in the plane of the triangle. This leads to a nonmagnetic doubly degenerate ground state, as in the trinuclear dysprosium system $[Dy_3(\mu_3\text{-}OH)_2L_3Cl(H_2O)_5]Cl_3\cdot 4H_2O\cdot 2MeOH\cdot 0.7MeCN$ (HL = *o*-vanillin) [42]. The magnetic data were fit by using a spin Hamiltonian which included isotropic exchange, local crystal field splitting up to second rank and the Zeeman interaction. Applying a transverse magnetic field leads to periodic quenching of the tunnel splitting, as observed in systems with high spin [50], or nonmagnetic ground states [51]. Further exchange coupled U^{5+} complexes are $[UO_2(dbm)_2K(18C6)]_2$ [52], $[\{UO_2(salen)\}_4(\mu^8\text{-}K)_2][\{(K(18C6)Py)\}_2]$ [53] and $[(Me_3SiOUO)_2(L)]$ (L = Schiff base macrocycle) [54]. In U^{5+}, the ligand field splitting is typically much larger than the thermal energy, even at room temperature. Hence, only the lowest Kramers doublet is populated, and the U^{5+} ion can be well modelled by an effective spin $S_{eff} = 1/2$.

Exchange interactions have also been observed between U^{4+} ions. Here, the analysis is complicated by the crystal field splitting, which is the dominant interaction, but not so strong, that excited crystal field levels can be neglected. In addition, in strictly octahedral symmetry, the ground state of the Russel–Saunders $^3H_4$ multiplet is a nonmagnetic singlet state. Hence, even in the absence of (antiferromagnetic) exchange interactions, the magnetic moment tends to zero on decreasing the temperature. Some indications for magnetic exchange in di- and trinuclear U^{4+} ethynylbenzene complexes were obtained by subtraction of the temperature dependence of the susceptibility for similar mononuclear complexes from the susceptibilities of the polynuclear compounds [55]. The magnetic interactions appeared to be weakly ferromagnetic, which would be expected for a meta-bridged system, but not for para-bridged systems. An isostructural heteronuclear $U^{4+}-Th^{4+}$, where the latter is diamagnetic, would allow more precise investigation of the magnetic properties of these clusters, but it proved to be synthetically not accessible. A second series of complexes in this respect is $[\{((t\text{-}BuArO)_3tacn)U\}_2(\mu\text{-}X)](X = O, S, Se)$ [56], which feature a linear bridging chalcogenide ion. These complexes show a clear maximum in the χ versus T curve, while similar complexes with different ancillary ligands do not. This was taken as an indication for the presence of significant exchange interactions, although it must be noted that the crystal field for the two families of complexes is also different (N_3O_3X vs NO_5X donor set). A similar example is $[U(Tren^{DMBS})(O)U\{N(CH_2CH_2NSiMe_2Bu^t)_2(CH_2CH_2NCOSiMe_2Bu^tCHO)\}]$ [57]. The relation between structure and the magnitude of the TIP of a series of U^{4+} fluoroketimides was investigated several years ago [58]. TIP arises when the applied magnetic field mixes ground and excited states, and is most clearly observable when the ground state is nonmagnetic [10].

There are no definitive examples of exchange coupling in polynuclear U^{3+} complexes. For this ion as well, the effect of exchange coupling is likely to be masked by the effects of the crystal field splitting on the magnetic susceptibility. Thus, though dinuclear $[U(BIPM^{TMS})(I)_2(\mu_2\text{-}\eta^6\text{:}\eta^6\text{-}C_6H_5CH_3)]$ (**6**) appears to possess an unusual delocalized electronic structure (as evidenced by very strong f→f electronic transitions, and supported by density functional theory calculations) it does not present any indications for superexchange between localized magnetic centres [34]. Further studies on similar systems appear to confirm this picture [59]. Furthermore, the magnetic properties of the U^{3+} complexes with the proposed compositions $[(UX)_2(L)]$ ($X = I^-$, BH_4^-; L = macrocyclic Schiff base ligand) were investigated, but the lack of structural data makes the analysis highly speculative [60]. Finally, the neptunyl complex **15** was discussed above [39].

An interesting development from a magnetic point of view is the recent preparation of large uranium-oxo clusters [61], reminiscent of the $\{M_{30}Mo_{72}\}$ ($M = Fe^{3+}$, V^{4+}, Cr^{3+}) polyoxometallates, that were the subject of many investigations on magnetism [62].

10.3.2 5f–4f Couplings

The magnetic properties of the 5f–4f heterotrinuclear species $(C_5Me_5)_2U^{IV}[-N{=}C(CH_2C_6H_5)(tpyYb^{III}(C_5Me_5)_2)]_2$ were investigated in detail, where the authors profited from the fact that both the U^{4+} mononuclear and the corresponding trinuclear $Yb^{III}{}_2Th^{IV}$ complexes were available. An elaborate subtraction procedure gave strong indications for the presence of exchange coupling between the paramagnetic centres [43, 63].

10.3.3 5f–3d Couplings

A larger number of examples of 5f–3d systems are known, and have been recently reviewed [43]. We reiterate here only the potentially enormously beneficial effect of sizable 5f–3d exchange couplings exploiting the large magnetic moments of first row transition metal ions and the anisotropy of actinides. Examples are the previously discussed U–Mn wheel and chain [36, 38].

10.3.4
5f–2p Couplings

The incorporation of radicals is a highly promising way to construct larger magnetically coupled clusters in 4f chemistry [44, 64], but this is yet to happen for the actinides [43]. A promising way forward seems to be the use of redox-active ligands [65].

10.4
Conclusions

The field of actinide molecular magnetism in general and of actinide single-molecule magnets is still in its infancy. The in-depth understanding of the magnetic properties of molecular actinide systems is clearly very challenging, because many interactions have to be considered simultaneously, especially for lower valency actinides. In other words, real progress in the advancement of science can be made in this area. We think that the recently sparked mutual interest of and fruitful collaborations between magnetochemists, spectroscopists and synthetic actinide chemists will help realize this potential in the years to come.

References

1. Gatteschi, D., Sessoli, R. and Villain, J. (2006) *Molecular Nanomagnets*, Oxford University Press, Oxford.
2. (a) Sessoli, R., Gatteschi, D., Caneschi, A. and Novak, M.A. (1993) *Nature (London)*, **365**, 141–143; (b) Sessoli, R. (1995) *Mol. Cryst. Liq. Cryst.*, **274**, 145–157.
3. Ishikawa, N., Sugita, M., Ishikawa, T., Koshihara, S. and Kaizu, Y. (2003) *J. Am. Chem. Soc.*, **125**, 8694–8695.
4. (a) Abragam, A. and Bleany, B. (1986) *Electron Paramagnetic Resonance of Transition Ions*, Dover Publications, Inc., New York, p. 911; (b) Morrish, A.H. (1966) *The Physical Principles of Magnetism*, John Wiley & Sons, Inc., New York.
5. Woodruff, D.N., Winpenny, R.E.P. and Layfield, R.A. (2013) *Chem. Rev.*, **113**, 5110–5148.
6. Rinehart, J.D. and Long, J.R. (2009) *J. Am. Chem. Soc.*, **131**, 12558–12559.
7. Orbach, R. and Stapleton, H.J. (1972) in *Electron Spin Resonance* (ed S. Geschwind), Plenum Press, New York.
8. Chiorescu, I., Wernsdorfer, W., Müller, A., Bogge, H. and Barbara, B. (2000) *Phys. Rev. Lett.*, **84**, 3454–3457.
9. Magnani, N., Apostolidis, C., Morgenstern, A., Colineau, E., Griveau, J.-C., Bolvin, H., Walter, O. and Caciuffo, R. (2011) *Angew. Chem. Int. Ed.*, **50**, 1696–1698.
10. Lueken, H. (1999) *Magnetochemie*, Teubner Verlag, Stuttgart, p. 360–367.
11. Liu, G. and Beitz, J.V. (2006) in *The Chemistry of the Actinide and Transactinide Elements*, vol. 3 (eds L.R. Morss, N.M. Edelstein, and J. Fuger), Springer, Dordrecht, p. 2063.
12. Shannon, R.D. (1976) *Acta Crystallogr. A*, **32**, 751–767.
13. Görller-Walrand, C. and Binnemans, K. (1996) in *Rationalization of Crystal Field Parametrization*, vol. 23 (eds K.A.

Gschneidner and L. Eyring), Elsevier, Amsterdam, p. 168.

14. Krupa, J.C. (1987) *Inorg. Chim. Acta*, **139**, 223–241.
15. Edelstein, N.M. and Lander, G.H. (2006) in *The Chemistry of the Actinide and Transactinide Elements*, vol. 4 (eds L.R. Morss, N.M. Edelstein, and J. Fuger).
16. Tsukerblat, B.S. (2006) *Group Theory in Chemistry and Spectroscopy*, Dover Publications, Inc., Mineola, NY.
17. (a) Boatner, L.A. and Abraham, M.M. (1978) *Rep. Prog. Phys.*, **41**, 87; (b) Lukens, W.W., Edelstein, N.M., Magnani, N., Hayton, T.W., Fortier, S. and Seaman, L.A. (2013) *J. Am. Chem. Soc.*, **135**, 10742–10754.
18. Cotton, S. (2006) *Lanthanide and Actinide Chemistry*, John Wiley & Sons, Ltd, Chichester.
19. Gourier, D., Caurant, D., Berthet, J.C., Boisson, C. and Ephritikhine, M. (1997) *Inorg. Chem.*, **36**, 5931–5936.
20. Lea, K.R., Leask, M.J.M. and Wolf, W.P. (1962) *J. Phys. Chem. Solids*, **23**, 1381–1405.
21. Marchand, R.L., Fish, G.E. and Stapleton, H.J. (1975) *Phys. Rev. B*, **11**, 23–26.
22. (a) Magnani, N., Amoretti, G., Blaise, A., Caciuffo, R., Dacheux, N., Genet, M., Stride, J.A. and Adroja, D.T. (2004) *J. Magn. Magn. Mater.*, **272–276**, E203–E204; (b) Kern, S., Lander, G.H., Soderholm, L., Loong, C.K., Trouw, F., West, M., Hoisington, D., Cort, B. and Welp, U. (1994) *J. Chem. Phys.*, **101**, 9338–9343.
23. (a) Carnall, W.T. (1992) *J. Chem. Phys.*, **96**, 8713–8726; (b) Amberger, H.D., Fischer, R.D. and Yünlü, K. (1986) *Organometallics*, **5**, 2109–2114.
24. Natrajan, L.S. (2012) *Coord. Chem. Rev.*, **256**, 1583–1603.
25. Baldoví, J.J., Cardona-Serra, S., Clemente-Juan, J.M., Coronado, E. and Gaita-Ariño, A. (2013) *Chem. Sci.*, **4**, 938–946.
26. Rinehart, J.D. and Long, J.R. (2011) *Chem. Sci.*, **2**, 2078–2085.
27. Maria, L., Paula Campello, M., Domingos, Â., Santos, I. and Andersen, R. (1999) *J. Chem. Soc., Dalton Trans.*, 2015–2020.
28. Zadrozny, J.M., Atanasov, M., Bryan, A.M., Lin, C.-Y., Rekken, B.D., Power, P.P., Neese, F. and Long, J.R. (2013) *Chem. Sci.*, **4**, 125–138.
29. (a) Rinehart, J.D., Meihaus, K.R. and Long, J.R. (2010) *J. Am. Chem. Soc.*, **132**, 7572–7573; (b) Meihaus, K.R., Rinehart, J.D. and Long, J.R. (2011) *Inorg. Chem.*, **50**, 8484–8489.
30. Ishikawa, N., Sugita, M., Ishikawa, T., Koshihara, S. and Kaizu, Y. (2004) *J. Phys. Chem. B*, **108**, 11265–11271.
31. Antunes, M.A., Pereira, L.C.J., Santos, I.C., Mazzanti, M., Marçalo, J. and Almeida, M. (2011) *Inorg. Chem.*, **50**, 9915–9917.
32. Coutinho, J.T., Antunes, M.A., Pereira, L.C.J., Bolvin, H., Marcalo, J., Mazzanti, M. and Almeida, M. (2012) *Dalton Trans.*, **41**, 13568–13571.
33. Rinehart, J.D. and Long, J.R. (2012) *Dalton Trans.*, **41**, 13572–13574.
34. Mills, D.P., Moro, F., McMaster, J., van Slageren, J., Lewis, W., Blake, A.J. and Liddle, S.T. (2011) *Nat. Chem.*, **3**, 454–460.
35. Moro, F., Mills, D.P., Liddle, S.T. and van Slageren, J. (2013) *Angew. Chem. Int. Ed.*, **52**, 3430–3433.
36. Mougel, V., Chatelain, L., Pécaut, J., Caciuffo, R., Colineau, E., Griveau, J.-C. and Mazzanti, M. (2012) *Nat. Chem.*, **4**, 1011–1017.
37. King, D.M., Tuna, F., McMaster, J., Lewis, W., Blake, A.J., McInnes, E.J.L. and Liddle, S.T. (2013) *Angew. Chem. Int. Ed.*, **52**, 4921–4924.
38. Mougel, V., Chatelain, L., Hermle, J., Caciuffo, R., Colineau, E., Tuna, F., Magnani, N., de Geyer, A., Pécaut, J. and Mazzanti, M. (2014) *Angew. Chem. Int. Ed.*, **53**, 819–823.
39. Magnani, N., Colineau, E., Eloirdi, R., Griveau, J.C., Caciuffo, R., Cornet, S.M., May, I., Sharrad, C.A., Collison, D. and Winpenny, R.E.P. (2010) *Phys. Rev. Lett.*, **104**, 197202.

40. Apostolidis, C., Morgenstern, A., Rebizant, J., Kanellakopulos, B., Walter, O., Powietzka, B., Karbowiak, M., Reddmann, H. and Amberger, H.-D. (2010) *Z. Anorg. Allg. Chem.*, **636**, 201–208.
41. Schelter, E.J., Wu, R., Veauthier, J.M., Bauer, E.D., Booth, C.H., Thomson, R.K., Graves, C.R., John, K.D., Scott, B.L., Thompson, J.D., Morris, D.E. and Kiplinger, J.L. (2010) *Inorg. Chem.*, **49**, 1995–2007.
42. (a) Tang, J., Hewitt, I., Madhu, N.T., Chastanet, G., Wernsdorfer, W., Anson, C.E., Benelli, C., Sessoli, R. and Powell, A.K. (2006) *Angew. Chem. Int. Ed.*, **45**, 1729–1733; (b) Luzon, J., Bernot, K., Hewitt, I.J., Anson, C.E., Powell, A.K. and Sessoli, R. (2008) *Phys. Rev. Lett.*, **100**, 247205.
43. Rinehart, J.D., Harris, T.D., Kozimor, S.A., Bartlett, B.M. and Long, J.R. (2009) *Inorg. Chem.*, **48**, 3382–3395.
44. (a) Rinehart, J.D., Fang, M., Evans, W.J. and Long, J.R. (2011) *J. Am. Chem. Soc.*, **133**, 14236–14239; (b) Rinehart, J.D., Fang, M., Evans, W.J. and Long, J.R. (2011) *Nat. Chem.*, **3**, 538–542.
45. Lukens, W.W. and Walter, M.D. (2010) *Inorg. Chem.*, **49**, 4458–4465.
46. Rosen, R.K., Andersen, R.A. and Edelstein, N.M. (1990) *J. Am. Chem. Soc.*, **112**, 4588–4590.
47. Cargill Thompson, A.M.W., Gatteschi, D., McCleverty, J.A., Navas, J.A., Rentschler, E. and Ward, M.D. (1996) *Inorg. Chem.*, **35**, 2701–2703.
48. (a) Spencer, L.P., Schelter, E.J., Yang, P., Gdula, R.L., Scott, B.L., Thompson, J.D., Kiplinger, J.L., Batista, E.R. and Boncella, J.M. (2009) *Angew. Chem. Int. Ed.*, **48**, 3795–3798; (b) Meskaldji, S., Zaiter, A., Belkhiri, L. and Boucekkine, A. (2012) *Theor. Chem. Acc.*, **131**, 1–10.
49. (a) Chatelain, L., Mougel, V., Pecaut, J. and Mazzanti, M. (2012) *Chem. Sci.*, **3**, 1075–1079; (b) Carretta, S., Amoretti, G., Santini, P., Mougel, V., Mazzanti, M., Gambarelli, S., Colineau, E. and Caciuffo, R. (2013) *J. Phys. Condens. Matter*, **25**, 486001.
50. Wernsdorfer, W. and Sessoli, R. (1999) *Science*, **284**, 133.
51. El Hallak, F., Rosa, P., Vidal, P., Sheikin, I., Dressel, M. and van Slageren, J. (2011) *Europhys. Lett.*, **95**, 57002.
52. Nocton, G., Horeglad, P., Pécaut, J. and Mazzanti, M. (2008) *J. Am. Chem. Soc.*, **130**, 16633–16645.
53. Mougel, V., Horeglad, P., Nocton, G., Pécaut, J. and Mazzanti, M. (2009) *Angew. Chem. Int. Ed.*, **48**, 8477–8480.
54. Arnold, P.L., Jones, G.M., Odoh, S.O., Schreckenbach, G., Magnani, N. and Love, J.B. (2012) *Nat. Chem.*, **4**, 221–227.
55. Newell, B.S., Rappé, A.K. and Shores, M.P. (2010) *Inorg. Chem.*, **49**, 1595–1606.
56. Lam, O.P., Heinemann, F.W. and Meyer, K. (2011) *Chem. Sci.*, **2**, 1538–1547.
57. Gardner, B.M., Stewart, J.C., Davis, A.L., McMaster, J., Lewis, W., Blake, A.J. and Liddle, S.T. (2012) *Proc. Natl. Acad. Sci. U.S.A.*, **109**, 9265–9270.
58. Schelter, E.J., Yang, P., Scott, B.L., Thompson, J.D., Martin, R.L., Hay, P.J., Morris, D.E. and Kiplinger, J.L. (2007) *Inorg. Chem.*, **46**, 7477–7488.
59. Vlaisavljevich, B., Diaconescu, P.L., Lukens, W.L., Gagliardi, L. and Cummins, C.C. (2013) *Organometallics*, **32**, 1341–1352.
60. Arnold, P.L., Potter, N.A., Magnani, N., Apostolidis, C., Griveau, J.-C., Colineau, E., Morgenstern, A., Caciuffo, R. and Love, J.B. (2010) *Inorg. Chem.*, **49**, 5341–5343.
61. (a) Biswas, B., Mougel, V., Pécaut, J. and Mazzanti, M. (2011) *Angew. Chem. Int. Ed.*, **50**, 5745–5748; (b) Falaise, C., Volkringer, C., Vigier, J.-F., Beaurain, A., Roussel, P., Rabu, P. and Loiseau, T. (2013) *J. Am. Chem. Soc.*, **135**, 15678–15681.
62. Kögerler, P., Tsukerblat, B. and Müller, A. (2010) *Dalton Trans.*, **39**, 21.
63. Schelter, E.J., Veauthier, J.M., Thompson, J.D., Scott, B.L., John, K.D., Morris, D.E.

and Kiplinger, J.L. (2006) *J. Am. Chem. Soc.*, **128**, 2198–2199.

64. Sessoli, R. and Powell, A.K. (2009) *Coord. Chem. Rev.*, **253**, 2328–2341.

65. (a) Cladis, D.P., Kiernicki, J.J., Fanwick, P.E. and Bart, S.C. (2013) *Chem. Commun.*, **49**, 4169–4171; (b) Kraft, S.J., Fanwick, P.E. and Bart, S.C. (2010) *Inorg. Chem.*, **49**, 1103–1110.

Index

Lanthanides and Actinides in Molecular Magnetism, First Edition.
Edited by Richard A. Layfield and Muralee Murugesu.